FINITE ELEMENT METHOD

Vs.

CLASSICAL METHODS

FINITE ELEMENT METHOD
Vs.
CLASSICAL METHODS

H.S. Govinda Rao

PUBLISHING FOR ONE WORLD

NEW AGE INTERNATIONAL (P) LIMITED, PUBLISHERS

(formerly Wiley Eastern Limited)

New Delhi • Bangalore • Chennai • Cochin • Guwahati • Hyderabad
Jalandhar • Kolkata • Lucknow • Mumbai • Ranchi

Visit us at **www.newagepublishers.com**

Published by New Age International (P) Ltd., Publishers
First Edition: 2007

ISBN : 81-224-2050-8

Rs. 195.00

C-06-12-1175

Printed in India at Ram Printoghaphs, Delhi.
Typeset at Bharati Composers, Delhi.

PUBLISHING FOR ONE WORLD
NEW AGE INTERNATIONAL (P) LIMITED, PUBLISHERS
(formerly Wiley Eastern Limited)
4835/24, Ansari Road, Daryaganj, New Delhi - 110002
Visit us at **www.newagepublishers.com**

Preface

This book is primarily intended to meet the requirements of senior undergraduate and post graduate students of Mechanical Engineering course at various Indian universities. Finite Element Method is a foundation course in Aerospace Engineering. The objective of this book is to present Finite Element Method in an easily understandable manner. This book is the outcome of extensive teaching of the subject at various levels by the author and his persuation by students and colleagues.

The material in this text has been divided into 9 chapters. A number of illustrations and worked examples have included to make the concepts clear. Exercise problems from other books and from past question papers have also been solved to provide confidence to the students and prepare them for the examination. I look forward to my colleagues and students for suggestions to improve this book.

Dr. H.S. Govinda Rao

Acknowledgements

First of all, I cannot find adequate words to express my gratitude Dr. P. Martin Jebaraj, Principal and Head of the Dept. of Mechanical Engg., Dr. Ambedkar Institute of Technology, Bangalore for his valuable suggestions and constant encouragement.

I would like to thank Mr. Saumya Gupta, Managing Director, New Age International Publications, Delhi, and Mr. Sudharshan Shetty of the same publication based in Bangalore. We are grateful to Mr. K.R. Krishnamurthy, AAO, Dr. Ambedkar Institute of Technology, Bangalore for his constant encouragement and support.

I take this opportunity to thank my parents for all that I am today and any credit for my success must go to them.

Dr. H.S. Govinda Rao

Contents

1

Fundamental Concepts

1.1 INTRODUCTION

Among the various numerical methods available, finite element method (FEM) is at present very widely used in every engineering analysis. Several engineering problems will be defined in terms of governing equations written in one, two and three dimensions. Usually these problems are expressed in the form of ordinary or partial differential equations. The problems of structural machanics such as deformation, trusses, stress analysis of automotive aircraft, building and bridge structures, magnetic flux, seepage etc. have been reduced to a system of linear simultaneous equations.

Solution of these equations gives us the approximate behaviour of the continuum. These facts suggest that we need to keep pace with the developments by understanding the basic theory, modeling techniques, and computational aspects of the finite element method. Applications range from problems relating to heat transfer, fluid flow, lubrication, soil machanics, electric and magnetic fields, structural engineering and discussions related to structural analysis problems.

1.2 A BRIEF HISTORY OF FEM

Although the name "finite element" is of recent origin, the basic concept has been used for centuries. The basic concept of finite elements originates from advances in aircraft structural analysis. One of the earlier problems dealt with this concept was by ancient mathematicians. In 1941, **Hrenikoff** found a solution of elasticity problems using the "framework method".

Courant introduced piecewise polynomial interpolation (or continuous trial functions) over triangular subregions to model tortion problems appeared in 1943.

In 1956, Turner et al. derived stiffness matrices for truss, beam, and other elements and presented their new findings.

Clough first introduced the term finite element, appeared in 1960. The two landmark papers to which the origin of the FEM is generally traced are those due to Turner, Clough, Martin and Topp and Argyris and Kelsey. The first book on finite elements by O.C. Zienkiewicz and Cheng was published in 1967. The finite element method was applied to the problems which are non-linear and large deformations in nature, appeared in the late 1960's and early 1970's. A book on nonlinear continua appeared in 1972. It is curious to note that the mathematicians continue to put the finite element method on sound theoretical ground whereas the engineers continue to find interesting extensions in various branches of engineering. In 1959, Greenstadt (279) utilised this technique to discretise cell. He defined the unknowns through a series of functions for each cell, proper variational principle for them and satisfied continuity requirements to tie together the cells giving fundamentalists of finite element technique. Thereafter in 1960s, Clough analysed plane elasticity problems further by this method. The method started getting its base when in 1963, it was recognised by Besseling and Melosh (174) as a form of Ritz's method of approximations. Zienkiewicz and Cheng (280), in 1965, discussed in a broad way and recognised that the method can handle all field problems which can be expressed in variational forms. After this, the technique was recognised by many and went on very fast.

Finite element method, today, is widely used in almost all fields of science and engineering, such as, aeroelasticity, aerodynamics, fluid flows, pipe and channel flows, thermodynamics, soil mechanics, foundation engineering, geotechnical engineering, structural engineering and structural dynamics, pile foundations, machine foundations, nuclear containment systems, lubrication, bearings, fluid and soil structure interactions, electrical technology, textile engineering, cable systems etc. Numerous software packages have been developed of which some of the most popular ones are SAP, ANSYS, ADINA, STAAD, STRUDL.

SAP and ANSYS are of the oldest softwares. SAP has got some of the best facilities for Dynamic analyses even till today. ANSYS is like an encyclopaedia of finite element packages. It has got almost all types of elements, with many facilities for many fields of machanics.

The accuracy of finite element method has always been a question to be answered. Based on Ritz's summation principle for approximating a solution with the help of parameters known at some selected points, the method always is of an approximate one. However, the accuracy can be controlled through

(i) Considering sufficient number of nodes and elements for a problem.
(ii) Choice of suitable approximating functions.
(iii) Choice of suitable numerical integration scheme.

Mathematicians and engineers have simultaneously been developing the approximation technique and some of the noted researchers in the field of FEM are listed below:

Researcher	*Period*	*Researcher*	*Period*
1. Rayleigh	1870	9. Mchenry	1943
2. Ritz	1909	10. Courant	1943
3. Richardson	1910	11. Prager and Synge	1947
4. Galerkin	1915	12. Newmark	1949
5. Liebman	1918	13. Morsh and Feshback	1953
6. Biezenokoch	1923	14. Mchahon	1953
7. Southwell	1940	15. Argyris	1955
8. Hrenikoff	1941	16. Turner and his group	1956

Researcher	*Period*	*Researcher*	*Period*
17. Clough, Martin and Topp	1956	20. Besseling	1963
18. Greenstadt	1959	21. Melosh	1963
19. Varga	1962	22. Zienkiewicz	1965

FEM Vs. Finite Difference Method (FDM)

1. In FEM, the final differential equation is surpassed, whereas it is the differential equation to start with in the finite difference method.
2. The approximation involved in FEM is physical in nature, as the actual continuum is replaced by finite elements. The element for mutation is mathematically exact whereas the finite difference technique involves the exact representation of the continuum in terms of the differential equation and on this actual physical system, the mathematical model is approximated.
3. FDM needs larger number of nodes to get good results while FEM needs fewer nodes.
4. FDM does not give the value at any point except at node points. It does not give any approximating function to evaluate the basic values (deflections, in case of solid mechanics) using the nodal values. FEM can give the values at the point. However, the values obtained at points other than nodes are by using suitable interpolation formulae.
5. FDM makes **pointwise approximation** to the governing equations i.e. it ensures continuity only at the node points. Continuity along the sides of grid lines are not ensured. FEM makes piecewise approximation i.e. it ensures the continuity at node points as well as along the sides of the element.
6. FEM caters irregular geometry whereas FDM makes stair type approximation to sloping and curved boundaries.

Finite Element Method Vs. Classical Methods

1. In FEM, we can obtain approximate solutions for the exact equations formed, whereas in classical methods exact equations are formed and exact solutions are obtained.
2. The FEM provides the solutions for all problems whereas classical method yields solutions for a few standard cases.
3. If structure consists of more than one materials, it is difficult to use classical method, but finite element can be used without any difficulty.
4. When material property is not isotropic, solutions for the problems become very difficult in classical method. Only few simple cases have been tried successfully by researchers. FEM can handle structures with anisotropic properties also without any difficulty.
5. FEM is superior to the classical methods in the sense that the problems which cannot be tackled by classical methods without making drastic assumptions.

1.3 NEED FOR STUDYING FEM

One may ask the question: "What is the need to study finite element method when there is a number of users-friendly packages available in the market?" This argument is not sound. The mathematicians continue to put the finite element method on sound theoretical ground whereas the engineers continue to find interesting extensions in various branches of engineering. Hence, the FE knowledge makes a good engineer better while just user without the knowledge of FE may produce more dangerous results. In order to use FEM packages properly, the user must know the following points clearly:

1. How to discretise the domain to get good results?
2. Identification of variables.
3. Which elements are to be used for solving the problems in hand?
4. Incorporation of boundary conditions.
5. Solution of simultaneous equations.
6. Choice of approximating functions.
7. Identification of variables.
8. How the element properties are developed and what are their limitations?

The FEM is a product of the computer age, and the application of the method to solve practical problems requires use of computer programs for analysis. Nowadays no such programs are developed as there are several commercial finite element packages available that can solve varieties of problems. Some of these packages are ANSYS, ALGOR, NISA, ABAQUS, NASTRAN etc. having pre- and postprocessors that give graphical pictures of the structure before and after loading.

Pre-processor helps in the generation of the finite element mesh and prepares the data for direct input into the analysis. This becomes essential as the modern structures have complicated shapes and geometry. As the geometry and loading become more and more complicated the discretization also becomes very cumbersome and nodal numbering is really a strenuous one.

Some of these packages also address questions like, "which type of analysis is the best for the problem?", "what is the best element type for the application?", "how can one combine different types of elements?" etc. The preprocessor graphically displays the structure to be analyzed. If the discretisation is not perfect then it can be modified before the analysis. Briefly, in pre-processing we build the model with defining geometry; specifying element type; defining material properties; creating meshes and nodes with numbering.

The post-processor is again a program that presents the results of the finite element analysis usually graphically and also performs further calculations as desired by the user on the results. This is a very attractive feature of the commercial packages.

Example: If at a node we have six degrees of freedom (3 displacements and 3 rotations) and 500 nodes, then nodal displacements will be 3000, i.e. we will get a print out of 3000 displacements. To get any physical idea of these will once again be a formidable task. On the other hand, a graphical picture incorporating these values and showing the structure in its displaced position will give a physical grasp of the problems. Briefly, in postprocessing we extract results such as displacements, stresses etc.; time–history relation wherever applicable; and graphical representation of the results.

The commercial package ANSYS is used to show what has been explained so far. Hence, it is necessary that the users of FEM package should have sound knowledge of FEM.

1.4 BASIC CONCEPTS OF ELASTICITY

The concept of elasticity provides a more sophisticated tool of stress analysis than strength of materials approach. It can be observed that when a wet ball impinges on a hard floor a circular path is observed. This shows that the ball is compressed at the point of concussion. When a body is compressed it is slightly deformed, and almost immediately tends to recover its original shape. Internal forces are called into play to restore the body to its original shape.

But, when a glass ball is dropped on a marble floor it rebounds and attains a height almost equal to the original one ; but if dropped on a wooden floor the height regained is much smaller. Again if balls of different materials are dropped on the same floor the heights regained are different. The velocities of all these balls will be the same when they reach the floor, but the velocities of the rebound are different and so the heights reached are also different.

The property by virtue of which the differences in the velocities of the rebound are caused, and the bodies tend to revert to the original shapes is called **elasticity**.

All bodies are more or less elastic. They possess the property of elasticity in different degrees.

Stresses and Equilibrium

In continuum mechanics, two distinct types of forces are considered: body forces and surface forces. Forces which are distributed over the volume of the body are called body forces. That is, body forces are forces that act on every element of a material and hence on the entire volume of the material. These forces are expressed as 'force per unit mass of the element'. Self weight of a structure, inertia force, magnetic force are examples of such body forces.

Forces distributed over the surface of the body are called surface forces. That is, surface forces act on the surface of a material. This surface may be either a part or the whole of the boundary surface, if any, of the material or an imaginary surface visualized in the interior of the material. In other words, surface forces are external forces that act on the boundary surface of the material. These are usually expressed as 'force per unit surface area of the element'. Hydrostatic pressure, frictional force, wind forces, and forces exerted by a liquid on a solid immersed in it are examples of such surface forces. **Stresses** are system's internal bond forces acting at molecular level at all points and in all directions. **Strains** are defined as rate of change of displacements (deformations) with respect to the original dimensions (positions). In a deformed body they are present at all points and in all directions.

Preliminary Definitions and Results in Elasticity

Definition (1) (Stress): The intensity of internally distributed forces that tend to resist change in the shape of a body is defined as stress.

$$\text{Stress } (\sigma) = \frac{\text{Load}}{\text{Cross-sectional area}} = \frac{P}{A}\,\frac{(\text{N})}{(\text{mm}^2)}$$

Definition (2) (Axial strain or linear strain or primary strain): Change in length per unit length of linear dimension of a body is defined as linear strain.

$$\text{Strain } (e \text{ or } \varepsilon) = \frac{\text{Change in length}}{\text{Original length}} = \frac{(\delta l)\ (\text{mm})}{(l)\ (\text{mm})}$$

Definition (3) (Lateral strain or secondary strain): It is defined as the ratio of change in lateral dimension to original dimension.

$$\text{Lateral strain} = \frac{\text{Change in diameter}}{\text{Original diameter}} = \frac{\delta d}{d}$$

Definition (4) (Volumetric strain): It is defined as the ratio of change in volume to original volume.

$$\text{Volumetric strain} = \frac{\text{Change in volume}}{\text{Original volume}} = \frac{\delta V}{V}$$

Definition (5) (Poisson's ratio): It is the ratio of lateral strain to linear strain.

$$\text{Poisson's ratio} = \frac{\text{Lateral strain}}{\text{Linear strain}}$$

Definition (6) (Body): A body is a portion of matter occupying a finite portion of space (and so is bounded).

Definition (7) (Mass, volume, density): Mass is the quantity of matter in the body. The amount of space occupied by a body is called its volume. Mass per unit volume of the body is called its density.

Definition (8) (Particle): A body of indefinitely small dimensions is a particle. Or, a particle is regarded as a mathematical point endowed with a mass.

Note: Very often the dimensions of a body are not considered, and we treat the body as if its whole mass is concentrated at a single point, or for purposes of discussion we treat it as a particle.

Definition (9) (Displacement): The displacement of a moving particle is the change of its position. It has both magnitude and direction.

Definition (10) (Strain energy): When a member is gradually loaded, within its elastic limit, it deforms. On removal of the load it regains its original shape. During deformation it absorbs the energy and gives back the energy while regaining the shape. The energy stored is called "strain energy".

Definition (11) (Work): A force acting on a particle is said to do work when the point of application of the force moves in the direction of the line of action of the force.

Definition (12) (Work done): Whenever a force *S* moves through a small distance, *dx*, then work is said to be done.

$$\begin{aligned}\text{Work done} &= \text{Force} \times \text{Distance moved}\\ &= S\,dx\end{aligned}$$

(i) If S and dx are in the same direction, then that work is said to be positive.

(ii) If S and dx are in opposite directions (such as frictional force), then the work is said to be negative.

Example: For a linearly elastic rod shown in Fig. 1.1, the force varies linearly from *O* to *P* when the displacement increases from *O* to Δ.

Small work done = $(+S)\ (+dx)$

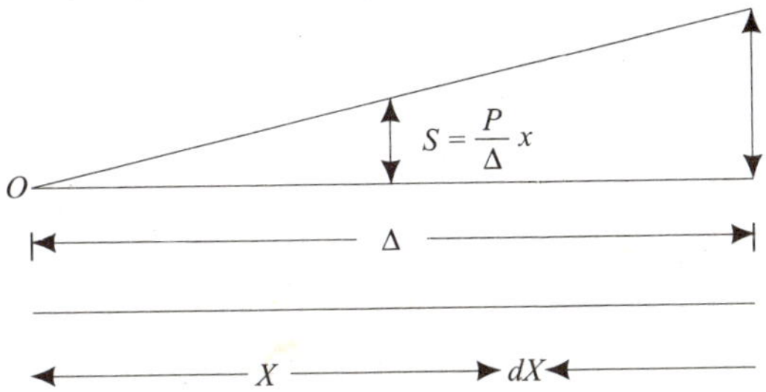

Fig. 1.1

We know $$S = \frac{P}{\Delta}x$$

Since in a linearly elastic material, force varies linearly with deformation

$$\text{Work done} = \int_O^\Delta S\,dx = \int_O^\Delta \frac{P}{\Delta}x\,dx$$

$$= \frac{1}{2}P\Delta \qquad \text{...[1]}$$

= (average force × distance moved)

Eq. (1) represents external work

$$\therefore \qquad \text{External work} = +\frac{1}{2}P\Delta$$

Definition (13) (Internal work): It is observed that at every section of Fig. 1.2, the material offers a resistance equal to the external force whose direction is opposite to that of the external force. But magnitude is same. This resistance undergoes the same displacement but with a difference. That is, in this case the force moves against the direction of displacement. Hence, work is negative.

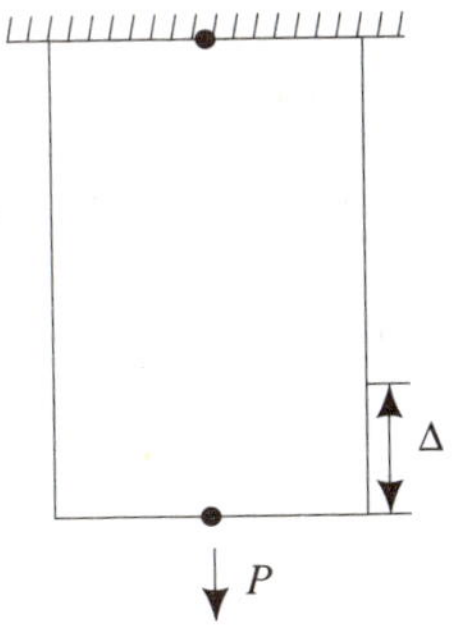

Fig. 1.2 Deformed configuration

$\therefore$
$$\text{Internal work} = -\frac{1}{2}P\Delta \qquad ...[2]$$

Definition (14) (Strain energy due to shear stress): Strain energy is the work done by shear force. It is given by the relation

$$U = \left(\frac{q^2}{2c} \times V\right)$$

where q = shear stress
c = modulus of rigidity
V = volume of the given material

Definition (15) (Column, strut): Any member subjected to compressive stress is called strut. Strut may be vertical, horizontal, or inclined. The vertical strut is called as column.

Definition (16) (Isotropic material): Materials having the same properties in all directions.

Definition (17) (Anisotropic material): Anisotropic materials possess different machanical properties in different directions with reference to their crystallographic planes.

Definition (18) (Orthotropic): If the material has three orthogonal planes of symmetry it is said to be orthotropic.

Definition (19) (Structure): Structure is defined as an assemblage of finite elements, interconnected at finite number of joints called nodes as shown in Fig. 1.3.

The deformations anywhere in the structure can be defined in terms of nodal displacements. For the truss shown in Fig. 1.3 if the displacements at the four joints are known, deformations at any other point can be uniquely determined.

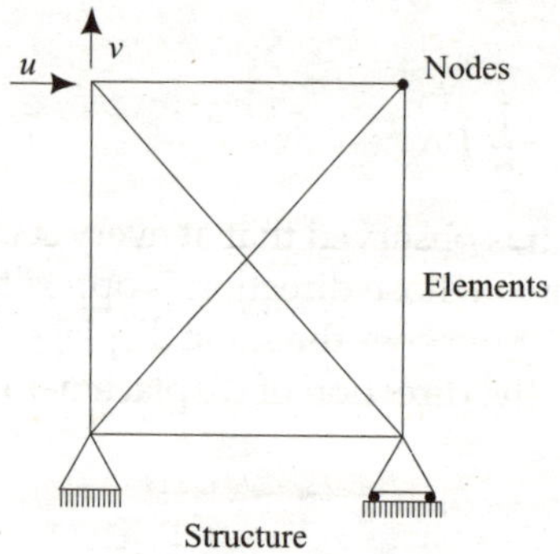

Fig. 1.1 Truss

Definition (20) (Degrees of freedom [Generalized coordinates]): These are the number of independent coordinates which must be specified to uniquely define all the displacements.

For the truss shown in Fig. 1.3 the degree of freedom is 5. This is also called kinematic degree of indeterminacy. The structure (finite element) has a fixed number of nodes. Each node is capable of deformation. The deformation includes displacements, rotations and/or strains. These are collectively known as Nodal degrees of freedom.

Boundary conditions

A structure cannot resist any external load or its own weight without proper boundary conditions. Because of all these the problems in structural machanics fall into boundary value problems essentially.

Work, Energy

Consider a structure in equilibrium under the action of forces $\{Q\}$ with a deformed configuration represented by generalized displacement coordinates $\{q\}$. Consider a conservative system and the loads are applied gradually. If W denotes the work done by the forces $\{Q\}$ on displacement $\{q\}$, then the external work W done by forces $\{Q\}$ on displacements $\{q\}$ (Fig. 1.4) is equal to the strain energy U stored in the structure (energy is conserved).

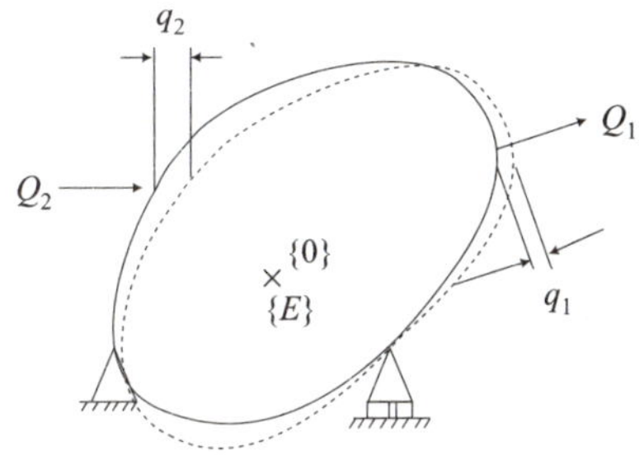

Fig. 1.4 Structure in equilibrium under the action of generalized forces

The structure shown in Fig. 1.5 in state A is the initial configuration. $\{\sigma\}$ are the internal stresses which are in equilibrium with the external loads $\{Q\}$ and the strains {E} are compatible with the displacement $\{q\}$.

Now, we shall study the response of the structure for small variation in displacements $\{\delta q\}$ and forces $\{\delta Q\}$ from initial position.

Due to small variation of displacements and forces there will be changes in work, complementary work, strain energy and complementary energy.

The small variation in displacements Δq takes the structure from equilibrium state A to state B. The incremental changes in work ΔW and strain energy ΔU are given by vertical strips of shaded area as shown in Fig. 1.6(a) (i) and (ii). Magnifying the Fig. 1.6(a) (i) and (ii) as in Figs. 1.6(b) (i) and (ii)

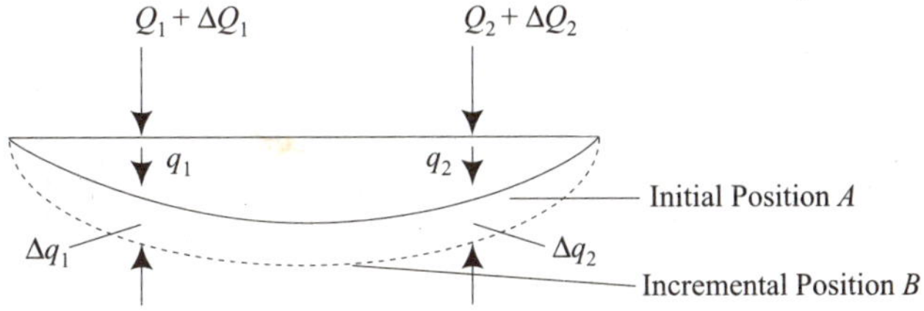

Fig. 1.5 Initial and deformed configuration of the structures

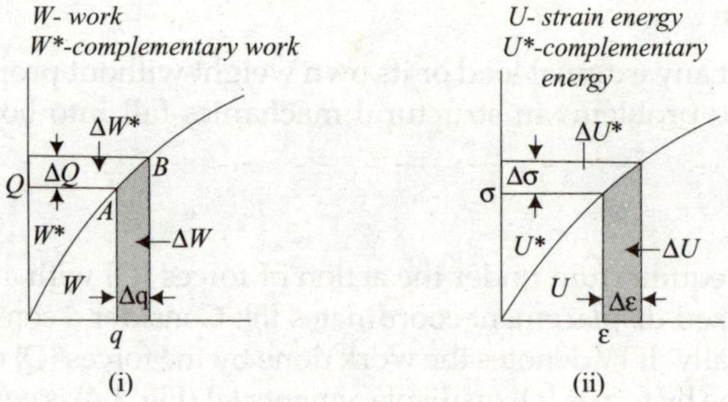

Fig. 1.6(a) Work and energy

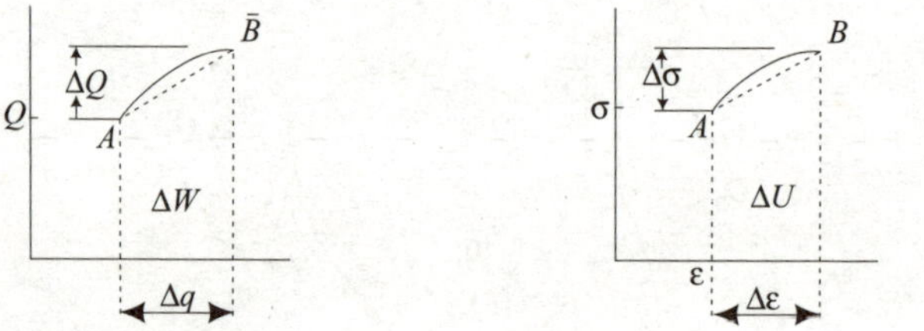

(i) Incremental work (ii) Incremental strain energy

Fig. 1.6(b)

$$\Delta W = \Sigma Q_j \, \Delta V_j + \frac{1}{2} \, \Sigma \, \Delta \, Q_j \, \Delta V_j + \text{higher order terms}$$

In matrix form

$$\Delta W = \{\Delta q\}^T \, \{Q\} + \frac{1}{2} \, \{\Delta q\}^T \, \{\Delta Q\} \quad \text{...[1]}$$

Similarly, ΔU can be written as

$$\Delta U = \int_v \{\Delta \varepsilon\}^T \, \{\sigma\} \, dv + \frac{1}{2} \int_v \{\Delta \varepsilon\}^T \, \{\Delta \sigma\} \, dv \quad \text{...[2]}$$

Recalling that the product of stress and strain represents energy/unit volume, if we assume that $\{Q\}$ and $\{\sigma\}$ remain unchanged while considering the variation of W and U which is equivalent as saying

$$\{\Delta \sigma\} = 0 \quad \text{...[3]}$$

$$\{\Delta Q\} = 0 \quad \text{...[4]}$$

In the study of continuous media, we are concerned with the manner in which forces are transmitted through a medium. If V is the volume occupied by a three-dimensional body bounded by a surface S, then any point in the body is represented by x, y, z coordinates.

The external loads acting on the body include point forces, distributed force per unit area T, also called traction. Under the loads the body deforms. The deformation of a point x (= $[xyz]^T$) is given by the three components of the displacement vector

$$U = [u, v, w]^T \quad \text{...[1]}$$

The external load (force vector) F consists of the following three types of loads that a body may normally be subjected to the vector (body force vector) f, given by

$$f = \begin{bmatrix} f_x \\ f_y \\ f_z \end{bmatrix} = [f_x f_y f_z]^T \qquad ...[2]$$

where f_x, f_y, f_z are the components in the x, y, z directions, respectively. The work done by these forces can be expressed as

$$\text{Work done} = \int_v u^T f\, dv \qquad ...[3]$$

The surface traction T can be expressed as

$$T = \begin{bmatrix} T_x \\ T_y \\ T_z \end{bmatrix} = [T_x\, T_y\, T_z]^T \qquad ...[4]$$

where T_x, T_y, T_z are the traction components in the x, y, z directions respectively. The work done by the forces will be

$$\text{Work done} = \int_A u^T T\, ds \qquad ...[5]$$

An external load P acting at a point i is represented by its three components (Fig. 1.7a)

$$P_i = [P_x, P_y, P_z]_i^T \qquad ...[6]$$

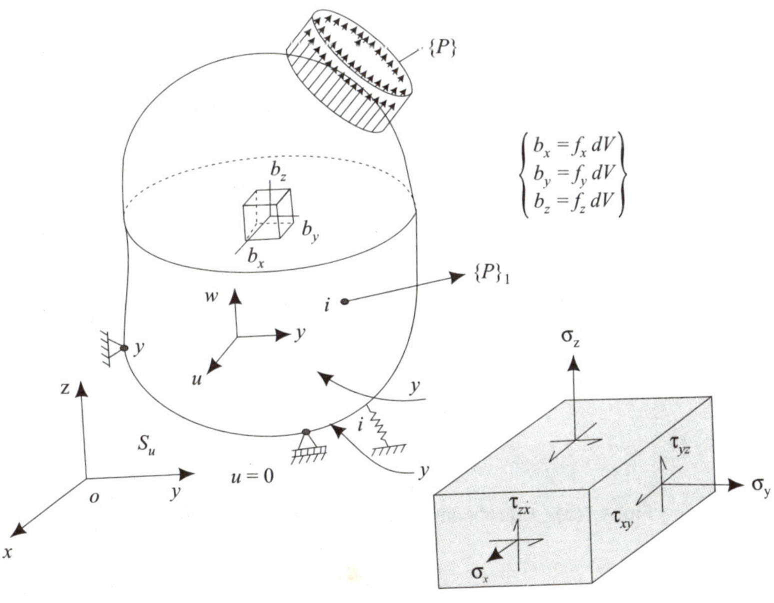

Fig. 1.7(a)

The state of stress in an elemental volume of a loaded body is defined in terms of six components of stress, expressed in a vector form as:

$$\sigma = \{\sigma\} = [\sigma_x \, \sigma_y \, \sigma_z \, \tau_{xy} \, \tau_{yz} \, \tau_{zx}] \quad ...[7]$$

where $\sigma_x, \sigma_y, \sigma_z,$ are normal stresses and $\tau_{yz}, \tau_{xz}, \tau_{xy}$ are shear stresses. The stresses must satisfy the equilibrium condition at each and every point within the body, considering the equilibrium of the elemental volume: $dV = dx\, dy\, dz,$ (Fig. 1.7b).

Writing: $\Sigma F_x = 0; \Sigma F_y = 0; \Sigma F_z = 0$, we get the equilibrium equations

$$\left\{\begin{aligned} \frac{\partial \sigma_x}{\partial x} + \frac{\partial \tau_{xy}}{\partial y} + \frac{\partial \tau_{xz}}{\partial z} + f_x = 0 \\ \frac{\partial \tau_{xy}}{\partial x} + \frac{\partial \sigma_y}{\partial y} + \frac{\partial \tau_{yz}}{\partial z} + f_y = 0 \\ \frac{\partial \tau_{xz}}{\partial x} + \frac{\partial \tau_{yz}}{\partial y} + \frac{\partial \sigma_z}{\partial z} + f_z = 0 \end{aligned}\right\} \quad ...[8]$$

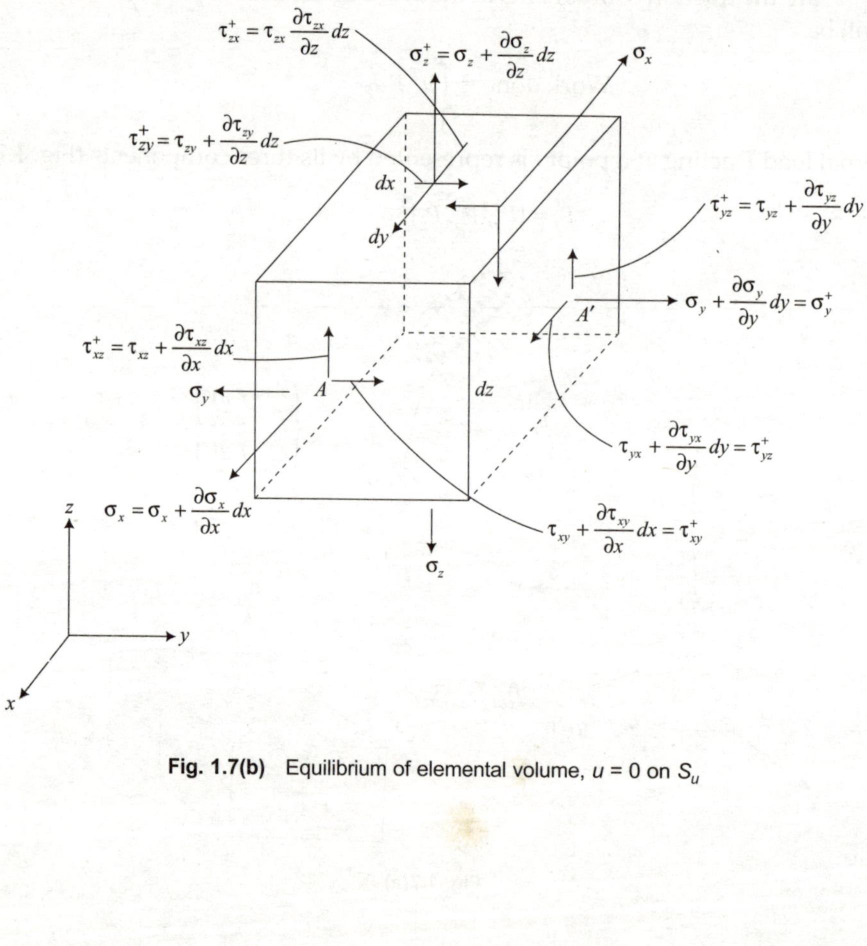

Fig. 1.7(b) Equilibrium of elemental volume, $u = 0$ on S_u

Sign convention: We shall employ the following sign convention for stresses.

A normal stress is positive when it produces tension; therefore, if it produces compression it is negative. Shear stresses shown in Fig. 1.8(a) is accompanied with two subscripts, the first indicating the plane given by its normal on which it acts and the second gives the direction of the component of the stress. To explain it further, if we consider the two shearing stress components acting on the plane which is perpendicular to *x*-axis, then the component which is in the direction of *y*-axis will be τ_{xy} and that in the *z*-direction will be τ_{xz}. The positive directions of the components of shearing stress on any side of the element are taken as the positive directions of the coordinate axes if a tensile stress on the same would have the positive direction of the corresponding axis. Thus if on any side, the tensile stress is opposite to the positive direction of the reference axis, the positive directions of the shearing stress components will then be reversed. All the stress components shown in Fig. 1.8(a) are positive, so that we can say that a stress is positive when it is on positive face in positive direction or on negative face in negative direction [Figs. 1.8(b) and (c)]. From the Fig. 1.8(a) it is clear that there are three components of normal stress—σ_x, σ_y and σ_z and six components of shear stress: τ_{xy}, τ_{yz}, τ_{zx}, τ_{yx}, τ_{zy} and τ_{xz}. Further it is known that $\tau_{xy} = \tau_{yx}$, $\tau_{xz} = \tau_{zx}$ and $\tau_{yz} = \tau_{zy}$. Therefore, to describe the stress at a point, the information of six quantities σ_x, σ_y, σ_z, σ_{xy}, σ_{yz}, σ_{zx} are all that is needed.

Boundary Conditions: There are two types of boundary conditions namely surface loading conditions and displacement boundary conditions, in addition to equations of equilibrium compatibility and constitutive law boundary conditions, to completely specify a problem in solid mechanics. If *u* is specified on part of the boundary denoted by S_u,

we have, $$u = 0 \text{ on } S_u. \quad ...[1]$$

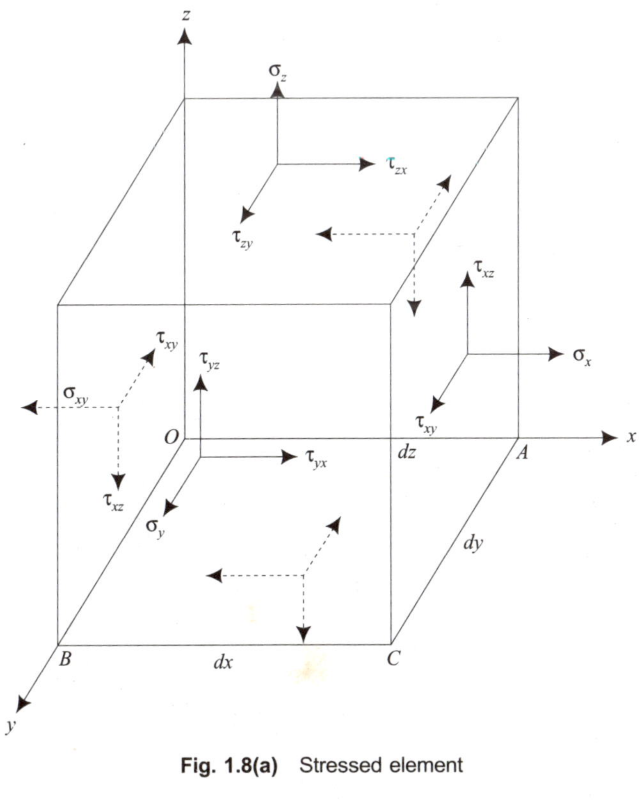

Fig. 1.8(a) Stressed element

Table 1.1

Face	*Stress on -ve Face*	*Stress on +ve Face*	*Remarks*
x	σ_x	$\sigma_x^+ = \sigma_x + \frac{\partial \sigma_x}{\partial x} dx$	Stress on positive face is equal to the stress on negative face plus rate of change of that stress multiplied by the distance between the faces. (See Fig. 1.9)
	τ_{xy}	$\tau_{xy}^+ = \tau_{xy} + \frac{\partial \tau_{xy}}{\partial x} dx$	
	τ_{xz}	$\tau_{xz}^+ = \tau_{xz} + \frac{\partial \tau_{xz}}{\partial x} dx$	
y	σ_y	$\sigma_y^+ = \sigma_y + \frac{\partial \sigma_y}{\partial y} dy$	
	τ_{yx}	$\tau_{yx}^+ = \tau_{yx} + \frac{\partial \tau_{yx}}{\partial y} dy$	
	τ_{yz}	$\tau_{yz}^+ = \tau_{yz} + \frac{\partial \tau_{yz}}{\partial y} dy$	
z	σ_z	$\sigma_z^+ = \sigma_z + \frac{\partial \sigma_z}{\partial z} dz$	
	τ_{zx}	$\tau_{zx}^+ = \tau_{zx} + \frac{\partial \tau_{zx}}{\partial z} dz$	
	τ_{zy}	$\tau_{zy}^+ = \tau_{zy} + \frac{\partial \tau_{zy}}{\partial z} dz$	

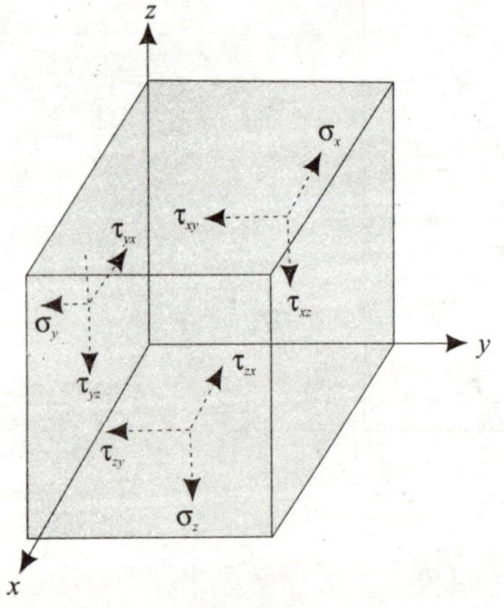

Fig. 1.8(b)

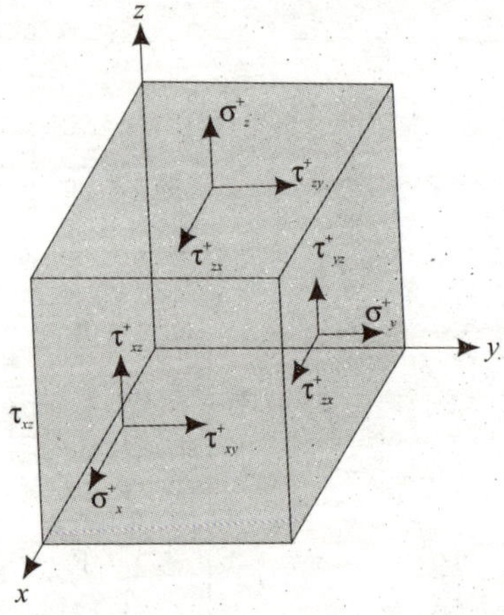

Fig. 1.8(c)

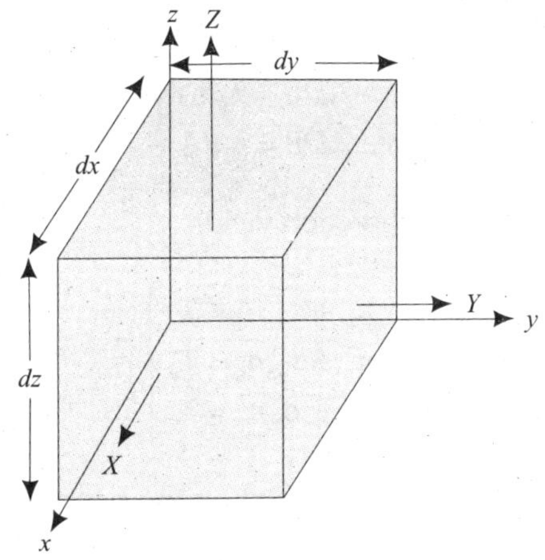

Fig. 1.9

Let us consider the equilibrium of an orthogonal tetrahedron element ABCD shown in Fig. 1.10 where *DA*, *DB* and *DC* are parallel to the *x*, *y* and *z* coordinate axes, respectively and *dA* = area of *ABC*, lies on the surface.

If $n = [n_x, n_y, n_z]^T$ is the outward drawn normal vector to *dA*, then

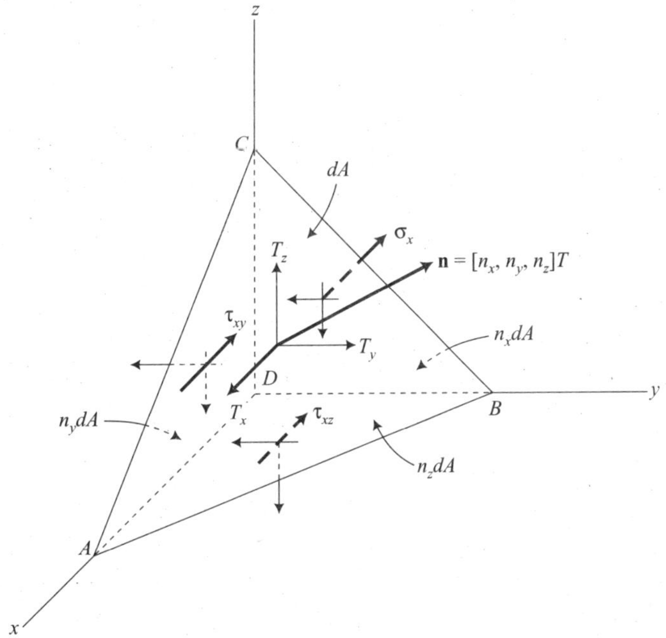

Fig. 1.10 An elemental volume at surface

$$\left\{\begin{aligned} &\text{area } BDC = n_x dA \\ &\text{area } ADC = n_y dA \\ \text{and } &\text{area } ADB = n_z dA \end{aligned}\right\} \quad \text{...[2]}$$

In equilibrium, along the three axes direction yields

$$\left\{\begin{aligned} \sigma_x n_x + \tau_{xy} n_y + \tau_{xz} n_z = T_x \\ \tau_{xy} n_x + \sigma_y n_y + \tau_{yz} n_z = T_y \\ \tau_{xz} n_x + \tau_{yz} n_y + \sigma_z n_z = T_z \end{aligned}\right\} \quad \text{...[3]}$$

The conditions [1], [2] and [3] must be satisfied on the boundary, S_T, where the tractions are applied.

Strain–Displacement Relations

The relations between the components of strain and the displacement components *u*, *v* and *w* at a point are

$$\varepsilon = [\varepsilon_x, \varepsilon_y, \varepsilon_z, \gamma_{yz}, \gamma_{xz}, \gamma_{xy}]^T \quad \text{...[1]}$$

where ε_x, ε_y, and ε_z are normal strains and γ_{yz}, γ_{xz} and γ_{xy} are the engineering shear strains.

Figure 1.11 yields the deformation of the *dx* – *dy* face for small deformations, which we consider here, and also considering the other faces, we get

$$\varepsilon = \left[\frac{\partial u}{\partial x}, \frac{\partial v}{\partial y}, \frac{\partial w}{\partial z}, \frac{\partial v}{\partial z} + \frac{\partial w}{\partial y}, \frac{\partial u}{\partial z} + \frac{\partial w}{\partial x}, \frac{\partial u}{\partial y} + \frac{\partial v}{\partial x}\right]^T \quad \text{...[2]}$$

where the strain and displacement components are given by:

$$\varepsilon_x = \frac{\partial u}{\partial x} + \frac{1}{2}\left[\left(\frac{\partial u}{\partial x}\right)^2 + \left(\frac{\partial v}{\partial x}\right)^2 + \left(\frac{\partial w}{\partial x}\right)^2\right]$$

$$\varepsilon_y = \frac{\partial v}{\partial y} + \frac{1}{2}\left[\left(\frac{\partial u}{\partial y}\right)^2 + \left(\frac{\partial v}{\partial y}\right)^2 + \left(\frac{\partial w}{\partial y}\right)^2\right]$$

$$\varepsilon_z = \frac{\partial w}{\partial z} + \frac{1}{2}\left[\left(\frac{\partial u}{\partial z}\right)^2 + \left(\frac{\partial v}{\partial y}\right)^2 + \left(\frac{\partial w}{\partial y}\right)^2\right] \quad \text{...[3]}$$

$$\gamma_{xy} = \frac{\partial v}{\partial x} + \frac{\partial u}{\partial y} + \frac{\partial u}{\partial x}\frac{\partial u}{\partial y} + \frac{\partial v}{\partial x}\frac{\partial v}{\partial y} + \frac{\partial w}{\partial x}\frac{\partial w}{\partial y}$$

$$\gamma_{yz} = \frac{\partial w}{\partial y} + \frac{\partial v}{\partial z} + \frac{\partial u}{\partial y}\frac{\partial u}{\partial z} + \frac{\partial v}{\partial y}\frac{\partial v}{\partial z} + \frac{\partial w}{\partial y}\frac{\partial w}{\partial z}$$

$$\gamma_{zx} = \frac{\partial u}{\partial z} + \frac{\partial w}{\partial x} + \frac{\partial u}{\partial z}\frac{\partial u}{\partial x} + \frac{\partial v}{\partial z}\frac{\partial v}{\partial x} + \frac{\partial w}{\partial z}\frac{\partial w}{\partial x}$$

$$\left.\begin{aligned} \varepsilon_x &= \frac{\partial u}{\partial x};\ \gamma_{xy} = \frac{\partial v}{\partial x} + \frac{\partial u}{\partial y} \\ \varepsilon_y &= \frac{\partial v}{\partial y};\ \gamma_{yz} = \frac{\partial w}{\partial y} + \frac{\partial v}{\partial z} \\ \varepsilon_z &= \frac{\partial w}{\partial z};\ \gamma_{zx} = \frac{\partial u}{\partial z} + \frac{\partial w}{\partial x} \end{aligned}\right\} \qquad \text{...[4]}$$

Stress–Strain Relations

The basic law of proportionality of stress and strain published by Robert Hooke in 1678 expressed in the form of law known as **Hooke's Law** which states that "within elastic limits of materials the elongation produced by the tensile force is proportional to the tensile force".

If the elongation is denoted by e and the tensile stress by T, then $T = Ee$, where E is a constant depending on the properties of the material. The constant E is called the modulus of tension (or Young's modulus).

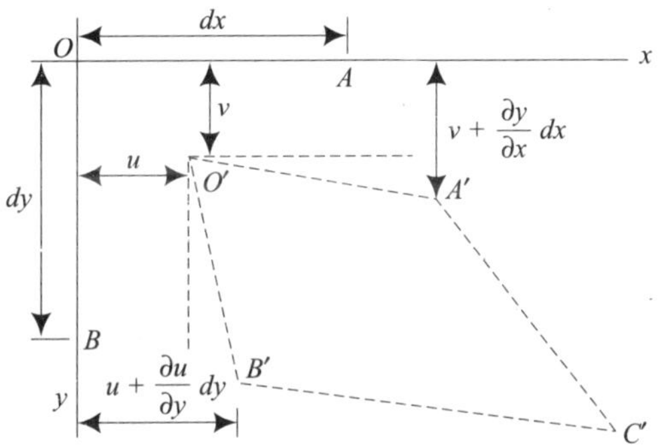

Fig. 1.11 Deformation of the element

Generalized Hooke's Law

The constitutive equations for a linear, elastic solid which relates all the stresses to all the strains is called the generalized Hooke's law. It has the form

$$\sigma_{ij} = C_{ijkm}\,\varepsilon_{km} \qquad \text{...[1]}$$

where C_{ijkm} are the *elastic constants* and in eq. [1] we have 81 such constants since C_{ijkm} is a tensor of fourth rank. The above equation is meant for an anisotropic material and the only requirement is that the material should be stressed within the elastic limits. However, due to symmetry of both stress and strain tensor there are utmost 36 distinct elastic constants only and the Hooke's law can be written as

$$\left.\begin{aligned}
\sigma_{11} &= C_{11}\varepsilon_{11} + C_{12}\varepsilon_{22} + C_{13}\varepsilon_{33} + C_{14}\varepsilon_{23} + C_{15}\varepsilon_{31} + C_{16}\varepsilon_{12} \\
\sigma_{22} &= C_{21}\varepsilon_{11} + C_{22}\varepsilon_{22} + C_{23}\varepsilon_{33} + C_{24}\varepsilon_{23} + C_{25}\varepsilon_{31} + C_{26}\varepsilon_{12} \\
\sigma_{33} &= C_{31}\varepsilon_{11} + C_{32}\varepsilon_{22} + C_{33}\varepsilon_{33} + C_{34}\varepsilon_{23} + C_{35}\varepsilon_{31} + C_{36}\varepsilon_{12} \\
\sigma_{23} &= C_{41}\varepsilon_{11} + C_{42}\varepsilon_{22} + C_{43}\varepsilon_{33} + C_{44}\varepsilon_{23} + C_{45}\varepsilon_{31} + C_{46}\varepsilon_{12} \\
\sigma_{31} &= C_{51}\varepsilon_{11} + C_{52}\varepsilon_{22} + C_{53}\varepsilon_{33} + C_{54}\varepsilon_{23} + C_{55}\varepsilon_{31} + C_{56}\varepsilon_{12} \\
\sigma_{21} &= C_{61}\varepsilon_{11} + C_{62}\varepsilon_{22} + C_{63}\varepsilon_{33} + C_{64}\varepsilon_{23} + C_{65}\varepsilon_{31} + C_{66}\varepsilon_{12}
\end{aligned}\right\} \quad \text{...[2]}$$

which can be written as

$$\sigma_i = C_{ij}\,\varepsilon_j \qquad (i, j = 1, 2, 3, 4, 5, 6) \quad \text{...[3]}$$

This way it is understood that

$$\sigma_{11} = \sigma_1;\ \sigma_{22} = \sigma_2;\ \sigma_{33} = \sigma_3;\ \sigma_{23} = \sigma_{32} = \sigma_4;\ \sigma_{31} = \sigma_{13} = \sigma_5;\ \sigma_{12} = \sigma_{21} = \sigma_6 \quad \text{...[4]}$$

and

$$\varepsilon_{11} = \varepsilon_1;\ \varepsilon_{22} = \varepsilon_2;\ \varepsilon_{33} = \varepsilon_3;\ 2\varepsilon_{23} = 2\varepsilon_{32} = \varepsilon_4;\ 2\varepsilon_{31} = 2\varepsilon_{13} = \varepsilon_5;\ 2\varepsilon_{12} = 2\varepsilon_{21} = \varepsilon_6 \quad \text{...[5]}$$

Constitutive Matrix: The system of equations [2] can be written in matrix form, for a linear elastic, anisotropic and homogeneous material.

$$\begin{Bmatrix} \sigma_x \\ \sigma_y \\ \sigma_z \\ \tau_{xy} \\ \tau_{yz} \\ \tau_{zx} \end{Bmatrix} = \begin{bmatrix} c_{11} & c_{12} & \cdots & c_{16} \\ c_{21} & c_{22} & \cdots & c_{26} \\ \cdots & \cdots & \cdots & \\ \cdots & \cdots & \cdots & \\ \cdots & \cdots & \cdots & \\ c_{61} & c_{62} & \cdots & c_{66} \end{bmatrix} \begin{Bmatrix} \varepsilon_x \\ \varepsilon_y \\ \varepsilon_z \\ \gamma_{xy} \\ \gamma_{yz} \\ \gamma_{zx} \end{Bmatrix}$$

or $\{\sigma\} = [C]\,\{\varepsilon\}$ ...[1]

where [C] is called the constitutive matrix and its inverse relation for strain is given by

$$\begin{Bmatrix} \varepsilon_x \\ \varepsilon_y \\ \varepsilon_z \\ \gamma_{xy} \\ \gamma_{yz} \\ \gamma_{zx} \end{Bmatrix} = \begin{bmatrix} d_{11} & d_{12} & \cdots & d_{16} \\ d_{21} & d_{22} & \cdots & d_{26} \\ \cdots & \cdots & \cdots & \\ \cdots & \cdots & \cdots & \\ \cdots & \cdots & \cdots & \\ d_{61} & d_{62} & \cdots & d_{66} \end{bmatrix} \begin{Bmatrix} \sigma_x \\ \sigma_y \\ \sigma_z \\ \tau_{xy} \\ \tau_{yz} \\ \tau_{zx} \end{Bmatrix}$$

Thus, from Hooke's law, for linear elastic materials, we find stress–strain relationship in three dimensions (and later we will reduce it to two dimensions).

$$\left.\begin{aligned}\varepsilon_x &= \frac{\sigma_x}{E} - \frac{\gamma\sigma_y}{E} - \frac{\gamma\sigma_z}{E}\\ \varepsilon_y &= \frac{\sigma_y}{E} - \frac{\gamma\sigma_x}{E} - \frac{\gamma\sigma_z}{E}\\ \varepsilon_z &= \frac{\sigma_z}{E} - \frac{\gamma\sigma_x}{E} - \frac{\gamma\sigma_z}{E}\end{aligned}\right\} \quad \text{...[1]}$$

where ε_x, ε_y and ε_z are linear strains. Now, the shearing strains are:

$$\gamma_{yz} = \frac{\tau_{yz}}{G}$$

$$\gamma_{xz} = \frac{\tau_{xz}}{G}$$

$$\gamma_{xy} = \frac{\tau_{xy}}{G}$$

Now,

$$G = \frac{E}{2(1+\gamma)} \quad \text{...[2]}$$

where γ is the Poisson's ratio of the material
E is the Young's modulus of elasticity
G is the shear modulus of elasticity

From [1], we have

$$\varepsilon_x + \varepsilon_y + \varepsilon_z = \frac{(1-2\nu)}{E}(\sigma_x + \sigma_y + \sigma_z), \quad \text{(From Hooke's law)} \quad \text{...[3]}$$

Substituting for $(\sigma_y + \sigma_z)$ and so on into equation [1],
we get,

$$\sigma = D\,\varepsilon \quad \text{...[4]}$$

where

$$\sigma = \begin{Bmatrix}\sigma_x\\ \sigma_y\\ \sigma_z\\ \tau_{xz}\end{Bmatrix} = [\sigma_x\,\sigma_y\,\sigma_z\,\tau_{xz}]^T$$

and

$$D = \frac{E}{(1+\nu)(1-2\nu)}\begin{bmatrix}1-\nu & \nu & \nu & 0 & 0 & 0\\ \nu & 1-\nu & \nu & 0 & 0 & 0\\ \nu & \nu & 1-\nu & 0 & 0 & 0\\ 0 & 0 & 0 & 0.5-\nu & 0 & 0\\ 0 & 0 & 0 & 0 & 0.5-\nu & 0\\ 0 & 0 & 0 & 0 & 0 & 0.5-\nu\end{bmatrix} \quad \text{...[5]}$$

Particularly any element spreading in a single direction is known as one-dimensional element. The normal stress σ and the corresponding normal strain ε are related by $\sigma = E\varepsilon$, which is the stress-strain relation in one dimension (Hooke's law).

If u be the displacement, then the linear strain is given by $\varepsilon_x = \dfrac{\partial u}{\partial x}$, which is called the strain–displacement relation. The stress is given by

$$\boxed{\sigma_x = E\varepsilon_x;\ \varepsilon_x = D\varepsilon_x}$$

where $D = E$, (Young's modulus).

1.5 POTENTIAL ENERGY AND EQUILIBRIUM

The variational or energy methods are of primary importance in the finite element analysis. In solid mechanics, we come across the problem of determining the displacement u of the body shown in Fig. 1.12, satisfying the equilibrium equations [8]. It may be noted that the stresses are related to strain, which, in turn, are related to displacements. Approximate solution methods normally employ variational or energy methods. The energy a body possesses by virtue of its position is known as potential energy.

POTENTIAL ENERGY: (Π) (a term applicable to conservative fields of force only).

The potential energy of the system arises because of two causes:

(i) Due to position of external loads

(ii) Due to deformation of the structure (i.e., energy stored within the body due to physical deformation known as strain energy. The concept of potential energy is fundamental and exceedingly important in the study of structural mechanics and FEM. The total potential energy is denoted by the Greek letter Π which is a function of displacements.

Definition: (Potential energy) is defined as the negative of the work done in displacing a particle from its standard position to any other position, and the total potential energy, Π of an elastic body is defined as the sum of total strain energy (U) and the work potential (W_p).

The potential energy of the external load P is

$$W_p = \text{(Load) (Displacement from zero potential state).}$$

$$= -Pu \qquad \text{...[1]}$$

where the zero potential state corresponds to $u = 0$. Hence the total potential energy π is

$$\boxed{\Pi = U + W_P} \qquad \text{...[2]}$$

where U is the strain energy, W_p is the potential of the applied loads (or work potential). The strain energy density (strain energy per unit volume in the body) for a linear elastic body is defined as

$$\boxed{dU = \frac{1}{2}\sigma^{\mathrm{T}}\varepsilon} \quad \Rightarrow \quad \boxed{\text{total energy} = U = \frac{1}{2}\int_V \sigma^T \varepsilon\, dV}$$

and the potential of the applied loads (or work potential) is given by (W_p).

$$W_p = -\int_V U^T f\, dV - \int_s U^T T . ds - \sum_i U_i^T P_i \qquad ...[3]$$

Hence the total potential becomes

$$\Pi = \frac{1}{2}\int_V \sigma^T \varepsilon\, dV - \int_V U^T f\, dV - \int_s U^T T\, ds - \sum_i U_i^T P_i \qquad ...[4]$$

Principle of virtual work: Since energy has to balance itself we equate the work done by the system to the work done by the system in opposition.

The application of forces displaces various points of the system deforming it. The internal stresses moving through the internal strains do another work.

Strain energy stored in the linear electric spring:

The spring constant (stiffness of the spring) is a constant denoted by *k* which is merely the force per unit displacement. Here, force and displacement are general terms.

If *D* be the displacement of the spring, then energy stored in the spring is

$$U = \left(\frac{1}{2}\right) k\, D^2$$

In case of flexural systems, the strain energy is expressed in two ways:

$$U = \int \frac{1}{2} M^2 \frac{dx}{EI}$$

Alternatively,

$$U = \frac{EI}{2} \int \left[\frac{d^2y}{dx^2}\right]^2 dx$$

Potential of a constant force : Each and every activeforce has a capacity to perform work. For example, it may be observed that from Fig. 1.12, a force is acting at a point A which is the reference point.

Let the potential or potential energy of the force at point *A* be zero. When the force *P* moves through a distance *D* along the line of action to a point *B* it loses some of its capacity to do work. Now the potential of the force at *B* is defined as

Potential at $B = 0 - (+P)(+D)$

$= -PD$

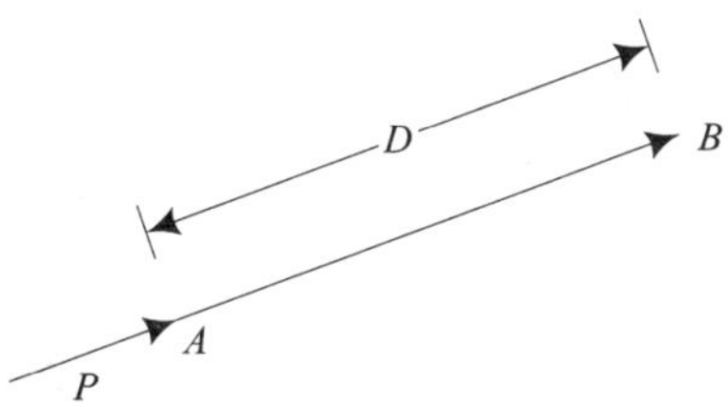

Fig. 1.12 Potential of a force

We may note two points:

(i) Both force and displacement are in the same direction. Hence (+) sign is attached to both of them.

(ii) When the force moves, it loses some of the energy. Hence (–) sign is attached in the beginning.

Potential Energy Function

The potential energy function is given by the equation

$$\Pi = U + W_P \quad ...[1]$$

where U is the strain energy of the material and W_p is the work potential.

Two spring systems are shown in Fig. 1.13.

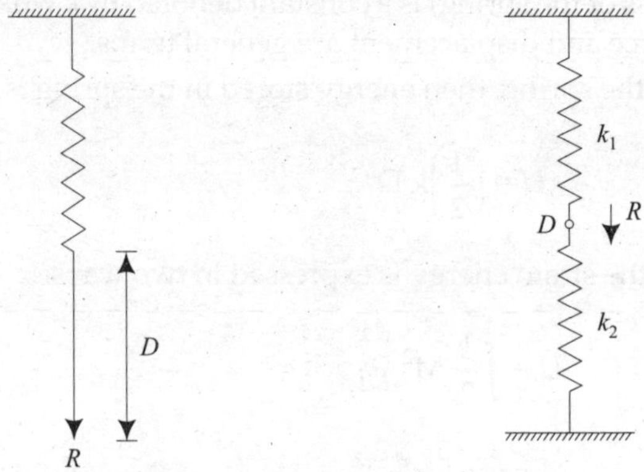

(a) Single spring (b) Springs in parallel with same displacement

Fig. 1.13

$$U = \frac{1}{2kD^2}$$

$$W_p = -RD$$

$$\pi = \frac{1}{2kD^2} - \frac{1}{2RD} \quad ...[2]$$

The P.E. function for spring system shown in Fig. 1.13 becomes

$$\pi = \frac{1}{2k_1D^2} + \frac{1}{2k_2D^2} - \text{RD} \quad ...[3]$$

Equations [1] and [2] are functions which depend on the displacement D.
Hence, they are called potential energy function.

Potential energy functional

Let us consider the potential energy for the beam shown in Fig. 1.14 below:

$$\Pi = U + W_p$$

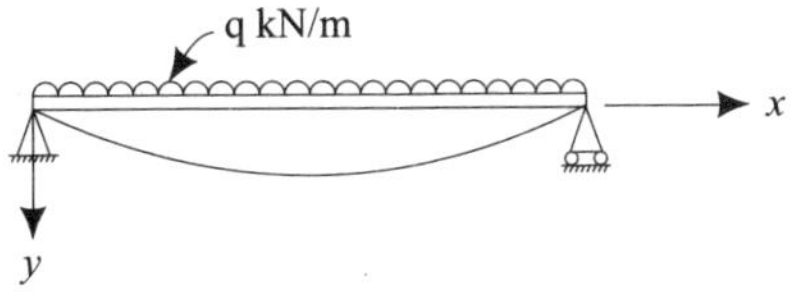

Fig. 1.14

$$U = \int_0^l \frac{EI}{2}\left[\frac{d^2y}{dx^2}\right]^2 dx$$

$$W_p = \int_0^l qy\,dx$$

$$\therefore \quad \pi = \int_0^l \left[\frac{EI}{2}\left(\frac{d^2y}{dx^2}\right)^2 - qy\right] dx \qquad ...[1]$$

($\because$ $\pi = U + W_p$)

[1] is known as a functional.

Note: Π depends not only on y and its derivatives at a point but upon their integrated effect over a region of interest. π becomes a functional when the system has infinite number of degrees of freedom.

Derivation of Principle of Stationary Potential Energy

It is known that the principle of virtual work assumes the following two systems of quantities:

(i) Equilibrium system:
σ with F in W_p
and T in S ...[1]

(ii) A system of virtual displacements u such that $\delta u = 0$

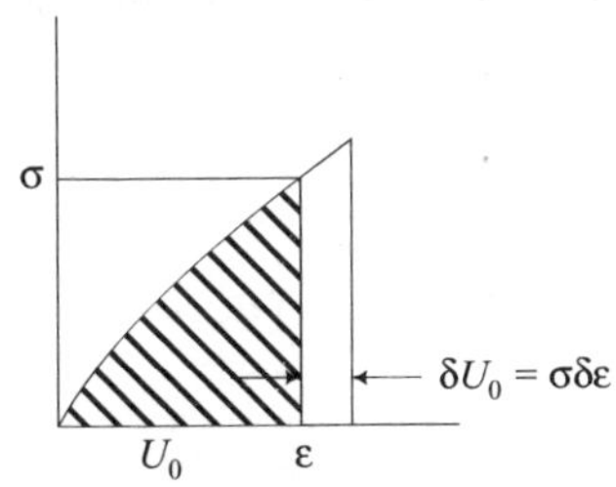

Fig. 1.15 Increment in strain energy

Another set of displacements u such that $u = \bar{u}$ on S_{ij} and u produce (by the principle of stationary potential energy) a set of ε which results in the stress σ through a single valued constitute relation.

If a unique strain energy density, "$U_0(\varepsilon)$" is associated with every set of strain ε, the material is said to be elastic. For such material

$$U = \int_0^{\varepsilon} \sigma\, \delta\, \varepsilon \qquad \text{...[2]}$$

energy in an elastic body is conserved and recovered and then

$$\frac{\partial U}{\partial \varepsilon} = \sigma \qquad \text{...[3]}$$

Now, if we consider the set of displacements, u, which produce the set of strains ε producing a set of stresses σ and if we now consider δ be infinitesimal increment in u resulting in incremental strain $\delta\varepsilon$ then the left hand side of the virtual work equation

$$\int_v \delta\, \varepsilon^T \sigma\, dv \quad = \quad \int \delta\, q^T \delta\, T ds \quad + \quad \int \delta\, q^T F\, dv$$

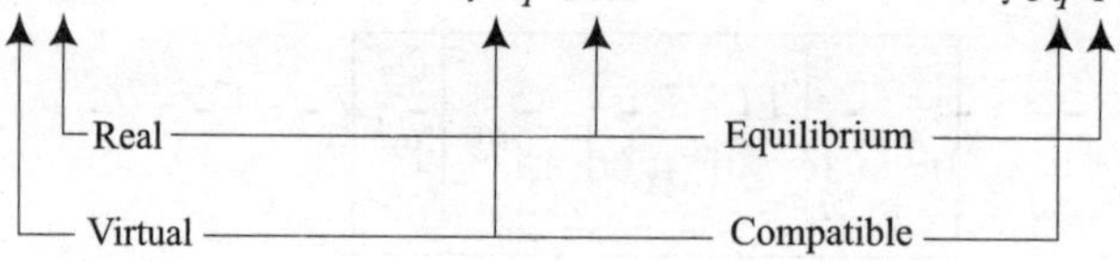

becomes
$$\int_V \sigma \delta \varepsilon\, dv = \int_V \frac{\partial U}{\partial \varepsilon}\, \delta \varepsilon\, dv = \int_V \delta\, U_0\, dv$$
$$= \delta \int U_0\, dv$$

Therefore, the principle of virtual work becomes

$$\boxed{\delta U = \int_V \delta q^T F\, dv + \int_s \delta q^T T\, ds} \qquad \text{...[4]}$$

which is now the equilibrium condition for stresses σ arising from the displacements q which are subjected to incremental displacements q, and which produces δu.

Eq. [4] is valid for an elastic body with non-conservative external forces. But it is more usual to consider the case of conservative forces for which it is possible to define the potential.

The total potential energy of the external forces V can be written as

$$W_p = -\left(\int_v q^T F\, dv + \int_s q^T T\, ds \right) \qquad \text{...[5]}$$

and the variation in potential is

$$\delta W_p = -\left(\int_v \delta q^T F\, dv + \int_s \delta q^T T\, ds\right) \quad ...[6]$$

Hence Eq. 6.17 may be written as

$$\delta U = -\delta W_p \quad \text{or,} \quad \delta (U + W_p) = 0, \quad ...[7]$$

Defining the total potential π as $U + W_p$

Then the equilibrium $\delta \Pi = 0$ for Principle of Stationary Potential energy

$$\boxed{\delta \Pi = \sum_{i=1}^{n} \frac{\partial \pi}{\partial q_i} \delta q_i = 0} \quad ...[8]$$

"Of all possible system of deformations, that are compatible and satisfy the displacement boundary conditions that system which also satisfies equilibrium makes the total potential energy stationary." That is,

$$\frac{\partial \pi}{\partial q_i} = 0 \text{ gives the condition of equilibrium.}$$

Principle of Minimum Potential Energy: It states that among all displacement fields that satisfy the prescribed constraint condition of the structure, those that satisfy the equilibrium condition make the total potential energy a minimum.

It may be observed that the structure (Fig. 1.16) is subjected to the generalized forces in the following two states *A* and *B* as q_i, $q_i + \delta q_i$ $\pi(q_i)$ and $\pi(q_i + \delta q_i)$ refer to the total potential in states *A* and *B* respectively. State *B* is in the vicinity of state *A* and is obtained by applying virtual displacement δq perturbing from states *A* to *B*. Total variation in potential energy in going from states *A* to *B* may be written as

$$\Delta \pi = \pi(q_i + \delta q_i) - \pi(q_i)$$

$$= \sum \frac{\partial \pi}{\partial q_i} \partial q_i + \frac{1}{2} \sum \sum \frac{\partial^2 \pi}{\partial q_i\, \partial q_j} \delta q_i \cdot \delta q_j \,.... \quad ...[1]$$

$$\Delta \pi = \delta \pi + \frac{1}{2} \delta^2 \pi + ...$$

For equilibrium $\delta \pi = 0$. If in addition we assume linear stress–strain relation (π is quadratic) hence $\delta^3 \pi$ ... and higher order terms are zero

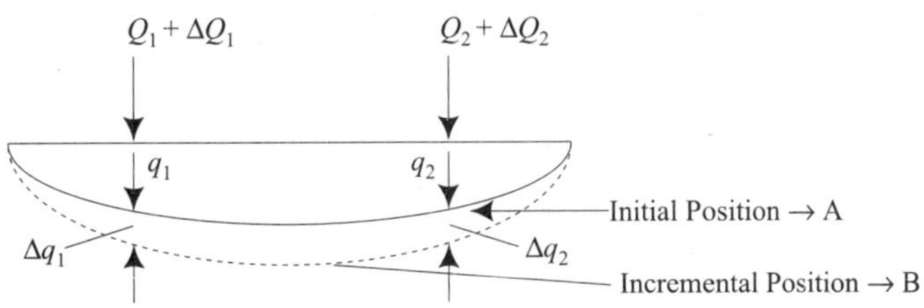

Fig. 1.16 Initial and deformed configurations of structure

$$\Delta\pi = \frac{1}{2}\sum\sum\frac{\partial^2 \pi}{\partial q_i\,\partial q_j}\,\delta q_i\;\delta q_j \qquad ...[2]$$

Moreover, $\delta^2\pi$ is not a function of Q because the potential of external forces is first order function in terms of displacement q

$$\Delta\pi = \frac{1}{2}\,\delta^2\,\pi = \frac{1}{2}\,\delta^2\,W_p \qquad [3]$$

Equation [3] is of course a familiar requirement that the slope of a function be zero at stationary point, from differential calculus.

A starting point must meet this condition but this is not sufficient to ascertain if we are at maximum, minimum or neutral points. In order to establish this the sign of curvature must be considered.

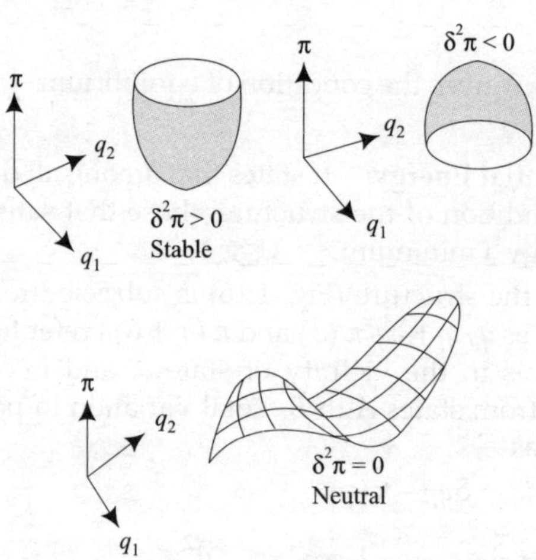

Fig. 1.17 States of equilibrium for two degree of freedom system

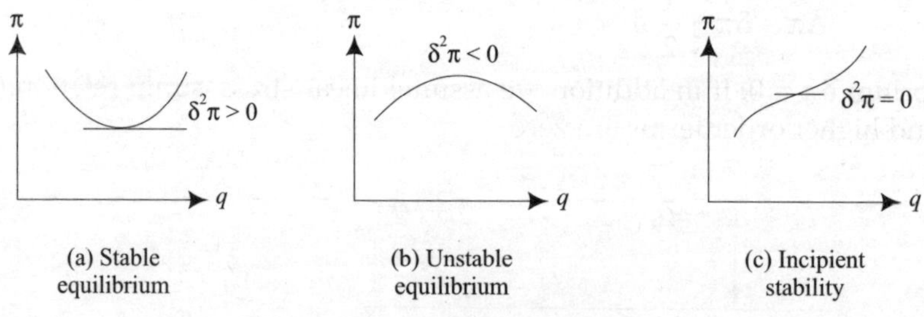

Fig. 1.18

$$\left\{\begin{array}{ll} \delta^2\pi > 0 & \text{then } \pi \text{ is minimum} \\ \delta^2\pi = 0 & \text{then } \pi \text{ is neutral} \\ \delta^2\pi < 0 & \text{then } \pi \text{ is maximum} \end{array}\right.$$

This is very clear from Figs. 1.17 and 1.18.

Note: For a linear elastic structure, we have

$$\pi = (U + W_p) \qquad \text{...(1)}$$

where U is always positive and W_p is always negative as shown in Fig. 1.19.

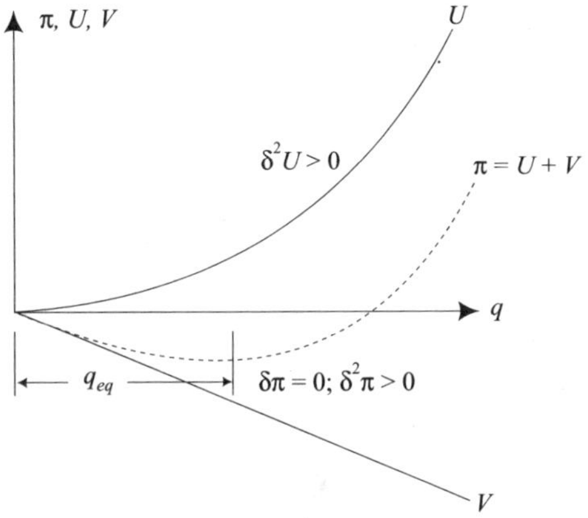

Fig. 1.19 Plot of π, U and V with respect to q

π is quadratic form and $\pi = 0$ locates a minimum point on the plot π against q.

Under these conditions principle of stationary potential energy becomes the principle of minimum potential energy.

It is seen that

(a) Minimum potential energy principle applies only to elastic systems with conservative forces.

(b) $\delta^2\,\pi$ is positive (if π is positive definite)—the theorem of stationary potential energy becomes the theorem of minimum potential energy.

Principle of Virtual Forces: The principle of virtual forces may be written as

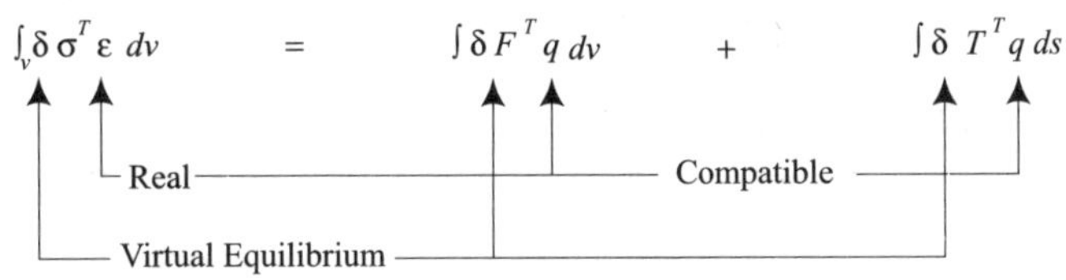

$$\int_v \delta\sigma^T \varepsilon\, dv = \int \delta F^T q\, dv + \int \delta\, T^T q\, ds$$

and the principle of stationary complementary potential may be stated as

$\delta\pi^* = 0$ for a compatible system

$$\delta\,(U^* + V^*) = 0 \qquad \text{...[2]}$$

Derivation of potential energy (Π) for the beam shown in Fig. 1.20.
(The system has a finite number of degrees of freedom.)

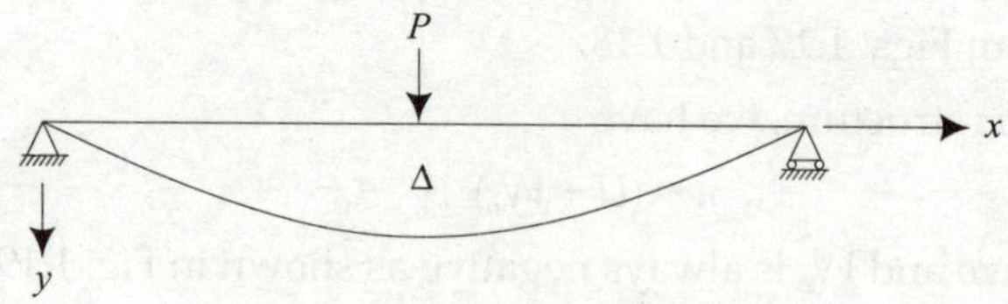

Fig. 1.20 Beam with a central load

Since $$\pi = U + W_p$$

$$U = \int_0^l \frac{EI}{2}\left(\frac{d^2y}{dx^2}\right)^2 dx$$

$$W_p = -P\Delta$$

$$\therefore \quad \pi = \int_0^l \frac{EI}{2}\left(\frac{d^2y}{dx^2}\right)^2 dx - P\Delta$$

Derivation of potential energy functional (Π) for a truss element subjected to uniform traction as shown in Fig. 1.21.

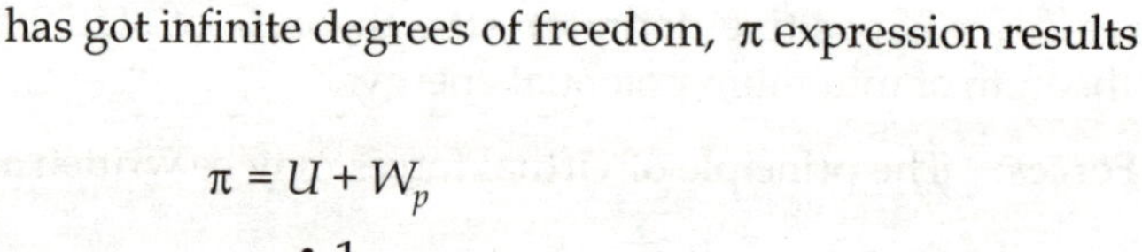

Fig. 1.21 Truss member with uniform traction

Since the system has got infinite degrees of freedom, π expression results in a functional.

Solution:

$$\pi = U + W_p \qquad \text{...[1]}$$

$$U = \int \frac{1}{2}\{\varepsilon\}^T\{\sigma\}\, dv \qquad \text{...[2]}$$

We know $$\sigma = E\varepsilon$$

$$dv = A\, dx$$

Potential Energy

$$\therefore \quad U = \int_0^l \frac{AE}{2}\varepsilon^2\, dx \qquad \text{...[3]}$$

But
$$\varepsilon = \frac{du}{dx}$$

$$\therefore \qquad U = \int_0^l \frac{AE}{2}\left(\frac{du}{dx}\right)^2 dx \qquad \text{...[4]}$$

$$W_p = -\int_0^l (q\,dx)\,u$$

$$W_p = -\int_0^l qu\,dx \qquad \text{...[5]}$$

Substituting Eqs. (4) and (5) in Eq. (1), we get

$$\Pi = \int_0^l \left[\frac{AE}{2}\left(\frac{du}{dx}\right)^2 - qu\right] dx$$

Example: Establish the P.E. functional for a beam column subjected to a uniformly distibuted load q kN/metre as shown in Fig. 1.22.
(Infinite degrees of freedom, hence, P.E. expression results in a functional.)

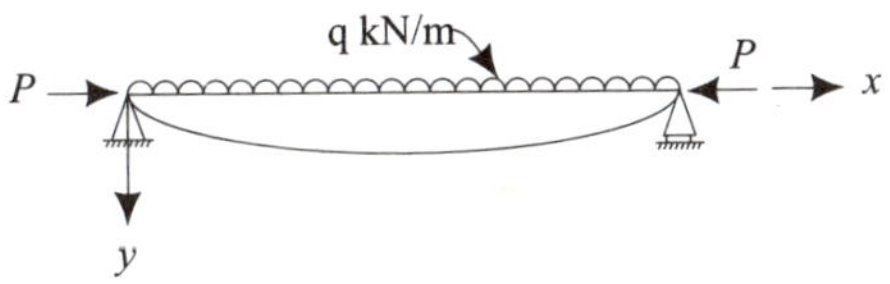

Fig. 1.22 A Beam Column

Solution: Since
$$\pi = U + W_P \qquad \text{...[1]}$$

$$U = \int_0^l \frac{EI}{2}\left(\frac{d^2y}{dx^2}\right)^2 dx \qquad \text{...[2]}$$

$$W_p = -P\lambda - qy \qquad \text{...[3]}$$

where

$$\lambda = \int_0^l \frac{1}{2}\left(\frac{dy}{dx}\right)^2 dx$$

Substituting Eqs. (2) and (3) in Eq. (1), we get

$$\pi = \int_0^1 \left[\frac{EI}{2}\left(\frac{d^2y}{dx^2}\right) - \frac{P}{2}\left(\frac{dy}{dx}\right)^2 - qy\right] dx$$

Computation of λ: From Fig. 1.23(b) originally straight *ds* has become curved *ds*. During this process a descent $d\lambda$ occurs as seen in Fig. 1.23(b).

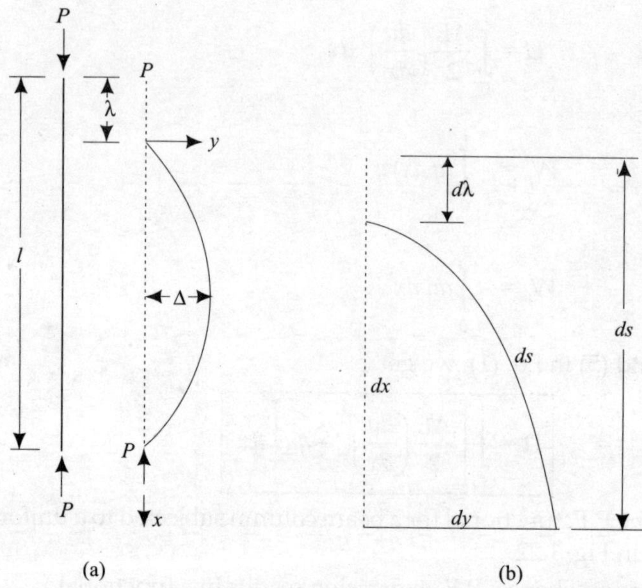

Fig. 1.23 (a) Two equilibrium shapes
(b) Originally straight *ds* has become curved *ds*

∴

$$d\lambda = ds - dx$$

$$= \sqrt{dx^2 + dy^2} - dx$$

$$= dx\left[1 + \left(\frac{dy}{dx}\right)^2\right]^{1/2} - dx$$

For small values of $\frac{dy}{dx}$, we know

$$\left[1 + \left(\frac{dy}{dx}\right)^2\right]^{1/2} = 1 + \frac{1}{2}\left(\frac{dy}{dx}\right)^2$$

$$d\lambda = dx\left[1 + \frac{1}{2}\left(\frac{dy}{dx}\right)^2\right] - dx$$

$$d\lambda = \frac{1}{2}\left(\frac{dy}{dx}\right)^2 dx \qquad ...[1]$$

Therefore, for the entire column Eq. (1) is integrated on both sides. Then

$$\lambda = \frac{1}{2}\int_0^l \left(\frac{dy}{dx}\right)^2 dx$$

Example: Figure 1.24 shows a system of springs. Assemble equations of equilibrium by direct approach. Prove that minimization of potential energy also yields the same result.

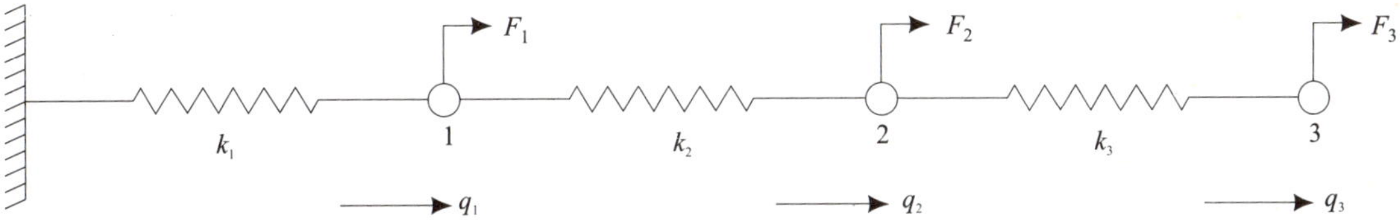

Fig. 1.24

Solution: (1) ***Direct approach:*** Let the nodes ①, ② and ③ be as shown in Fig. 1.25.

If q_1, q_2, q_3 be the displacements of nodes, then the extensions of springs 1, 2 and 3 are:

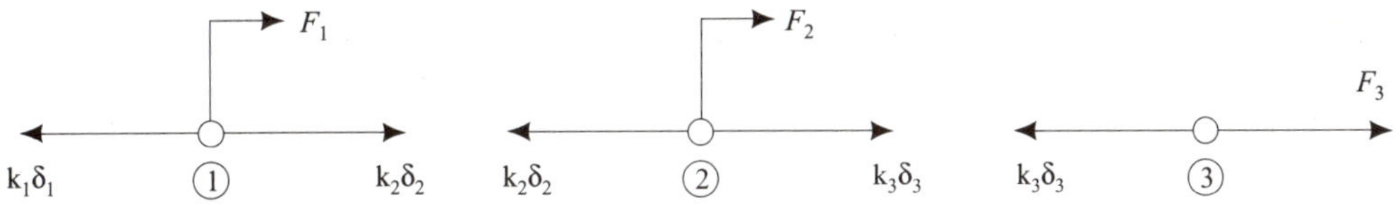

Fig. 1.25

$$\left\{\begin{aligned} \delta_1 &= q_1; \\ \delta_2 &= q_2 - q_1; \\ \delta_3 &= q_3 - q_2; \end{aligned}\right\} \qquad \text{...[1]}$$

The equilibrium equations are:

$$\left\{\begin{aligned} -k_1\delta_1 + k_2\delta_2 + F_1 &= 0 \\ -k_2\delta_2 + k_3\delta_3 + F_2 &= 0 \\ -k_3\delta_3 + F_3 &= 0 \end{aligned}\right\} \qquad \text{...[2]}$$

From first and second equations of eq. [2], we get

$$\left\{\begin{aligned} -k_1q_1 + k_2(q_2 - q_1) + F_1 &= 0 \\ -k_2(q_2 - q_1) + k_3(q_3 - q_2) + F_2 &= 0 \\ -k_3(q_3 - q_2) + F_3 &= 0 \end{aligned}\right\} \qquad \text{...[3]}$$

$$\left\{\begin{aligned} (k_1 + k_2)q_1 - k_2q_2 &= F_1 \\ -k_2q_1 + (k_2 + k_3)q_2 - k_3q_3 &= F_2 \\ -k_3q_2 + k_3q_3 &= F_3 \end{aligned}\right\} \qquad \text{...[4]}$$

Matrix form of system [4] is

$$\begin{bmatrix} (k_1+k_2) & -k_2 & 0 \\ -k_2 & k_2+k_3 & -k_3 \\ 0 & -k_3 & k_3 \end{bmatrix} \begin{Bmatrix} q_1 \\ q_2 \\ q_3 \end{Bmatrix} = \begin{Bmatrix} F_1 \\ F_2 \\ F_3 \end{Bmatrix} \quad \text{...[5]}$$

(2) ***Potential energy approach:***

$$\pi = \frac{1}{2}k_1\delta_1^2 + \frac{1}{2}k_2\delta_2^2 + \frac{1}{2}k_3\delta_3^2 - F_1q_1 - F_2q_2 - F_3q_3$$

$$= \frac{1}{2}k_1q_1^2 + \frac{1}{2}k_2(q_2-q_1)^2 + \frac{1}{2}k_3(q_3-q_2)^2 - F_1q_1 - F_2q_2 - F_3q_3$$

$$\therefore \quad \frac{\partial \pi}{\partial q_1} = 0 \Rightarrow \left\{ \begin{aligned} k_1q_1 + k_2(q_2-q_1)(-1) - F_1 &= 0 \\ k_1q_1 - k_2(q_2-q_1) - F_1 &= 0 \\ (k_1+k_2)q_1 - k_2q_2 &= F_1 \end{aligned} \right\} \quad \text{...[6]}$$

$$\frac{\partial \pi}{\partial q_2} = 0 \Rightarrow k_2(q_2-q_1) + k_3(q_3-q_2)(-1) = F_2$$

$$-k_2q_1 + (k_2+k_3)q_2 - k_3q_3 = F_2 \quad \text{...[7]}$$

and
$$\frac{\partial \pi}{\partial q_3} = 0 \Rightarrow k_3(q_3-q_2) - F_3 = 0$$

$$-k_3q_2 + k_3q_3 = F_3$$

Matrix form is given by

$$\begin{bmatrix} (k_1+k_2) & -k_2 & 0 \\ -k_2 & k_2+k_3 & -k_3 \\ 0 & -k_3 & k_3 \end{bmatrix} \begin{Bmatrix} q_1 \\ q_2 \\ q_3 \end{Bmatrix} = \begin{Bmatrix} F_1 \\ F_2 \\ F_3 \end{Bmatrix} \quad \text{...[8]}$$

Equations [5] and [8] are same. This completes the proof.

Example: The potential energy for the linear elastic rod with body force neglected is

$$\pi = \frac{1}{2}\int_0^L EA\left(\frac{du}{dx}\right)dx - 2u_1$$

where $u_1 = u(1)$.

Compute the stress in the bar.

Solution: We assume a quadratic trial function

$$u = C_1 + C_2x + C_3x^2 \quad \text{...[1]}$$

Now, when $x = 0$, $u(0) = 0$ yields $C_1 = 0$ and when $x = 2$, $u(2) = 0$ yields

$$0 = C_1 + 2C_2 + 4C_3$$

$$\Rightarrow C_2 = -2\,C_3$$

$\therefore$ [1] becomes $\quad u = C_3\,(-2x + x^2)$

$$\Rightarrow u_1 = -C_3 \quad (\text{at } x = 1)$$

$$\therefore \quad \frac{du}{dx} = 2C_3\,(-1 + x) \text{ and } \pi = \frac{1}{2}\int_0^2 4C_3^2(-1 + x)^2\,dx - 2(-C_3)$$

$$= 2C_3^2 \int_0^2 (1 - 2x + x^2)\,dx + 2C_3$$

$$= 2C_3^2\left(\frac{2}{3}\right) + 2C_3$$

$$\therefore \quad \frac{\partial \pi}{\partial C_3} = 4C_3\left(\frac{2}{3}\right) + 2 = 0.$$

$$\Rightarrow C_3 = -\,0.75;$$

$$u_1 = -C_3 = 0.75$$

Hence, the stress in the bar is given by

$$\sigma = E\,\frac{du}{dx}$$

$$= 2(-0.75)\,(-1 + x)$$

$$\Rightarrow \quad \boxed{\sigma = 1.5(1 - x)}$$

1.6 RAYLEIGH–RITZ'S METHOD

The Rayleigh–Ritz method was proposed independently by Lord Rayleigh (1842–1919) and Walter Ritz (1878–1909). Stated simply, "a function is assumed in terms of some unknown coefficients that would represent a solution of equation $\frac{d^2T}{dx^2} = -\frac{Q}{K}$. An approximate solution can be obtained for a differential equation using the Rayleigh–Ritz method whereby an approximating function is substituted into the variational function. The approximating function must satisfy the boundary conditions for the problem being studied.

The Rayleigh–Ritz method and the Ritz method are entirely different. The distinction is that the Ritz method always refers to the nonvariational integral formulation whereas the Rayleigh–Ritz method makes use of variational approach.

Illustration:

Let $\quad I = \int_a^b F(x, y, \dot{y}, \ddot{y})\,dx \quad$ be a functional. ...[1]

Choosing $$F = F(x, y, \dot{y}, \ddot{y}) = -\frac{1}{2}\left(\frac{dy}{dx}\right)^2 + 1000\,x^2y \qquad ...[2]$$

$\therefore$ Equation [1] becomes

$$I = \int_0^1 \left(1000\,x^2y - \frac{1}{2}\left(\frac{dy}{dx}\right)^2\right)dx \qquad ...[3]$$

[3] maximizes or minimizes I.

In Rayleigh–Ritz method for continuous system we deal with the functional :

$$I = \int_a^b F(x, y, \dot{y}\,)\,dx \qquad ...[4]$$

Discription of the Method

Rayleigh–Ritz method is a method for determining approximate solutions of functional equations through the expedient of replacing them by finite systems of equations.

It is one of the most useful approximate methods stemming from variational considerations, wherein, for the present we employ the following approximate displacement field components for expressing the total potential energy:

$$\left\{\begin{aligned} U_n &= \phi_0(x, y, z) + \sum_{i=1}^{l} a_i\phi_i(x, y, z) \\ V_n &= \psi_0(x, y, z) + \sum_{j=l+1}^{m} b_j\psi_j(x, y, z) \\ W_n &= \gamma_0(x, y, z) + \sum_{k=m+1}^{n} c_k\gamma_k(x, y, z) \end{aligned}\right\}_{n>m>l} \qquad ...[1]$$

where a_i, b_j and c_k are $3n$ linearly independent parameters, which behave as generalised coordinates. In order to determine these coefficients $3n$ equations/conditions are sought: ϕ_i, ψ_j and γ_k are chosen as to satisfy kinematic boundary conditions.

Finite element technique may be thought of as an extension of Ritz's analysis of a continuous system. It tries to represent any continuous system with the help of parameters at some chosen points only. It was first taken up by Rayleigh and then elaborated by Ritz. The method is also known as Rayleigh–Ritz method.

For a most stable equilibrium, the total potential energy is minimized.

We may write the system of equations [1] as

$$\left\{\begin{aligned} U_n &= \sum a_i \phi_i ; \\ V_n &= \sum b_j \psi_j ; \\ W_n &= \sum c_k \gamma_k ; \end{aligned}\right\} \qquad ...[2]$$

$$\therefore \quad \left\{\begin{aligned} \delta U_n &= \sum \phi_i . \delta a_i \\ \delta V_n &= \sum \psi_j . \delta b_j \\ \delta W_n &= \sum \gamma_k . \delta c_k ; \end{aligned}\right\} \qquad ...[3]$$

The total energy can be expressed as variations in a_i, b_j and c_k means variations in energy.

$$\therefore \quad \delta U = \sum \left(\frac{\partial U}{\partial a_i} . \delta a_i + \frac{\partial U}{\partial b_j} . \delta b_j + \frac{\partial U}{\partial c_k} . \delta c_k \right) = 0 \qquad ...[4]$$

This gives $3n$ equations as:

$$\frac{\partial U}{\partial a_i} = 0 = \frac{\partial U}{\partial b_j} = \frac{\partial U}{\partial c_k} \qquad ...[5]$$

a_i, b_j and c_k can be solved from these equations. Figure 1.26 shows the type of approximations it follows:

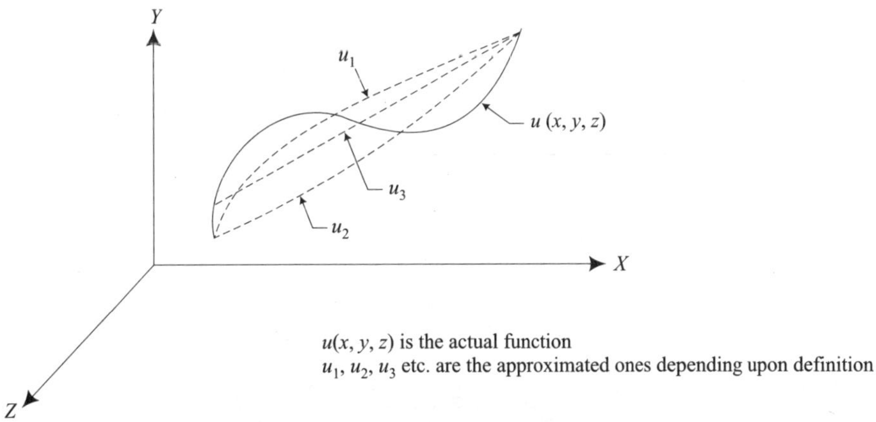

Fig. 1.26 Ritz's Approximation

Illustration:

We know that

$$\frac{\partial^2 q}{\partial x^2} + \frac{\partial^2 q}{\partial y^2} + f(x, y) = 0 \qquad ...[1]$$

is a Poisson's equation. Its solution can be approximated in the form:

$$q(x, y) = \Sigma M_i(x, y)\, q_i \qquad ...[2]$$

where q_i are the parameters defined at only some selected points i (i = 1, 2, 3, ... n), M_i are the functions to be designed/chosen in a way to satisfy all the boundary conditions related to the problem.

The governing functional can be expressed as

$$j(q) = \frac{1}{2}\iint \left[\left(\frac{\partial q}{\partial x}\right)^2 + \left(\frac{\partial q}{\partial y}\right)^2 - 2f(x, y)q\right] dx\, dy \qquad ...[3]$$

In order that $J(q)$ is minimum,

$$\frac{\partial J}{\partial q_i} = 0$$

$$\text{Therefore, } \frac{\partial J}{\partial q_i} = \iint \left[\frac{\partial q}{\partial x} \cdot \frac{\partial}{\partial q_i}\left(\frac{\partial q}{\partial x}\right) + \frac{\partial q}{\partial y} \cdot \frac{\partial q}{\partial q_i}\left(\frac{\partial q}{\partial y}\right) - f(x, y) \cdot \frac{\partial q}{\partial q_i}\right] dx\, dy = 0 \qquad ...[4]$$

The equation [4] yields n equations for the unknowns giving approximate solution to $q(x, y)$ as $q'(x, y)$, if $M_i = N_i$

OR,

$$\iint \left[\frac{\partial}{\partial x}[N](q) \cdot \frac{\partial}{\partial q_i}\left(\left[\frac{\partial N}{\partial x}\right]\{q\}\right)\right] + \frac{\partial}{\partial y}[N](q) \cdot \frac{\partial}{\partial q_i}\left(\left[\frac{\partial N}{\partial y}\right]\{q\}\right)$$

$$- \left[f(x, y) \cdot \frac{\partial}{\partial q_i}[N]\{q\}\right] dx \,.\, dy = 0 \qquad ...[5]$$

OR,

$$\iint \left(\left[\frac{\partial N}{\partial x}\right]\{q\} \cdot \frac{\partial N_i}{\partial x} + \left[\frac{\partial N}{\partial y}\right]\right)(q) \cdot \frac{\partial N_i}{\partial y} - \int (x, y) \,.\, N_i \Bigg) dx \,.\, dy = 0 \qquad ...[6]$$

OR,

$$\iint \left(\left[\frac{\partial N}{\partial x}\right] \cdot \frac{\partial N_i}{\partial x} + \left[\frac{\partial N}{\partial y}\right]\frac{\partial N_i}{\partial y}\right) .\, dx \,.\, dy \,.\, \{q\} + \iint - f(x, y) \,.\, N_i \,.\, dx \,.\, dy = 0 \qquad ...[7]$$

OR,

$$[K]\{q\} - \{F\} = \{0\} \qquad ...[8]$$

where

$$[K] = \iint \left(\left[\frac{\partial N}{\partial x}\right] \cdot \frac{\partial N_i}{\partial x} + \left[\frac{\partial N}{\partial y}\right] \cdot \frac{\partial N_i}{\partial y}\right) .\, dx \,.\, dy$$

$$\{F\} = \iint f(x, y) \,.\, N_i \,.\, dx \,.\, dy$$

The accuracy of the method is dependent on number of points selected and the trial functions taken. The larger the number of points the nearer would be the solution to the exact one.

Comments on the Rayleigh-Ritz Method

The Rayleigh–Ritz method consists of assuming the form of the unknown solution in terms of known functions (trial functions) with unknown adjustable parameters. The assumed fields must be admissible and they should satisfy the geometric boundary conditions of the system. It should have a minimum of one Ritz parameter.

Since the Ritz method is a method involving functional, i.e. since the method is not derived after the extremising process in the variational principle the admissible function need not satisfy the natural boundary conditions but only the geometric or forced boundary conditions. But, a physical problem may contain natural boundary conditions as well. In as far as the approximating functions satisfying only the forced conditions, Ritz method will be an approximate method only. If the approximating function satisfies all the boundary conditions then Rayleigh–Ritz method will yield good results. Note that the basic idea of Rayleigh–Ritz method is to approximate the continuous structure by a system having a finite number of degrees of freedom. Reducing the degree of freedom of a structure is equivalent to introducing additional geometric constraints. This renders the structure stiffer than the original. This is the reason for getting an approximate result.

In choosing the approximate function the following are to be noted. In method like Rayleigh–Ritz , the approximate function need to satisfy only the geometric boundary conditions and not all boundary conditions. If the chosen function satisfies this criterion then it is called *admissible function*, while an approximate solution satisfying all the boundary conditions is called a *comparison function*.

The approximating functions used normally are polynomials, sometimes sine and cosine terms are also used. Beyond this, there are no explicit answers to the following questions:

(i) Whether this is the only function which can be used ?

(ii) What kind of assumed pattern is best ?

(iii) How many terms in the function are to be used ?

(iv) How good is the approximate solution ?

It should be noted that the completeness demands that low-order terms be included. For example, assume that the bar carries only an axial load P as shown in Fig. 1.27. We found that the exact displacement field for this case is $u = a_1x$. If we omit the term a_1x in the polynomial

$$u = a_2x^2 + a_3x^3 + ...$$

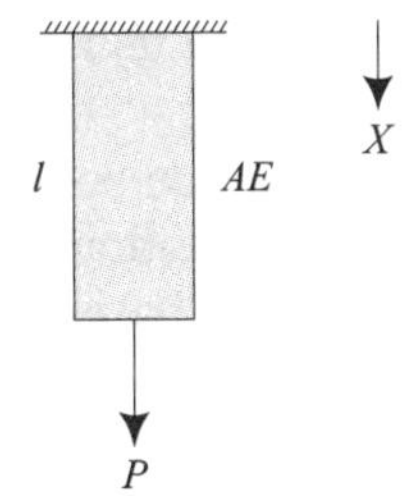

Fig. 1.27

then completeness is destroyed, and we cannot get the exact answer no matter how high is the degree of the polynomial. On the other hand we retain this as shown below:

$$u = a_1x + a_2x^2 + ...$$

and solve the problem, the solution does not change.

Notes: (i) Fourier series are complete.

(ii) Although Rayleigh–Ritz method yields good results for the function, still it cannot be reliably used for getting other values especially if they have to be obtained after differentiation.

Conditions for Convergence

In the Rayleigh–Ritz method, it is observed as the number of approximating functions increases the solution converges to the true solution provided the functions satisfy the following sufficient conditions:

(1) The infinite set of chosen approximate functions must be capable of representing the true solution exactly. This is called the condition of *completeness*.

(2) The function must give rise to only finite terms in the functional to be minimized.

(3) Each function must be an acceptable solution in the physical sense, i.e. it must not violate the material continuity conditions implicit in the derivation of the differential equation.

(4) The chosen function must at least satisfy the essential boundary conditions even though it is preferable that natural boundary conditions are satisfied as well.

In general, to ensure good convergence it is better to choose an approximate function having continuity of values and derivatives upto an order one less than the highest order derivative appearing in the equation to be minimized.

Important characteristics of Ritz's method are listed as follows.

1. Ordinarily the accuracy of the assumed displacement is increased with the increase in the number of the parameters used

$$W = \sum_{n=1}^{\alpha} a_n \phi_n(x)$$

2. Although Ritz's method may lead to fairly accurate expressions for displacements, the corresponding stress may differ significantly from their exact values. This is due to the fact that stresses generally depend upon derivatives of the displacements. Evidently, derivatives of approximate functions are usually less accurate approximations than the functions themselves.
3. The differential equations of equilibrium do not enter the analysis. Equilibrium is satisfied in an average sense through effecting variation due to stationary property of the potential energy. Thus, stresses computed on the basis of approximate displacements do not, in general, satisfy the equilibrium conditions.
4. Since the Ritz's method approximates the system having infinite degrees of freedom to finite degrees of freedom, the approximate system is more stiff than the actual system. Hence, the method overestimates the stiffness of a structure.

5. It follows from the above that the buckling load computed by this method will always be greater than or equal to the exact analysis. This is analogous to the argument advanced for the higher value of the Rayleigh frequency in vibration theory.

Similarity in Rayleigh–Ritz's Method and FEM.

In both methods

(1) We use approximating functions as trial solution.

(2) We take linear combinations of trial functions.

(3) Completeness condition of the function should be satisfied.

(4) Solution is sought by making a functional stationary.

Difference between FEM and Ritz's Method.

The following table provides the differences between FEM and Ritz's method.

Finite Element Method	*Ritz's Method*
1. Trial functions need not be defined over the whole domain.	1. Trial functions are to be defined over the whole domain.
2. Trial functions need not satisfy boundary conditions.	2. Trial functions are chosen so as to satisfy boundary conditions.
3. Trial functions have to satisfy continuity and compatibility only.	3. Compatibility and continuity are obvious.
4. There is no limitation for geometry of the system except that for each element.	4. Since the trial functions have to satisfy boundary conditions and are to be defined over whole domain solutions for complex geometry are difficult.
5. This method is a special case of Ritz's method.	5. It forms basis for all FEM and weighted residual approaches.

Comparison and Contrast between Potential Energy Method and Rayleigh–Ritz Method.

In both the cases the starting point is the computation of total potential energy, $\Pi = U + W_p$.

Potential Energy Method	*Rayleigh–Ritz Method*
1. Π is a function of generalized coordinates.	1. Π is a function of Ritz parameters.
2. Number of generalised coordinates is equal to the number of degrees of freedom.	2. A displacement function in terms of Ritz parameter is chosen which can be from 1 to ∞.
3. Is an exact method.	3. Usually approximate method. For simple structure, sometimes it may lead to exact value.
4. Variation is effected w.r.t. generalised coordinates using the stationary property.	4. It is w.r.t. Ritz parameters.
5. The stress resultants are obtained using the constitutive laws such as the slope deflection-deflection equations.	5. Using the derivatives of approximate displacement field various quantities are found.
6. Solution of equations is needed in both the methods.	

Distinction between Boundary Condition Parameter and Ritz Parameter.

The trial function or displacement function for a cantilever beam shown in Fig. 1.28 can be assumed as :

$$u = a_0 + a_1x + a_2x^2 \quad ...[1]$$

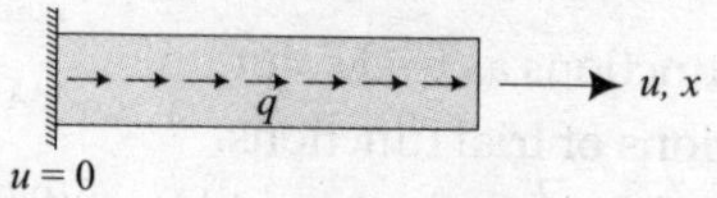

Fig. 1.28

The function $u = a_0 + a_1x + a_2x^2$

or, u = [Constant term] + [Linear term] + [Quadratic term]

has two types of parameters known as (i) boundary condition parameter and (ii) Ritz parameter.

Now, we eliminate the boundary condition parameter from [1]. So, when

$$x = 0; u = 0$$

$\therefore$ $u = a_0 + a_1x + a_2x^2$ becomes

$$0 = a_0 + 0 + 0$$

or, $\boxed{a_0 = 0}$ (b.c. parameter)

where a_1 and a_2 are Ritz parameters. These can be found by using the stationary condition of the potential energy. The number of Ritz parameters may be any number from one onwards.

Another condition is that the function should start from the lower order term.

Example: Find an expression for the stress in the system shown in Fig. 1.29.

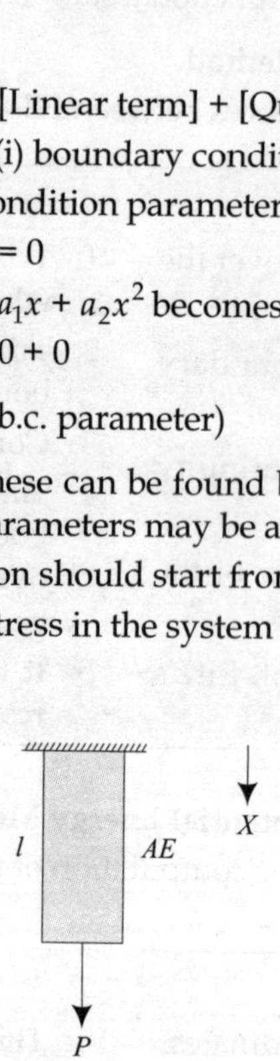

Fig. 1.29

Solution: Let the trial function be $u = a_0 + a_1x$ ...[1]

$$\left\{\begin{array}{c}\text{When } x = 0, u = 0 \\ \therefore a_0 = 0; \text{ the polynomial in [1]} \\ \text{where } a_1 \text{ is Ritz parameter.}\end{array}\right\}$$

$$\frac{du}{dx} = \varepsilon = a_1$$

We know $$U = \int \frac{1}{2} \sigma \varepsilon \, dv$$

$$\sigma = E\varepsilon$$

$$dv = A\,dx$$

$$U = \int_0^l \frac{1}{2} E\varepsilon^2 A\,dx = \frac{1}{2} . AEl\, a_1^2$$

Now, potential energy $= - Pu_1 = W_p$

When $x = 1,\ u_1 = a_1 l$

$\therefore$ $$\Pi = U + W_p = \frac{1}{2} AEl\, a_1^2 - Pa_1 l$$

Now effecting variation with respect to a_1

$$\delta\pi = \frac{d\pi}{da_1} \delta a_1 = 0$$

$$[AEl\, a_1 - Pl]\, \delta a_1 = 0$$

$\therefore$ $$a_1 = \frac{P}{AE}.$$

Therefore, displacement function becomes

$$u = \frac{P}{AE} x$$

Now strain $$= \varepsilon = \frac{du}{dx} = \frac{P}{AE}.$$

Stress $$= \sigma = E\varepsilon = E \frac{P}{AE} = \frac{P}{A}.$$

Example: Find the stress at the support and in the middle for the system shown in Fig. 1.30. Use the following displacement field

$$u = \frac{4hx(l - x)}{l^2}$$

where h is the unknown displacement in the middle.

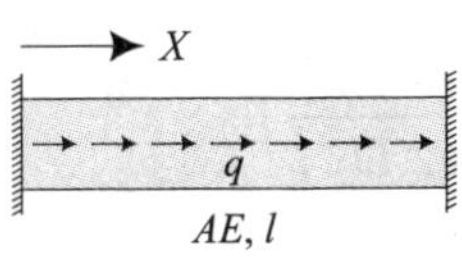

Fig. 1.30

Solution: Let $u = a_0 + a_1 x + a_2 x^2$ be the trial function. The given displacement field satisfies the geometric boundary conditions and hence is a suitable polynomial.

$$u = \frac{4hx(l-x)}{l^2}$$

where h is Ritz parameter.

$$\frac{du}{dx} = \frac{4h}{l^2}(l-2x)$$

$$U = \frac{1}{2}\int \sigma\varepsilon\, dv$$

$$U = \frac{1}{2}\int_0^l AE\varepsilon^2\, dx$$

$$U = \frac{1}{2} AE \int_0^l \frac{16h^2}{l^4}(l-2x)^2\, dx$$

$$U = \frac{8}{3}\frac{AE\,h^2}{l}$$

$$W_p = -\int_o^l \frac{q\,4hx\,(l-x)\,dx}{l^2}$$

$$W_p = -\frac{4}{6}\,qhl$$

$$\therefore \quad \pi = W_p + U = \left[-\frac{4}{6}\,qhl + \frac{8}{3}\frac{AE}{l}h^2\right]$$

Effecting variation with respect to h

$$\delta\pi = \frac{d\pi}{dh}\,\delta h = 0$$

$$= \left[\frac{-4}{6}\,ql + \frac{16}{3}\frac{A\,Eh}{l}\right]\delta h = 0$$

Since δh is not zero

$$h = \frac{3\,ql^2}{24\,AE}$$

$$\therefore \quad u = \frac{4\times 3\times ql^2}{24\,AEl^2}\,[lx - x^2]$$

$$\varepsilon = \frac{du}{dx} = \frac{q}{2AE}\,[l-2x]$$

$$\sigma = E\varepsilon = \frac{q}{2A}[l - 2x]$$

when $x = 0$, $\sigma = \frac{ql}{2A}$ (Tension)

when $x = l/2$, $\sigma = 0$

when $x = l$, $\sigma = -\frac{ql}{2A}$ (compressive)

In the application of Rayleigh–Ritz method to solve the problems relating to flexural systems, the only internal energy which contributes to the strain energy U is due to bending. The other energies are assumed to be negligible.

$$U = \frac{1}{2}\int \frac{M^2 dx}{EI} \quad \text{...[1]}$$

We know $EI\frac{d^2y}{dx^2} = M$

Substituting in the above $U = \frac{EI}{2}\int_0^l \left(\frac{d^2y}{dx^2}\right)^2 dx.$ ...[2]

Example: Using Rayleigh–Ritz method determine the expressions for displacement and stress for the system shown in Fig. 1.31, the intensity of loading varies linearly. Assume the following displacement fields and $A = E = l = C =$ unity.

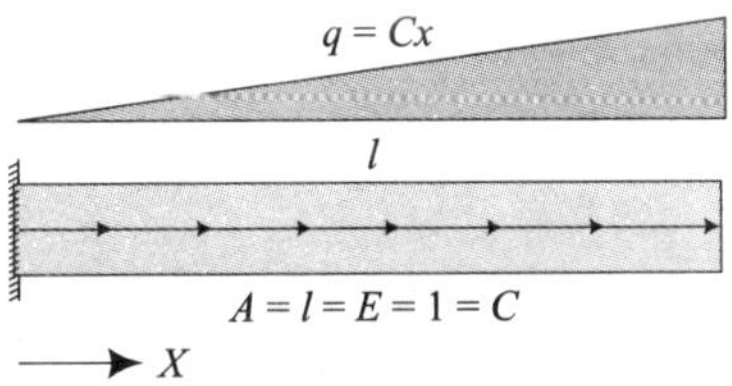

Fig. 1.31

(i) $u = a_0 + a_1x$
(ii) $u = a_0 + a_1x + a_2x^2$
(iii) $u = a_0 + a_1x + a_2x^2 + a_3x^3$

Solution (i):

Given: $u = a_0 + a_1x;\ q = Cx$

When $x = 0,\ u = 0,\ \therefore\ a_0 = 0$

$u = a_1x$ ...[1]

where a_0 is boundary condition parameter and a_1 is Ritz parameter.

Now, $$\boxed{\Pi = U + W_p}$$ ...[2]

where $$W_p = -\int_0^l qu\,dx = -\int_0^l a_1 x x\,dx$$

$$= -\frac{a_1}{3}; \quad (\therefore q = x; \text{ since } C = 1)$$

Now $$U = \frac{1}{2}\int \sigma^T \varepsilon\, dv$$

$$= \frac{AE}{2}\int_0^l a_1^2\,dx$$

$$= \frac{AE\,a_1^2}{2}$$

Hence $$\Pi = -\frac{a_1}{3} + \frac{AE\,a_1^2}{2}, \text{ (using [2])}$$

$$\delta\Pi = \frac{\partial \pi}{\partial a_1}\delta a_1 = 0$$

$$\Rightarrow \quad \delta\Pi = \left[-\frac{1}{3} + \frac{2AE}{2}a_1\right]\delta a_1 = 0$$

$$\Rightarrow \quad a_1 = \frac{1}{3} \quad (\because A = 1 = E)$$

$\therefore$ Equation [1] that is, $$u = a_1 x = \frac{1}{3}x$$

$$\therefore \quad \frac{du}{dx} = \frac{1}{3} = \varepsilon$$

$$\sigma = E\left(\frac{1}{3}\right) = \frac{1}{3} \quad (\because E = 1)$$

$$\sigma = \frac{1}{3}$$

Solution (ii):

$$u = a_1 x + a_2 x^2$$

where a_1 and a_2 are Ritz parameters

$$\frac{du}{dx} = \varepsilon = a_1 + 2a_2 x$$

$$U = \frac{1}{2}\int \sigma^T \varepsilon\, dv$$

$$U = \frac{1}{2}\int_0^1 [a_1 + 2a_2 x]^2\,dx$$

$$U = \frac{1}{2}\left[a_1^2 + \frac{4a_2^2}{3} + 2a_1a_2\right]$$

$$W_p = -\int_0^1 qu\,dx = -\int_0^1 x[a_1x + a_2x^2]\,dx$$

$$W_p = -\left[\frac{a_1}{3} + \frac{a_2}{4}\right]; \text{Now, } \boxed{\pi = U + W_p}$$

$$\Rightarrow \qquad \pi = -\left[\frac{a_1}{3} + \frac{a_2}{4}\right] + \frac{1}{2}\left[a_1^2 + \frac{4}{3}a_2^2 + 2a_1a_2\right]$$

$$\delta\pi = \frac{\partial \pi}{\partial a_1}\delta a_1 = 0$$

$$\frac{1}{2}[2a_1 + 2a_2] - \frac{1}{3} = 0 \qquad \text{...[1]}$$

Similarly, $\frac{1}{2}\left[2a_1 + \frac{8}{3}a_2\right] - \frac{1}{4} = 0$ ...[2]

Solving equations [1] and [2] simultaneously,

$$a_1 = \frac{7}{12}, a_2 = -\frac{1}{4}$$

$$\therefore \qquad u = \frac{7}{12}x - \frac{1}{4}x^2$$

$$\frac{du}{dx} = \frac{7}{12} - \frac{x}{2} = \varepsilon$$

$$\sigma = E\varepsilon = \left(\frac{7}{12} - \frac{x}{2}\right); \quad \text{(Since } E = 1\text{)}$$

when $x = 0$, $\sigma = \frac{7}{12}$ instead of $\frac{1}{2}$.

when $x = 1$, $\sigma = \frac{1}{12}$ instead of zero.

Solution (iii):

$$u = a_1x + a_2x^2 + a_3x^3 \qquad (\because a_0 = 0 \text{ because of boundary conditions})$$

We know that $q = x$

$$\therefore \qquad W_p = -\int_0^1 qu\,dx$$

$$= -\int_0^1 x(a_1x + a_2x^2 + a_3x^3)\,dx$$

$$= -\left[\frac{a_1}{3} + \frac{a_2}{4} + \frac{a_3}{5}\right]$$

and $$U = \frac{1}{2}\int \sigma^T \varepsilon\, dv$$

$$= \frac{1}{30}\left[15a_1^2 + 20a_2^2 + 27a_3^2 + 30a_1a_2 + 30a_1a_3 + 45a_2a_3\right]$$

$$\boxed{\Pi = U + W_p}$$

$$= \frac{1}{30}\left[15a_1^2 + 20a_2^2 + 27a_3^2 + 30a_1a_2 + 30a_1a_3 + 45a_2a_3\right] - \left[\frac{a_1}{3} + \frac{a_2}{4} + \frac{a_3}{5}\right]$$

Now, effecting variation, we get the following matrix:

$$\frac{1}{30}\begin{bmatrix} 30 & 30 & 30 \\ 30 & 40 & 45 \\ 30 & 45 & 54 \end{bmatrix}\begin{bmatrix} a_1 \\ a_2 \\ a_3 \end{bmatrix} = \begin{bmatrix} 1/3 \\ 1/4 \\ 1/5 \end{bmatrix}$$

$$\left.\begin{aligned} a_1 &= 1/2 \\ a_2 &= 0 \\ a_3 &= -1/6 \end{aligned}\right\} \text{Ritz parameters.}$$

∴ The displacement function

$$u = \frac{1}{2}x - \frac{1}{6}x^3$$

$$\varepsilon = \frac{du}{dx} = \frac{1}{2} - \frac{1}{2}x^2$$

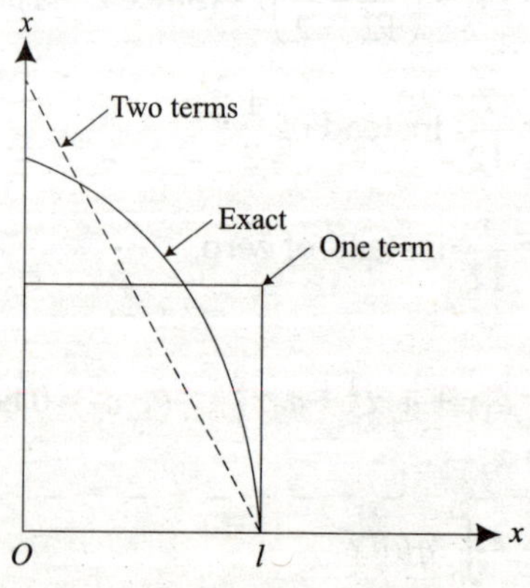

Fig. 1.32 Stress pattern

and $$\sigma = E\varepsilon = \left(\frac{1}{2} - \frac{1}{2}x^2\right); \qquad (\because E = 1)$$

$$\left\{\begin{array}{ll} \text{when } x = 0; & \sigma = \frac{1}{2}; \\ \text{when } x = 1, & \sigma = 0; \end{array}\right\}$$

This shows that the b.c. is satisfied. Hence the given u (displacement function) is exact one whereas the former two are approximate functions.

Example: Use Rayleigh–Ritz method to find the value of the central deflection of a uniform fixed beam shown in Fig. 1.33 that carries a central concentrated load P. Assume the following two trial functions:

(1) $y = c[x^4 + l^2x^2 - 2lx^3]$

(2) $y = y_0\left[1 - \cos\frac{2x\pi}{l}\right]$

where c and y_0 are Ritz parameters.

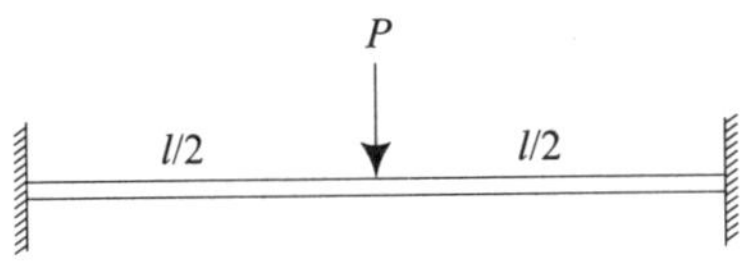

Fig. 1.33

Solution (i): *Central deflection*

$$\left\{\begin{array}{l} y = c[x^4 + l^2x^2 - 2lx^3] \\ y' = c[4x^3 + 2xl^2 - 6lx^2] \\ y'' = e[12x^2 + 2l^2 - 12lx] \end{array}\right\} \qquad \text{...[1]}$$

$\therefore$ $$\boxed{U = \frac{EI}{2}\int_0^l (y'')^2\, dx} \qquad \text{...[2]}$$

Substituting for y'' and integrating, we get

$$U = \frac{48}{120}\, c^2 l^5$$

Now the central deflection under the load is

$$y_{max} = \frac{cl^4}{16}$$

$$\therefore \qquad \pi = (\pi = U + W_p) = U - \frac{Pcl^4}{16}$$

$$\therefore \qquad \pi = EI\,\frac{48}{120}\,c^2 l^5 - \frac{Pcl^4}{16}$$

$$\delta\pi = \frac{d\pi}{dc}\,\delta c = 0$$

$$= \left[\frac{48}{60}\,cl^5 - \frac{Pl^4}{16}\right]\delta c = 0$$

$$c = \frac{15}{192}\,\frac{P}{lEI}$$

$$\therefore \qquad y_{max} = \frac{15}{16}\,\frac{Pl^3}{192\,EI}$$

The exact value is $\dfrac{Pl^2}{192EI}$.

Solution (ii): *Central deflection*

$$\left\{\begin{aligned} y &= y_0\left[1 - \cos\frac{2\pi x}{l}\right] \\ y'' &= \frac{4\pi^2}{l^2}\cos\frac{2\pi\, x}{l} \end{aligned}\right\} \qquad \text{...[1]}$$

$$\boxed{U = \frac{EI}{2}\int_0^l (y'')^2\,dx} \qquad \text{...[2]}$$

Substituting for y'' and simplifying

$$U = \frac{4\pi^4 y_0^2 EI}{l^3}$$

The central deflection $\quad y_{max} = 2y_0$

$$\therefore \qquad \Pi = U + W_p = U - P2y_0$$

$$\Pi = \frac{4\pi^4 y_0^2\,EI}{l^3} - 2Py_0$$

$$\delta\Pi = \frac{\partial\pi}{\partial y_0}\,\delta y_0 = 0$$

$$= \left[\frac{8\pi^4 y_0\,EI}{l^3} - 2P\right]\delta y_0 = 0$$

$$\therefore \qquad y_0 = \frac{Pl^3}{4\pi^4 EI}$$

The central deflection

$$y_{max} = 2y_0 = \frac{Pl^3}{2\pi^4 EI}$$

which is very close to the exact value $\dfrac{Pl^3}{192EI}$

Example: Find the deflection at the centre of a simply supported beam subjected to a concentrated load P at its midpoint as shown in Fig. 1.34. Find the maximum stress developed in the beam.

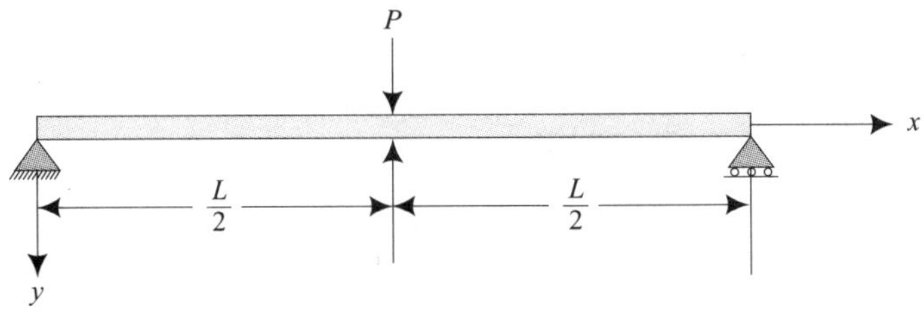

Fig. 1.34

Solution: By data, the beam is simply supported, the boundary conditions are

$$y = 0 \text{ and } \frac{d^2y}{dx^2} = 0, \text{ at } x = 0 \text{ and } x = L. \qquad \text{...[1]}$$

Let us take for $y(x)$ an expression of the form

$$y = Ax + Bx^2 \qquad \text{...[2]}$$

To determine A and B, substitute eq. [2] into the geometric boundary conditions.
At $x = 0$, the condition $y = 0$ is automatically satisfied. At $x = L$, $y = 0$ yields $A = -BL$.
Hence,
$y = -BLx + Bx^2$ which can be written in general form as

$$y = Cx\,(L - x) \qquad \text{...[3]}$$

We have deliberately chosen equation [3] to satisfy only the forced boundary condition as required by Rayleigh-Ritz method. Eq. [3] is an admissible function substituting it into energy expressions,

Strain energy, $$U = \frac{EI}{2}\int_0^L \left(\frac{d^2y}{dx^2}\right)^2 dx = \frac{EI}{2}\int_0^L (-2C)^2\,dx = 2EILC^2 \qquad \text{...[4]}$$

and the load is displaced by an amount $\delta = y_{\frac{L}{2}} = C\left(\frac{L}{2}\right)\left(L - \frac{L}{2}\right) = \frac{CL^2}{4}$

Hence, the work done by the load, $W_p = P\delta = P\left(\frac{CL^2}{4}\right)$ ...[5]

Therefore, the total potential is, $I = 2EILC^2 - \frac{PCL^2}{4}$ ...[6]

Since I is a function of C, taking the derivative of I with respect to C and equating it to zero we get

$$\frac{dI}{dC} = 4EILC - \frac{PL^2}{4} = 0$$

or $$C = \frac{PL}{16EI}$$

Hence, according to Rayleigh–Ritz method

$$y = \frac{PL}{16EI}x(L - x) \text{ is the deflection function.} \quad ...[7]$$

Now, y_{max} at centre is

$$(y_{max})_{\frac{L}{2}} = \frac{PL^3}{64EI} \text{ as against } \frac{PL^3}{48EI} \text{ which is the exact solution.}$$

Choose $$y = C\sin\frac{\pi x}{L} \quad ...[8]$$

which satisfies all the boundary conditions and is a comparison function. Substituting eq. [8] into the strain energy expression gives

$$U = \frac{EI}{2}\int_0^L \left(-\frac{C\pi^2}{L^2}\sin\frac{\pi x}{L}\right)^2 dx = \frac{\pi^4 EI}{4L^3}C^2 \quad [9]$$

and the work done by external load is

$$W_p = P(y)_{x=\frac{L}{2}} = PC \quad ...[10]$$

Therefore, the total potential $I = \frac{\pi^4 EI}{4L^3}C^2 - PC$ ...[11]

Differentiating eq. [11] with respect to C gives

$$\frac{dI}{dC} = \frac{\pi^4 EI}{2L^3}2C - P = 0 \Rightarrow C = 0.0205\frac{PL^3}{EI}$$

Hence, the deflection function

$$y = 0.0205 \frac{PL^3}{EI} \sin \frac{\pi x}{L} \qquad ...[12]$$

At the centre the maximum deflection is $y_{max} = \dfrac{PL^3}{48.78\,EI}$ which is very close to the exact value.

Maximum Stress Developed in the Beam

We now have an expression for the deflection as

$$y = 0.0205 \frac{PL^3}{EI} \sin \frac{\pi x}{L}$$

The moment $\qquad M = -EI \dfrac{d^2y}{dx^2}$

$$\therefore \qquad M = 0.0205\, PL\, \pi \sin \frac{\pi x}{L}$$

This is maximum at the centre where $x = L/2$, i.e.

$$M_{max} = 0.0205\, \pi\, PL$$

If the cross-section of the beam is rectangular, then the maximum bending stress will be

$$\sigma_{max} = \frac{Mh}{I} = 0.0205 \frac{d}{2I} PL = 0.3865 \frac{PL}{bd^2} \text{ as against } \frac{3PL}{bd^2},$$

where b is the width and d the depth of the rectangle and h the distance to the extreme fibre from the neutral axis.

Example: The governing equation for the bending of a simply supported beam is given as

$$\frac{d^2M}{dx^2} = -q$$

where M is the bending moment. Using the Rayleigh–Ritz method find the maximum moment at the centre.

Solution: Given $\left(\dfrac{d^2M}{dx^2} + q\right) = 0 \qquad ...[1]$

Let us write the D.E. in [1] of the given problem as the Euler equation of some variational problem as follows:

Differential equation of the given boundary value problem $D(y) = 0$	$\Rightarrow$	To identify $D(y) = 0$ as Euler-Lagrange equation
		$\Downarrow$
Use direct method like Rayleigh–Ritz to get an approximate solution for Y	$\Leftarrow$	Find the corresponding functional $I(Y)$

The variational function Y is then an approximation to the exact y.

To obtain this, multiply eq. [1] by δM and integrate between $x = 0$ and $x = L$, i.e.

$$\int_0^L \left[\frac{d^2M}{dx^2} + q\right] \delta M \, dx = 0 \qquad ...[2]$$

Integrating the first term by parts

$$\left|\frac{dM}{dx}\delta M\right|_0^L - \int_0^L \frac{dM}{dx}\frac{d(\delta M)}{dx}\,dx + \int_0^L q\delta M\,dx = 0 \qquad ...[3]$$

Since $\delta M = 0$ at $x = 0$ and $x = L$, eq. [3] reduces to

$$\left[\int_0^L \left\{\frac{dM}{dx}\frac{d(\delta M)}{dx} \quad q\delta M\right\}\right] \qquad ...[4]$$

The above can be written as the variation of a functional, I, as

$$-\delta\left[\frac{1}{2}\int_0^L \left\{\left(\frac{dM}{dx}\right)^2 - 2qM\right\} dx\right] = 0 \qquad ...[5]$$

i.e.

$$I = \int_0^L \left[\left(\frac{dM}{dx}\right)^2 - 2qM\right] dx \qquad ...[6]$$

To apply Rayleigh–Ritz method we have to use eq. [6]. Here $M = 0$ at $x = 0$ and $x = L$. Hence, an approximating function for M is assumed as

$$M = Ax\,(L - x) \qquad ...[7]$$

Substituting eq. [7] into eq. [6]

$$I = \int_0^L [A^2(L - 2x)^2 - 2qAx(L - x)]\,dx \qquad ...[8]$$

Integrating and minimizing with respect to A, i.e. setting $\frac{\partial I}{\partial A} = 0$ gives $A = \frac{q}{2}$. Hence

$$M = \frac{q}{2}x\,(L - x) \qquad ...[9]$$

At $x = L/2$

$$(M)_{\frac{L}{2}} = \frac{qL^2}{8},$$ the well-known result.

Note: This is an interesting and informative problem having the differential equation. However, Rayleigh–Ritz method cannot be applied directly to differential equation. We have tried to find out the functional through which differential equation is obtained.

Example: A fixed beam with varying moment of inertia is as shown in Fig. 1.35. Find out a suitable trial function. Hence compute the value of central deflection.

How to choose the trial field ?

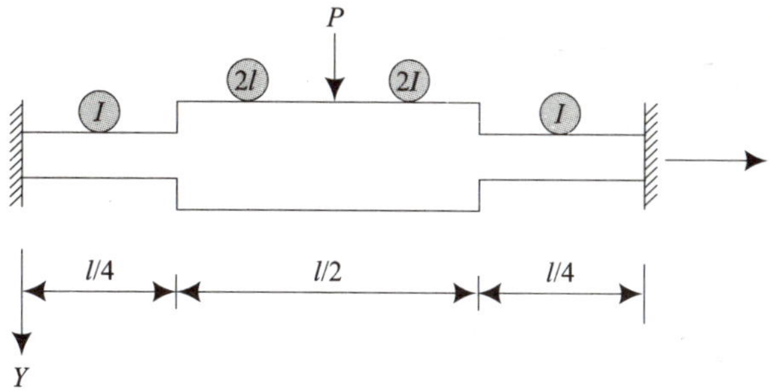

Fig. 1.35

Solution:

Method (1): Here, we assume the polynomial as

$$y = a_0 + a_1x + a_2x^2 + a_3x^3 + a_4x^4 \qquad ...[1]$$

as the condition is that the polynomial must have at least one Ritz parameter and as many boundary condition parameters as there are geometric boundary conditions.

The trial function must have minimum of five parameters. These parameters can be evaluated besides it should satisfy the symmetric condition of the profile.

Clearly, cubic, quadratic and linear polynomials do not satisfy the condition.

Method (2): Since there is a symmetry, we can consider half the portion.

Hence, $\left\{\begin{aligned} \text{when } x = 0, y = 0 \\ x = 0, y' = 0 \\ x = l/2, y' = 0 \end{aligned}\right\}$

We have 3 b.c.'s in this portion.

Let us consider strain energy for half the portion and double it to get for the entire structure.

For this portion, we have

$$y = a_0 + a_1x + a_2x^2 + a_4x^4$$

which yields 3 b.c. parameters and one Ritz parameter

$$\left\{\begin{aligned} \text{when } x = 0, y = 0 \\ x = 0, y' = 0 \\ \Rightarrow \quad a_0 = 0; a_1 = 0 \end{aligned}\right\}$$

The conditions

$$\frac{dy}{dx} = 0, \text{ when } x = l/2$$

$$y = a, \text{ when } x = l/2$$

give
$$a_2 = -\frac{3a_3 l}{4}$$

$$a_3 = -\frac{16a}{l^2}$$

So the polynomial becomes

$$y = \frac{4ax^2}{l^3}(3l - 4x)$$

where a is Ritz parameter, valid for $0 \le x \le l/2$)

When $x = l/2, y = y_{max} = a$

Now
$$U = 2\left(\frac{EI}{2}\right)\int_0^{l/4} (y'')^2\, dx + 2\left(\frac{2EI}{2}\right)\int_{l/4}^{l/2} (y'')^2\, dx$$

Putting down for y'' and simplifying

$$U = \frac{144EIa^2}{l^3}$$

$$W_p = -Pa$$

$\therefore$
$$(\pi = U + W_p)$$

$$\pi = \left[\frac{144EIa^2}{l^3} - Pa\right]$$

$$\delta\pi = \frac{d\pi}{da}\delta a = 0 \quad \text{(Stationary condition)}$$

$\Rightarrow$
$$\left[\frac{288EIa}{l^3} - P\right]\delta a = 0$$

$$a = (y)_{max} = \frac{Pl^3}{288EI} = 0.00347\frac{Pl^3}{EI}$$

Exact $(y')_{max}$ is

$$y_{max} = \frac{11Pl^3}{3072EI} = 0.00358\frac{Pl^3}{EI} \quad \textbf{Ans.}$$

Example: Use the Rayleigh–Ritz method to obtain an approximate solution for the one-dimensional heat conduction problem $k \cdot \frac{d^2T}{dx^2} + Q = 0$. Assume a rod of length L, constant area, and constant heat source along the length of the rod. Assume boundary conditions of $T(0) = T(L) = 0$. Compare the approximate solution with the exact solution.

Solution:

Given: $$k \cdot \frac{d^2T}{dx^2} + Q = 0 \qquad \text{...[1]}$$

$$\Rightarrow \qquad \frac{d^2T}{dx^2} = -\frac{Q}{k} \qquad \text{...[2]}$$

Integrating equation [2] twice w.r.t. x, we get

$$T = -\frac{Q}{2k}x^2 + C_1 x + C_2 \qquad \text{...[3]}$$

From the given boundary conditions, we have

$$T(0) = 0$$

Equation [3] yields : $C_2 = 0,$

and $$T(L) = 0 \;\Rightarrow\; -\frac{Q}{2k}L^2 + C_1 L = 0$$

$$\Rightarrow \qquad C_1 = \frac{QL^2}{2kL} = \frac{QL}{2k}$$

∴ [3] becomes

$$T = \frac{Q(Lx - x^2)}{2k} \qquad \text{...[4]}$$

We assume a quadratic trial function

$$T = C_1 + C_2 x + C_3 x^2 \qquad \text{...[5]}$$

$$\left\{ \begin{array}{lll} \text{Now, when} & x = 0; & T(0) = 0 \text{ gives } C_1 = 0 \\ \text{when} & x = L, & T(L) = 0 \text{ gives } C_2 = -C_3 L \end{array} \right\}$$

∴ Eq. [5] becomes $$T = C_3 (x^2 - Lx) \qquad \text{...[6]}$$

$$\therefore \qquad \frac{dT}{dx} = C_3(2x - L) \qquad \text{...[7]}$$

Substituting equations [6] and [7] into the equation:

$$J(T) = \int_V \left[\frac{1}{2}k\left(\frac{dT}{dx}\right)^2 - QT \right] dV$$

and replacing $$\int_V dv \text{ with } A\int_V^L dx$$

gives $$J = \int_0^L \frac{1}{2}\left[kC_3^2(4x^2 - 4Lx + L^2) - 2QC_3(x^2 - Lx)\right]A\,dx$$

or, integrating and substituting limits,

$$J = \frac{AkC_3^2L^3}{6} + \frac{AC_3QL^3}{6} \qquad ...[8]$$

We find C_3 as follows (which makes J a minimum)

$$\frac{\partial J}{\partial C_3} = \frac{2AkC_3L^3}{6} + \frac{AQL^3}{6} = 0$$

and

$$C_3 = -\frac{Q}{2k} \qquad ...[9]$$

Putting down equation [9] into equation [6], we get

$$T = \frac{Q(Lx - x^2)}{2k} \qquad ...[10]$$

Thus, equations [4] and [10] are identical.

⇒ approximate solution is identical to the exact solution in this application.

Example: Using the Rayleigh–Ritz method, obtain the value of the critical load, for a uniform two-hinged column which is subjected to a compressive load P at the two ends (take trial function as $y = \frac{4hx(l-x)}{l^2}$, h = Ritz parameter).

Solution:

Given:

$$y = \frac{4hx(l-x)}{l^2}$$

∴

$$y' = \frac{4h}{l^2}[l - 2x]$$

$$y'' = -\frac{8h}{l^2}$$

Substitute for y', y'' in the potential energy functional given by

$$\pi = \int_0^l \frac{EI}{2}\left(\frac{d^2y}{dx^2}\right)^2 dx - \int_0^l \frac{P}{2}\left(\frac{dy}{dx}\right)^2 dx,$$

Or,

$$\pi = \frac{EI}{2} \times \frac{16h^2}{3l} - \frac{P}{2} \times \frac{64}{l^3} \quad \text{(on integration)}$$

Effecting variation

$$\delta\pi = \left[\frac{16EI}{6l} - \frac{64P}{l^3}\right] h\delta h = 0$$

h and δh are not zero.

Hence

$$\frac{64P}{l^3} = \frac{16EI}{6l}$$

$$P_{cr} = \frac{12EI}{l^3}$$

1.7 GALERKIN'S METHOD

We shall now discuss an important method for finding an approximate solution to a differential equation, namely the Galerkin's method. This method of approximate analysis is classified as a method of weighted residuals. The analysis is based upon assuming an approximate solution for a differential equation.

In Galerkin's method, the trial functions $N_i(x)$ themselves are used as the weighting functions, W_i.

Consider the linear equation

$$Lu = f \qquad \text{...[1]}$$

on a region V, where the b.c.'s are homogeneous.

In one-dimensional case, we consider the governing equation as the differential equation

$$\frac{d}{dx}\left(EA\frac{du}{dx}\right) = 0$$

$$\Rightarrow \qquad \frac{d}{dx}EA\frac{d}{dx}() \quad \text{operating on } u.$$

Suppose, we have a differential equation

$$Lu = f \text{ in } \Omega \qquad \text{...[1]}$$

with boundary conditions

$$\varphi(u) = g \quad \text{in } \Gamma_1 \text{ (geometric)} \qquad \text{...[2]}$$

$$\psi(u) = h \quad \text{in } \Gamma_2 \text{ (natural)} \qquad \text{...[3]}$$

We assume a trial function $\bar{u}$ and substitute in equations [1] to [3]. Since $\bar{u}$ does not satisfy any of the equations [1] to [3] there will be errors (residual) both in the domain and on the boundary, denoted by

$$\varepsilon(x) = L\bar{u} - f \qquad \text{...[4]}$$

The approximate methods revolve around setting the residual relative to weighting function W_i to zero.

$$\int_V W_i(L\bar{u} - f)\,dv = 0 \qquad i = 1 \text{ to } n \qquad \text{...[5]}$$

The choice of the weighting function W_i leads to various approximation methods.

The weighting functions W_i are chosen from the basis function used for constructing $\bar{u}$ in the Galerkin's method.

Let
$$\bar{u} = \sum_{i=1}^{n} Q_i G_i$$

where G_i, $i = 1, 2, \ldots n$ are basis functions or polynomials in (x, y, z).

Here, the trial functions $N_i(x)$ are used as weighted function. It is a linear combination of basis functions.

The following example shows how Galerkin's method can be utilised.

If we consider a two-dimensional transient heat flow problem, where the coefficient of thermal conductivity K is independent of temperature and the material is homogeneous and isotropic

$$\frac{\partial}{\partial x}\left(K_x \frac{\partial T}{\partial x}\right) + \frac{\partial}{\partial y}\left(K_y \frac{\partial T}{\partial y}\right) + Q_s = C.\frac{\partial T}{\partial t} \qquad \text{...[1]}$$

where C = specific heat

T = temperature

t = time

and Q_S = external heat flux applied.

If temperature function T is approximated as

$$T' = \sum M_i \, . \, T_i$$

where M_i are the interpolation or weighting polynomials for point i. If the exact solution gives $f[T(x, y)] = q$ and after Galerkin's distribution of error $e_r = f(T') - q \to 0$.

$$T \simeq \sum M_i T_i$$

According to definition

$$\int M_i . \, f\left(\sum M_i T_i\right).dA = 0$$

Galerkin assumes M_i as equal to the shape function N_i of the finite elements itself and so

$$\int N_i f(T). \, dA = 0$$

$$\int N_i \left[\frac{\partial}{\partial x}\left(K_x \frac{\partial T}{\partial x}\right) + \frac{\partial}{\partial y}\left(K_y \frac{\partial T}{\partial y}\right) + Q_s - C.\frac{\partial T}{\partial t}\right] dA = 0 \qquad \text{...[2]}$$

Integrating by parts, we get

$$1^{\text{st}} \text{ term} = \int N_i . \left[\frac{\partial}{\partial x}\left(K_x \frac{\partial T}{\partial x}\right)\right].dA$$

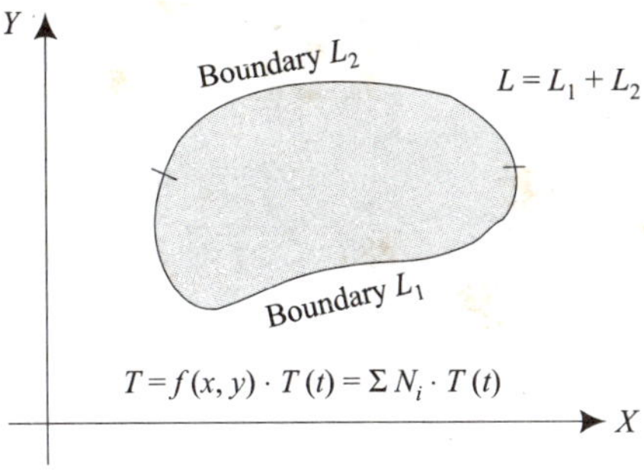

Fig. 1.36 A two-dimensional transient heat flow problem

$$= \int K_x \cdot \frac{\partial T}{\partial x} \cdot N_i \cdot dy - \iint K_x \cdot \frac{\partial T}{\partial x} \cdot \frac{\partial N_i}{\partial x} \cdot dx \cdot dy$$

$$= \int K_x \cdot \frac{\partial T}{\partial x} \cdot nx \cdot N_i \cdot dL - \iint K_x \cdot \frac{\partial T}{\partial x} \cdot \frac{\partial N_i}{\partial x} \cdot dx \cdot dy$$

where, nx is the direction cosine in x direction and dL is differential arc length along the boundary.

Similarly, for others and the above equation can be written as

$$\int_{S_2} \left(K_x \cdot \frac{\partial T}{\partial x} \cdot nx + K_y \cdot \frac{\partial T}{\partial y} \cdot ny \right) \cdot N_i \, dL + \iint N_i \cdot Q_s \cdot dx \cdot dy$$

$$- \iint \left[K_x \cdot \frac{\partial T}{\partial x} \cdot \frac{\partial N_i}{\partial x} + K_y \cdot \frac{\partial T}{\partial y} \cdot \frac{\partial N_i}{\partial y} \right] \cdot dx \cdot dy - \int N_i \cdot C \cdot \frac{\partial T}{\partial t} \cdot dx \cdot dy = 0 \quad \text{...[3]}$$

Assuming time dependency to be separate and not coupled with geometry, i.e.,

$$T(x, y, t) = \sum N_i(x, y) \,.\, T_i(t)$$

$$= [N] \{T\} \quad \text{...[4]}$$

$$\iint \left(K_x \cdot \left[\frac{\partial N}{\partial x} \right] \{T\} \cdot \frac{\partial N_i}{\partial x} + K_y \cdot \left[\frac{\partial N}{\partial y} \right] \cdot \{T\} \cdot \frac{\partial N_i}{\partial y} \right) . \, dx \cdot dy - \iint N_i \cdot Q_s \cdot dx \cdot dy$$

$$+ \iint N_i \cdot C[N] \cdot \left(\frac{\partial T}{\partial t} \right) \cdot dx \cdot dy - \int \left[K_x \cdot \frac{\partial T}{\partial x} \cdot nx + K_y \cdot \frac{\partial T}{\partial y} \cdot ny \right] \cdot Ni \cdot dL = 0 \quad \text{...[5]}$$

or

$$\iint \left[K_x \cdot \frac{\partial N_i}{\partial x} \cdot \frac{\partial N_j}{\partial x} + K_y \cdot \frac{\partial N_i}{\partial y} \cdot \frac{\partial N_j}{\partial y} \right] \cdot dx \cdot dy \, \{T\}$$

$$- \iint N_i \cdot Q_s \cdot dx \cdot dy + \int C \cdot N_i \cdot N_j \cdot dx \cdot dy \cdot \left[\frac{\partial T}{\partial t} \right] - \{bc\} = 0 \quad \text{...[6]}$$

or

$$[H] \{T\} + [K] \left[\frac{\partial T}{\partial t} \right] - (Q) - (bc) = 0 \quad \text{...[7]}$$

where $$[H] = \iint \left[K_x \cdot \frac{\partial N_i}{\partial x} \cdot \frac{\partial N_j}{\partial x} + K_y \cdot \frac{N_i}{y} \cdot \frac{N_j}{y} \right] \cdot dx \cdot dy \, \{T\}$$

$$\{Q\} = \iint N_i \cdot Q_s \cdot dx \cdot dy$$

$$[K] = \iint C \cdot N_i \cdot N_j \cdot dx \cdot dy \cdot \left[\frac{\partial T}{\partial t} \right]$$

$$\{bc\} = \int \left(K_x \cdot \frac{\partial T}{\partial t} \cdot nx + K_y \cdot \frac{\partial T}{\partial y} \cdot ny \right) N_i \cdot dL$$

Galerkin's Method Applied to Elasticity Problems

A two-dimensional theory of elasticity problem contains three unknown σ_x, σ_y and τ_{xy}. Only two equilibrium equations are available. They are given by

$$\left\{ \begin{aligned} \frac{\partial \sigma_x}{\partial x} + \frac{\partial \tau_{xy}}{\partial y} + P_b = 0 \\ \frac{\partial \tau_{xy}}{\partial x} + \frac{\partial \sigma_y}{\partial y} + Q_b = 0 \end{aligned} \right\} \qquad \text{...[1]}$$

The equilibrium equations of three-dimensional problem are the extension of equations [1]. They are given by the equations

$$\left\{ \begin{aligned} \frac{\partial \sigma_x}{\partial x} + \frac{\partial \tau_{xy}}{\partial y} + \frac{\partial \tau_{xz}}{\partial z} + P_b = 0 \\ \frac{\partial \tau_{xy}}{\partial x} + \frac{\partial \sigma_y}{\partial y} + \frac{\partial \tau_{yz}}{\partial z} + Q_b = 0 \\ \frac{\partial \tau_{xz}}{\partial x} + \frac{\partial \tau_{yz}}{\partial y} + \frac{\partial \sigma_z}{\partial z} + R_b = 0 \end{aligned} \right\} \qquad \text{...[2]}$$

where the stresses $\{\sigma\}$ are expressed in terms of displacements $\{u\}$.

We find from Fig. 1.37 that if α is the angle made by the normal $\vec{n}$ to the surface 'S' with x-axis, and P_s and Q_s are the surface forces, then from the equilibrium of the element (t = thickness of the element),

$$\left\{ \begin{aligned} P_s RQt &= \sigma_x \, RPt + \tau_{xy} PQt \\ \Rightarrow \quad P_s &= \sigma_x \left(\frac{RP}{RQ} \right) + \tau_{xy} \left(\frac{PQ}{RQ} \right) \\ &= \sigma_x \cos \alpha + \tau_{xy} \sin \alpha \\ &= \sigma_x l + \tau_{xy} m \end{aligned} \right\} \qquad \text{...[3]}$$

where $l = \cos \alpha$, $m = \sin \alpha$ are the d.c.'s of the normal to the surface S.

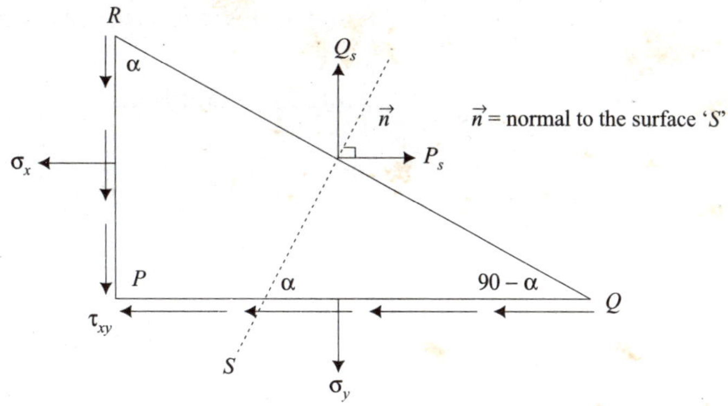

Fig. 1.37 Stressed element

Also, equilibrium of the forces on the element in y-direction, we get

$$\left\{\begin{aligned} & Q_s.RQt = \tau_{xy}RPt + \sigma_y PQt \\ \Rightarrow \quad & Q_s = \tau_{xy}\left(\frac{RP}{RQ}\right) + \sigma_y\left(\frac{PQ}{RQ}\right) \\ & \quad = \tau_{xy}\cos\alpha + \sigma_y \sin\alpha \\ & \quad = \tau_{xy}l + \sigma_y m \end{aligned}\right\} \qquad ...[4]$$

Hence, the surface forces on the element are:

$$\left\{\begin{aligned} P_s &= \sigma_x l + \tau_{xy} m \\ Q_s &= \tau_{xy} l + \sigma_y m \end{aligned}\right\} \qquad ...[5]$$

We shall now consider the integral:

$$\Pi = \iint \left[\left(\frac{\partial \sigma_x}{\partial x} + \frac{\partial \tau_{xy}}{\partial y} + P_b\right)\delta u + \iint\left(\frac{\partial \tau_{xy}}{\partial x} + \frac{\partial \sigma_y}{\partial y} + Q_b\right)\delta v\right] dx\, dy$$

where δu and δv are the elemental displacements in x and y directions.

We note that the integral $\Pi \equiv 0$,

Since $$\frac{\partial \sigma_x}{\partial x} + \frac{\partial \tau_{xy}}{\partial y} + P_b = 0$$

and $$\frac{\partial \tau_{xy}}{\partial x} + \frac{\partial \sigma_y}{\partial y} + Q_b = 0 \quad \text{(Using [1])}$$

$$\Rightarrow \quad \iint \left[\left(\frac{\partial \sigma_x}{\partial x} + \frac{\partial \tau_{xy}}{\partial y} + P_b\right)\delta u + \left(\frac{\partial \tau_{xy}}{\partial x} + \frac{\partial \sigma_y}{\partial y} + Q_b\right)\delta v\right] dx\, dy = 0 \qquad ...[6]$$

L.H.S of integral given in equation [6] can be expanded using Green's theorem in a plane.

Green's theorem: in a plane states that if $M = M(x, y)$ and $N = N(x, y)$ are continuous and

$$\left(\frac{\partial M}{\partial y}, \frac{\partial^2 M}{\partial y^2}, \frac{\partial N}{\partial x}, \frac{\partial^2 N}{\partial x^2}\right) \text{ are also continuous,}$$

then,

$$\iint \left[\frac{\partial M}{\partial x}\frac{\partial N}{\partial x} + \frac{\partial M}{\partial y}\frac{\partial N}{\partial y}\right] dx\, dy$$

$$= -\iint M\left[\frac{\partial^2 N}{\partial x^2} + \frac{\partial^2 N}{\partial y^2}\right] dx\, dy + \int M\left[\frac{\partial N}{\partial x} l + \frac{\partial N}{\partial y} m\right] ds \qquad ...[7]$$

Setting $M = \sigma_x$; $\frac{\partial N}{\partial x} = \partial u$ and $\frac{\partial N}{\partial y} = 0$, we get from the integral in the first terms of equation [7], that :

$$\iint \frac{\partial \sigma_x}{\partial x} \partial u\, dx\, dy = -\iint \sigma_x \frac{\partial(\delta u)}{\partial x} dx\, dy + \int \sigma_x\, l\, \delta u\, ds$$

Along the similar lines,

$\iint \frac{\partial \sigma_y}{\partial y} \delta u\, dx\, dy$ can be found by setting $M = \sigma_y$, $\frac{\partial M}{\partial x} = 0$; $\frac{\partial M}{\partial y} = \delta v$

we obtain the following:

$$\left.\begin{aligned}
&\iint \frac{\partial \sigma_y}{\partial y} \delta u\, dx\, dy = -\iint \sigma_y \frac{\partial}{\partial y}(\delta u)\, dx\, dy + \int \sigma_y\, m\, \delta v\, ds \\
\text{and similarly } &\iint \frac{\partial \tau_{xy}}{\partial x} \delta v\, dx\, dy = -\iint \tau_{xy} \frac{\partial}{\partial x}(\delta v) dx\, dy + \int \tau_{xy}\, l\, \delta v\, ds \\
&\iint \frac{\partial \tau_{xy}}{\partial y} \delta u\, dx\, dy = -\iint \tau_{xy} \frac{\partial}{\partial y}(\delta u) + \int \tau_{xy}\, m\, \delta u\, ds
\end{aligned}\right\} \qquad ...[8]$$

Hence equation [6] is equal to

$$-\iint \left[\sigma_x \frac{\partial}{\partial x}(\delta u) + \sigma_y \frac{\partial}{\partial y}(\delta v) + \tau_{xy}\frac{\partial}{\partial x}(\delta v) + \tau_{xy}\frac{\partial}{\partial y}(\delta u)\right] dx\, dy + \int [\sigma_x\, l\, \delta u + \sigma_y\, m\, \delta v + \tau_{xy}\, l\, \delta v$$

$$+ \tau_{xy}\, m\, \delta u]\, ds + \iint P_b\, \delta u\, dx\, dy + \iint Q_b\, \delta v\, dx\, dy = 0$$

Regrouping the term, we get,

$$-\iint \left\{\sigma_x \frac{\partial}{\partial x}(\delta u) + \sigma_y \frac{\partial}{\partial y}(\delta v) + \tau_{xy}\frac{\partial}{\partial x}(\delta v) + \frac{\partial}{\partial y}(\delta u)\right\} dx\, dy + \iint (P_b\, \delta u + Q_b \delta v) dx\, dy$$

$$+ \int (\sigma_x l + \tau_{xy} m)\delta u\, ds + \int (\tau_{xy}\, l + \sigma_y m)\, \delta v\, ds = 0 \qquad ...[9]$$

Explanation of the term: $\iint (P_b\delta u + Q_b\delta v)dx\,dy$ in equation [9]

where P_b is the body force along x-direction and δu is arbitrary (virtual) displacement in x-direction. Also, Q_b and δv are the body force and virtual displacements in y-directions.

We note that $\iint (P_b\delta u + Q_b\delta v)\,dx\,dy$ represents virtual work done by the body forces.

Explanation of the term: $\int [(\sigma_x l + \tau_{xy} m)\,\delta u + (\tau_{xy} l + \sigma_y m)\,\delta v]\,ds$ in equation [9]

We have $\quad (P_s = \sigma_x l + \tau_{xy} m;\ Q_s = \tau_{xy} l + \sigma_y m)$

$$\therefore \quad \int [\sigma_x l + \tau_{xy} m)\delta u + (\tau_{xy} l + \sigma_y m)\delta v]ds = \int [P_s\delta u + Q_s\delta v]ds \qquad \text{...[10]}$$

We note that equation [10] represents the virtual work done by the surface forces. Except the first term the other terms in equation [9] represent the virtual work done by external forces due to virtual displacements δu and δv.

Explanation of the term: $-\iint \left[\sigma_x \frac{\partial}{\partial x}(\delta u) + \sigma_y \frac{\partial}{\partial y}(\delta v) + \tau_{xy} \frac{\partial}{\partial x}(\delta v) + \frac{\partial}{\partial y}(\delta u)\right] dx\,dy$ in equation [9].

We can give the physical meaning for the above expression. The rectangular element of unit thickness is shown in Fig. 1.38. If δu be the virtual displacement given to the element

$$\therefore \quad (\text{Work done by } \sigma_x) = \sigma_x dy\left[\delta u + \frac{\partial}{\partial x}(\delta u)\,dx\right] - \sigma_x\,dy\,\delta u$$

$$= \sigma_x \frac{\partial}{\partial x}(\delta u)\,dx\,dy \qquad \text{...[11]}$$

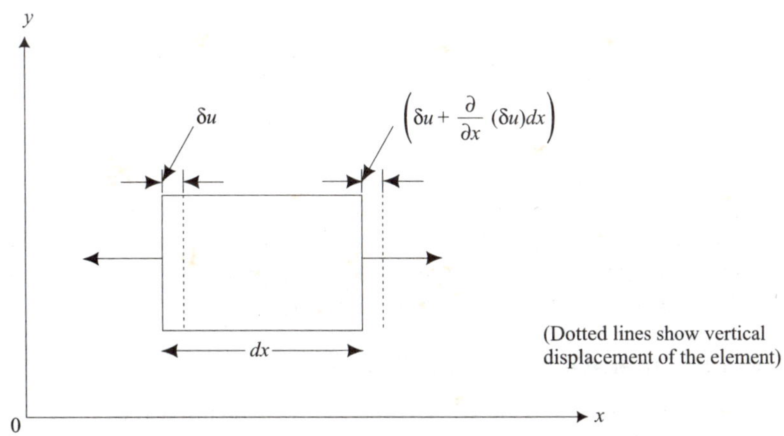

Fig. 1.38

If $\delta\varepsilon_x$ is the change in strain due to virtual displacement δu, then

$$\delta\varepsilon_x = \frac{\left[\delta u + \frac{\partial}{\partial x}(\delta u)\,dx\right] - \delta u}{dx}$$

$$= \frac{\partial}{\partial x}(\delta u)$$

Now, substituting in equation [11], we get

(Virtual work done by σ_x) = $\sigma_x(\delta\varepsilon_x)\,dx\,dy$ ...[12]

Along the similar lines, we can show that $\iint \sigma_y \frac{\partial}{\partial y}(\delta v)\,dx\,dy$ and $\iint \tau_{xy}\left\{\frac{\partial}{\partial y}(\delta v) + \frac{\partial}{\partial x}(\delta u)\right\}dx\,dy$

or, $\iint \sigma_y \delta\varepsilon_y dx\,dy$ and $\iint \tau_{xy}\delta\gamma_{xy}\,dx\,dy$

represent the work done by σ_y and τ_{xy}.

∴ The term $\iint\left\{\sigma_x \frac{\partial}{\partial x}(\delta u) + \sigma_y \frac{\partial}{\partial y}(\delta v) + \tau_{xy}\frac{\partial}{\partial x}(\delta v) + \frac{\partial}{\partial y}(\delta u)\right\}dx\,dy$ in equation [9] becomes:

$$\delta U = \iint (\sigma_x\,\delta\varepsilon_x + \sigma_y\delta\varepsilon_y + \tau_{xy}\delta\gamma_{xy})\,dx\,dy \quad \text{...[13]}$$

Hence, the first term in equation [9] represents the work done by internal forces with negative sign.

∴ Equation [9] can be put in the compact form as

$$-\delta U + \delta W_E = 0$$

⇒ $$\boxed{\delta U = \delta W_E}$$ ...[14]

where U = internal work done

W_E = external work done

Hence, the Galerkin's method in elasticity problems turns out to be the principle of virtual work.

"The principle of virtual work states that the work done by the external forces due to virtual displacements for a structure (a deformable body) in equilibrium is equal to the work done by the internal forces". Here, the forces and deformations considered for the derivation, $\delta U = \delta W_E$, are not related. The work done considered in the derivation, $\delta U = \delta W_E$, is called virtual, δu and δv are arbitrary displacements. The principle of virtual work is independent of the material properties (it holds good for linearly elastic as well as non-linear materials).

Galerkin's Method in Three-Dimensional Problems

In case of 3-dimensional problems, virtual work done by the external force is

$$\delta W_E = \iiint_V (P_b\delta u + Q_b\delta v + R_b\delta w)\,dv + \iint_s (P_s\delta u + Q_s\delta v + R_s\delta w)\,ds \quad \text{...[1]}$$

where S represents the surface on which forces are acting.

The matrix form of equation [1] is

$$\delta W_E = \iiint \{\delta u\}^T \{f\} dv + \iint \{\delta u\}^T \{P\} ds$$

where $\{\delta u\}^T = [\delta u \; \delta v \; \delta w]$

and $\{f\}^T = [P_b \; Q_b \; R_b]$

$$\{P\}^T = [P_s \; Q_s \; R_s]$$

The internal work expression is

$$\delta U = \iiint_V (\sigma_x \delta \varepsilon_x + \sigma_y \delta \varepsilon_y + \sigma_z \delta \varepsilon_x + \tau_{xy} \delta \gamma_{xy} + \tau_{yz} \delta \gamma_{yz} + \tau_{xz} \delta \gamma_{xz}].dv$$

or

$$\delta U = \iiint_V \{\delta \varepsilon\}^T \{\sigma\} \qquad ...[2]$$

where

$$\{\delta \varepsilon\}^T = [\delta \varepsilon_x \; \delta \varepsilon_y \; \delta \varepsilon_z \; \gamma_{xy} \; \gamma_{yz} \; \gamma_{xz}]$$

and $\{\sigma\}^T = [\sigma_x \; \sigma_y \; \sigma_z \; \tau_{xy} \; \tau_{yz} \; \tau_{zx}]$

from the principle of virtual work,

$$\boxed{\delta U = \delta W_E}$$

Comments on Galerkin's Method

In solid machanics it turns out to be virtual work method. This method can be applied to any problem involving solution of a system (set) of equations subject to specified values.

Example (1): Use the Galerkin's method and the trial function $a_1 x(10 - x)$ to obtain an approximate solution of

$$y'' + 100 = 0; \; (0 \le x \le 10)$$

with the boundary conditions $y(0) = 0$; $y(10) = 0$. Obtain the parameter a_1.

Solution: Given the trial function $T_R = a_1 x(10 - x)$, we choose the trial function as the weighting function.

$$T_R = a_1(10x - x^2)$$

$$\frac{dT_R}{dx} = a_1(10 - 2x)$$

$$\frac{d^2 T_R}{dx^2} = -2a_1$$

Substituting in the given differential equation

$$R = R(x, a_1) = -2a_1 + 100; \; W_i = a_1(10x - x^2)$$

Galerkin's method gives

$$\therefore \qquad \int_0^{10} W_i R(x; a_1) dx = 0$$

$\Rightarrow \quad \int_0^{10} a_1(10x - x^2)\,[-2a_1 + 100]\,dx = 0$

$\Rightarrow \quad 2a_1 = 100$

$\Rightarrow \quad a_1 = 50$

$\therefore$ The trial function is

$$T_R = 50x\,(10 - x)$$

Example (2): Use the Galerkin's method to obtain the approximate solution of

$$y'' + 1000\,x^2 = 0 \qquad \text{for } (0 \le x \le 1)$$

the boundary conditions are

$$y(0) = 0;\ y(1) = 0$$

given $\quad y = a_1(x - x^4)$.

Find the value of the parameter a_1.

Solution: Let us verify that the trial function satisfies the boundary conditions.

Given trial functions $\quad y = a_1(x - x^4) \quad$ ..[1]

when $\quad x = 0;\ y = 0$

$\quad x = 1;\ y = 0$

Galerkin's method requires

$$\int_0^1 W_i(x)\,R(x, a_i)\,dx = 0$$

$$W_i = a_1(x - x^4)$$

$$\therefore \int_0^1 a_1(x - x^4)\,(-12a_1 + 1000)x^2 dx = 0$$

$\Rightarrow \quad 12a_1 = 1000$

$\Rightarrow \quad a_1 = \dfrac{1000}{12}$

$\therefore \quad y = \dfrac{1000}{12}\,(x - x^4)$

Example (3): Use Galerkin's method and the trial function $T_R = a_0 + a_1x + a_2x^2$ to obtain the approximate solution of $\dfrac{d^2T}{dx^2} = \dfrac{Q}{k}$.

With the boundary conditions $T(0) = 0$; $T(L) = 0$ obtain the parameter a_1.

Solution: In Galerkin's method we assume trial function itself as the weighting function $W_i(x)$:

$$T_R = a_0 + a_1x + a_2x^2, \text{ (trial function).}$$

with boundary conditions

$$\left.\begin{aligned} T(0) &= 0 = a_0 \\ \text{and} \quad T(L) &= 0 = a_0 + a_1L + a_2L^2 \end{aligned}\right\}$$

$$\Rightarrow \qquad a_0 = 0; a_2 = -\frac{a_1}{L}$$

$\therefore$ the trial function (assumed solution) is

$$T_R = a_1\left[x - \frac{x^2}{L}\right] \qquad \text{...[1]}$$

Substituting into the governing equation as

$$R = \frac{d^2T_R}{dx^2} - \frac{Q}{k}$$

$$\Rightarrow \qquad -2\frac{a_1}{L} - \frac{Q}{k} = R$$

Since there is only one parameter a_1, there will be only one weighted residual equation. Galerkin's method gives

$$\int_{\Omega} W_i(x)\, R(x; a_1)\, dx = \int_0^L W_i(x)\, R(x; a_1)\, dx = 0$$

$$\Rightarrow \quad \int_0^L \left(x - \frac{x^2}{L}\right)\left(-2\frac{a_1}{L} - \frac{Q}{k}\right) dx = 0$$

$$\Rightarrow \quad \left(-\frac{a_1}{L} - \frac{Q}{k}\right)\left[\frac{x^2}{2} - \frac{x^3}{3L}\right]_0^L = 0$$

$$\Rightarrow \quad \left(-\frac{a_1}{L} - \frac{Q}{k}\right)\left(\frac{L^2}{2} - \frac{L^3}{3L}\right) = 0$$

Integrating, we obtain: $a_1 = -QL/2k$.

$$\therefore \qquad a_1 = -\frac{QL}{2k}$$

Substituting into equation [1] gives the exact solution;

$$T_R = \frac{Q}{2k}[x^2 - xL]$$

Example (4): Use Galerkin's method to obtain the approximate solution of the differential equation :

$$x^2\frac{d^2y}{dx^2} + 2x\frac{dy}{dx} = 6x$$

With the boundary conditions

$$y(1) = y(2) = 0.$$

In Galerkin's method, we assume the trial function as the weighting function with single parameter a_1 as:

$$T_R = a_1(x-1)(x-2) = a_1(x^2 - 3x + 2) \qquad \text{...[1]}$$

which satisfies the given b.c.'s.

Note that the given differential equation can be written as follows:

$$-\frac{d}{dx}\left(x^2\frac{dy}{dx}\right) = -6x \qquad \text{...[2]}$$

where $$L = -\left(\frac{d}{dx}x^2\frac{d}{dx}\right).$$

Since there is only one parameter a_1, there will be only one weighted residual equation. Galerkin's method gives

$$\int_\Omega W_i(x)\, R(x, a_1)\, dx = -\int_1^2 \left\{\frac{d}{dx}\left[x^2\frac{dy}{dx}\right]y - 12yx\right\}dx = 0$$

$$= -\int_1^2 \left(x^2 y\frac{d^2y}{dx^2} + 2xy\frac{dy}{dx} - 12xy\right)dx = 0 \qquad \text{...[3]}$$

We now choose a single coordinate function ϕ and write the trial functions assumed in [1] as:

$$T_R = a_1\phi_1 = a_1(x^2 - 3x + 2) \qquad \text{...[4]}$$

Substituting T_R into [2], we get

$$\bar{I} = -\int_1^2 \left[x^2\phi_1\left(\frac{d^2\phi_1}{dx^2}\right)a_1^2 + 2x\phi_1\frac{d\phi_1}{dx}a_1^2 - 12x\phi_1 a_1\right]dx$$

Extremising $\bar{I}$ w.r.t. a_1 we get, on cancelling –2 from the result,

$$\int_1^2 \left(x^2\phi_1 . \frac{d^2\phi_1}{dx^2}a_1 + 2x\phi_1\frac{d\phi_1}{dx}a_1 - 6x\phi_1\right)dx = 0 \qquad \text{...[5]}$$

$$\Rightarrow \qquad \int_1^2 [(3x^4 - 12x^3 + 15x^2 - 6x)a_1 - 3(x^3 - 3x^2 + 2x)]dx = 0$$

(on substituting for ϕ_1)

Integrating, we get

$$\left[\left(\frac{3}{5}x^5 - 3x^4 + 5x^3 - 3x^2\right)a_1 - \left(\frac{3}{4}x^4 - 3x^3 + 3x^2\right)\right]_1^2 = 0$$

$\therefore \qquad a_1 = 1.875$

$\therefore$ approximate solution is $T_R = 1.875\,(x^2 - 3x + 2)$.

Example (5): Use the Galerkin's method to obtain the approximate solution of the differential equation

$$\frac{d^2v}{dx^2} - \frac{kV}{T} + \frac{f}{T} = 0,$$

with the boundary conditions:

$$V(0) = V(L) = 0$$

Solution: Let us assume the trial function as:

$$U_R = a_0 + a_1 x + a_2 x^2 + a_4 x^4 \qquad ...[1]$$

$$\frac{d^2 v}{dx^2} - \frac{kv}{T} + \frac{f}{T} = 0 \qquad v(0) = v(L) = 0 \qquad ...[2]$$

The trial solution that satisfies the boundary conditions becomes

$$U_R = a_2(x^2 - xL) + a_4(x^4 - xL^3) \qquad ...[3]$$

Substituting Eq. (3) into Eq. (2) gives the residual

$$2a_2 + 12 a_4 x^2 - \left(\frac{k}{T}\right)[a_2(x^2 - xL) + a_4(x^4 - xL^3)] + \frac{f}{T} = R$$

There are two weighted equations corresponding to Eq. $\int_\Omega W_i(x) R(x; a_i)\, dv = 0$

$$\left\{\begin{aligned} &\int_0^L \left\{2a_2 + 12a_4x^2 - \left(\frac{k}{T}\right)[a_2(x^2 - xL) + a_4(x^4 - xL^3)] + \frac{f}{T}\right\}(x^2 - xL)\, dx \qquad ...[4] \\ &\int_0^L \left\{2a_2 + 12a_4x^2 - \left(\frac{k}{T}\right)[a_2(x^2 - xL) + a_4(x^4 - xL^3)] + \frac{f}{T}\right\}(x^4 - xL^3)\, dx \qquad ...[5] \end{aligned}\right\}$$

Integrating, [4] and [5] become

$$\left\{\begin{aligned} &a_2\left(-\frac{1}{3} + \frac{kL^2}{30T}\right) + a_4\left(-\frac{3\,L^2}{5} - \frac{5kL^4}{84T}\right) = \frac{f}{6T} \\ &a_2\left(-\frac{6}{10} - \frac{5kL^2}{84T}\right) + a_4\left(-\frac{12L^2}{7} - \frac{kL^4}{9T}\right) = \frac{3f}{10T} \end{aligned}\right\}$$

Substitute $L = 120$, $T = 600$, $f = 2$, and $k = \frac{1}{2}$ and solve for $a_2 = -\,7.1603(10^{-4})$ and $a_4 = -\,1.5816\,(10^{-9})$. The approximate solution is

$$U_R = -\,7.1603(10^{-4})\,(x^2 - xL) - 1.5816\,(10^{-9})\,(x^4 - xL^3)$$

Example (6): Given a differential equation

$$\nabla^2 u = 0; \quad \text{on } |x| \le 1,\ |y| \le 1$$

$$u = 0; \quad \text{on the boundary.}$$

Use the following trial function

$$T_R = T_R(x, y) = (1 - x^2)\,(1 - y^2)\,[a_1 + a_2(x^2 + y^2) + ...]$$

Solution: Given, $T_R = (1 - x^2)(1 - y^2)[a_1 + a_2(x^2 + y^2) + ...]$

Substituting in the differential equation, we get

$$E(x, y, a_1) = 1 + 2a_1(2 - x^2 - y^2) + ...$$

Galerkin's method yields

$$\Rightarrow \qquad \int_{-1}^{1}\int_{-1}^{1} [1 + 2a_1(2 - x^2 - y^2)]\,(1 - x^2)\,(1 - y^2)\,dx\,dy = 0$$

where we have retained one parameter.

Integrating and simplifying, we get

$$a_1 = -\frac{5}{16}.$$

∴ The approx. solution is

$$W(x, y) = -\frac{5}{16}(1 - x^2)\,(1 - y^2)$$

Example (7): Find a one parameter Galerkin solution of the boundary value problem

$$\nabla^2 u = x^2 - 1, \text{ for } |x| \le 1, \ |y| \le \frac{1}{2}$$

$$u = 0, \text{ on the boundary.}$$

Solution: We assume the approximate solution as:

$$U_R = \text{U}\,(x, y) = a_0(1 - x^2)\left(\frac{1}{4} - y^2\right)$$

which satisfies the boundary conditions.

The error in satisfying the differential equation is given by

$$E(x, y, a_0) = \left(\frac{\partial^2}{\partial x^2} + \frac{\partial^2}{\partial y^2}\right)\left[a_0\,(1 - x^2)\left(\frac{1}{4} - y^2\right)\right] - (x^2 - 1)$$

$$= -\left[2a_0\left(\frac{5}{4} - x^2 - y^2\right) + (x^2 - 1)\right]$$

Galerkin's method gives the equation

$$\int_{-1/2}^{1/2}\int_{-1}^{1}\left[(x^2 - 1) + 2a_0\left(\frac{5}{4} - x^2 - y^2\right)\right](1 - x^2)\left[\frac{1}{4} - y^2\right]dx\,dy = 0$$

or

$$4\int_{0}^{1/2}\int_{0}^{1}\left[(x^2 - 1) + 2a_0\left(\frac{5}{4} - x^2 - y^2\right)\right](1 - x^2)\left(\frac{1}{4} - y^2\right)dx\,dy = 0$$

$$\Rightarrow \qquad -\frac{2}{45} + \frac{a_0}{9} = 0$$

$$\Rightarrow \qquad a_0 = \frac{2}{5}.$$

⇒ Approximate solution:-

$$U_R = U(x, y) = \frac{2}{5}(1 - x^2)\left(\frac{1}{4} - y^2\right)$$

1.8 SAINT VENANT'S PRINCIPLE

The principle of Saint Venant states that "if a loading system be replaced by a statically equivalent system then the response is unaffected at a distance which is of the order of length over which the statically equivalent system is applied".

We can also state this principle as the distribution of stress over the cross-section is uniform except at the extreme ends of a bar carrying direct load, i.e., no change of deformation occurs at sections which are away from the point of application of load.

Consider a member subjected to compressive load as shown in planes normal to the axis of the member (Fig. 1.39). Planes normal to the axis of the member are subjected to the stress $\sigma_{avg} = \frac{P}{A}$.

A section near the point of application of force is subjected to the stress which increases from minimum value at the surface to a peak value at the centre, which is greater than σ_{avg} as in the section-XX. Variation of stress is not seen in the section-YY which is away from the load.

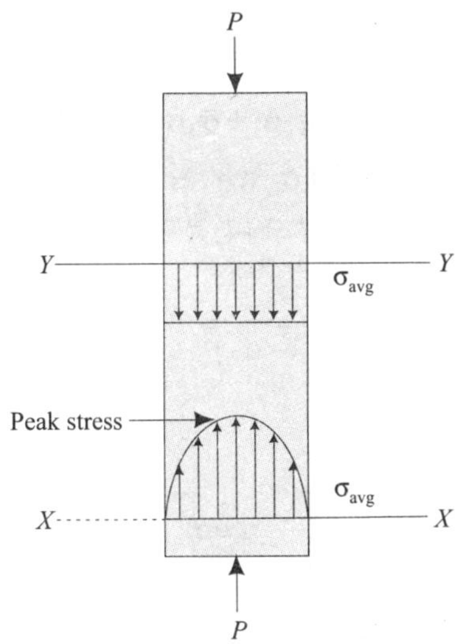

Fig. 1.39

Principle of Superposition

Sometimes, a body is subjected to a number of forces acting on its outer edges as well as at some other sections, along the length of the body. In such a case, the forces are split up, and their effects are considered on individual sections. The resulting deformation of the body is equal to the algebraic sum of the deformations of the individual sections. Such a principle, of finding out the resultant deformation, is called the principle of superposition.

The relation for the resulting deformation is modified as:

$$\delta = \frac{Pl}{AE} = \frac{1}{AE}(P_1l_1 + P_2l_2 + P_3l_3 + ...)$$

where P = force acting on section 1,

l_1 = length of section 1,

P_2l_2 = corresponding values of section 2, and so on.

1.9 VON MISES STRESS

The Von Mises stress is denoted by σ_{VM} and is defined as

$$\sigma_{VM} = \sqrt{I_1^2 - 3I_2} \quad ...[1]$$

where I_1 and I_2 are the first two invariants of the stress tensor and $\sigma_{VM} \leq \sigma_Y$ (yield stress of the material).

Generally, the state of stress is given by equation:

$$\sigma = [\sigma_x\, \sigma_y\, \sigma_z\, \tau_{yz} \tau_{xz} \tau_{xy}]^T \quad ...[2]$$

where $\sigma_x, \sigma_y, \sigma_z$ are normal stresses and $\tau_{yz}, \tau_{xz}, \tau_{xy}$ are shear stresses.

I_1 and I_2 are given by

$$\left.\begin{aligned} I_1 &= \sigma_x + \sigma_y + \sigma_z \\ I_2 &= \sigma_x\sigma_y + \sigma_y\sigma_z + \sigma_z\sigma_x - \tau_{yz}^2 - \tau_{xz}^2 - \tau_{xy}^2 \end{aligned}\right\} \quad ... [3]$$

Or, in terms of principal stresses σ_1, σ_2 and σ_3 we can write I_1 and I_2 as

$$I_1 = \sigma_1 + \sigma_2 + \sigma_3;\ I_2 = \sigma_1\sigma_2 + \sigma_2\sigma_3 + \sigma_3\sigma_1$$

Hence, σ_{VM} can be expressed as:

$$\sigma_{VM} = \frac{1}{\sqrt{2}}\sqrt{(\sigma_1 - \sigma_2)^2 + (\sigma_2 - \sigma_3)^2 + (\sigma_3 - \sigma_1)^2} \quad ...(4)$$

For plane stress and plane strain:

$$\left\{\begin{aligned} I_1 &= \sigma_x + \sigma_y \\ I_2 &= \sigma_x\sigma_y - \tau_{xy}^2 \end{aligned}\right\} \quad ...[5]$$

and

$$I_1 = \sigma_x + \sigma_y + \sigma_z$$

$$I_2 = \sigma_x\sigma_y + \sigma_y\sigma_z + \sigma_z\sigma_x - \tau_{xy}^2 \quad ...[6]$$

where $\sigma_z = \nu(\sigma_x + \sigma_y)$

EXERCISES

1. Define 'Potential energy'.
2. State and explain the principle of minimum potential energy.
3. Derive Euler-Lagrange equation for an integral function using variational principle.
4. What is an exact solution? What is an approximate solution? Does Rayleigh-Ritz method yield an exact or approximate solution?
5. Explain the principle of Rayleigh-Ritz method.
6. Write a short note on variational principles.
7. What are comparison and admissible functions?
8. Write a note on the conditions for convergence of Rayleigh-Ritz method.
9. Find the Galerkin finite element solution of the bvp

$$\nabla^2 u = -12xy, \quad x, y > 0, \quad x^2 + y^2 < 1$$

$$u = 0, \text{ on the boundary}$$

with $h = 1/2$ and a triangular network.

[Exact solution is : $u = xy(1 - x^2 - y^2)$]

2

Matrix Algebra and Gaussian Elimination

2.1 INTRODUCTION

Matrix theory was invented by the British mathematicians Arthur Cayley and J. J. Sylvester in 1858. Matrices came into existence while solving a system of linear equations. They play a very important role in solving many problems in Engineering, Statistics and also in structural analysis where matrices are used as a mathematical tool.

A matrix is an array of real numbers along horizontal lines and vertical lines (or a set of 'mn' numbers arranged in the form of m rows and n columns. If these are 'm' rows and 'n' columns in a matrix , then it is called an "$m \times n$ matrix over the set of real numbers". The matrices are denoted by A, B, C etc. and its elements $a_{11}, a_{12}, \ldots\ldots b_{11}, b_{12}, \ldots\ldots c_{11}, c_{12}, \ldots\ldots$ These elements are enclosed within the brackets of the type [ ] or ( ) or || ||.

2.2 MATRIX ALGEBRA

In this chapter, we study the various types of matrices. The study of matrices here is largely motivated from the need to solve systems of simultaneous equations of the form

$$\left\{\begin{array}{c} a_{11}x_1 + a_{12}x_2 + \ldots + a_{1n}x_n = b_1 \\ a_{21}x_1 + a_{22}x_2 + \ldots + a_{2n}x_n = b_2 \\ \ldots \quad \ldots \quad \ldots \\ a_{n1}x_1 + a_{n2}x_2 + \ldots + a_{nn}x_n = b_n \end{array}\right\} \qquad \ldots[1]$$

where $x_1, x_2, \ldots x_n$ are the unknowns. The matrix form of equation (1) can be conveniently expressed as:

$$Ax = b \qquad \ldots[2]$$

where A is a square matrix of dimension $(n \times n)$, and x, b are vectors of dimension $(n \times 1)$, given as

$$A = \begin{bmatrix} a_{11} & a_{12} & \cdots & a_{1n} \\ a_{21} & a_{22} & \cdots & a_{2n} \\ \cdots & \cdots & \cdots & \cdots \\ a_{n1} & a_{n2} & \cdots & a_{nn} \end{bmatrix}; x = \begin{Bmatrix} x_1 \\ x_2 \\ \vdots \\ x_n \end{Bmatrix}, b = \begin{Bmatrix} b_1 \\ b_2 \\ \vdots \\ b_n \end{Bmatrix}$$

Thus, a matrix is a rectangular array of numbers [elements].

It is denoted as $[A]$. An element located at the ith row and j^{th} column of A is denoted by a_{ij}.

Examples:

$$\begin{bmatrix} 2 & 5 & -1 \\ 3 & 2 & 5 \end{bmatrix} \text{ is a } 2 \times 3 \text{ matrix}$$

$$\begin{bmatrix} 1 & 2 \\ 3 & 5 \\ 4 & 7 \end{bmatrix} \text{ is a } 3 \times 2 \text{ matrix}$$

$$\begin{bmatrix} 1 & 2 & 3 \\ 3 & 2 & 1 \\ 4 & 1 & 5 \end{bmatrix} \text{ is a } 3 \times 3 \text{ matrix}$$

In general, a matrix is given by

$$[B] = \begin{bmatrix} b_{11}\ b_{12} \ldots b_{1j} \ldots b_{1n} \\ b_{21}\ b_{22} \ldots b_{2j} \ldots b_{2n} \\ \vdots \\ \vdots \\ b_{m1}\ b_{m2} \ldots b_{mj} \ldots b_{mn} \end{bmatrix}_{m \times n} \qquad \ldots[1]$$

Each number is known as an element. This matrix is represented by the general element i.e. the element in the ith row, j^{th} column $[b_{ij}]$.

All the elements $[b_{ij}]$ are written in compact form $[B]$. There are m rows and n columns in the matrix given by equation [1] and the matrix is said to be of size m × n.

Note: A beginner may confuse between a matrix and a determinant. A matrix merely represents an array of numbers without implying any relationship among the elements and the number of rows and columns need not be same. The number of rows and columns are same in the case of determinant and there is a relationship between the elements which results in a unique numerical value.

Types of Matrices

Definition (1): (Row matrix or Row vector) When the elements of the matrix lie in one row ($m = 1$), the matrix is a row matrix or also known as row vector and will be denoted as $\{B\}$.

For example, $(5 \quad 3 \quad 2)$

Definition (2): (Column matrix or column vector) When the elements are in one column ($n = 1$), the matrix is known as a column matrix or a column vector and will be denoted as $\{B\}$.

For example, $$\begin{bmatrix} -3 \\ 2 \\ 0 \end{bmatrix}$$

Definition (3): (Square matrix) A matrix with an equal number of rows and columns ($m = n$) is called a square matrix. The diagonal whose elements are a_{ii} ($i = 1, 2, 3, \ldots, n$) is called the principal diagonal.

For example, $$A = \begin{bmatrix} a_{11} & a_{12} & a_{13} \\ a_{21} & a_{22} & a_{23} \\ a_{31} & a_{32} & a_{33} \end{bmatrix}$$

Definition (4): (Diagonal matrix) A square matrix with non-zero elements only on the principal diagonal, i.e. $a_{ij} \neq 0$ when $i = j$ and $a_{ij} = 0$ when $i \neq j$, is called a diagonal matrix.

For example, $$\begin{bmatrix} a & 0 & 0 \\ 0 & b & 0 \\ 0 & 0 & c \end{bmatrix}$$

Definition (5): (Scalar matrix) A diagonal matrix in which all the principal diagonal elements are equal is called a scalar matrix.

For example, $$\begin{bmatrix} a & 0 & 0 \\ 0 & a & 0 \\ 0 & 0 & a \end{bmatrix}$$

Definition (6) : (Unit matrix or Identity matrix) A diagonal matrix with all elements in the principal diagonal as unity is called an identity matrix and is denoted by $[I]$

i.e., $$a_{ii} = 1 \ (i = 1, 2, \ldots n),$$

For example, $$\begin{bmatrix} 1 & 0 & 0 \\ 0 & 1 & 0 \\ 0 & 0 & 1 \end{bmatrix}_{3\times 3}, \ [I] = \begin{bmatrix} 1 & 0 \\ 0 & 1 \end{bmatrix}_{2\times 2}$$

Definition (7): (Symmetric matrix) A square matrix which is symmetric about principal diagonal is called a symmetric matrix, i. e. $a_{ij} = a_{ji}$ for all values of i and j.

For example, $\begin{bmatrix} 1 & -1 & 3 \\ -1 & 2 & 2 \\ 3 & 2 & 7 \end{bmatrix}$ is symmetric.

Definition (8): (Skew and skew symmetric matrices) A matrix whose a $a_{ij} = -a_{ji}$ and $a_{ii} \neq 0$ for all values of i and j is called a skew matrix. A skew matrix which has $a_{ii} = 0$ $(i = 1, 2, \dots n)$ is called a skew-symmetric matrix.

For example,

$$\begin{bmatrix} 0 & 1 \\ -1 & 0 \end{bmatrix} \text{ is a skew-symmetric matrix.}$$

$$\begin{bmatrix} 0 & 0 \\ 0 & 0 \end{bmatrix}, \quad \text{since } 0^T = 0 = -0 \text{ when } 0 \text{ is square.}$$

∴ It is symmetric and skew symmetric.

Definition (9): (Null matrix) A matrix of any order whose all elements are zero is referred to as null matrix and is designated as {0}.

For example,

$$0_3 = \begin{bmatrix} 0 & 0 & 0 \\ 0 & 0 & 0 \\ 0 & 0 & 0 \end{bmatrix} \text{ is a null matrix of order 3.}$$

Definition (10): (Triangular matrix) A square matrix having all elements as zero on one side of the principal diagonal is called a triangular matrix.

Thus $A = [a_{ij}]$, $a_{ij} \in C$ is called a triangular matrix if $a_{ij} = 0$ when $i > j$ or $i < j$.

(a) When $i > j$: Here every element below the diagonal is zero.
In this case, the matrix is known as upper-triangular matrix.

For example, $\begin{bmatrix} 1 & 2 & 3 \\ 0 & 4 & 5 \\ 0 & 0 & 6 \end{bmatrix}$ is an upper-triangular matrix.

(b) When $i < j$: Here every element above the diagonal is zero.
In this case, the matrix is known as lower-triangular matrix.

For example, $\begin{bmatrix} 1 & 0 & 0 \\ 2 & 3 & 0 \\ 4 & 5 & 6 \end{bmatrix}$ is lower-triangular matrix.

Definition (11): (Banded Matrix) A square matrix whose only non-zero coefficients are located on or near the main diagonal is called banded matrix as shown below.

$$\overset{\longleftarrow \quad nbh \quad \longrightarrow}{\begin{bmatrix} k_{11} & k_{12} & k_{13} & k_{14} & 0 & 0 & 0 & 0 \\ k_{21} & k_{22} & k_{23} & k_{24} & k_{25} & 0 & 0 & 0 \\ k_{31} & k_{32} & k_{33} & k_{34} & k_{35} & k_{36} & 0 & 0 \\ k_{41} & k_{42} & k_{43} & k_{44} & k_{45} & k_{46} & k_{47} & 0 \\ 0 & k_{52} & k_{53} & k_{54} & k_{55} & k_{56} & k_{57} & k_{58} \\ 0 & 0 & k_{63} & k_{64} & k_{65} & k_{66} & k_{67} & k_{68} \\ 0 & 0 & 0 & k_{74} & k_{75} & k_{76} & k_{77} & k_{78} \\ 0 & 0 & 0 & 0 & k_{85} & k_{86} & k_{87} & k_{88} \end{bmatrix}}$$

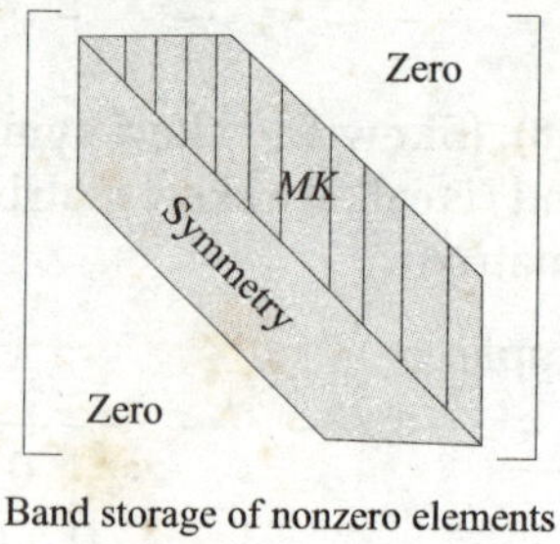

Band storage of nonzero elements

The above matrix has non-zero coefficients on either side of the main diagonal joining the element k_{11} and k_{88}. These are within the band whose width is given by *nhb* called the half-band width. The half-band width or a semi-band width is defined as the *greatest number of coefficient of stiffness in any row of the stiffness matrix from and including the leading diagonal to the outside right-hand non-zero coefficient*. Its value can be calculated as follows:

For a row *i* find the column number *j* of the last non-zero entry. Then, the half-band width is given by

$$(nhb) = 1 + (j - i) \qquad \text{...[2]}$$

nhb, for the problem is the maximum of $(nhb)_i$ over the rows. For example in eq. [2]

$$\left\{\begin{array}{l} \text{for } i = 1 \text{ the value of } j = 4. \text{ Hence, } (nhb)_1 = 1 + (4 - 1) = 4 \\ \text{for } i = 2 \text{ the value of } j = 5. \text{ Hence, } (nhb)_2 = 1 + (5 - 2) = 4 \\ \text{for } i = 4 \text{ the value of } j = 7. \text{ Hence, } (nhb)_4 = 1 + (7 - 4) = 4 \end{array}\right\} \qquad \text{...[3]}$$

Thus, the half-band width is 4.

Alternatively, the half-band width can be calculated as follows.

$$nhb = (R + 1) \times (\text{Number of degrees of freedom})$$

(R = largest difference between the numbers of any two nodes)

Half band width = (largest difference between the numbers of any two nodes + 1 degree of freedom per node – 1. Thus, apart from storing the half band, another effort required is proper node numbering. This governs the computational time as tremendously.

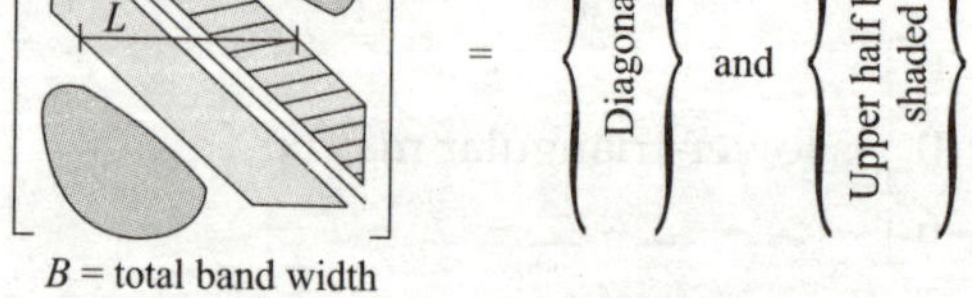

B = total band width

Fig. 2.1 Band width of a matrix

Condensation. There occur many a times, situations when some of the degrees of freedoms are not required. With the help of some matrix operations required degrees of freedom can be eliminated. This reduces the size of matrices to be handled and can be done as follows.

Let the matrix be

$$\begin{bmatrix} a_{11} & a_{12} & a_{13} & a_{14} & a_{15} & a_{16} \\ a_{21} & a_{22} & a_{23} & a_{24} & a_{25} & a_{26} \\ a_{31} & a_{32} & a_{33} & a_{34} & a_{35} & a_{36} \\ a_{41} & a_{42} & a_{43} & a_{44} & a_{45} & a_{46} \\ \cdots & \cdots & \cdots & \cdots & \cdots & \cdots \\ a_{51} & a_{52} & a_{53} & a_{54} & a_{55} & a_{56} \\ a_{61} & a_{62} & a_{63} & a_{64} & a_{65} & a_{66} \end{bmatrix} \begin{bmatrix} x_1 \\ x_2 \\ x_3 \\ x_4 \\ \cdots \\ x_5 \\ x_6 \end{bmatrix} = \begin{bmatrix} F_1 \\ F_2 \\ F_3 \\ F_4 \\ \cdots \\ F_5 \\ F_6 \end{bmatrix} \quad \text{...[1]}$$

or

$$\begin{bmatrix} [A_{11}] & [A_{12}] \\ [A_{21}] & [A_{22}] \end{bmatrix} \begin{bmatrix} [x_1] \\ [x_2] \end{bmatrix} = \begin{bmatrix} (F_1) \\ (F_2) \end{bmatrix} \quad \text{...[2]}$$

From the above we can solve as

$$([A_{11}] - [A_{12}]\,[A_{22}]^{-1}\,[A_{21}]\,[x_1] = (F_1) - [A_{12}]\,[A_{22}]^{-1}\,[F_2]$$

or

$$[k']\,[x_1] = [F'] \quad \text{...[3]}$$

where,

$$[k'] = ([A_{11}] - [A_{12}]\,[A_{22}]^{-1}\,[A_{21}])$$

$$[F'] = [A_{12}]\,[A_{22}]^{-1}\,[F_2]$$

With the help of these matrix operations all unwanted degrees of freedom can be eliminated.

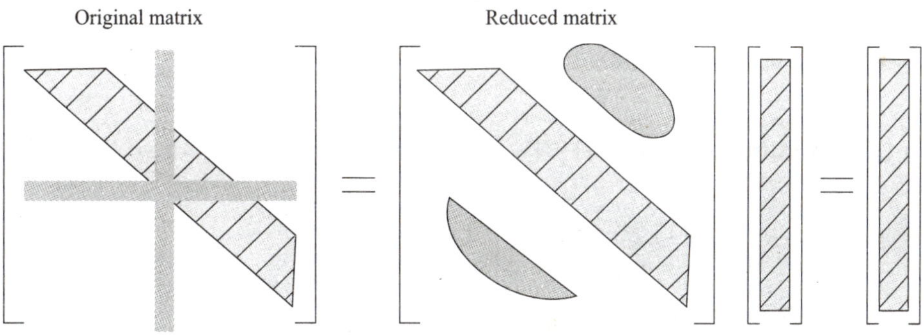

Fig. 2.2 Elements in green band are to be eliminated—row and column elimination.

Definition (12): (Transpose of a matrix) A matrix $[B]$ is said to be transpose of matrix $[A]$ if $[B]$ is resulted from the interchange of corresponding rows and columns of $[A]$ and is denoted by $[B] = [A]^T$.

For example,

$$[A] = \begin{bmatrix} 1 & -2 & 6 \\ 1 & 8 & 9 \end{bmatrix}$$

$$[B] = [A]^T = \begin{bmatrix} 1 & 1 \\ -2 & 8 \\ 6 & 9 \end{bmatrix}$$

Note: $([ABC]^T = C^T B^T A^T$, in general).

Definition (13): (Orthogonal matrix) A square matrix $[A]$ is called an orthogonal matrix.

if $$[A]\,[A]^T = [I].$$

For example, $\begin{bmatrix} 1/9 & 8/9 & -4/9 \\ 4/9 & -4/9 & -7/9 \\ 8/9 & 1/9 & 4/9 \end{bmatrix}$ is orthogonal matrix.

We need only to show $AA^T = I$:

$$AA^T = \begin{bmatrix} 1/9 & 8/9 & -4/9 \\ 4/9 & -4/9 & -7/9 \\ 8/9 & 1/9 & 4/9 \end{bmatrix} \begin{bmatrix} 1/9 & 4/9 & 8/9 \\ 8/9 & -4/9 & 1/9 \\ -4/9 & -7/9 & 4/9 \end{bmatrix} = \frac{1}{81} \begin{bmatrix} 1+64+16 & 4-32+28 & 8+8-16 \\ 4-32+28 & 16+16+49 & 32-4-28 \\ 8+8-16 & 32-4-28 & 64+1+16 \end{bmatrix}$$

$$= \frac{1}{81} \begin{bmatrix} 81 & 0 & 0 \\ 0 & 81 & 0 \\ 0 & 0 & 81 \end{bmatrix} = \begin{bmatrix} 1 & 0 & 0 \\ 0 & 1 & 0 \\ 0 & 0 & 1 \end{bmatrix} = I.$$

Definition (14): (Equality of matrices) Two matrices $[A]$ and $[B]$ can be equal if and only if

(a) They are of same order and

(b) $a_{ij} = b_{ij}$ for all values of i and j.

$(1 \le i \le m)$, $(1 \le j \le n)$.

Definition (15): (Addition of matrices) Two or more matrices can be added if they are of same order.

For example, $$\underset{m \times n}{[Y]} = \underset{m \times n}{[A]} + \underset{m \times n}{[B]} + \underset{m \times n}{[C]}$$

where $$y_{ij} = a_{ij} + b_{ij} + c_{ij}$$

This rule is valid for all values of i and j.

Definition (16): (Subtraction of matrices) Two matrices of the same order can be subtracted.

For example, $$\underset{m \times n}{[X]} = \underset{m \times n}{[A]} - \underset{m \times n}{[B]}$$

where $$x_{ij} = a_{ij} - b_{ij}$$

Definition (17): (Scalar multiplication) When a matrix is multiplied by a scalar quantity, the resulting matrix is obtained by multiplying every element of the matrix by that number.

For example,
$$4\begin{bmatrix} 7 & 2 \\ -4 & 3 \end{bmatrix} = \begin{bmatrix} 28 & 8 \\ -16 & 12 \end{bmatrix}$$

Definition (18): (Matrix division rules) Matrices cannot be divided directly. The operation is carried by premultiplying the matrix with the inverse of the matrix by which the given matrix is to be divided.

$$[k]\,\{x\} = [m]\,[x]\,[p^2]$$

to carry out division by $[m]$

$$[m]^{-1}\,[k]\,[x] = [x]\,[p^2] \qquad \text{...[1]}$$

Definition (19): (Matrix multiplication) The product of an $(m \times n)$ matrix A and an $(n \times p)$ matrix B results in an $(m \times p)$ matrix C. That is,

$$\underset{(m\times n)}{\mathrm{A}} \quad \underset{(n\times p)}{\mathrm{B}} \quad = \underset{(m\times p)}{\mathrm{C}}$$

In other words, two matrices A and B can be multiplied if and only if the number of columns in A is equal to the number of rows in B. The product matrix is denoted by AB.

In this case we say matrices A and B are conformable for multiplication and the product matrix AB has the same number of rows as A and the same number columns as B; $(AB = C)$.

The $(ij)^{\text{th}}$ component of C is obtained by taking the det product

$$C_{ij} = (i^{\text{th}} \text{ row of } A) \cdot (j^{\text{th}} \text{ column of } B)$$

For example,

$$\underset{(2\times 3)}{\begin{bmatrix} 1 & 2 & 3 \\ 3 & 2 & 1 \end{bmatrix}} \underset{(3\times 2)}{\begin{bmatrix} 4 & 5 \\ 2 & 3 \\ 1 & 2 \end{bmatrix}} = \begin{bmatrix} 11 & 17 \\ 17 & 23 \end{bmatrix}$$

(***Note:*** $AB \neq BA$; in fact BA may not even be defined, since the number of columns of B may not equal the number of rows of A).

Differentiation and Integration

The components a_{ij} of a matrix $[A]$ are functions of a variable x, then matrix differentiation is defined as

$$\frac{d}{dx}A(x) = \left[\frac{d}{dx}a_{ij}(x)\right] \qquad \text{...[1]}$$

$$\int A(x)\,dx\,dy = \left[\int a_{ij}\,dx\,dy\right] \qquad \text{...[2]}$$

In this regard, matrices may be differentiated and integrated.

Imp. case: In the formula [1], let A be $(n \times n)$ matrix of constants, and $x = [x_1 x_2 \ldots x_n]^T$ be a column vector of n variables.

Then $$\frac{d}{dx_p} A(x) = a^p \quad \text{(derivative w.r.t. } x_p) \qquad \ldots[3]$$

where a^p is the pth column of A.

Matrix Inversion

If A is a non-singular matrix, then the inverse of A is defined to be a matrix denoted by A^{-1} such that

$$A^{-1}A = AA^{-1} = I$$

where I is the identity matrix of the same order as that of A.

If det $A = 0$, then we say that A is singular, for which the inverse is not defined. The minor M_{ij} of a square matrix A is the determinant of the $(n-1 \times n-1)$ matrix obtained by eliminating the ith row and the jth column of A. The cofactor C_{ij} of matrix A is given by

$$C_{ij} = (-1)^{i+j} M_{ij}$$

Matrix C with elements C_{ij} is called the cofactor matrix.

The adjoint of matrix A is defined as

$$Adj\ A = C^T$$

The inverse of a square matrix A is given as

$$A^{-1} = \frac{\text{adj } A}{\det A}$$

For example,

$$\begin{bmatrix} a_{11} & a_{12} \\ a_{21} & a_{22} \end{bmatrix}^{-1} = \frac{1}{\det A} \begin{bmatrix} a_{22} & -a_{12} \\ -a_{21} & a_{11} \end{bmatrix}$$

Quadratic Forms

Let A be an $(m \times n)$ matrix and x be an $(n \times 1)$ vector, then the scalar quantity $X^T AX$ is called a quadratic form, since upon expansion we obtain the quadratic expression

$$\begin{aligned} & a_{11}x_1x_1 + a_{12}x_1x_2 + \ldots + a_{1n}x_1x_n \\ & a_{21}x_1x_2 + a_{22}x_2x_2 + \ldots + a_{2n}x_2x_n \\ & + \ldots\ldots\ldots\ldots \\ & + a_{n1}x_1x_n + a_{n2}x_2x_n + \ldots + a_{nn}x_nx_n \end{aligned}$$

This quadratic form can be written as

$$\sum_{i=1}^{m} \sum_{j=1}^{n} a_{ij} x_i x_j$$

$$= [x_1 x_2 \dots x_n] \begin{bmatrix} a_{11} & a_{12} & \dots & a_{1n} \\ a_{21} & a_{22} & \dots & a_{2n} \\ \dots & \dots & \dots & \dots \\ a_{n1} & a_{n2} & \dots & a_{nn} \end{bmatrix} \begin{bmatrix} x_1 \\ x_2 \\ \vdots \\ x_n \end{bmatrix} = X^T A X$$

where $X = \begin{bmatrix} x_1 \\ x_2 \\ \vdots \\ x_n \end{bmatrix}$ and $A = [a_{ij}]$.

The matrix A of the coefficient is called matrix of the quadratic form.

Note: The matrix A of the quadratic form must be symmetric. If it is not so, we make it symmetric by defining a new matrix B as follows:

$$(a_{ij} = a_{ji}) \text{ and } b_{ji} = b_{ij} = \frac{1}{2}(a_{ij} + a_{ji})$$

Example: Write the symmetric matrix of the quadratic form

$$u = 3x_1^2 - 4x_1x_2 + 6x_1x_3 - x_2^2 + 5x_3^2$$

Method to write matrix of a quadratic form

In the first row, write the coefficients of x_1^2, $\frac{1}{2}$ coeff. of x_1x_2, $\frac{1}{2}$ coeff. of x_1x_3 etc.

In the second row, write $\frac{1}{2}$ coeff. of x_1x_2, coeff. of x_2^2, $\frac{1}{2}$ coeff. of x_2x_3 etc.

In the third row, write $\frac{1}{2}$ coeff. of x_1x_3, $\frac{1}{2}$ coeff. of x_2x_3, coeff. of x_3^2 etc.

$$\therefore \quad u = [x_1\, x_2\, x_3] \begin{bmatrix} 3 & -2 & 3 \\ -2 & -1 & 0 \\ +3 & 0 & 5 \end{bmatrix} \begin{Bmatrix} x_1 \\ x_2 \\ x_3 \end{Bmatrix}$$

$$= X^T A X$$

Eigen Values and Eigen Vectors

Let $(A)_{n \times n} (X)_{n \times 1} = \lambda (X)_{n \times 1}$ ...[1]

holds for some scalar λ, where A is a square matrix, $(n \times n)$.

We wish to find a non-trivial solution. That is, we wish to find a nonzero eigen vector X and the corresponding eigen value λ that satisfy the equation [1].

If we write equation [1] as

$$(A - \lambda I)\, X = 0 \qquad \text{...[2]}$$

we see that a nonzero solution for X will occur when $A - \lambda I$ is a singular matrix,

or

$$\det (A - \lambda I) = 0 \qquad \text{...[3]}$$

Equation [3] is called the characteristic equation. We can solve eq. [3] for n roots or eigen values $\lambda_1, \lambda_2, \ldots \lambda_n$. For each eigen value λ_i, the associated eigen vector X^i is then obtained from equation [2].

$$(A - \lambda_i I)\, X^i = 0 \qquad \text{...[4]}$$

It should be noted that the eigen vector X^i is determined only to within a multiplicative constant since $(A - \lambda_i I)$ is a singular matrix.

Example:

$$A = \begin{bmatrix} 2 & -1 \\ 0 & 1 \end{bmatrix}$$

then

$$(A - \lambda I) = \begin{bmatrix} 2 & -1 \\ 0 & 1 \end{bmatrix} - \lambda \begin{bmatrix} 1 & 0 \\ 0 & 1 \end{bmatrix}$$

$$= \begin{bmatrix} 2-\lambda & -1 \\ 0 & 1-\lambda \end{bmatrix}$$

The characteristic equation is

$$\det \begin{bmatrix} 2-\lambda & -1 \\ 0 & 1-\lambda \end{bmatrix} = 0$$

which yields $\quad (\lambda - 1)(\lambda - 2) = 0$

On solving, we get $\quad \lambda_1 = 1, \lambda_2 = 2$

To get the eigen vector $\quad X_1 = [x_1, x_2]^T$

Corresponding to the eigen value λ_1, we substitute $\lambda_1 = 1$ into equation [4],

that is,

$$\begin{bmatrix} 1 & -1 \\ 0 & 0 \end{bmatrix} \begin{Bmatrix} x_1 \\ x_2 \end{Bmatrix} = \begin{Bmatrix} 0 \\ 0 \end{Bmatrix}$$

Thus, the components of X_1 satisfy the equation $x_1 - x_2 = 0 \quad \therefore \quad x_1 = x_2 = c$

$\therefore \quad X_1 = \begin{Bmatrix} c \\ c \end{Bmatrix}$ is the eigen vector associated with $\lambda = 1$, where c is any arbitrary number $(c \neq 0)$.

Again, we substitute $\lambda_2 = 2$, into equation [4], we get

$$\begin{bmatrix} 0 & -1 \\ 0 & -1 \end{bmatrix} \begin{Bmatrix} x_1 \\ x_2 \end{Bmatrix} = \begin{Bmatrix} 0 \\ 0 \end{Bmatrix}$$

i.e. $-x_2 = 0, -x_2 = 0,$ i.e. $x_2 = 0, x_1 = c$

$\therefore$ $X_2 = \begin{Bmatrix} c \\ 0 \end{Bmatrix}$ is the eigen vector associated with $\lambda_2 = 2$.

Note: Eigen value problems in *FEA* are of the type $AX = \lambda BX$ or $Ay = \lambda By$.

Positive Definite Matrix

(a) Quadratic form

If X is a column vector of order $n \times 1$ and there is a symmetric matrix

$$A = [a_{ij}]_{n \times n} \quad \text{i.e.} \quad a_{ij} = a_{ji}$$

then

$$Q = (X^T)_{1 \times n} (A)_{n \times n} (X)_{n \times 1}$$

is called a quadratic form.

Classification: If we apply L.T.

say $X = PY$, then

$$Q = Y^T (P^T AP) Y$$

If

$$P^T AP = I_n$$

where P is a non-singular matrix and $f(A) = n$, then Q reduces to

$$Q = \sum_{i=1}^{n} Y_i^2 \Rightarrow Q > 0, \text{ unless all } y\text{'s are zero and } y\text{'s cannot be zero}$$

unless all X's are zeros.

Definition:

If $Q > 0 \ \forall \ x \neq 0$ then

$Q = X^T AX$ is called a positive definite quadratic form and A is called positive definite matrix.

Alternatively, a symmetric matrix A of dimension $(n \times n)$ is positive definite if, for any non-zero vector

$$\left.\begin{aligned} X &= [x_1 x_2 \ldots, x_n]^T, \\ X^T AX &> 0 \end{aligned}\right\} \qquad \ldots[1]$$

Note: (i) A symmetric matrix is said to be positive definite if all its eigen values are strictly positive (greater than zero).

Or the positive definite quadratic form must necessarily possess positive eigen values.

Example: Let $A = \begin{bmatrix} 4 & -2.236 \\ -2.236 & 8 \end{bmatrix}$; (symmetric matrix)

has eigen values $\lambda_1 = 3 > 0$; $\lambda_2 = 9 > 0$ and hence is positive definite.

2.3 GAUSSIAN ELIMINATION

Consider a set of linear equations to be solved in the finite element analysis.

$$\left\{\begin{array}{l} k_{11}^0 x_1 + k_{12}^0 x_2 + k_{13}^0 x_3 + k_{14}^0 x_4 = b_1^0 \\ k_{21}^0 x_1 + k_{22}^0 x_2 + k_{23}^0 x_3 + k_{24}^0 x_4 = b_2^0 \\ k_{31}^0 x_1 + k_{32}^0 x_2 + k_{33}^0 x_3 + k_{34}^0 x_4 = b_3^0 \\ k_{41}^0 x_1 + k_{42}^0 x_2 + k_{43}^0 x_3 + k_{44}^0 x_4 = b_4^0 \end{array}\right\} \qquad \text{...[1]}$$

The above system of equations are solved using Gaussian elimination, a procedure wherein the matrix [*K*] is transformed into upper triangular form which can be solved directly for unknowns as direct approach. The matrix equation of [1] is given by

$$\underset{K}{\begin{bmatrix} k_{11}^0 & k_{12}^0 & k_{13}^0 & k_{14}^0 \\ k_{21}^0 & k_{22}^0 & k_{23}^0 & k_{24}^0 \\ k_{31}^0 & k_{32}^0 & k_{33}^0 & k_{34}^0 \\ k_{41}^0 & k_{42}^0 & k_{43}^0 & k_{44}^0 \end{bmatrix}} \underset{X}{\begin{bmatrix} x_1 \\ x_2 \\ x_3 \\ x_4 \end{bmatrix}} \underset{=}{=} \underset{B}{\begin{bmatrix} b_1^0 \\ b_2^0 \\ b_3^0 \\ b_4^0 \end{bmatrix}} \qquad \text{...[2]}$$

1. Perform: $k_{ij}^0 / k_{11}^0 = C$ to get k_{ij}^1

   or $k_{ij}^1 = k_{ij}^0 / k_{11}^0 = C$

   where $j = 2, 3, 4$. Check whether the leading diagonal is either zero or negative.

   If it is so, elimination will not be performed.

   Thus, the step yields $[1\ A_{12}^1\ A_{13}^1\ A_{14}^1]$ ...[1]

2. In [1], choose the row below 1 on which elimination is to be performed.

$$\left\{\begin{array}{l} k_{22}^1 = k_{22}^0 - k_{21}^0 k_{12}^1 \\ k_{2j}^1 = k_{2j}^0 - k_{21}^0 k_{1j}^0 \\ k_{ij}^1 = k_{ij}^0 - k_{i1}^0 k_{ij}^1 \\ k_{ij}^1 = k_{ij}^0 - C k_{1i}^1 \end{array}\right\} \qquad \text{...[2]}$$

   If $k_{1i}^0 = k_{i1}^0 = 0$, stop the procedure.

3. We have to perform the operations (as in step [2])

$$\left\{\begin{array}{l} \text{for} \quad i = 2, 3, 4. \\ \text{by symmetry, } (j = i \text{ to } 4 \text{ and } i = 2 \text{ to } 4) \\ k_{32}^1 = k_{32}^0 - k_{31}^0 k_{12}^1 = k_{23}^0 - k_{21}^0 k_{13}^0 / k_{11}^0 = k_{23}^1 \end{array}\right\} \qquad \text{...[3]}$$

4. $k_{ij}^1 = A_{ji}^1$ (By symmetry)

5. R. H. S. can be modified as:

$$\begin{Bmatrix} b_2^1 = b_2^0 - k_{21}^0 b_1^0 / k_{11}^0 \\ b_2^1 = b_2^0 - k_{12}^1 b_1^0 \\ b_1^1 = b_i^0 - k_{1i}^0 b_1^0 \end{Bmatrix}$$

6. R. H. S. of the first row can be modified as

$$b_1^1 = b_1^0 / k_{11}^0$$

After performing the operations involved in the above six steps, we get the resulting matrix as

$$\begin{bmatrix} 1 & k_{12}^1 & k_{13}^1 & k_{14}^1 \\ 0 & k_{22}^1 & k_{23}^1 & k_{24}^1 \\ 0 & k_{32}^1 & k_{33}^1 & k_{34}^1 \\ 0 & k_{42} & k_{43}^1 & k_{44}^1 \end{bmatrix} \begin{Bmatrix} X_1 \\ X_2 \\ X_3 \\ X_4 \end{Bmatrix} = \begin{Bmatrix} b_1^1 \\ b_2^1 \\ b_3^1 \\ b_4^1 \end{Bmatrix}$$

Repeating the same type of operations from equations 2 to 4, we get

$$\begin{bmatrix} 1 & k_{23}^{11} & k_{24}^{11} \\ 0 & k_{33}^{11} & k_{34}^{11} \\ 0 & k_{43}^{11} & k_{44}^{11} \end{bmatrix} \begin{Bmatrix} X_2 \\ X_3 \\ X_4 \end{Bmatrix} = \begin{Bmatrix} b_2^{11} \\ b_3^{11} \\ b_4^{11} \end{Bmatrix}$$

Reducing equations 3 and 4, we get,

$$\begin{bmatrix} 1 & k_{34}^{111} \\ 0 & k_{44}^{111} \end{bmatrix} \begin{Bmatrix} X_3 \\ X_4 \end{Bmatrix} = \begin{Bmatrix} b_3^{111} \\ b_4^{111} \end{Bmatrix}$$

It is seen that general operations to be performed will be three times when there are four equations. If there are 'n' equations, six steps have to be performed $(n - 1)$ times. The reduced equations are:

$$\begin{bmatrix} 1 & k_{12}^1 & k_{13}^1 & k_{14}^1 \\ 0 & 1 & k_{23}^{11} & k_{24}^{11} \\ 0 & 0 & 1 & k_{34}^{111} \\ 0 & 0 & 0 & k_{44}^{111} \end{bmatrix} \begin{Bmatrix} X_1 \\ X_2 \\ X_3 \\ X_4 \end{Bmatrix} = \begin{Bmatrix} b_1^1 \\ b_2^{11} \\ b_3^{111} \\ b_4^{111} \end{Bmatrix}$$

Now, the stiffness matrix $[K]$ has been transformed to upper-triangular matrix.

By back substitution, we can solve for unknowns as:

$$X_4 = b_4^{111} / k_{44}^{111}$$

$$X_3 = b_3^{111} - k_{34}^{111} X_4$$

X can be stored in the same location of b.

The above can be stated in a general form as follows:

Step 1. Divide the first row with coefficient of x_1.

Step 2. Eliminate the variable x_1 from the second row by multiplying first row with coefficient of x_1 from second row and subtracting this from the second row. similarly, eliminate the variable x_1 from the third row by multiplying first row with coefficient of x_1 from third row and subtracting this from the third row. Similarly repeat this procedure with all the subsequent rows.

Step 3. Similarly eliminate the variable x_2 from all the equations after the second; eliminate the variable x_3 from all the equations after the third and so on till the last equation contains only the variable x_n. This completes what is called the forward elimination.

Step 4. From step 3 find the variable x_n.

Step 5. By back substitution into the last but one equation find value of x_{n-1}. Repeat this process till the value of x_1 is obtained. The above can be programmed and the listing is given below:

```
C     SOLUTION OF AX = B USING GAUSSIAN ELIMINATION METHOD
C     PROGRAM BY DR. J. RAAMACHANDRAN
      DIMENSION A(50,51), X(50)
      READ (*, 1) N
1     FORMAT (I5)
      M = N + 1
      L = N - 1
2     READ (*, 3) ((A(I, J), J = 1, M), I = 1, N)
3     FORMAT(F 10.0)
      DO 10 K = 1, L
      JJ = K
      BIG = ABS(A (K, K))
      KP1= K + 1
      DO 5I = KP1, N
      AB = ABS(A (I, K))
      IF (BIG-AB) 4, 5, 5
4     BIG = AB
      JJ = I
5     CONTINUE
      IF (JJ-K) 6, 8, 6
6     DO 7 J = K, M
      TEMP = A(JJ, J)
      A(JJ, J) = A(K, J)
7     A(K, J) = TEMP
8     DO 9 I = KP1, N
      QUOT = A(I, K)/A(K, K)
      DO 9 J = KP1, M
9     A(I, J) = A(I, J)-QUOT *A(K, J)
      DO 10 I = KP1, N
```

```
10      A(I, K) = 0
        X(N) = A(N, M) / A(N, N)
        DO 12 NN = 1, L
        SUM = 0
        I = N-NN
        IP1 = I + 1
        DO 11 J = IP1,N
11      SUM = SUM + A(I, J) *X(J)
12      X(I) = (A(I, M) - SUM/ A(I, I)
        DO 14 I = 1, N
        WRITE(*, 13) X (I)
13      FORMATE ('',E 14.8)
14      CONTINUE
        STOP
        END
```

Illustration

Consider the simultaneous equations

$$3x_1 - 2x_2 + x_3 = 6 \quad \text{...[1]}$$

$$x_1 + 10x_2 - x_3 = 2 \quad \text{...[2]}$$

$$-3x_1 - 2x_2 + x_1 = 0 \quad \text{...[3]}$$

The above can be put in the form

$$\begin{bmatrix} A_{11} & A_{12} & A_{13} \\ A_{21} & A_{22} & A_{23} \\ A_{31} & A_{32} & A_{33} \end{bmatrix} \begin{Bmatrix} x_1 \\ x_2 \\ x_3 \end{Bmatrix} = \begin{Bmatrix} 6 \\ 2 \\ 0 \end{Bmatrix} \quad \text{...[4]}$$

As a first step we will reduce the coefficient A_{11} to 1 by dividing the eq. ($A1$) by A_{11}. This results in

$$x_1 - \frac{2}{3}x_2 + \frac{1}{3}x_3 = \frac{6}{3} = 2 \quad \text{...[5]}$$

The second step is to eliminate x_1 from eqs. [1] and [3]. To this end we multiply eq. [5] by A_{21} which in this case is 1 and subtract the resulting equation from eq. [2]

$$[A_{21} \times \text{eq. [5]} - \text{eq. [2]]} \Rightarrow \frac{32}{3}x_2 - \frac{4}{3}x_3 = 0 \quad \text{...[6]}$$

Similarly multiply eq. [5] by A_{31} which in this case is -3 and subtract the resulting equation from eq. [3]. This yields

$$[A_{31} \times \text{eq. [5]} - \text{eq. [2]]} \Rightarrow -4x_2 + 2x_3 = 6 \quad \text{...[7]}$$

Next, we reduce the coefficient of x_2 in eq. [6] to 1 by dividing eq. [6] by 32/3. This results in

$$\left[\frac{\text{eq. [6]}}{\text{coeff. of } x_2}\right] \Rightarrow x_2 - \frac{1}{8}x_3 = 0 \qquad \text{...[8]}$$

The next step is to eliminate x_2 from eq. [8]. Therefore, we multiply eq. [8] by the coefficient of x_2 from eq. [7] that is –4 now and the resulting equation is subtracted from eq. [8]. This yields

$$[\text{eq. [8]} - (-4)(\text{eq. [7]}] \Rightarrow \frac{3}{2}x_3 = 6 \qquad \text{...[9]}$$

Thus after reduction of the columns we have the equations

$$x_1 - \frac{2}{3}x_2 + \frac{1}{3}x_3 = 2 \qquad \text{...[10]}$$

$$0 - x_2 - \frac{1}{8}x_3 = 0 \qquad \text{...[11]}$$

$$0 - 0 + \frac{3}{2}x_3 = 6 \qquad \text{...[12]}$$

The above eqs. [10] to [12] can be put in the form

$$\begin{bmatrix} 1 & -\frac{2}{3} & \frac{1}{3} \\ 0 & -1 & -\frac{1}{8} \\ 0 & 0 & \frac{3}{2} \end{bmatrix} \begin{Bmatrix} x_1 \\ x_2 \\ x_3 \end{Bmatrix} = \begin{Bmatrix} 2 \\ 0 \\ 6 \end{Bmatrix} \qquad \text{...[13]}$$

Comparing eqs. [4] and [13] we find eq. [4] is reduced to one having coefficients only in upper triangle. Hence the name triangular elimination method.

From. eq. [12] we find $x_3 = 4$. Next step is to back substitute and find the remaining unknowns. Substituting for x_3 in eq. [11] yields for $x_2 = ½$. From these values and eq. [10] we find $x_1 = 1$.

Skyline Storage

In the following matrices, it may be observed that the line separating the top zeros from the first non-zero element is called the skyline (Fig. 2.3). For the matrix given below the skyline is indicated. After assembly of all element matrices we may see that the global characteristic matrices have the following properties:

(i) The matrices have large number of coefficients as zero quantities (in context of coefficients of matrices these will be called, now onwards, as zero or non-zero elements) and all such null/ zero elements occur in some specific zones of the matrices.

(ii) Some of the zones have both zero and non-zero elements but they are arranged in a way that after a particular range all the elements are zero. Thus we see that global matrices have non-zero elements in some band form parallel to the main diagonal. Beyond the band all the elements are zeros.

$$\begin{bmatrix} a_{11} & a_{12} & 0 & a_{14} & 0 & 0 & 0 & 0 \\ & a_{22} & a_{23} & 0 & a_{25} & 0 & 0 & 0 \\ & & a_{33} & a_{34} & 0 & a_{36} & 0 & 0 \\ & & & a_{44} & 0 & a_{46} & 0 & a_{48} \\ & & & & a_{55} & a_{56} & a_{57} & 0 \\ & & & & & a_{65} & a_{67} & a_{68} \\ & & & & & & a_{77} & a_{78} \\ & & & & & & & a_{88} \end{bmatrix}$$

Skyline

Fig. 2.3(a)

$$\begin{bmatrix} 24 & -12 & -24 & -12 & 0 & 0 & 0 & 0 & 0 \\ -12 & 8 & 12 & 4 & 0 & 0 & 0 & 0 & 0 \\ -24 & 12 & 48 & 0 & -24 & -12 & 0 & 0 & 0 \\ -12 & 4 & 0 & 16 & 12 & 4 & 0 & 0 & 0 \\ 0 & 0 & -24 & 12 & 48 & 0 & -24 & -12 & 0 \\ 0 & 0 & -12 & 4 & 0 & 16 & 12 & 4 & 0 \\ 0 & 0 & 0 & 0 & -24 & 12 & 48 & 0 & -24 \end{bmatrix}$$

Fig. 2.3(b) Skyline

$$\begin{bmatrix} 23 & -12 & -24 & -12 & 0 & 0 & 0 & 0 & 0 & 0 \\ -12 & 8 & 12 & 4 & 0 & 0 & 0 & 0 & 0 & 0 \\ -24 & 12 & 48 & 0 & -24 & -12 & 0 & 0 & 0 & 0 \\ -12 & 4 & 0 & 16 & 12 & 4 & 0 & 0 & 0 & 0 \\ 0 & 0 & -24 & 12 & 48 & 0 & -24 & -12 & 0 & 0 \\ 0 & 0 & -12 & 4 & 0 & 16 & 12 & 4 & 0 & 0 \\ 0 & 0 & 0 & 0 & -24 & 12 & 48 & 0 & -24 & -12 \\ 0 & 0 & 0 & 0 & -12 & 4 & 0 & 16 & 12 & 4 \\ 0 & 0 & 0 & 0 & 0 & 0 & -24 & 12 & 24 & 12 \\ 0 & 0 & 0 & 0 & 0 & 0 & -12 & 4 & 12 & 8 \end{bmatrix}$$

Fig. 2.3(c) A typical Global Matrix

In this system of storage, if there are zeros at the top of a column, only the elements starting from non-zero value need to be stored.

Consider the matrix shown in Fig. 2.4. The zero elements such as a_{37}, a_{77} etc. within the matrix may be changed to nonzero elements in the solution process. The matrix elements will form a skyline as shown in Fig. 2.4. Therefore, we need to allocate storage allocations to zero elements within the skyline but do not need to store zero elements that are outside the skyline.

The matrix is stored as shown in Fig. 2.5. The locations of the diagonal elements are given by

$$< 1\ 2\ 4\ 7\ 11\ 14\ 18\ 25\ 33\ 42\ 52\ 61 >.$$

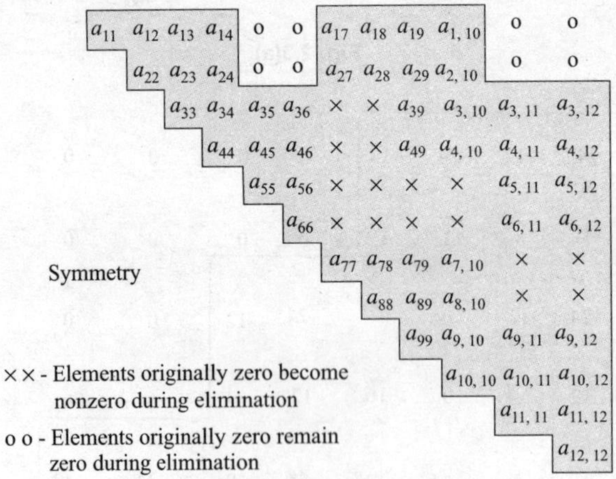

Fig. 2.4 Skyline of the matrix

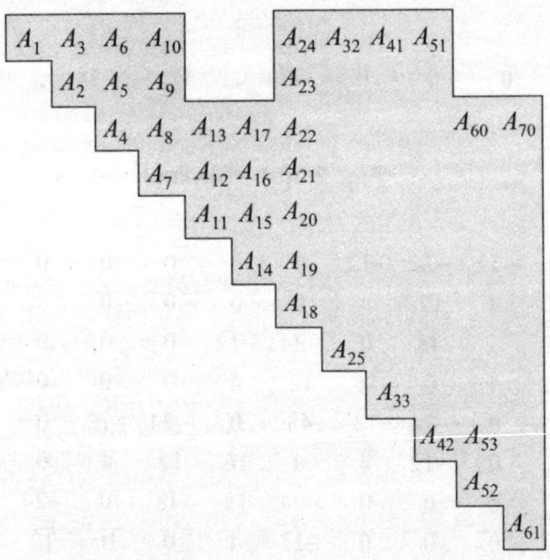

Fig. 2.5 Skyline storage location

Frontal Solver

Wave front equation or frontal solvers carry out the solution at the same time as assembly, that is total matrix is not completely assembled before the solution starts. This limits the high speed storage to the square of m and eliminates one of the "tape storage" of the blocking scheme as explained in band solvers. Such an idea comes from a simple observation on the inner loop of the equation

$$a_{ij} = a_{ij} - a_{in}\, a_{nj}/a_{nn} \qquad \text{...[1]}$$

Observation:

Irrespective of the subtraction of the second term before or after the first term is formed, the second term can be correctly evaluated as soon as the terms in the pivot row are correct (i.e. as soon as a_{nn}, a_{ni} and a_{nj} are known). Irons (1970) developed this concept specifically for the solution of finite element equations.

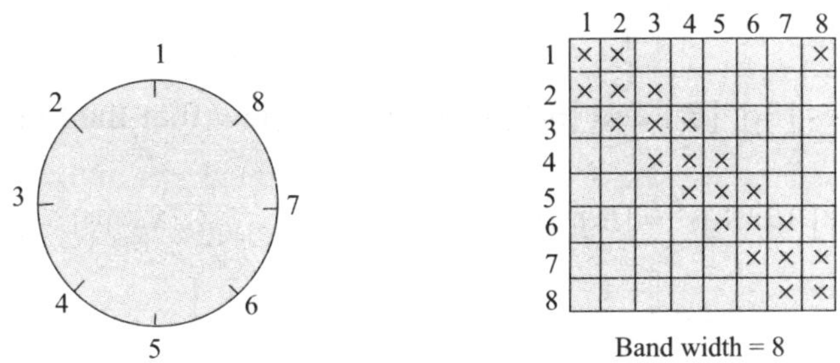

(a) Ring Structure (b) Connectivity Matrix

Fig. 2.6

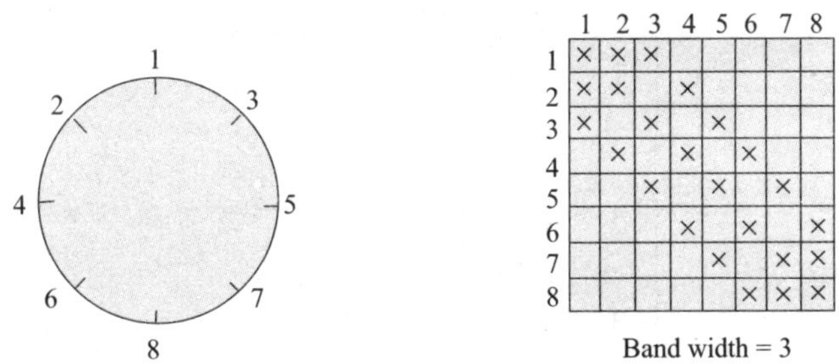

(a) (b)

Fig. 2.7 (a) Renumbered ring structure (b) Connectivity matrix

EXERCISES

1. Find the product of [A] {B}.

$$[A] = \begin{bmatrix} 4 & 0 & 2 & -1 \\ 3 & -2 & 4 & 3 \end{bmatrix}, \{B\} = \begin{Bmatrix} 5 \\ -2 \\ 1 \\ 4 \end{Bmatrix}$$

2. Obtain the matrix [C] in order that the following equation is satisfied:

$$-10[A] + 3[C] = 2[B]$$

$$[A] = \begin{bmatrix} 7 & -2 \\ 4 & 5 \end{bmatrix}, [B] = \begin{bmatrix} -3 & 4 \\ 2 & 5 \end{bmatrix}$$

3. Using matrices $[B_1]$, $[B_2]$ and $[B_3]$ given below, show that the following relationship is satisfied:

$$([B_1]\,[B_2]\,[B_3])^T = [B_3]^T\,[B_2]^T\,[B_1]^T$$

$$[B_1] = \begin{bmatrix} -4 & 1 \\ 2 & 3 \end{bmatrix} \qquad [B_2] = \begin{bmatrix} 2 & 1 \\ 0 & -3 \end{bmatrix}$$

and $$[B_3] = \begin{bmatrix} 3 & -2 \\ -1 & 4 \end{bmatrix}$$

4. Solve $$\begin{bmatrix} 5 & -2 & 4 \\ 1 & 2 & 3 \\ -4 & 1 & -1 \end{bmatrix} \begin{Bmatrix} x_1 \\ x_2 \\ x_3 \end{Bmatrix} = \begin{Bmatrix} 20 \\ 20 \\ -9 \end{Bmatrix}$$

5. Find the inverse of $$\begin{bmatrix} 3 & -6 & 3 \\ -6 & 1 & 6 \\ 3 & 6 & 1 \end{bmatrix}$$

6. Find the solution of the following simultaneous equations by inverting the matrix.

$$2.15x - 1.10y + 2.50z = 10.25$$

$$2.30x + 4.60y - 3.10z = 7.20$$

$$5.92x + 3.05y + 2.15z = 6.90$$

7. Find $\dfrac{d[A]}{dx}$ where

$$[A] = \begin{bmatrix} 5x^2 - 3 & 9x^3 - 6x^2 & 4x^2 + 3x + 2 \\ 4 & 7x + 6 & 8 \\ 3x^2 + 7x + 1 & 8x^2 & 9x^2 + 7x + 4 \end{bmatrix}$$

8. Find $\int_0^1 [A]\,dx$

where
$$[A] = \begin{bmatrix} 4x^3 + 7x + 3 & x^2 + 1 & x + 2 \\ 8x^2 + 9x + 7 & x^3 + 1 & 8x + 3 \\ 2x^2 + 4x + 3 & x^4 + 1 & 9x^2 + 4 \end{bmatrix}$$

9. Find $\int_0^2 \int_0^2 [A]\,dx\,dy$

$$[A] = \begin{bmatrix} x^2 + y^2 & 0 & x^2y + xy^2 \\ x + y & x^3 + y^3 & x^3y^2 + x^2y^3 \\ xy & x^2y^2 & xy + x^2y^2 \end{bmatrix}$$

10. Using the cofactor approach, determine the inverse of the matrix

$$\begin{bmatrix} 2 & 1 & 1 \\ 1 & 2 & 1 \\ 1 & 1 & 2 \end{bmatrix}$$

11. Given that

$$A = \begin{bmatrix} 8 & -2 & 0 \\ -2 & 4 & -3 \\ 0 & -3 & 3 \end{bmatrix} \quad D = \begin{Bmatrix} 2 \\ -1 \\ 3 \end{Bmatrix}$$

Determine
(a) $I - DD^T$
(b) det $\boldsymbol{A}$
(c) the eigen values and eigen vectors of $\boldsymbol{A}$. Is $\boldsymbol{A}$ positive definite?
(d) the solution to $\boldsymbol{Ax} = \boldsymbol{d}$ using Algorithms 1 and 2, by hand calculation.

12. Given that

$$N = [\xi, 1 - \xi^2]$$

find

(a) $\int_{-1}^{1} N d\xi$

(b) $\int_{-1}^{1} N^T N d\xi$

13. Given that the area of a triangle with corners at (x_1, y_1), (x_2, y_2) and (x_1, y_3) can be written in the form

$$\text{Area} = \frac{1}{2}\det\begin{bmatrix} 1 & x_1 & y_1 \\ 1 & x_2 & y_2 \\ 1 & x_3 & y_3 \end{bmatrix}$$

determine the area of the triangle with corners at (1, 1), (4, 2) and (2, 4).

14. For the triangle in Fig. P2.14 the interior point P at (2, 2) divides it into three areas A_1, A_2 and A_3, as shown. Determine A_1/A, A_2/A and A_3/A.

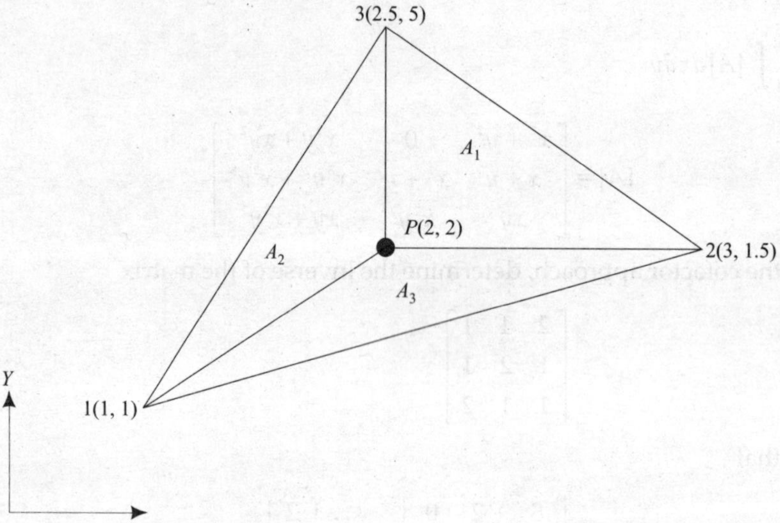

Fig. P2.14

15. Express $q = x_1 - 6x_2 + 3x_1^2 + 5x_1x_2$ in the matrix form $\frac{1}{2}\,x^TQx + c^Tx$.

16. A symmetric matrix $[A]_{n \times n}$ has a bandwidth *nbw* and is stored in the matrix $[B]_{n \times nbw}$.
(a) Find the location in **B** that corresponds to $A_{11,14}$.
(b) Find the location in **A** that corresponds to $B_{6,1}$.
For a symmetric (10 × 10) matrix with all nonzero elements, determine the number of locations needed for banded and skyline storage methods.

3

One-dimensional Problems

3.1 INTRODUCTION

Several engineering problems will be defined in terms of governing equations written in one dimension. Elements having spread in a single direction are known as one-dimensional elements. Such elements have a particular dimension quite longer compared to its other dimensions. Such elements are beams, pipes, shafts, truss members, frame skeleton members, cables, reinforcements, poles, small diameter chimneys, thin towers, piles etc.

Examples:

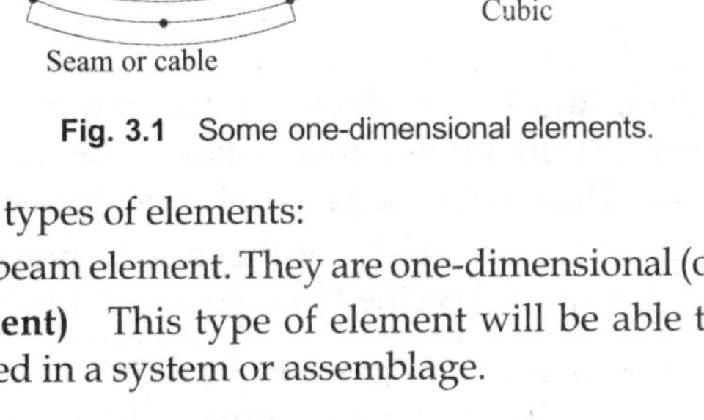

Fig. 3.1 Some one-dimensional elements.

Let us consider the two types of elements:

A truss element, and A beam element. They are one-dimensional (or unidimensional) elements.

Definition (i) (Truss element) This type of element will be able to carry either a tensile or a compressive force when used in a system or assemblage.

(Truss element)

It is also called a linear element or axial element.

Definition (ii) (Beam element) This type of element will be able to resist an axial force, then force and bending moment.

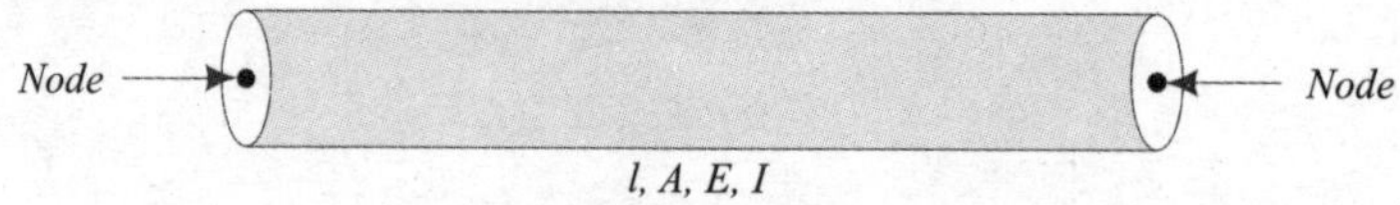

(Beam element)

It is seen that these two elements have got two nodes one at each end.

A one-dimensional problem in elasticity is given by the balance of forces in an elastic rod in terms of normal stress σ, are A and axial body force f. The force in the rod is σA, and the change in the force is balanced by the external body force:

$$\frac{d}{dx}[\sigma A] + fA = 0 \qquad \text{...[1]}$$

For one-dimensional problems, the stress, strain, displacement and loading depend only on the variable x.

$u = u(x)$; (displacement vector) $\sigma = \sigma(x)$; (stress vector) $\varepsilon = \varepsilon(x)$; (strain vector) $T = T(x)$; (traction force vector) $f = f(x)$; (body force vector)
$\sigma = E\varepsilon$; (stress-strain relation)
$\varepsilon = \frac{du}{dx}$; (strain-displacement relation)
$dv = Adx$; (differential volume for one-dimensional problem) A = area of elastic rod

Definition: (Degree of freedom, (dof)) The degree of freedom is a variable that describes the behaviour of a node in an element. In a one-dimensional problem, every node is permitted to displace only in the $\pm x$ direction. Thus, each node has only one degree of freedom.

Note: Degree of freedom shows the possibility of the joint to move. These are the number of independent co-ordinates which must be specified to uniquely define all the displacements.

Displacement function Each node is free to move in $\pm x$ direction so that there is one degree of freedom per node. Therefore, for 2 nodes there are 2 degrees of freedom. The displacement function for u is represented by a linear polynomial. Total number of constant terms to be associated with the polynomial will be two, because the number of constant terms equals the total degrees of freedom for the element. Therefore, the displacement vector is given by $\{f\} = \{u\}$.

Let us take the displacement polynomial describing state of displacement variation within the element as

$$u = \alpha_1 + \alpha_2 x$$

∴ For the two nodes, we get

$$u_1 = \alpha_1 + \alpha_2 x_1 \qquad ...[1]$$

$$u_2 = \alpha_1 + \alpha_2 x_2 \qquad ...[2]$$

or
$$\begin{Bmatrix} u_1 \\ u_2 \end{Bmatrix} = \begin{bmatrix} 1 & x_1 \\ 1 & x_2 \end{bmatrix} \begin{Bmatrix} \alpha_1 \\ \alpha_2 \end{Bmatrix}$$

or
$$\begin{Bmatrix} \alpha_1 \\ \alpha_2 \end{Bmatrix} = \begin{bmatrix} 1 & x_1 \\ 1 & x_2 \end{bmatrix}^{-1} \begin{Bmatrix} u_1 \\ u_2 \end{Bmatrix} \qquad ...[3]$$

$$= \frac{1}{(x_2 - x_1)} \begin{bmatrix} x_2 & -x_1 \\ -1 & 1 \end{bmatrix} \begin{Bmatrix} u_1 \\ u_2 \end{Bmatrix}$$

∴ The displacement polynomial can be written as $u = [(u_1 x_2 - u_2 x_1) + (u_2 - u_1)] \times \frac{1}{(x_2 - x_1)} . x$

Since $(x_2 - x_1) = L$ = Length of the member

$$u = \left[\frac{u_1(x_2 - x)}{L} + \frac{u_2(x - x_1)}{L} \right] \qquad ...[4]$$

or
$$u = [N_1 u_1 + N_2 u_2] \qquad ...[5]$$

where
$$N_1 = \frac{(x_2 - x)}{L}; N_2 = \frac{(x - x_1)}{L}$$

Strain: In one dimension for axial elements we will have only one strain (e_x)

$$e_x = \frac{\partial u}{\partial x}$$

∴
$$\{e_x\} = \frac{\partial u}{\partial x} = \frac{\partial}{\partial x} [N_1 u_1 + N_2 u_2] = \frac{\partial}{\partial x} [N_1 \; N_2] \begin{Bmatrix} u_1 \\ u_2 \end{Bmatrix} \quad (\because u = N_1 u_1 + N_2 u_2)$$

or,
$$\{e_x\} = \begin{bmatrix} \frac{\partial N_1}{\partial x} & \frac{\partial N_2}{\partial x} \end{bmatrix} \begin{Bmatrix} u_1 \\ u_2 \end{Bmatrix}$$

$$= \begin{bmatrix} \frac{-1}{L} & \frac{1}{L} \end{bmatrix} \begin{Bmatrix} u_1 \\ u_2 \end{Bmatrix}$$

∴
$$\boxed{N_1 = \frac{(x_2 - x)}{L}; N_2 = \frac{(x - x_1)}{L}}$$

$$\therefore \qquad \boxed{\frac{\partial N_1}{\partial x} = -\frac{1}{L}; \frac{\partial N_2}{\partial x} = \frac{1}{L}}$$

$$\therefore \qquad \{e_x\} = \frac{1}{L}[-1\ 1]\begin{Bmatrix} u_1 \\ u_2 \end{Bmatrix} \qquad \text{...[1]}$$

or
$$\{B\} = \frac{1}{L}[-1\ 1] = \text{strain matrix} \qquad \text{...[2]}$$

Stress: In one dimension, only one component of stress is there in the axial element.

$$\therefore \qquad \sigma_x = E.e_x$$

or
$$\sigma_x = \frac{E}{L}[-1\ 1]\{u\} \qquad \text{...[1]}$$

$$\left(\because e_x = \frac{1}{L}[-1\ 1]\begin{Bmatrix} u_1 \\ u_2 \end{Bmatrix}\right)$$

where $\{u\}$ is displacement vector.

3.2 FINITE ELEMENT MODELING

Element Division: Finite element analysis by virtue of its basic principle starts with discretisation and dividing/differentiating the whole system into a number of small-small elements. Let us consider a bar as in Fig. 3.2, the first step is to model the bar as a stepped shaft, constituting of a discrete number of elements, each having a uniform cross-section.

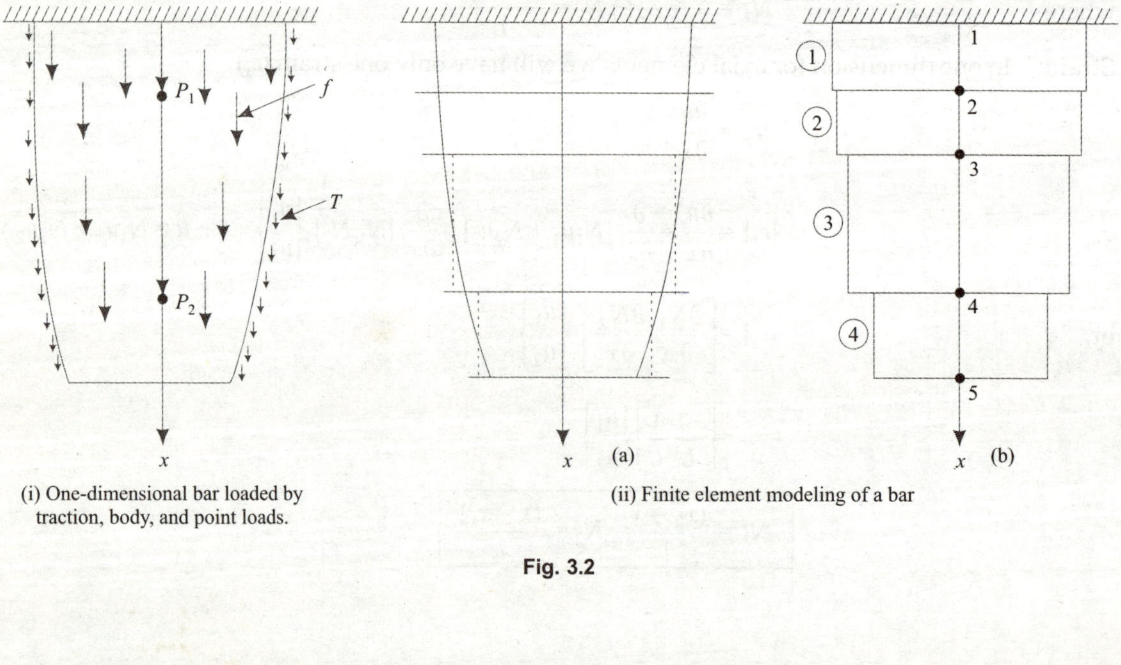

(i) One-dimensional bar loaded by traction, body, and point loads.

(ii) Finite element modeling of a bar

Fig. 3.2

A simple method for doing this is to divide the bar into four regions, as in Fig. 3.2(ii)(*a*). The resulting four elements, five node finite element model is shown in Fig. 3.2(ii)(*b*), the element numbers are circled to distinguish them from node numbers. The smaller the element the larger is the number of elements and so the better will be the accuracy of results.

But under the limitation of time and money we have to see as what minimum number of elements would be adopted to give a satisfactory result. However, cross-sectional area, traction, and body forces can differ in magnitude from element to element.

Global and Local Numbering of Nodes

When a body is discretised into elements certain numbering must be made to identify the elements. Initially numbering will be done with respect to overall system.

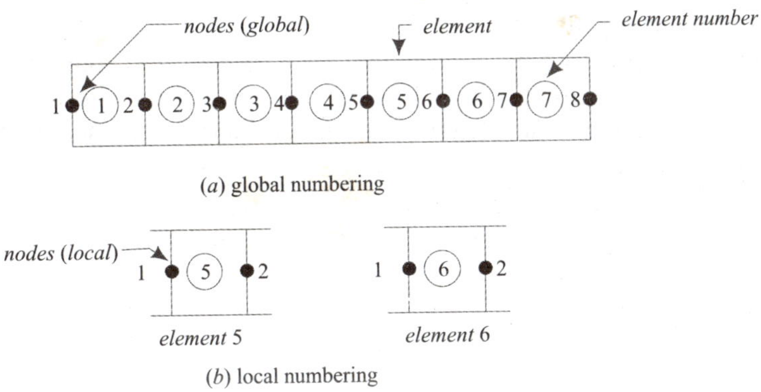

Fig. 3.3

Connectivity: Connectivity relates to the manner in which one element in a finite element model is connected with an adjacent element. It is the nodal information for the individual elements with details how they fit together to form the complete original system. Referring to Fig. 3.3, we see that local number 1 of element 5 refers to global number 5. Thus, we can establish all the relations. For the system shown in Fig. 3.3 we have the following element connectivity table (Table 3.1).

Table 3.1 Element connectivity table

Element	*Nodes* 1	2	*← local nodes*
①	1	2	global nodes
②	2	3	
③	3	4	
④	4	5	
⑤	5	6	
⑥	6	7	
⑦	7	8	

From Table 3.1, if we consider, for example, element 5, then its local node 1 refers to global node 5 and its local node 2 refers to global node 6.

3.3 COORDINATES AND SHAPE FUNCTIONS

I. COORDINATE SYSTEMS

In order to orient a truss in space we need a coordinate system. First remove the supports. We define the displacements at each joint as positive in the positive coordinate directions. The following coordinate systems are commonly referred in FEM.

(i) Global coordinates,

(ii) Local coordinates, and

(iii) Natural coordinates or intrinsic coordinates.

Apart from these coordinates, there is another term 'generalised coordinates'. These are the number of independent coordinates which must be specified or uniquely define all the displacements. This has nothing to do with the 'coordinates' terms used here to define the location of points in the element.

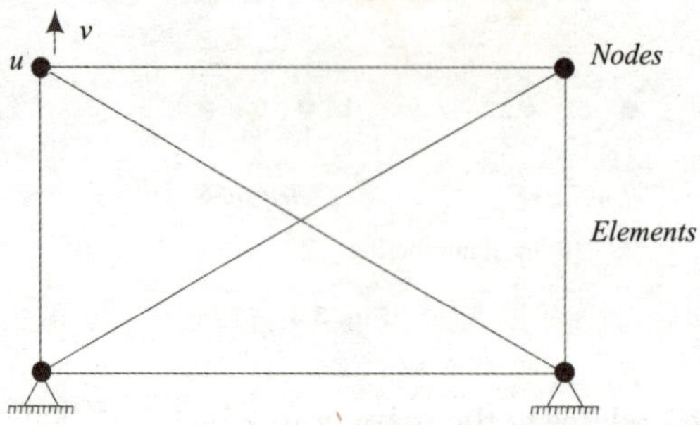

Fig. 3.4 Structure

For the truss shown in Fig. 3.4, the degree of freedom is 5. This is also called Kinematic degree of indeterminacy .

(i) Global Coordinates

The coordinate system used to define the points of individual element and the complete model (entire structure) to which all other coordinate systems are referred is called global coordinate system.

Example: Figure 3.5 shows the Cartesian global coordinate system used for some of the typical cases.

(ii) Local Coordinates

In order to derive conveniently the element properties in FEM many times, for each element a element a separate coordinate system is used.

Example: For example, for typical elements shown in Fig. 3.6(a) the local coordinates may be as shown in Fig. 3.6(b). However, the final equations are to be formed in the common coordinate system, i.e., global coordinate system only.

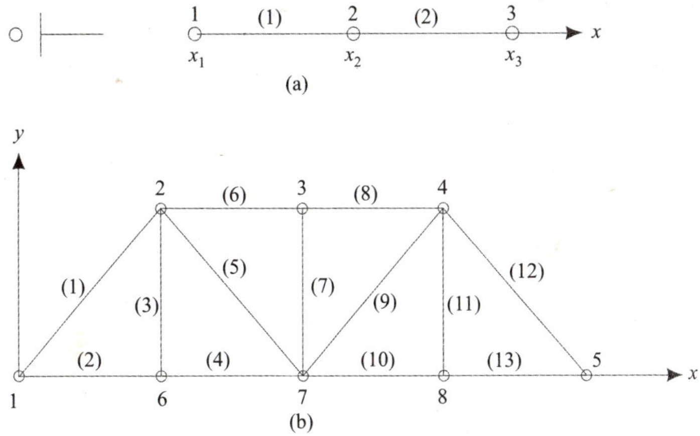

Fig. 3.5 Global coordinate system

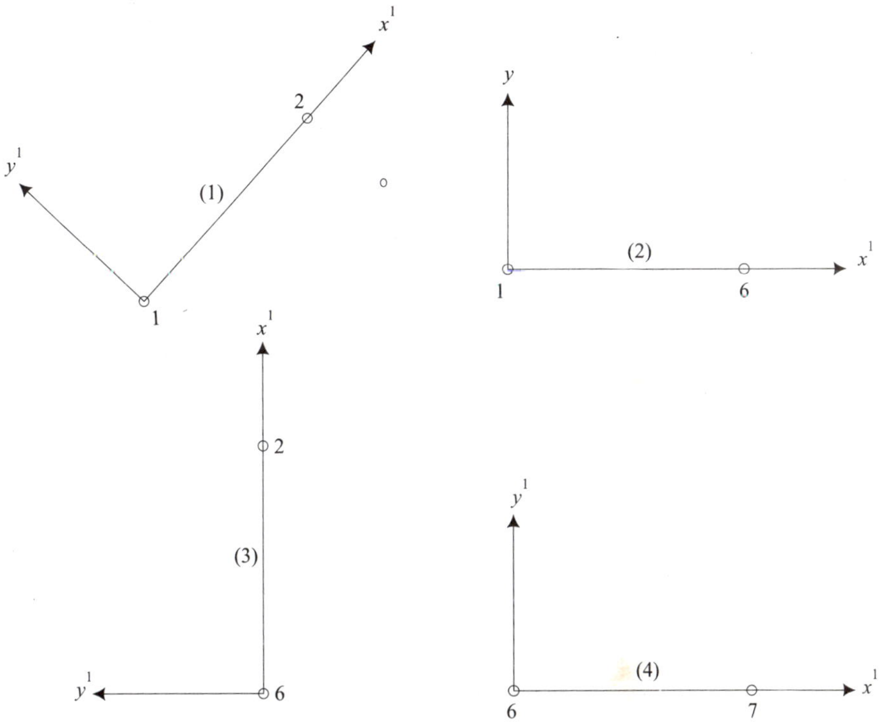

Fig. 3.6 Local coordinate system

(iii) Natural Coordinates (or Intrinsic Coordinates)

A natural coordinate system is nothing but a local coordinate system which is dimensionless and whose maximum value can be unity, i.e, it is a local coordinate within the element which varies normally between 0 and 1 or –1 and 1 depending upon the element type. This system will normally be represented by ξ in one dimension and by (ξ, η) in two dimensions etc. They are related to global coordinates by a mapping procedure.

The natural coordinates are obtained by assigning weightages to the nodal coordinates in defining the coordinate of any point inside the element. Hence such system has the property that ith coordinate has unit value at node i of the element and zero value at all other nodes.

The use of natural coordinate system is advantageous in assembling element properties (stiffness matrices), since closed form integrations formulae are available when the expressions are in natural coordinate systems.

Illustration

Natural coordinate: L in one-dimensional problems.

Consider Fig. 3.7 let the natural coordinate of point P be (L_1, L_2) and the Cartesian coordinate be x. Node 1 and node 2 have the Cartesian coordinates x_1 and x_2.

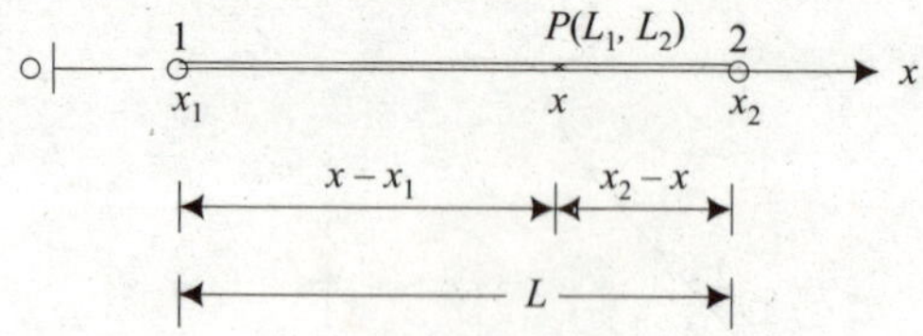

Fig. 3.7

$$L_1 + L_2 = 1 \quad \text{...[1]}$$

and also

$$L_1 x_1 + L_2 x_2 = x \quad \text{...[2]}$$

i.e., in matrix form

$$\begin{bmatrix} 1 & 1 \\ x_1 & x_2 \end{bmatrix} \begin{Bmatrix} L_1 \\ L_2 \end{Bmatrix} = \begin{Bmatrix} 1 \\ x \end{Bmatrix}$$

$$\therefore \begin{Bmatrix} L_1 \\ L_2 \end{Bmatrix} = \begin{bmatrix} 1 & 1 \\ x_1 & x_2 \end{bmatrix}^{-1} = \begin{Bmatrix} 1 \\ x \end{Bmatrix}$$

$$\begin{Bmatrix} L_1 \\ L_2 \end{Bmatrix} = \frac{1}{x_2 - x_1} \begin{bmatrix} x_2 & -x_1 \\ -1 & 1 \end{bmatrix}^{\tau} \begin{Bmatrix} 1 \\ x \end{Bmatrix}$$

$$= \frac{1}{x_2 - x_1} \begin{bmatrix} x_1 & -1 \\ -x_1 & 1 \end{bmatrix} \begin{Bmatrix} 1 \\ x \end{Bmatrix} = \frac{1}{x_2 - x_1} \begin{Bmatrix} x_2 - x \\ -x_1 + x \end{Bmatrix}$$

Noting that $x_2 - x_1$ is the length of the element, say l, we can write

$$\begin{Bmatrix} L_1 \\ L_2 \end{Bmatrix} = \begin{Bmatrix} \dfrac{x_2 - x}{l} \\ \dfrac{x - x_1}{l} \end{Bmatrix} \quad \text{...[3]}$$

The variation of L_1 and L_2 is shown in Fig. 3.8. L_1 is 1 at node 1 and is zero at node 2 whereas L_2 is zero when referred to node 1 and is one when referred to node 2. The variation is linear.

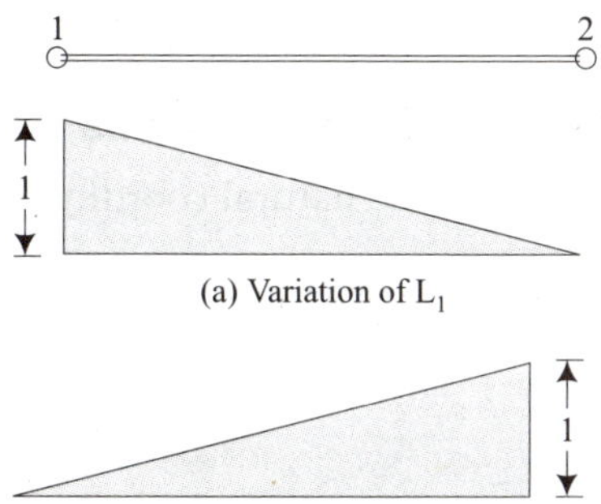

(a) Variation of L_1

(b) Variation of L_2

Fig. 3.8 Variation of natural coordinates L_1, L_2

The standard closed form integration over entire length is

$$\int_{x_1}^{x_2} L_1^p L_2^q \, dx = \frac{p!\,q!}{(p+q+1)!} l$$

Example: Integrate the following over the entire length l of the element

(i) $\int_0^l L_1^2 \, dx$ (ii) $\int_0^l L_1^3 L_2 \, dx$

Solution: (i) $\int_0^l L_1^2 \, dx$

Using the standard formula, $\int_0^l L_1^p L_2^q \, dx = \dfrac{p!\,q!}{(p+q+1)!}$

We note, $p = 2$, $q = 0$.

Hence $$\int_0^l L_1^2 \, dx = \frac{2!\,0!}{(2+0+1)!} l$$

$$= \frac{2}{2 \times 3} l = \frac{1}{3}$$

(ii) $$\int_o^l L_1^3 L_2 \, dx = \frac{3!1!}{(3+1+1)!} l$$

$$= \frac{3 \times 2 \times 1}{5 \times 4 \times 3 \times 2} l = \frac{l}{20}$$

Natural Coordinate ξ in One-dimensional Problems

In one-dimensional problems, we define the natural coordinate ξ for any point in the element as $\xi = \dfrac{PC}{\left(\dfrac{x_2 - x_1}{2}\right)}$, where *P* is a point within the element and *C* is the centre of nodes 1 and 2 (Fig. 3.9).

Since $$\xi = \frac{PC}{\left(\frac{l}{2}\right)} = \frac{x - x_c}{\left(\frac{l}{2}\right)}$$

$(l = x_2 - x_1)$ = length of *e*)

$$x_c = \frac{x_2 + x_1}{2}$$

$$= \frac{2}{l}\left(x - \frac{x_2 + x_1}{2}\right)$$

$$= \frac{2}{l}\left(x - \frac{x_2 - x_1 + 2x_1}{2}\right)$$

$$= \frac{2}{l}\left(x - \frac{l + 2x_1}{2}\right)$$

$$= \frac{2}{l}\left[x - x_1 - \frac{l}{2}\right]$$

$$\frac{l}{2}\xi = x - x_1 - \frac{l}{2}$$

$$\frac{l}{2}(1 + \xi) = x - x_1$$

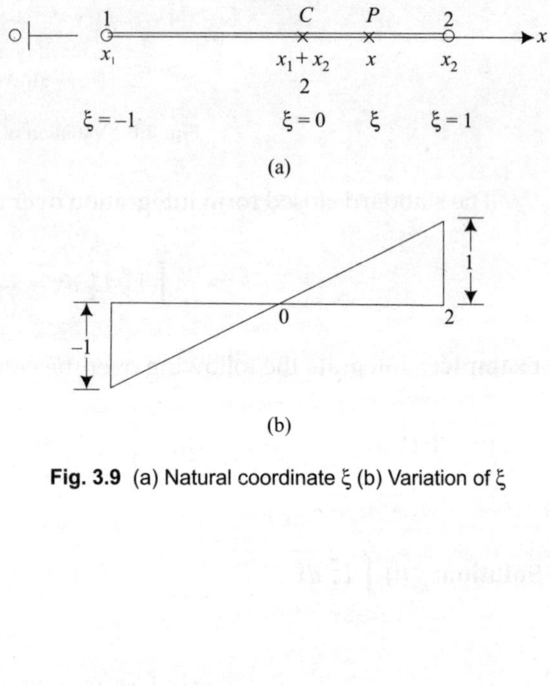

Fig. 3.9 (a) Natural coordinate ξ (b) Variation of ξ

It may be noted that, at node 1 where $x = x_1$,

$$\frac{l}{2}(1 + \xi) = x - x_1 = 0$$

$$\therefore \xi = -1$$

and at node 2, where $x = x_2$, we get

$$\frac{l}{2}(1 + \xi) = x_2 - x_1 = l$$

or $$\xi = 1$$

The variation of local coordinate ξ is as shown in Fig. 3.9(b). The integration formula for integration over entire length is

$$\int_{x_1}^{x_2} \xi^p dx = \int_{-1}^{1} \xi^p \frac{l}{2} d\xi \text{ since } dx = \frac{l}{2} d\xi \text{ (using equation } \frac{l}{2}(1 + \xi) = x - x_1)$$

$$= \frac{l}{2} \frac{1}{p+1} \left[\xi^{p+1}\right]_{-1}^{1}$$

$$= 0 \text{ if } p \text{ is odd} \qquad ...[2]$$

$$= \frac{l}{2} \frac{1}{p+1} (2) = \frac{l}{p+1}, \text{ if } p \text{ is even} \qquad ...[3]$$

Note: To evaluate $\int_{x_1}^{x_2} \xi^p dx$, we substitute

$$x = \frac{(x_2 - x_1)}{2} \xi + \frac{(x_2 + x_1)}{2}$$

$$\therefore \quad dx = \frac{(x_2 - x_1)}{2} d\xi$$

$$dx = \frac{l}{2} d\xi;$$

$$(l = x_2 - x_1)$$

$$\therefore \quad \int_{x_1}^{x_2} \xi^p dx = \frac{l}{2} \int_{-1}^{1} \xi^p d\xi$$

$$= \frac{l}{2} \left[\frac{\xi^{p+1}}{p+1}\right]_{-1}^{1}$$

$$= 0, \text{ if } p \text{ is odd}$$

$$= \frac{l}{2}\frac{1}{p+1}(2) = \frac{l}{p+1}, \text{ if } p \text{ is even.}$$

Generalized Coordinate Form of Displacement Models

In this system displacement/parametric model is taken in the form of a polynomial where unknown coefficients of the variables are named as generalized coordinates. Polynomial is the most common form of displacement model. This form is widely used because

(i) It is easy to handle mathematics of polynomial

(ii) A polynomial of arbitrary order permits a recognizable approximation to the solution.

Example:

$$u(x) = a_1 + a_2x + a_3x^2 + ... + a_{n+1}x^n \quad ...[1]$$

is a polynomial of order 'n' and $a_1, ... a_{n+1}$ are called generalized coordinates. The approximations of the polynomial in comparison with the true solution taking one, two and three terms are shown in Fig. 3.10 (For one dimension).

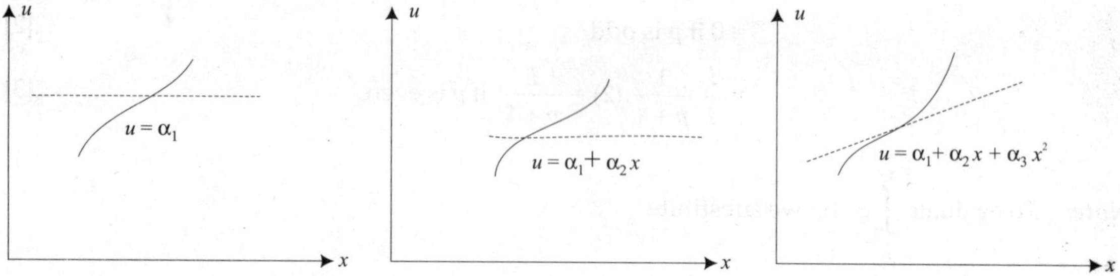

Fig. 3.10

The generalized coordinates $a_1a_2 ...a_{n+1}$ do not represent displacements rather they represent l.c. of displacements and slopes, as well.

$$u(x) = [N]\{a\}; [N] = [1 \times x^2 ... x^n] \quad \text{(i)}$$

and $$[a] = [a_1a_2 .... a_{n+1}] \quad \text{(ii)}$$

II. SHAPE FUNCTIONS

Finite elements cannot be taken randomly with any displacement variation and of any shape. They have to represent the behaviour of the system truly. As soon as we limit the number of elements, the true minimization of virtual work/energy may not be reached. Yet to obtain a convergence to actual some requirements are to be satisfied by shape functions we choose.

Field variable within an element is normally expressed in terms of nodal values. The shape function dictates the size of these nodal contributions. One shape function is associated with each node of the element. The shape function is usually denoted by the letter N and is usually the coefficient that appears in the interpolation polynomial. A shape function is written for each individual node of a finite element and has the property that its magnitude is 1 at that node and 0 for all other nodes is that element. The terminology is often interchanged between interpolation polynomial and shape function.

Example: Using generalized coordinate approach, find shape functions for two noded bar/truss elements.

Solution:

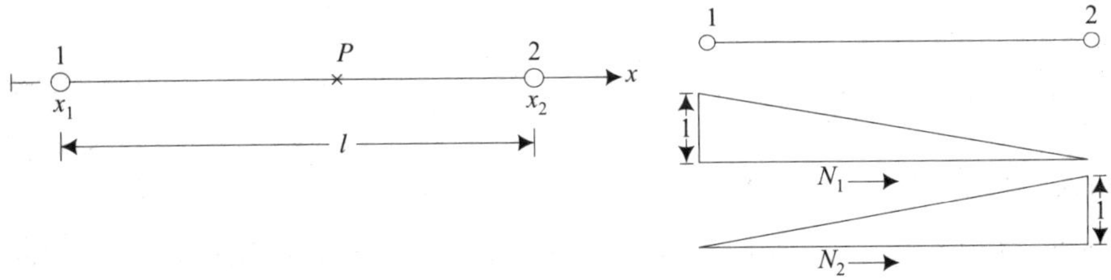

Fig. 3.11 Bar/Truss element with two nodes

A typical truss element is shown in Fig. 3.11.

Let $u = \alpha_1 + \alpha_2 x$ be a linear polynomial over the element (α_1, α_2 are generalized coordinates). The polynomial satisfies compatibility and completeness requirement.

$$u = \alpha_1 + \alpha_2 x = [1\ x] \begin{Bmatrix} \alpha_1 \\ \alpha_2 \end{Bmatrix}$$

Since $u = u_1$ at node 1 and equal to u_2 at node 2, we have,

$$\{q\} = \begin{Bmatrix} u_1 \\ u_2 \end{Bmatrix} = \begin{bmatrix} 1 & x_1 \\ 1 & x_2 \end{bmatrix} \begin{Bmatrix} \alpha_1 \\ \alpha_2 \end{Bmatrix}$$

$$\Rightarrow \qquad \begin{Bmatrix} \alpha_1 \\ \alpha_2 \end{Bmatrix} = \begin{bmatrix} 1 & x_1 \\ 1 & x_2 \end{bmatrix}^{-1} \begin{Bmatrix} u_1 \\ u_2 \end{Bmatrix}$$

$$\Rightarrow \qquad \begin{Bmatrix} \alpha_1 \\ \alpha_2 \end{Bmatrix} = \frac{1}{x_2 - x_1} \begin{bmatrix} x_2 & -1 \\ -x_1 & 1 \end{bmatrix}^T \begin{Bmatrix} u_1 \\ u_2 \end{Bmatrix}$$

$$= \frac{1}{l} \begin{bmatrix} x_2 & -x_1 \\ -1 & 1 \end{bmatrix} \begin{Bmatrix} u_1 \\ u_2 \end{Bmatrix}$$

$$\therefore \qquad u = [1\ x] \begin{Bmatrix} \alpha_1 \\ \alpha_2 \end{Bmatrix} = [1\ x] . \frac{1}{l} \begin{bmatrix} x_2 & -x_1 \\ -1 & 1 \end{bmatrix} \begin{Bmatrix} u_1 \\ u_2 \end{Bmatrix}$$

$$= \frac{1}{l} [x_2 - x \quad - x_1 + x] \begin{Bmatrix} u_1 \\ u_2 \end{Bmatrix}$$

$$= \left[\frac{x_2 - x}{l} \quad \frac{x - x_1}{l}\right]\begin{Bmatrix} u_1 \\ u_2 \end{Bmatrix}$$

$$= [N_1\ N_2]\begin{Bmatrix} u_1 \\ u_2 \end{Bmatrix} = N_1 u_1 + N_2 u_2$$

where $\quad N_1 = \dfrac{x_2 - x}{l}$ and $N_2 = \dfrac{x - x_1}{l}$

The shape function $[N]$ is

$$[N] = [N_1\ N_2] = \left[\frac{x_2 - x}{l} \quad \frac{x - x_1}{l}\right]$$

Variations of N_1 and N_2 are shown in Fig. 3.12.

Example: Derive the expression for shape function for a two-noded bar element taking natural coordinate ξ as varying from –1 to 1.

Solution:

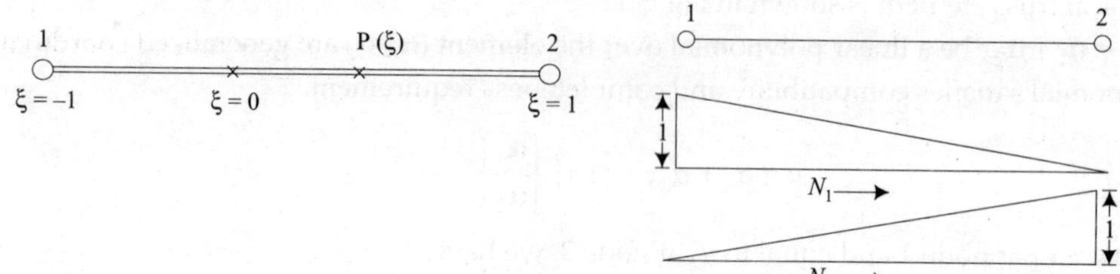

Fig. 3.12

Let displacement at any point P (ξ) be

$$u = \alpha_1 + \alpha_2 \xi$$

$$= [1\ \xi]\begin{Bmatrix} \alpha_1 \\ \alpha_2 \end{Bmatrix}$$

$$\begin{Bmatrix} u_1 \\ u_2 \end{Bmatrix} = \begin{bmatrix} 1 & -1 \\ 1 & 1 \end{bmatrix}\begin{Bmatrix} \alpha_1 \\ \alpha_2 \end{Bmatrix}$$

Since $\xi = -1$ at node 1 and $\xi = 1$ at node 2

$$\therefore \quad \begin{Bmatrix} \alpha_1 \\ \alpha_2 \end{Bmatrix} = \begin{bmatrix} 1 & -1 \\ 1 & 1 \end{bmatrix}^{-1}\begin{Bmatrix} u_1 \\ u_2 \end{Bmatrix}$$

$$= \frac{1}{1+1}\begin{bmatrix} 1 & -1 \\ 1 & 1 \end{bmatrix}^{T}\begin{Bmatrix} u_1 \\ u_2 \end{Bmatrix}$$

$$= \frac{1}{2}\begin{bmatrix} 1 & 1 \\ -1 & 1 \end{bmatrix}\begin{Bmatrix} u_1 \\ u_2 \end{Bmatrix}$$

$\therefore$
$$u = [1 \;\; \xi]\begin{Bmatrix} \alpha_1 \\ \alpha_2 \end{Bmatrix}$$

$$= [1 \;\; \xi]\frac{1}{2}\begin{bmatrix} 1 & 1 \\ -1 & 1 \end{bmatrix}\begin{Bmatrix} u_1 \\ u_2 \end{Bmatrix}$$

$$= \frac{1}{2}\,[1-\xi \quad 1+\xi]\begin{Bmatrix} u_1 \\ u_2 \end{Bmatrix}$$

$$= \begin{bmatrix} \frac{1-\xi}{2} & \frac{1+\xi}{2} \end{bmatrix}\begin{Bmatrix} u_1 \\ u_2 \end{Bmatrix}$$

$$= [N_1 \; N_2]\begin{Bmatrix} u_1 \\ u_2 \end{Bmatrix}$$

where
$$N_1 = \frac{1-\xi}{2} \text{ and } N_2 = \frac{1+\xi}{2} \qquad \text{...[1]}$$

Example: For a two-noded bar element, determine the shape functions. Use natural coordinate system.

Solution:

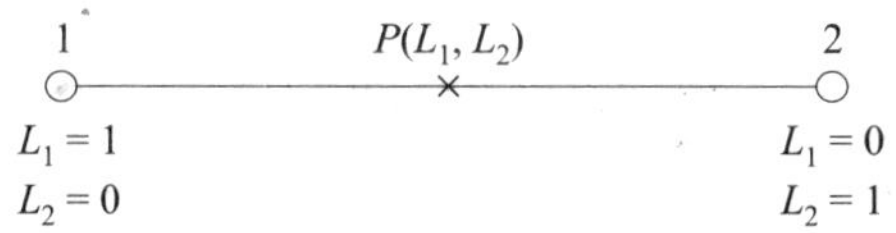

Fig. 3.13

Let
$$u = \alpha_1 L_1 + \alpha_2 L_2 = [L_1 L_2]\begin{bmatrix} \alpha_1 \\ \alpha_2 \end{bmatrix}$$

$$\therefore \quad \begin{Bmatrix} u_1 \\ u_2 \end{Bmatrix} = \begin{bmatrix} 1 & 0 \\ 0 & 1 \end{bmatrix}\begin{Bmatrix} \alpha_1 \\ \alpha_2 \end{Bmatrix};$$ (since $L_1 = 1$; $L_2 = 0$ at node 1 and $L_1 = 0$ and $L_2 = 1$ at node 2)

$$\therefore \quad \begin{Bmatrix} \alpha_1 \\ \alpha_2 \end{Bmatrix} = \begin{bmatrix} 1 & 0 \\ 0 & 1 \end{bmatrix}^{-1}\begin{Bmatrix} u_1 \\ u_2 \end{Bmatrix} = \frac{1}{1}\begin{bmatrix} 1 & 0 \\ 0 & 1 \end{bmatrix}\begin{Bmatrix} u_1 \\ u_2 \end{Bmatrix}$$

$$\therefore \quad u = [L_1 \; L_2]\begin{Bmatrix} \alpha_1 \\ \alpha_2 \end{Bmatrix} = [L_1 \; L_2]\begin{bmatrix} 1 & 0 \\ 0 & 1 \end{bmatrix}\begin{Bmatrix} u_1 \\ u_2 \end{Bmatrix} = [L_1 \; L_2]\begin{Bmatrix} u_1 \\ u_2 \end{Bmatrix}$$

Since $$u = [N_1\ N_2] \begin{Bmatrix} u_1 \\ u_2 \end{Bmatrix}; \text{ (by def}^{\,n} \text{ of shape function)}$$

$$\therefore \quad N_1 = L_1 \text{ and } N_2 = L_2 .$$

3.4 THE POTENTIAL ENERGY APPROACH

The potential energy Π of the system is given by

$$\Pi = \frac{1}{2} \int_{Area} \varepsilon^T D\varepsilon\, t\, dA - \int_{Area} u^T f\, t\, dA - \int_{Line} u^T T\, t\, dl - \sum_i u_i^T P_i \qquad \text{...[1]}$$

As usual f, T and P_i refer to body force, surface traction and point load applied at point i, respectively. Here, $P_i = [P_x\ P_y]^T_i$. The summation on i gives the total potential energy due to all point loads. The first term on RHS of eq. (i) represents the total strain energy of the system. Since this total strain energy is the sum of strain energy of all the elements then the above can be written as

$$\Pi = \sum_{NE} U_e - \sum_{NE} \int_{Area} u^T f\, t\, dA - \int_{Line} u^T T\, t\, dl - \sum_i u_i^T P_i \qquad \text{...[2]}$$

where NE is the number of elements and

$$U_e = \frac{1}{2} \int_{Area} \varepsilon^T D\, \varepsilon\, t\, dA \qquad \text{...[3]}$$

is the element strain energy

Element stiffness: We substitute for ε from (ε = Bq) into equation [3], the latter becomes:

$$U_e = \frac{1}{2} \int_{Area} q^T B^T D\, Bq\, t\, dA \qquad \text{...[4]}$$

If we assume that the element has a constant thickness throughout and since all terms in **B** and **D** matrices are constant then the above eq. [4] will be reduced to

$$U_e = \frac{1}{2} q^T B^T D\, B\, t_e \left(\int_{Area} dA \right) q = \frac{1}{2} t_e A_e q^T B^T D\, B \qquad \text{...[5]}$$

which can be put in the form

$$U_e = \frac{1}{2} q^T k\, q \qquad \text{...[6]}$$

where the element stiffness k is given by

$$k = t_e A_e B^T D\, B \qquad \text{...[7]}$$

Element Stiffness Matrix for the Axial Element

Now, consider an individual element of length l of an axial element subjected to an axial force as shown in Fig. 3.14.

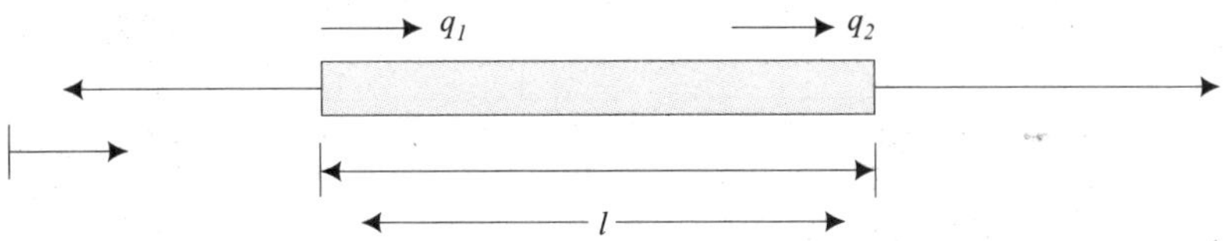

Fig. 3.14

It is assumed, the displacement u varies linearly along the length of the element,

letting $$u = a_0 + a_1 x \qquad \text{...[4]}$$

where a_0 and a_1 are constants. Since q_1 and q_2 are the displacements at the ends 1 and 2, we set

$$q_1 = a_0 + a_1 x_1$$
$$q_2 = a_0 + a_1 x_2$$

Solving for a_0 and a_1,

we find $$a_0 = \frac{(q_1 x_2 - q_2 x_1)}{l}$$

and $$a_1 = \frac{(q_2 - q_1)}{l}.$$

Hence, $$u = \frac{x_2 - x}{l} q_1 + \frac{x - x_1}{l} q_2 \qquad \text{...[5]}$$

We can write [2] as: $$u = N_1 q_1 + N_2 q_2 \qquad \text{...[6]}$$

Matrix form of the above is

$$u = [N_1 N_2] \begin{Bmatrix} q_1 \\ q_2 \end{Bmatrix} = Nq$$

where $$N_1 = \frac{x_2 - x}{l};\ N_2 = \frac{x - x_1}{l} \qquad \text{...[7]}$$

N_1 and N_2 are the shape functions.
$N = [N_1\ N_2]$ is a row matrix.

$q = \begin{Bmatrix} q_1 \\ q_2 \end{Bmatrix} = [q_1\ q_2]^T$ is called the element displacement vector.

Now, the strain in the element is given by $\varepsilon = \dfrac{du}{dx} = \dfrac{-q_1 + q_2}{l}$; (using equation [5]) or $\varepsilon = Bq$, where $B = \frac{1}{l}[-1 \;\; 1]$ is the element strain-displacement matrix.

For the single element considered here, the strain energy is given by $U = \frac{1}{2}\int_{x_1}^{x_2} \sigma^T \varepsilon A\, dx$ ...[8]

Note: $\varepsilon = Bq$ is the strain nodal parameter relationship of the axial element

Since, $u = Nq$ and $\varepsilon = Bq$

and by Hooke's law $\sigma = E\varepsilon$

then $\sigma = EBq$.

∴ Equation [8] becomes

$$U = \frac{EA}{2}\int_{x_1}^{x_2} \varepsilon^2 dx \qquad ...[9]$$

Let us assume EA is a constant within the element. Substituting for ε from eqn. [5] into eqn. [9] we find

$$U = \frac{EA}{2}\int_{x_1}^{x_2}\left[\frac{q_2 - q_1}{l}\right]^2 dx$$

$$= \frac{EA}{2l^2}(q_2 - q_1)^2 (x_2 - x_1) = \frac{EA}{2l}(q_2 - q_1)^2 \qquad ...[10]$$

Matrix form of equation [10] is:

$$U = \frac{EA}{2l}[q_1 \;\; q_2]\begin{bmatrix} 1 & -1 \\ -1 & 1 \end{bmatrix}\begin{Bmatrix} q_1 \\ q_2 \end{Bmatrix} = \frac{1}{2}\{q\}^T [k] \{q\} \qquad ...[11]$$

where $\{q\} = \begin{Bmatrix} q_1 \\ q_2 \end{Bmatrix}$ is the displacement vector

and $[k] = \frac{AE}{l}\begin{bmatrix} 1 & -1 \\ -1 & 1 \end{bmatrix}$ is the stiffness matrix.

Force Terms

We have seen that the work done by the body force is given by $\int_e u^T f A\, dx$ term can be written as

Substituting $u = N_1 q_1 + N_2 q_2$, we have

$$\int_e u^T f A\, dx = A_e f \int_e (N_1 q_1 + N_2 q_2)\, dx \qquad \text{...[1]}$$

Equation [1] can be written as:

$$\int_e u^T f A\, dx = q_1 \left(A_e f \int_e N_1 dx \right) + q_2 \left(A_e f \int_e N_2 dx \right) \qquad \text{...[2]}$$

$$= q^T \begin{Bmatrix} A_e f \int_e N_1 dx \\ A_e f \int_{-e} N_2 dx \end{Bmatrix} \qquad \text{...[3]}$$

The integrals of the shape functions can be evaluated by the substitution:

$$dx = \left(\frac{l_e}{2} \right) d\xi$$

Hence,

$$\begin{Bmatrix} \int_e N_1 dx = \frac{l_e}{2} \int_{-1}^{1} \frac{1-\xi}{2}\, d\xi = \frac{l_e}{2} \\ \int_e N_2 dx = \frac{l_e}{2} \int_{-1}^{1} \frac{1-\xi}{2}\, d\xi = \frac{l_e}{2} \end{Bmatrix} \qquad \text{...(4)}$$

Since $\int_e N_1\, dx$ is simply the area under the N_1 curve as shown in (Fig. 3.15)

$$\int_e N_1 dx = \text{Area} = \frac{1}{2} \cdot l_e - 1 = \frac{l_e}{2}.$$

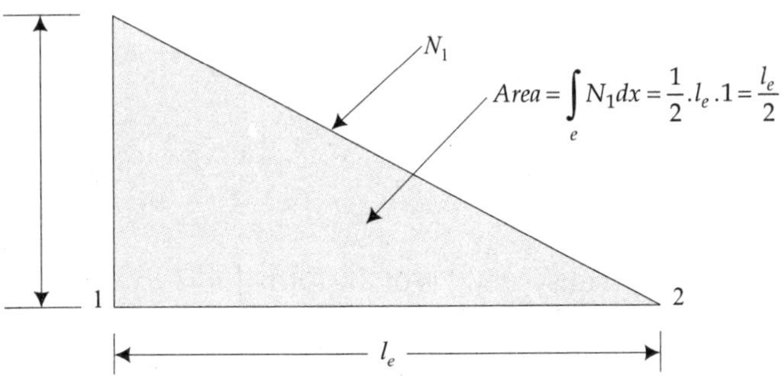

Fig. 3.15

Similarly, $$\int_e N_2\, dx = \frac{1}{2} . l_e . 1 = \frac{l_e}{2}$$

The body force term in equation (3) reduces to

$$\int_e u^T fA dx = q^T \frac{A_e}{2} l_e f \begin{Bmatrix} 1 \\ 1 \end{Bmatrix} \quad ...[5]$$

or, $$\int_e u^T fA dx = q^T f^e \quad ...[6]$$

(r.h.s. of (6) is of the form : Displacement × Force)

Thus, the element body force vector f^e, is defined as

$$f^e = \frac{A_e l_e f}{2} \begin{Bmatrix} 1 \\ 1 \end{Bmatrix} \quad ...[7]$$

Note: The factor $\frac{1}{2}$ in equation [7] tells us that this body force is equally distributed to the two nodes of the element

$A_e l_e$ = volume of the element

f = body force per unit volume

$A_e l_e f$ is the total body force acting on the element.

Traction Force Term

We have

$$\int_e u^T T\, dx = \int_e (N_1 q_1 + N_2 q_2) T\, dx \quad ...]1]$$

Since traction force T is constant within the element, we have

$$\int_e u^T T\, dx = q^T \div \begin{Bmatrix} T\int_e N_1 dx \\ T\int_e N_2 dx \end{Bmatrix} \quad ...[2]$$

where $\left(\int_e N_1 dx = \int_e N_2 dx = \frac{l_e}{2} \right)$. Thus, eq. [2] is of the form $\int_e u^T T\, dx = q^T T^e$

where element traction force vector T^e is given by

$$T^e = \frac{T l_e}{2} \begin{Bmatrix} 1 \\ 1 \end{Bmatrix} \quad ...[3]$$

Total Potential and System Equation

If we now assemble all the elements then we get for the total strain energy

$$U = \sum_{NE} U_e = \sum_{NE} \frac{1}{2} q^T k q = \frac{1}{2} Q^T \boldsymbol{k} Q \qquad ...[1]$$

where $\boldsymbol{k}$ is the global stiffness matrix. The global stiffness matrix will be symmetric; banded and sparse. Similarly, the total work done by the external loads can be put as

$$W = Q^T F \qquad ...[2]$$

where $F = \sum_{NE} (f^e + T^e) + \boldsymbol{P}$ ...[3]

Hence, the total potential of the system is

$$\Pi = \frac{1}{2} Q^T K Q - Q^T F \qquad ...[4]$$

Minimizing eq. [4] leads to the system equation

$$KQ = F \qquad ...[5]$$

Eq. [5] is to be solved after substituting for the boundary conditions.

3.5 THE GALERKIN APPROACH

We represent a virtual displacement field by $\phi = \phi(x)$ with the prescribed boundary conditions, then the corresponding strain will be represented by

$$\varepsilon(\phi) = \frac{d\phi}{dx} \qquad ...[1]$$

Principle of virtual work in the virtual displacement field A structure in equilibrium under the action of external forces for arbitrary virtual displacement ϕ (with corresponding compatible strain $\delta\phi$) form a state of compatible deformation q with compatible strain ε, the virtual w is δw equal to the virtual strain energy

$$\delta w = \delta \phi$$

The Galerkin's variation is given by

$$\int_V \sigma^T \varepsilon(\phi)\, dV - \int_V \phi^T f dV - \int_S \phi^T T\, ds - \sum_i \phi^T P = 0 \qquad ...[2]$$

For one-dimensional case [2] becomes

$$\int_l \sigma^T \varepsilon(\phi) A dx - \int_l \phi^T f A dx - \int_l \phi^T T\, dx - \sum_{i=1} \phi_i P_i = 0 \qquad \text{...[3]}$$

The first term on the LHS represents the strain energy or the internal work done, while other terms represent the external work done by various loads. If we discretize the region then equation [3] will assume the form

$$\sum_{NE} \int_e \varepsilon^T E \varepsilon(\phi) A dx - \sum_{NE} \int_e \phi^T f A dx - \sum_{NE} \int_e \phi^T T dx - \sum_{i=1} \phi_i P_i = 0 \qquad \text{...[4]}$$

where ε is the actual strain and $\varepsilon(\phi)$ is the virtual strain. Now, representing the above in terms of shape function expression and strain displacement matrix that is,

$$\phi = N\psi \qquad \text{...[5]}$$

$$\varepsilon(\phi) = B\psi \qquad \text{...[6]}$$

where $\psi = [\psi_1\ \psi_2]^T$ is the nodal displacement vector.

Element Stiffness Matrix

Hence, the first term equation [4] will take the form:

$$\int_e \varepsilon^T E \varepsilon(\phi)\, A dx = \int_e q^T B^T E B\psi\, A\, dx \qquad \text{...[7]}$$

where $\varepsilon = Bq$ is used for the actual strain. We know that for an axial element, B is a constant matrix and

$$dx = \frac{l_e}{2} d\xi.$$

$\therefore$ The eq. [7] becomes

$$\int_e \varepsilon^T E \varepsilon(\phi)\, A dx = \int_e q^T B^T E B\psi\, A\, dx$$

$$= q^T \left[EA \frac{l_e}{2} B^T B \int_{-1}^{1} d\xi \right] \psi$$

$$= q^T K^e \psi = \psi^T k^e q \qquad \text{...[8]}$$

where $k^e = EA\, l_e B^T B$ is the stiffness matrix

We substitute for B from the equation: $B = \frac{1}{l}$ [-1 1] and obtain:

$$k^e = \frac{EA}{l_e} \begin{bmatrix} 1 & -1 \\ -1 & 1 \end{bmatrix} \qquad \text{...[9]}$$

Force Terms

Consider the second term in equation [2] and using $\phi = N\psi$, $dx = \frac{l_e}{2} d\xi$, and body is assumed constant, we have

$$\int_V \phi^T f\, dV = \int_e \phi^T A\, dx \quad \text{...[1]}$$

$$= \int_{-1}^{1} \psi^T N^T f A_e . \frac{l_e}{2}\, d\xi \quad \text{...[2]}$$

$$= \psi^T f^e \quad \text{...[3]}$$

where
$$f^e = \frac{A_e l_e f}{2} \begin{Bmatrix} \int_{-1}^{1} N_1 d\xi \\ \int_{-1}^{1} N_2 d\xi \end{Bmatrix} \quad \text{...[4]}$$

is called the element body force vector.

We substitute, $N_1 = \frac{(1-\xi)}{2}$, $N_2 = \frac{(1+\xi)}{2}$

and obtain $\int_{-1}^{1} N_1 d\xi = 1$. Alternatively $\int_{-1}^{1} N_1 d\xi$ is the area under the N_1 curve $= \frac{1}{2} \times 2 \times 1 = 1$.

Similarly, $\int_{-1}^{1} N_2 d\xi = 1$.

Thus
$$f^e = \frac{A_e l_e f}{2} \begin{Bmatrix} 1 \\ 1 \end{Bmatrix} \quad \text{...[5]}$$

and the element traction term reduces to

$$\int_e \phi^T T\, dx = \psi^T T^e \quad \text{...[6]}$$

where T^e, element traction force, is given by

$$\boxed{T^e = \frac{T l_e}{2} \begin{Bmatrix} 1 \\ 1 \end{Bmatrix}} \quad \text{...[7]}$$

Note: Since k^e, f^e and T^e are obtained, also

$$\psi = [\psi_1, \psi_2]^T \text{ for element 1,}$$
$$\psi = [\psi_2, \psi_3] \text{ for element 2,}$$

the variational form

$$\sum_e \psi^T k^e q - \sum_e \psi^T f^e - \sum_e \psi^T T^e - \sum_i \psi_i P_i = 0 \qquad \text{...[8]}$$

$$\Rightarrow \quad \psi^T (KQ - F) = 0 \qquad \text{...[9]}$$

for every ψ with the constant boundary conditions.

3.6 ASSEMBLY OF THE GLOBAL STIFFNESS MATRIX AND LOAD VECTOR

We know that the stiffness matrix for an individual element (has been already developed) is given by

$$[K] = \frac{AE}{l}\begin{bmatrix} 1 & -1 \\ -1 & 1 \end{bmatrix} \qquad \text{...[1]}$$

Since the system contains many number of elements then the sum of the individual stiffness over the number of elements will give the global stiffness matrix. Similarly we get global load and global displacement vectors. Now, the potential energy of the complete system will be the sum of the potential energy of the individual elements and if there are n-number of elements then we have for the potential energy of the entire system

$$\Pi = \sum_{e=1}^{n}\left(\frac{1}{2} q^T k q - q^T F\right) \qquad \text{...[2]}$$

$$= \frac{1}{2} Q^T K Q - Q^T F \qquad \text{...[3]}$$

where K is the global stiffness matrix ; Q the global displacement vector and F the global load vector. Now, recall that k is 2×2 matrix. If two elements are joined then one node will be common and hence the stiffness matrix for two elements will be a 3×3 matrix corresponding to 3 nodes. Thus, in the global system for an axial element the global stiffness matrix will be of size $n \times n$, where n is the number of nodes. In general, the global stiffness matrix will be of order $mn \times mn$, where m is the degree of freedom (dof) at each node and n the number of nodes. In the case of axial element $m = 1$.

Let us consider the strain energy in, say element 3 (Fig. 3.16), we have

$$U_3 = \frac{1}{2} q^T k^3 q \qquad \text{...[4]}$$

We substitute for k_3, $U_3 = \frac{1}{2} q^T \frac{E_3 A_3}{l_3}\begin{bmatrix} 1 & -1 \\ -1 & 1 \end{bmatrix} q$ ...[5]

For element (3), $q = [Q_3, Q_4]^T$.

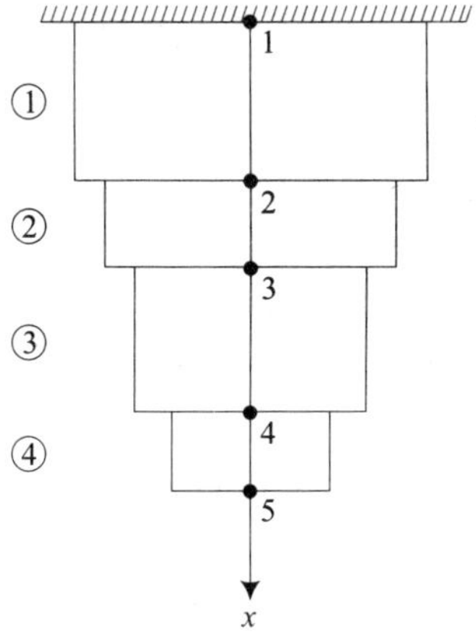

Fig. 3.16

Hence

$$U_3 = \frac{1}{2}[\,Q_1, Q_2, Q_3, Q_4, Q_5]\begin{bmatrix} 0 & 0 & 0 & 0 & 0 \\ 0 & 0 & 0 & 0 & 0 \\ 0 & 0 & \dfrac{E_3A_3}{l_3} & \dfrac{-E_3A_3}{l_3} & 0 \\ 0 & 0 & \dfrac{-E_3A_3}{l_3} & \dfrac{E_3A_3}{l_3} & 0 \\ 0 & 0 & 0 & 0 & 0 \end{bmatrix}\begin{Bmatrix} Q_1 \\ Q_2 \\ Q_3 \\ Q_4 \\ Q_5 \end{Bmatrix}$$

[6]

Example: Consider the bar shown in Fig. 3.17, which is subjected to all the three types of forces. Determine the global force vector assuming the bar is divided into three elements.

Solution: Since the bar is divided into three elements (Fig. 3.17(b)) each element is subjected to a traction T_i per unit length assumed to be constant over that element. The body force f per unit volume is, however, the same for all the elements. Now, if we consider element 1, then the total body

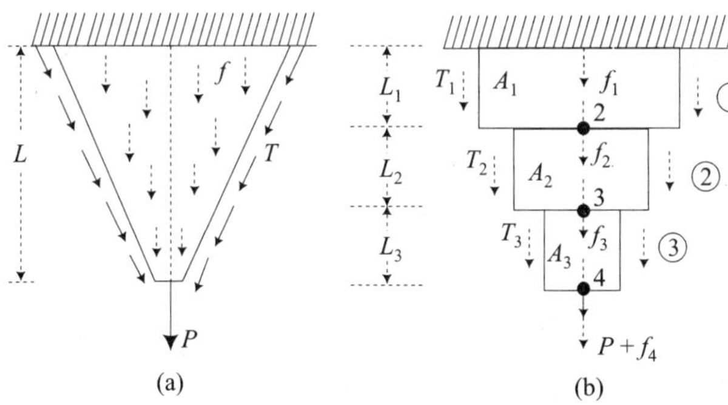

Fig. 3.17

force acting over this element will be fA_1L_1 which is equally divided between nodes 1 and 2 and marked as f_1 and f_2 and each is equal to $\frac{fA_1L_1}{2}$.

The traction acting over this element will be T_1L_1 and again will be equally divided between nodes 1 and 2 and will be equal to $\frac{T_1L_1}{2}$.

There are no concentrated load acting either at the node 1 or 2.

Thus, the total load F_1 for node 1

$$F_1 = f_1 + t_1 + P_1 = \frac{fA_1L_1}{2} + \frac{T_1L_1}{2}$$

For node 2 there will be contribution from the second element which can be calculated exactly the same way as already explained. For node 2, we get

$$F_2 = f_2 + t_2 + P_2 = \frac{fA_1L_1}{2} + \frac{T_1L_1}{2} + \frac{fA_2L_2}{2} + \frac{T_2L_2}{2} + 0$$

Proceeding and assembling all the loads gives the global load vector as

$$F = \begin{Bmatrix} \frac{fA_1L_1}{2} + \frac{T_1L_1}{2} \\ \frac{fA_1L_1}{2} + \frac{T_1L_1}{2} + \frac{fA_2L_2}{2} + \frac{T_2L_2}{2} \\ \frac{fA_2L_2}{2} + \frac{T_2L_2}{2} + \frac{fA_3L_3}{2} + \frac{T_3L_3}{2} \\ \frac{fA_3L_3}{2} + \frac{T_3L_3}{2} \end{Bmatrix} + \begin{Bmatrix} 0 \\ 0 \\ 0 \\ P \end{Bmatrix} \quad \text{...[1]}$$

Note that in structural mechanics problems, normally the body force will be neglected and surface traction may appear in a particular case. The only load that will be encountered often is the concentrated or distributed external load P.

3.7 PROPERTIES OF [K]

(1) Stiffness matrix is always a symmetric matrix i.e., $(K_{ij}) = (K_{ji})$

(2) Since we know that there is one shape function corresponding to each dof and hence the dimension of [K] is $(n \times n)$, n is the number of nodes.

(3) The entries along the main diagonal ($k_{ii} > 0$) are always positive.

(4) Sum of all the elements along any column is zero i.e., $k_{1i} + k_{2i} + \cdots + k_{ni} = 0$ and sum of all the elements along any row is zero i.e., $k_{i1}\, k_{i2} + k_{i3} + \cdots + k_{in} = 0$.

(5) It is a banded matrix.

$$K_{banded} = E\begin{bmatrix} \frac{A_1}{L_1} & -\frac{A_1}{L_1} \\ \frac{A_1}{L_1}+\frac{A_2}{L_2} & -\frac{A_2}{L_2} \\ \frac{A_2}{L_2}+\frac{A_3}{L_3} & -\frac{A_3}{L_3} \\ \frac{A_3}{L_3}+\frac{A_4}{L_4} & -\frac{A_4}{L_4} \\ \frac{A_4}{L_4} & 0 \end{bmatrix}$$ (all elements outside the band are zero).

> ***Note:*** Dimension of $K_{banded} = [N \times NBW]$ where NBW is the half bandwidth.

3.8 THE FINITE ELEMENT EQUATIONS

(Treatment of Boundary Conditions)

After using a discretization scheme to model the continuum we have obtained an expression for the total potential energy in the body as:

$$\Pi = \frac{1}{2} Q^T K Q - Q^T F \qquad \text{...[1]}$$

where K is the structural stiffness matrix, F is the global load vector, and Q is the global displacement vector.

The finite element formulation used here is based on the theorem of minimum potential energy, which states that:

"Of all the displacements satisfying the given boundary conditions, those that satisfy the equations of equilibrium are obtained by the stationary value of the potential energy."

According to the above principle, for minimum potential the displacements must be such that

$$\frac{\partial \Pi}{\partial q_1} = \frac{\partial \Pi}{\partial q_2} = 0$$

$$\Rightarrow \qquad \frac{\partial \Pi}{\partial \{q\}} = 0 \qquad \text{...[2]}$$

$\therefore$ [1] can be written as

$$\Pi = \frac{1}{2} \{q\}^T [k] \{q\} - \{q\}^T \{F\}$$

After differentiating, we get

$$\frac{\partial \Pi}{\partial \{q\}^T} = [k]\{q\} - \{F\} = 0 \qquad \text{...[3]}$$

$$\Rightarrow \qquad \boxed{F = kq}$$

or expanding

$$\left(\begin{Bmatrix} F_1 \\ F_2 \end{Bmatrix} = \frac{AE}{l}\begin{bmatrix} 1 & -1 \\ -1 & 1 \end{bmatrix}\begin{Bmatrix} q_1 \\ q_2 \end{Bmatrix}\right)$$

Consequently, the equations of equilibrium can be obtained by minimizing, w.r.t. Q, the *P.E.* $\Pi = Q^T k Q - Q^T F$ subject to boundary conditions.

Boundary conditions are given by

$$Q_{P_1} = a_1;\ Q_{P_2} = a_2;\ \ldots\ Q_{P_r} = a_r \qquad \ldots[4]$$

That is, the displacements along dof's P_1, P_2, ... P_r are specified to be equal to a_1, a_2, a_r, respectively.

Boundary conditions eliminate the probability of the structure moving as a rigid body.

Two approaches will be discussed for handling specified displacement boundary conditions of the type given in equation [4]:

(a) The Elimination approach and

(b) The Penalty approach.

(a) Elimination Approach

After obtaining the global stiffness and load matrix, we obtained the total potential for the complete system as given by eq. (39) which is reproduced here

$$\Pi = \frac{1}{2} Q^T K Q - Q^T F \qquad \ldots[1]$$

The equation of equilibrium is obtained by minimizing the total potential energy with respect to Q and subject to the boundary conditions. The handling of boundary condition can be the elimination method (which we shall see in this problem).

Now, assume that

$$Q = [q_1\, q_2\, q_3 \ldots q_n]^T \qquad \ldots[2]$$

$$F = [F_1\, F_2\, F_3 \ldots F_n]^T \qquad \ldots[3]$$

$$K = \begin{bmatrix} K_{11} & K_{12} & K_{13} \ldots & K_{1N} \\ K_{21} & K_{22} & K_{23} \ldots & K_{2N} \\ \ldots & \ldots & \ldots & \ldots \\ \ldots & \ldots & \ldots & \ldots \\ \ldots & \ldots & \ldots & \ldots \\ \ldots & \ldots & \ldots & K_{NN} \end{bmatrix} \qquad \ldots[4]$$

Now write the total potential expression in an expanded form for clarity. Then,

$$\Pi = \frac{1}{2}\,(q_1 K_{11} q_1 + q_1 K_{12} q_2 + q_1 K_{13} q_3 \cdots + q_1 K_{1N} q_N$$

$$+ (q_2 K_{21} q_1 + q_1 K_{22} q_2 + q_2 K_{23} q_3 \cdots + q_2 K_{2N} q_N$$

$$+ (q_3 K_{31} q_1 + q_2 K_{32} q_2 + q_2 K_{33} q_3 \cdots + q_3 K_{3N} Q_N \quad \text{...[5]}$$

....

$$+ (q_N K_{N1} q_1 + q_2 K_{N2} q_2 + q_1 K_{N3} q_3 \cdots + q_2 K_{NN} q_N)$$

$$- (q_1 F_1 + q_2 F_2 + q_3 F_3 \cdots + q_N F_N)$$

The above expression must be minimized with respect to $q_1, q_2, \ldots q_N$. Now before this operation we can substitute the boundary condition thereby eliminating the particular displacement. For example, if the boundary condition is such that $q_1 = \alpha$, then eq. [5] will become

$$\Pi = \frac{1}{2} (\alpha K_{11} q_1 + \alpha K_{12} q_2 + \alpha K_{13} q_3 \cdots + \alpha K_{1N} q_N$$

$$+ q_2 K_{21} \alpha + q_2 K_{22} q_2 + q_2 K_{13} q_3 \cdots + q_2 K_{2N} q_N$$

$$+ q_3 K_{31} \alpha + q_2 K_{32} q_2 + q_2 K_{33} q_3 \cdots + q_2 K_{3N} q_N \quad \text{...[6]}$$

....

$$+ (q_N K_{N1} \alpha + q_2 K_{N2} q_2 + q_2 K_{N3} q_3 \cdots + q_2 K_{NN} q_N$$

$$- (\alpha F_1 + q_2 F_2 + q_3 F_3 \cdots + q_N F_N)$$

Eq. [6] is to be minimised with respect to all q_j for $j = 2, 3, 4 \ldots N$, *i.e.*,

$$\frac{\partial \Pi}{\partial q_j} = 0 \quad j = 2, 3, 4 \ldots N \quad \text{...[7]}$$

Carrying this operation with respect to each q_j's we get the following simultaneous equations

$$K_{22} q_2 + K_{23} q_3 + K_{24} q_4 + \cdots + K_{2N} q_N = F_2 - K_{21}\alpha$$

$$K_{32} q_2 + K_{33} q_3 + K_{34} q_4 + \cdots + K_{3N} q_N = F_3 - K_{31}\alpha$$

$$K_{42} q_2 + K_{43} q_3 + K_{44} q_4 + \cdots + K_{4N} q_N = F_4 - K_{41}\alpha$$

....

$$K_{N2} q_2 + K_{N3} q_3 + K_{N4} q_4 + \cdots + K_{NN} q_N = F_N - K_{N1}\alpha \quad \text{...[8]}$$

and can be put in the form

$$\begin{bmatrix} K_{22} & K_{23} & K_{24} \ldots K_{2N} \\ K_{32} & K_{33} & K_{34} \ldots K_{3N} \\ \cdots & & \\ \cdots & & \\ K_{N2} & K_{N3} & K_{N4} \ldots K_{NN} \end{bmatrix} \begin{Bmatrix} q_2 \\ q_3 \\ q_4 \\ \cdots \\ q_N \end{Bmatrix} = \begin{Bmatrix} F_2 - K_{21}\alpha \\ F_3 - K_{31}\alpha \\ F_4 - K_{41}\alpha \\ \cdots \\ F_N - K_{N1}\alpha \end{Bmatrix} \quad \text{...[9]}$$

or

$$K_R Q_R = F_R \qquad \text{...[10]}$$

where K_R Q_R and F_R are the reduced stiffness, displacement and load matrices respectively.

The above equation is of the order $(n - 1) \times (n - 1)$ as q_1 is eliminated. This approach is called *elimination* method. The solution of the equation will yield all other q's. Normally, the Gaussian elimination scheme will be used for solving eq. [9]. Now, looking at the stiffness expression in eq. [9], it may be noticed that eq. [10] can be simply obtained by deleting the row and column containing the known boundary condition. Since q_1 is known and if we delete first row and column in eq. [9] we get the same expression as obtained in eq. [9]. In general, to form the equilibrium equation it is enough if we delete all rows and column corresponding to the specified support or the dof.

Example: The tapered bar is subjected to all the three types of forces as shown in Fig. 3.18 assuming $f = T = 0$ and $P = 1$ kN, $E = 200$ kN/mm^2, find the nodal displacements and the reaction at the support.

Also find the stresses in the elements (the bar is divided into there elements).

Solution: The global stiffness and load matrix for this problem (Fig. 3.18) has been obtained earlier as

$$\mathbf{K} = E\begin{bmatrix} \frac{A_1}{L_1} & -\frac{A_1}{L_1} & 0 & 0 \\ -\frac{A_1}{L_1} & \frac{A_1}{L_1}+\frac{A_2}{L_2} & -\frac{A_2}{L_2} & 0 \\ 0 & -\frac{A_2}{L_2} & \frac{A_2}{L_2}+\frac{A_3}{L_3} & -\frac{A_3}{L_3} \\ 0 & 0 & -\frac{A_3}{L_3} & \frac{A_3}{L_3} \end{bmatrix}$$

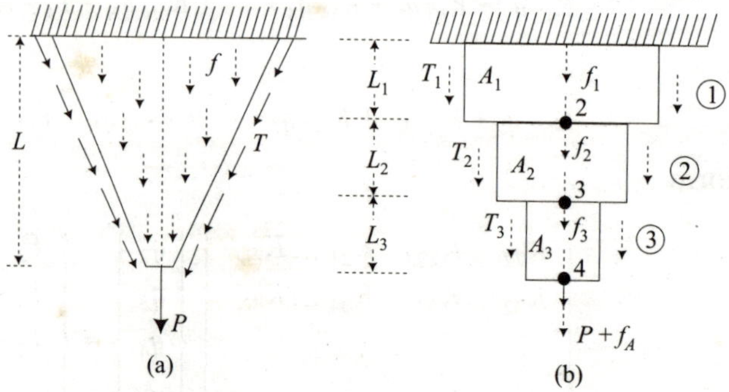

Fig. 3.18

and

$$F = \begin{Bmatrix} \dfrac{fA_1L_1}{2} + \dfrac{T_1L_1}{2} \\ \dfrac{fA_1L_1}{2} + \dfrac{T_1L_1}{2} + \dfrac{fA_2L_2}{2} + \dfrac{T_2L_2}{2} \\ \dfrac{fA_2L_2}{2} + \dfrac{T_2L_2}{2} + \dfrac{fA_3L_3}{2} + \dfrac{T_3L_3}{2} \\ \dfrac{fA_3L_3}{2} + \dfrac{T_3L_3}{2} \end{Bmatrix} + \begin{Bmatrix} 0 \\ 0 \\ 0 \\ P \end{Bmatrix}$$

Now, q_1 is known and is equal to zero. Hence, removing first row and first column in **K** matrix and the first row in **Q** and **F** matrices we obtain the equilibrium equation as

$$E = \begin{bmatrix} \dfrac{A_1}{L_1} + \dfrac{A_2}{L_2} & -\dfrac{A_2}{L_2} & 0 \\ -\dfrac{A_2}{L_2} & \dfrac{A_2}{L_2} + \dfrac{A_3}{L_3} & -\dfrac{A_3}{L_3} \\ 0 & -\dfrac{A_3}{L_3} & \dfrac{A_3}{L_3} \end{bmatrix} \begin{Bmatrix} q_2 \\ q_3 \\ q_4 \end{Bmatrix} = \begin{Bmatrix} \dfrac{fA_1L_1}{2} + \dfrac{T_1L_1}{2} + \dfrac{fA_2L_2}{2} + \dfrac{T_2L_2}{2} \\ \dfrac{fA_2L_2}{2} + \dfrac{T_2L_2}{2} + \dfrac{fA_3L_3}{2} + \dfrac{T_3L_3}{2} \\ \dfrac{fA_3L_3}{2} + \dfrac{T_3L_3}{2} \end{Bmatrix} + \begin{Bmatrix} 0 \\ 0 \\ P \end{Bmatrix}$$

These are the equilibrium equations to be solved for all q_i's. Substituting the numerical values, the above reduces to

$$200 \times 10^7 \begin{bmatrix} 6.0 & -2.5 & 0 \\ -2.5 & 4.0 & -1.5 \\ 0 & -1.5 & 1.5 \end{bmatrix} \begin{Bmatrix} q_2 \\ q_3 \\ q_4 \end{Bmatrix} = \begin{Bmatrix} 0 \\ 0 \\ 1000 \end{Bmatrix}$$

Solving the above, we will find

$$q_2 = 0.142857 \times 10^{-6} \text{ cm}; \; q_3 = 0.342857 \times 10^{-6} \text{ cm}; \; q_4 = 0.67619 \times 10^{-6} \text{ cm}.$$

To obtain reaction R_1 at node 1, we use the original equation including the reaction. Thus, at node 1 we have

$$K_{11}q_1 + K_{12}q_2 + K_{13}q_3 + K_{14}q_4 = F_1 + R_1$$

In this case $F_1 = 0$. Hence,

$$\begin{aligned} R_1 &= K_{11}q_1 + K_{12}q_2 + K_{13}q_3 + K_{14}q_{41} \\ &= 0 - 2 \times 10^9 \times 3.5 \times 0.142857 \times 10^{-6} + 0 + 0 = -1000 \text{ N} \end{aligned}$$

as $q_1 = K_{13} = K_{14} = 0$. The –ve sign indicates that the reaction acts opposite to the load.

The stresses can be determined using the relation: $\varepsilon = Bq$ and Hooke's law.

We have,

$$\sigma = E\,\mathbf{B}\,\mathbf{q}.$$

For element 1,

$$\sigma_1 = E\frac{1}{L_1}[-1\ 1]\begin{Bmatrix} q_1 \\ q_2 \end{Bmatrix} = 2\times 10^9 \frac{1}{40}[-1\ 1]\begin{Bmatrix} 0 \\ 0.1428\times 10^{-6} \end{Bmatrix}$$

from which we find $\sigma_1 = 7.14\ \text{N/cm}^2$

Similarly,

$$\sigma_2 = E\frac{1}{L_2}[-1\ 1]\begin{Bmatrix} q_2 \\ q_3 \end{Bmatrix} = 2\times 10^9 \frac{1}{40}[-1\ 1]\begin{Bmatrix} 0.1428\times 10^{-6} \\ 0.3428\times 10^{-6} \end{Bmatrix} = 10.0\ \text{N/cm}^2$$

$$\sigma_3 = E\frac{1}{L_3}[-1\ 1]\begin{Bmatrix} q_3 \\ q_4 \end{Bmatrix} = 2\times 10^9 \frac{1}{40}[-1\ 1]\begin{Bmatrix} 0.3428\times 10^{-6} \\ 0.6762\times 10^{-6} \end{Bmatrix} = 16.7\ \text{N/cm}^2$$

(b) Penalty Approach

This is another method of treating the boundary conditions which is much more elegant and easy to implement in the computer program.

This method is as explained below:

Suppose we have the boundary given by $q_1 = \alpha$. Then, wherever we have the specified boundary conditions at those points let us introduce an imaginary spring with large value of spring constant C_1, where i is the degree of freedom. In this case $i = 1$. The physical situation will be as shown in Fig. 13.19.

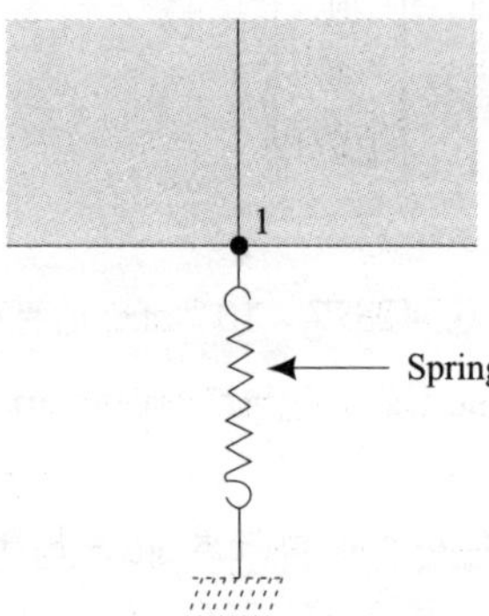

Fig. 13.19

The net displacement which the spring experiences is $= q_1 - \alpha$

Hence, the work done by the spring will be $= \frac{1}{2}C_1(q_1 - \alpha)^2$

Therefore the total potential of the entire system will be

$$\Pi = \frac{1}{2}Q^T KQ - Q^T F + \frac{1}{2}C_1(q_1 - \alpha)^2 \qquad ...[1]$$

Now, we can minimize eq. (1) with respect to all q_1's where $i = 1, 2, ...N$. This will result in the following equation.

$$\begin{bmatrix} K_{11}+C_1 & K_{12} & K_{13} \dots K_{1N} \\ K_{21} & K_{22} & K_{23} \dots K_{2N} \\ \dots & & \\ \dots & & \\ K_{N1} & K_{N2} & K_{N3} \dots K_{NN} \end{bmatrix} \begin{Bmatrix} q_1 \\ q_2 \\ \dots \\ \dots \\ q_N \end{Bmatrix} = \begin{Bmatrix} F_1 + C_1\alpha \\ F_2 \\ \dots \\ \dots \\ F_N \end{Bmatrix} \quad ...[2]$$

Now, taking the first equation we get

$$(K_{11} + C_1)\, q_1 + K_{12}q_2 + K_{13}q_3 + \cdots + K_{1N}q_N = F_1 + C_1\alpha \quad ...[3]$$

Dividing [3] throughout by C_1 we find

$$\left(\frac{K_{11}}{C_1} + 1\right) q_1 + \frac{K_{12}}{C_1} q_2 + \frac{K_{13}}{C_1} q_3 + \cdots + \frac{K_{1N}}{C_1} q_N = \frac{F_1}{C_1} + \alpha \quad ...[4]$$

If we have to get the value $q_1 = \alpha$, the original boundary condition, then all the terms in eq. (4) whose denominator contains C_1 must be zero. This will happen only if C_1 is very large. So choosing a large value for C_1 will give us the necessary equations. Now, how large C_1 should be? Theoretically ∞. But in practice it should be larger than any of the stiffness values. Since the same value of C can be used for any degree of freedom we can dispense away with the subscript for C. The following guideline may be useful in choosing the value for C.

$$C = \max |K_{ij}| \times 10^4 \quad (i = 1, 2 \dots N; j = 1, 2 \dots N) \quad ...[5]$$

Example: Find the global stiffness matrix for the tapered bar with dimensions as shown in Fig. 3.20 by the penalty method.

(Assume P =1 kN; E = 200 kN/mm^2)

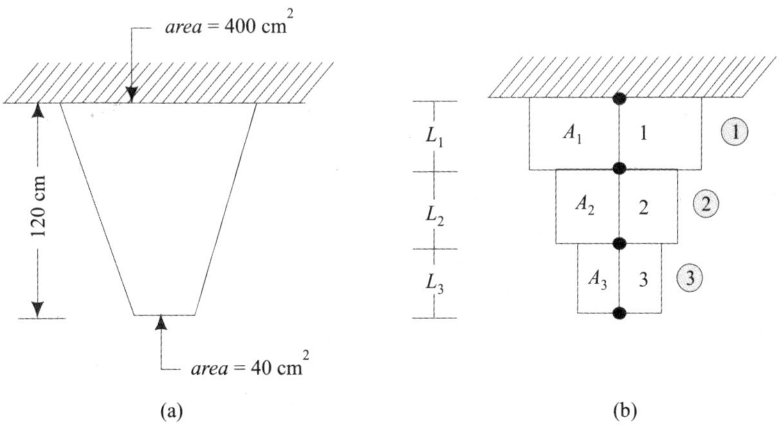

Fig. 3.20

Solution:

We have

$$K = E\begin{array}{c} \begin{array}{cccc} \quad 1 \quad & \quad 2 \quad & \quad 3 \quad & \quad 4 \quad \end{array} \\ \begin{bmatrix} \dfrac{A_1}{L_1} & -\dfrac{A_1}{L_1} & 0 & 0 \\ -\dfrac{A_1}{L_1} & \dfrac{A_1}{L_1}+\dfrac{A_2}{L_2} & -\dfrac{A_2}{L_2} & 0 \\ 0 & -\dfrac{A_2}{L_2} & \dfrac{A_2}{L_2}+\dfrac{A_3}{L_3} & -\dfrac{A_3}{L_3} \\ 0 & 0 & -\dfrac{A_3}{L_3} & \dfrac{A_3}{L_3} \end{bmatrix} \end{array} \qquad \text{...[1]}$$

Dividing the bar into three elements of equal length, we find $L_1 = L_2 = L_3 = 40$ cm; $A_1 = 140$ cm^2; $A_2 = 100$ cm^2; $A_3 = 60$ cm^2. Substituting these values in equation [1], we get

$$K = E\begin{bmatrix} \dfrac{140}{40} & -\dfrac{140}{40} & 0 & 0 \\ -\dfrac{140}{40} & \dfrac{140}{40}+\dfrac{100}{40} & -\dfrac{100}{40} & 0 \\ 0 & -\dfrac{100}{40} & \dfrac{100}{40}+\dfrac{60}{40} & -\dfrac{60}{40} \\ 0 & 0 & -\dfrac{60}{40} & \dfrac{60}{40} \end{bmatrix} = E\begin{bmatrix} 3.5 & -3.5 & 0 & 0 \\ -3.5 & 6.0 & -2.5 & 0 \\ 0 & -2.5 & 4.0 & -1.5 \\ 0 & 0 & -1.5 & 1.5 \end{bmatrix}$$

Or,
$$K = 2\times 10^9\begin{bmatrix} 3.5 & -3.5 & 0 & 0 \\ -3.5 & 6.0 & -2.5 & 0 \\ 0 & -2.5 & 4.0 & -1.5 \\ 0 & 0 & -1.5 & 1.5 \end{bmatrix}$$

We find the largest value of the stiffness is 12×10^9. Hence we can assume for C the value

$$C = 12\times 10^9 \times 10^4 = 12\times 10^{13} = 2\times 10^9\ [6\times 10^4]$$

Since, in this problem the node 1 is given a prescribed displacement = 0, then to the stiffness K_{11} the value of C must be added. The modified stiffness matrix will be

$$K = 2\times 10^9\begin{bmatrix} 3.5+6\times 10^4 & -3.5 & 0 & 0 \\ -3.5 & 6.0 & -2.5 & 0 \\ 0 & -2.5 & 4.0 & -1.5 \\ 0 & 0 & -1.5 & 1.5 \end{bmatrix}$$

Hence, the equilibrium equation will be

$$2 \times 10^9 \begin{bmatrix} 3.5+6\times10^4 & -3.5 & 0 & 0 \\ -3.5 & 6.0 & -2.5 & 0 \\ 0 & -2.5 & 4.0 & -1.5 \\ 0 & 0 & -1.5 & 1.5 \end{bmatrix} \begin{Bmatrix} q_1 \\ q_2 \\ q_3 \\ q_4 \end{Bmatrix} = \begin{Bmatrix} 0 \\ 0 \\ 0 \\ 1000 \end{Bmatrix}$$

We find $q_1 = 0.83326 \times 10^{-12}$ cm; $q_2 = 0.14285 \times 10^{-6}$ cm; $q_3 = 0.34284 \times 10^{-6}$ cm; $q_4 = 0.67618 \times 10^{-6}$ cm. The value of q_1 being very small can be taken equal to zero.

3.9 QUADRATIC SHAPE FUNCTIONS

The shape functions have already been derived for a two-noded bar element and they are given by $N_1(\xi) = \frac{1-\xi}{2}$ and $N_2(\xi) = \frac{1+\xi}{2}$.

They are called linear shape functions as the variation of $N_i(\xi)$ is linear within the element. However, such a linear assumption may not give good result in certain problems and is better to have different variation rather than linear. One such is *quadratic variation* within the element achieved by introducing a third node within element called the internal node.

Illustration

Consider a typical three-node quadratic element as shown in Fig. 3.21.

The relation between x and ξ is given by

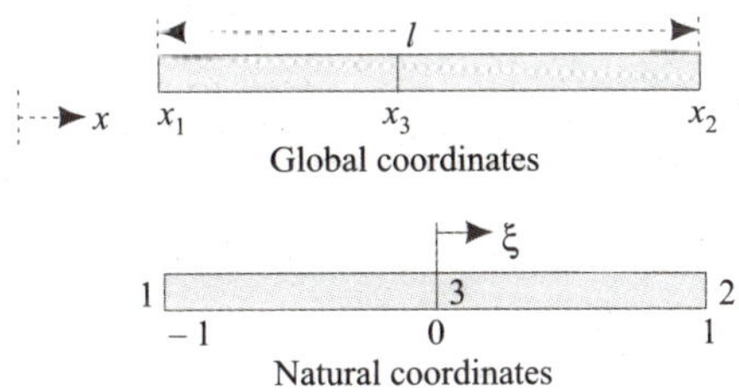

Fig. 3.21

$$\xi = \frac{2(x - x_3)}{x_2 - x_1} \quad \text{...[1]}$$

The element displacement vector is given by

$$q = \begin{Bmatrix} q_1 \\ q_2 \\ q_3 \end{Bmatrix} = [q_1\ q_2\ q_3]^T \quad \text{...[2]}$$

The shape functions N_1, N_2, N_3 will be obtained considering that N_i will be equal to unity at the node i but zero at the other nodes.

$$\left\{\begin{aligned} N_1 &= -\frac{1}{2}\,\xi(1-\xi) \quad &...[3]\\ N_2 &= \frac{1}{2}\,\xi(1+\xi) \quad &...[4]\\ N_3 &= (1+\xi)\,(1-\xi) \quad &...[5]\end{aligned}\right.$$

The shape functions given by eqs. [3], [4] and [5] are called the *Lagrange shape functions.*

If we now try to plot these functions the following will be obtained (Fig. 3.22). This displacement within the element is now given by

$$u = N_1q_1 + N_2q_2 + N_3q_3 \qquad ...[6]$$

which can be put in the form

$$u = \mathbf{N}\,\mathbf{q} \qquad ...[7]$$

where **N** is a 1 × 3 vector shape function given as

$$\mathbf{N} = [N_1 \;\; N_2 \;\; N_3] \qquad ...[8]$$

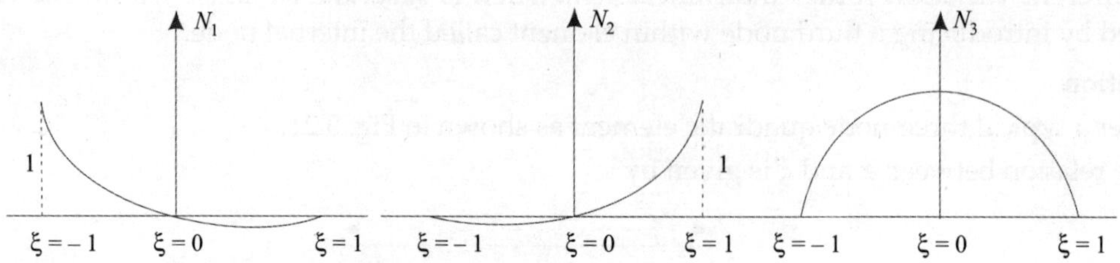

Fig. 3.22

The deflection pattern given by eq. [6] will appear as shown in Fig. 3.23.

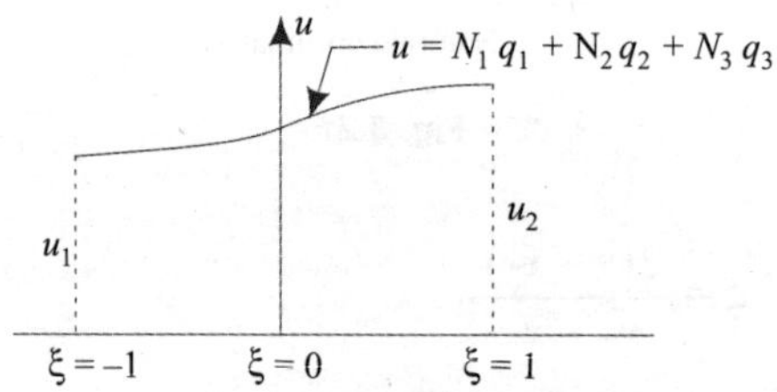

Fig. 3.23

Note (i): In the case of axial element the use of quadratic shape function is very much limited. Only when the force in the element varies along the length such an element is useful. The rotating member is one example where the centrifugal force varies along the length.

Since the displacement field within the element is written in terms of the nodal displacements as

$$u = N_1 q_1 + N_2 q_2 + N_3 q_3 \qquad \text{...[9]}$$

or

$$u = Nq. \qquad \text{...[10]}$$

where $N = [N_1, N_2, N_3]$ is a (1×3) vector of shape functions and $q = [q_1, q_2, q_3]^T$ is the (3×1) element displacement vector. At node 1, $N_1 = 1$, $N_2 = N_3 = 0$ and hence $u = q_1$. Similarly, $u = q_2$ at node 2, and $u = q_3$ at node 3. Thus [2] is a quadratic interpolation passing through q_1, q_2 and q_3 (Fig. 3.23).

Now,

$$\varepsilon = \frac{du}{dx}; \text{ (strain–displacement relation)}$$

$$= \frac{du}{d\xi}\frac{d\xi}{dx}; \text{ (chain rule)}$$

$$= \frac{2}{x_2 - x_1}\frac{du}{d\xi}; \qquad \text{...[11]}$$

$$\left(\text{using equation } \xi = \frac{2(x - x_1)}{(x_2 - x_1)}\right)$$

$$= \frac{2}{x_2 - x_1}\left[\frac{dN_1}{d\xi}, \frac{dN_2}{d\xi}, \frac{dN_3}{d\xi}\right].q;$$

using equation:

$u = Nq$

$u = N_1 q_1 + N_2 q_2 + N_3 q_3$

Using equations $N_1(\xi) = -\frac{1}{2}\xi(1 - \xi)$; $N_2(\xi) = \frac{1}{2}\xi(1 + \xi)$; $N_3(\xi) = (1 + \xi)(1 - \xi)$, we have

$$\varepsilon = \frac{2}{x_2 - x_1}\left[-\frac{1 - 2\xi}{2}, \frac{1 + 2\xi}{2}, -2\xi\right].q; \qquad [12]$$

or

$$\varepsilon = Bq; \qquad \text{...[13]}$$

where B is given by

$$B = \frac{2}{x_2 - x_1}\left[-\frac{1 - 2\xi}{2}, \frac{1 + 2\xi}{2}, -2\xi\right]; \qquad [14]$$

Using Hooke's law, we have

$$\sigma = EBq \qquad \text{...[15]}$$

B is linear in ξ, N_i are quadratic shape functions.

This implies that the strain and stress can vary linearly within the element.

Now, from equation $\xi = \frac{2(x - x_3)}{x_2 - x_1}$, we have

$$d\xi = \frac{2}{x_2 - x_1}dx = \left(\frac{2}{l_e}\right)dx$$

$$\Rightarrow \qquad dx = \left(\frac{l_e}{2}\right) d\xi; \ (x_2 - x_1 = l_e)$$

Note(ii): When using linear shape functions, the strain and stress came out to be constant within the element.

Quadratic Shape Functions in Terms of Line Coordinates

Let us define variation of the field variable u as

$$u = \alpha_1 + \alpha_2 L_2 + \alpha_3 L_2^2 \qquad \text{...[1]}$$

where α_1, α_2 and α_3 are constants to be determined and L_2 (as given in eqn. [2]) varies between 0 and 1. From Fig. 3.24

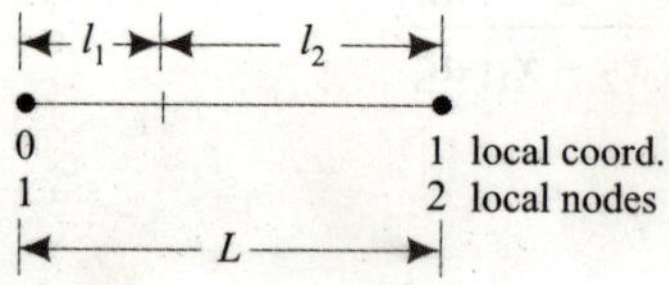

Fig. 3.24

$$L_1 = \frac{l_1}{L} \text{ and } L_2 = \frac{l_2}{L} \qquad \text{...[2]}$$

$$u = u_1 \text{ at } \quad L_2 = 0$$

$$u = u_3 \text{ at } \quad L_2 = 0.5$$

$$u = u_2 \text{ at } \quad L_2 = 1.0$$

Substituting the above into eq. [1] yields three equations solving which the constants are obtained as

$$\alpha_1 = u_1$$

$$\alpha_2 = \frac{-3u_1 + 4u_3 - u_2}{L} \qquad \text{...[3]}$$

$$\alpha_3 = \frac{2u_1 - 4u_3 + 2u_2}{L^2}$$

Substituting equation [2] into eq. $N = [N_1 \ N_2 \ N_3]$

$$u = u_1 + \frac{-3u_1 + 4u_3 - u_2}{L} L_2 + \frac{2u_1 - 4u_3 + 2u_2}{L^2} L_2^2$$

and can be rearranged as

$$u = N_1 u_1 + N_2 u_2 + N_3 u_3$$

where the shape functions N_1, N_2 and N_3 are given by

$$N_1 = L_1(2L_1 - 1) \quad \text{...[4a]}$$

$$N_2 = L_2(2L_2 - 1) \quad \text{...[4b]}$$

$$N_3 = 4L_1L_2 \quad \text{...[4c]}$$

Note that at node 1, $L_1 = 1$ and $L_2 = 0$; at node 2, $L_2 = 1$ and $L_1 = 0$; and at node 3, $L_1 = L_2 = 0.5$.

3.10 ONE-DIMENSIONAL STEADY STATE HEAT CONDUCTION FORMULATION

In the present chapter, we discuss one-dimensional steady state problems in which a temperature gradient exists along only one coordinate axis and the temperature at each point does not change with time.

Two different approaches are considered, and they are

(i) Functional approach

(ii) Galerkin's approach.

We may derive the equations of one-dimensional steady state condition using the balance of energy and a constitutive equation. Since the balance of energy states that the change in the heat flux q is balanced by an external heat source Q; we have

$$\frac{d}{dx}\{q(x)\,A(x)\} = Q\,(x)A(x) \quad \text{...[1]}$$

where $A(x)$ = area; negative Q implies that the heat is being removed from the system.

The constitutive equation is given by

$$q\,(x) = -\,k(x)\,\frac{dT(x)}{dx}; \qquad \text{(Fourier's Law)} \quad \text{...[2]}$$

where T = Temperature; k = Thermal conductivity.

Equations [1] and [2] yield the differential equation:

$$\frac{d}{dx}\left[k(x)A(x)\frac{dT(x)}{dx}\right] + Q(x)A(x) = 0 \quad \text{...[3]}$$

Notes: (1) Boundary conditions on T and natural b.c's can be specified for $q(x)$ = heat flux.

(2) The units of measurements are

$q(x)$ = heat flux in E/tL^2,

$A(x)$ = area = L^2,

$Q(x)$ = external heat source = E/tL^3,

$T(x)$ = temperature in T,

$k(x)$ = thermal conductivity in E/tLT.

Let us consider a linear variation of temperature in the element. The temperature at the two nodal points, denoted by T, are the unknowns within the element e. The temperature field is approximated using shape functions N_1 and N_2 as

$$T = T_1 N_1 + T_2 N_2$$

$$= [N_1\ N_2] \begin{Bmatrix} T_1 \\ T_2 \end{Bmatrix} = [N]\{T\}^e \qquad \text{...[1]}$$

where $N_1 = \dfrac{(1-\xi)}{2}$; $N_2 = \dfrac{(1+\xi)}{2}$, ξ varies from –1 to + 1, $N = [N_1\ N_2]$; $T^e = [T_1\ T_2]$.

Now, using
$$\xi = \frac{2}{x_2 - x_1}(x - x_1) - 1,\ (x - x_1) - 1 = \frac{2}{x_2 - x_1}\, dx \qquad \text{...[2]}$$

we have
$$\frac{dT}{dx} = \frac{dT}{d\xi} \cdot \frac{d\xi}{dx}$$

$$= \frac{2}{(x_2 - x_1)} \frac{dN}{d\xi} \,.\, T^e \qquad \text{...[3]}$$

$$= \frac{1}{(x_2 - x_1)} [-1, 1]\, T^e \qquad \text{...[4]}$$

or
$$\frac{dT}{dx} = B_T\, T^e \qquad \text{...[5]}$$

where
$$B_T = \frac{1}{(x_2 - x_1)} [-1, 1] \qquad [6]$$

Functional Approach (for Heat Conduction)

To solve the heat conduction equation:

$$\frac{d}{dx}\left(k \frac{dT}{dx}\right) + Q = 0 \qquad \text{...[1]}$$

with boundary condition $T(0) = 0$;

$$q(L) = h(T_L - T_\infty) \qquad \text{...[2]}$$

$$\left(\because\ q(x) = -k(x)\frac{dT(x)}{dx}\right)$$

We assume the solution of [1], equivalent to minimizing a functional, which is the energy form

$$\Pi_T = \int_O^L \frac{1}{2} k \left(\frac{dT}{dx}\right)^2 dx - \int_O^L QT\, dx + \frac{1}{2} h (T_L - T_\infty)^2 \qquad \text{...[3]}$$

subject to $T(0) = T_0$.

Let $Q = Q_e$; $k = k_e$ be the source and thermal conductivity, and are constant within the element.

using
$$dx = (l_e/2)\, d\xi;$$

$$T = NT^e \quad \text{and} \quad \frac{dT}{dx} = B_T T^e.$$

$\therefore$ From [3], we get

$$\Pi_T = \sum_e \frac{1}{2} T^{e^T} \left[\frac{k_e l_e}{2} \int_{-1}^{1} B_T^T B_T d\xi\right] T^e - \sum_e \left[\frac{Q_e l_e}{2} \int_{-1}^{1} N d\xi\right] T^e + \frac{1}{2} h (T_L - T_\infty)^2 \quad \text{...[4]}$$

We find from [4], that $\quad k_T = \dfrac{k_e l_e}{2} \displaystyle\int_{-1}^{1} B_T^T B_T \, d\xi$ ...[5]

Or $\quad k_T = \dfrac{k_e}{l_e} \begin{bmatrix} 1 & -1 \\ -1 & 1 \end{bmatrix}$ ...[6]

and the element heat rate vector due to heat source is

$$r_Q{}^T = \frac{Q_e l_e}{2} \int_{-1}^{1} N \, d\xi \quad \text{...[7]}$$

Or, since $\displaystyle\int_{-1}^{1} N_i d\xi = L$, we write

$$r_Q = \frac{Q_e l_e}{2} \begin{Bmatrix} 1 \\ 1 \end{Bmatrix} \quad \text{...[8]}$$

We note that global matrices k_T and R are assembled from element matrices k_T and r_Q in the usual manner using element connectivity information.

$\therefore$ The term $\frac{1}{2} h (T_L - T_\infty)^2$ in equation (4) becomes

$$\frac{1}{2} T_L h T_L - (h T_\infty) T_L + \frac{1}{2} h T_\infty{}^2 \quad \text{...[9]}$$

and $2hT_\infty{}^2$ is a constant which gives zero contribution while minimizing Π_∞.

Finally, we find the finite element equations as

$$\begin{bmatrix} (K_{11} + C) & K_{12} & \cdots & K_{1L} \\ K_{21} & K_{22} & \cdots & K_{2L} \\ \vdots & & & \\ K_{L1} & K_{L2} & \cdots & (K_{LL} + h) \end{bmatrix} \begin{bmatrix} T_1 \\ T_2 \\ \vdots \\ T_L \end{bmatrix} = \begin{bmatrix} (R_1 + (T_o) \\ R_2 \\ \vdots \\ (R_L + h T_\infty \end{bmatrix} \quad \text{...[10]}$$

Note (1): In [10], h gets added to the $(L, L)^{th}$ location of K_T and $(h T_\infty)$ gets added to the Lth row of R.

Example: A composite wall consists of three materials, as shown in Fig. 3.26(a). The outer temperature is $T_o = 20°C$. Convection heat transfer takes palce on the inner surface of the wall with $T_\infty = 800°C$ and $h = 25 \text{ W/m}^2.°C$. Determine the temperature distribution in the wall.

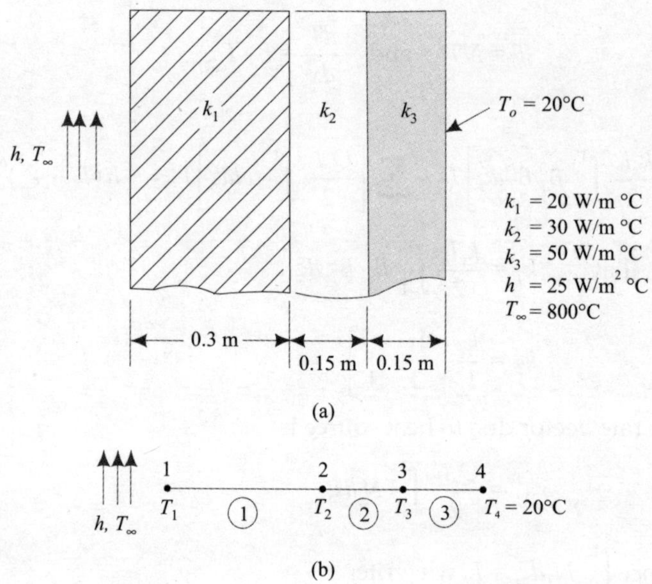

Fig. 3.26

Solution:

A three-linear element model of the composite wall is as shown in Fig. 3.26. The element conductivity matrices are:

$$\mathbf{k}_T^{(1)} = \frac{20}{0.3}\begin{bmatrix} 1 & -1 \\ -1 & 1 \end{bmatrix} \quad \mathbf{k}_T^{(2)} = \frac{30}{0.15}\begin{bmatrix} 1 & -1 \\ -1 & 1 \end{bmatrix} \quad \mathbf{k}_T^{(3)} = \frac{50}{0.15}\begin{bmatrix} 1 & -1 \\ -1 & 1 \end{bmatrix} \qquad ...[1]$$

$\therefore$

$$\mathbf{K} = \Sigma\, k_T$$

$$\mathbf{K} = 66.7\begin{bmatrix} 1 & -1 & 0 & 0 \\ -1 & 4 & -3 & 0 \\ 0 & -3 & 8 & -5 \\ 0 & 0 & -5 & 5 \end{bmatrix}$$

Now, since convection occurs at node 1, the constant h = 25 is added to the (1, 1) location of **K**. This results in

$$\mathbf{K} = 66.7\begin{bmatrix} 1.375 & -1 & 0 & 0 \\ -1 & 4 & -3 & 0 \\ 0 & -3 & 8 & -5 \\ 0 & 0 & -5 & 5 \end{bmatrix}$$

Since no heat generation Q occurs in this problem, the heat rate vector **R** consists only of $h\,T_\infty$ in the first row. That is,

$$\mathbf{R} = [25 \times 800, 0, 0, 0]^T$$

The specified temperature boundary condition, T_4 = 20°C, will now be handled by the penalty approach. We choose C based on

$$C = \max |K_{ij}| \times 10^4$$
$$= 66.7 \times 8 \times 10^4$$

Now, C gets added to (4, 4) location of **K**, while CT_4 is added to the fourth row of **R**. The resulting equations are

$$66.7\begin{bmatrix} 1.375 & -1 & 0 & 0 \\ -1 & 4 & -3 & 0 \\ 0 & -3 & 8 & -5 \\ 0 & 0 & -5 & 80\,005 \end{bmatrix}\begin{bmatrix} T_1 \\ T_2 \\ T_3 \\ T_4 \end{bmatrix} = \begin{Bmatrix} 25 \times 800 \\ 0 \\ 0 \\ 10\,672 \times 10^4 \end{Bmatrix}$$

The solution is

$$\mathbf{T} = [304.6, 119.0, 57.1, 20.0]^T \,^{\circ}\mathrm{C}$$

Galerkin's Approach (for Heat Conduction)

To solve the heat conduction equation:

$$\frac{d}{dx}\left(k\frac{dT}{dx}\right) + Q = 0 \qquad \text{...[1]}$$

with b.c's

$$T(0) = 0;$$
$$q(L) = h(T_L - T_\infty) \qquad [2]$$

$$\boxed{q(x) = -k(x)\frac{d}{dx}T(x)}$$

We assume ϕ as an approximation to T, where T is the exact solution of [1], i.e., $\phi \cong T$, then Galerkin's approach requires to solve the (integral) equation

$$\int_O^L \phi\left[\frac{d}{dx}\left(k\frac{dT}{dx}\right) + Q\right] dx = 0 \qquad \text{...[3]}$$

with $\phi(0) = 0$; Integrating equation [3] by parts, we get

$$\left(\text{Integrate } \frac{d}{dx}\left(k\frac{dT}{dx}\right) \text{ first}\right)$$

$$= \left\{\phi\left(k.\frac{dT}{dx}\right)\right\}_0^L - \int_0^L \left(\frac{d\phi}{dx}\right)\left(k.\frac{dT}{dx}\right) dx + \int_0^L \phi\, Q\, dx = 0 \qquad \text{...[4]}$$

Consider the term $\left\{\phi\, k\frac{dT}{dx}\right\}_0^L$

$$= \phi(L) \,.\, K(L)\frac{d}{dx}T(L) - \phi(0)\,k(0)\frac{d}{dx}(0)$$

$$= -\phi(L)\, h(T_L - T_\infty) \qquad ...[5]$$

$\therefore$ $\phi(0) = 0;$ b.c. given

and $q = -k(L)\dfrac{d}{dx}T(L) = h(T_L - T_\infty)$ (Fourier's law)

h = convection heat transfer coefficient

q = convective heat flux

$q = h(T_s - T_\infty)$

T_s = surface temperature

T_∞ = Fluid temperature

Equation [4] becomes

$$-\phi(L)\, h(T_L - T_\infty) - \int_0^L k\frac{d\phi}{dx}\frac{dT}{dx}dx + \int_0^L \phi Q\, dx = 0 \qquad ...[6a]$$

Now we know that, by isoparametric concept,

$$T = \sum_{i=1}^{2} N_i(\xi) T_i$$

$$= N_1 T_1 + T_2 T_2 \qquad ...[6b]$$

$$= [N_1\ N_2]\begin{bmatrix} T_1 \\ T_2 \end{bmatrix}$$

$$T = NT^e \qquad ...[7]$$

$$\xi = \frac{2}{(x_2 - x_1)}(x - x_1) - 1 \text{ for one dimension} \qquad ...[8]$$

$$\frac{dT}{dx} = \frac{dT}{d\xi}\frac{d\xi}{dx}$$

$$= \frac{2}{(x_2 - x_1)}\frac{dN}{d\xi}\,.\,T^e$$

$$\frac{dT}{dx} = \frac{2}{(x_2 - x_1)}[-1, 1]\, T^e \qquad ...[9]$$

Or $$\frac{dT}{dx} = B_T T^e \qquad ...[10]$$

$$B_T = \frac{1}{(x_2 - x_1)}[-1, 1]$$

$$= \frac{1}{l}[-1, 1]$$

For global system, let us denote the temperature vector

$$\psi = [\psi_1\ \psi_2 \ldots\ \psi_2]^T$$

We can write the approximation, ϕ

as $$\phi = N\psi \quad \text{...[11]}$$

and $$\frac{dT}{dx} = B_T T^e$$

$$\Rightarrow \quad \frac{d\phi}{dx} = B_T \psi \qquad (\because \phi \cong T)$$

$\therefore$ Equation [6a] becomes

$$-\psi_L h(T_L - T_\infty) - \Sigma\psi^T\left(\frac{k_e l_e}{2}\int_{-1}^{1} B_T^T B_T \, d\xi\right)T^e + \sum_e \psi^T Q_e \frac{l_e}{2}\int_{-1}^{1} N^T d\xi = 0 \quad \text{...[12]}$$

where ψ is a global virtual temperature vector.

Or $$-\psi_L hT_L + \psi_L hT_\infty - \psi^T k_T T + \psi^T R = 0 \quad \text{...[13]}$$

Equation [13] satisfies for all ψ with $\psi_1 = 0$.

The global matrices K_T and R are assembled from element matrices k_T as r_Q.

We choose ψ as :

$$\psi = [\psi_1 \, \psi_{2,} \ldots]$$

where
$$\psi_1 = [0, 1, 0, \ldots 0]^T$$
$$\psi_2 = [0, 0, 1, 0, \ldots 0]^T$$
$$\psi_n = [0, 0, \ldots 0, 1]^T$$

and $T_1 = T_0$, then equation [13] becomes

$$\begin{bmatrix} K_{22} & K_{23} & \ldots & K_{2L} \\ K_{32} & K_{33} & \ldots & K_{3L} \\ \vdots & \vdots & \vdots & \vdots \\ K_{L2} & K_{L3} & \ldots & (K_{LL}+h) \end{bmatrix} \begin{Bmatrix} T_2 \\ T_3 \\ \vdots \\ T_L \end{Bmatrix}$$

$$= \begin{bmatrix} R_2 - K_{21}T_0 \\ R_3 - K_{31}T_0 \\ \vdots \\ (R_L + hT_\infty) - K_{L1}T_0 \end{bmatrix} \quad \text{...[14]}$$

We can solve [14] for $T_2, T_3, \ldots T_L$

at node 1, $T = T_0$

Example: Assume a rod of length 0.2 m with the temperature at $x = 0$ maintained at 100°C. Assume a constant heat source of $Q = 3(10^6)$ W/m^3 along the length of the rod, a boundary heat flux of $q = 3\,(10^6)$ W/m^2 along the length of the rod, a boundary heat flux of $q = 1.8(10^6)$ W/m^2 removing heat at $x = 0.2$ m, and thermal conductivity $k = 6000$ W/(m.K).

Assume the area of the rod is 0.4 (10^{-3}) m^2. Use 5 linear elements of equal length and compute the temperature distribution at each element node.

Solution: A five linear element model of the rod is as shown in Fig. 3.27.

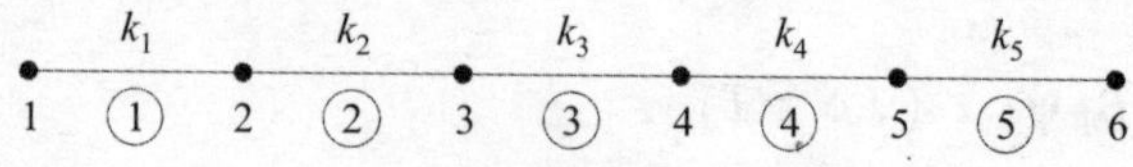

Fig. 3.27

With $\frac{AE}{L}$ replaced by Ak/L and f by Q, each element length is 0.2 m/5 = 0.04 m and $\frac{Ak}{L}$ = 0.4 (10^{-3}) (6000)/0.04 = 60 W/K

and

$$AQL = 0.400(10^{-3})(3)(10^6)(0.04)$$
$$= 48 \text{ W.}$$

The local stiffness matrix is

$$\begin{bmatrix} Ak/L & -Ak/L \\ -Ak/L & Ak/L \end{bmatrix} \begin{Bmatrix} T_1 \\ T_2 \end{Bmatrix} = \begin{Bmatrix} AQL/2 \\ AQL/2 \end{Bmatrix} \quad \text{...[1]}$$

Or,

$$60 \begin{bmatrix} 1 & -1 \\ -1 & 1 \end{bmatrix} \begin{Bmatrix} T_1 \\ T_2 \end{Bmatrix} = \frac{48}{2} \begin{Bmatrix} 1 \\ 1 \end{Bmatrix} \quad \text{...[2]}$$

The global stiffness matrix is :

$$\begin{bmatrix} 60 & -60 & 0 & 0 & 0 & 0 \\ -60 & 120 & -60 & 0 & 0 & 0 \\ 0 & -60 & 120 & -60 & 0 & 0 \\ 0 & 0 & -60 & 120 & -60 & 0 \\ 0 & 0 & 0 & -60 & 120 & -60 \\ 0 & 0 & 0 & 0 & -60 & 60 \end{bmatrix} \begin{Bmatrix} T_1 \\ T_2 \\ T_3 \\ T_4 \\ T_5 \\ T_6 \end{Bmatrix} = \begin{Bmatrix} 24 \\ 48 \\ 48 \\ 48 \\ 48 \\ 24 \end{Bmatrix} \quad \text{...[3]}$$

The flux b.c'. at node 6 is specified as $q = 1.8(10^6)$ W/m^2.
We compute surface boundary integral

$$\int_s [N]^T q ds \quad \text{...[4]}$$

In this problem, since the element 5 defines the node that contains the surface, node 6 is the location of the surface. Node 5 corresponds to a location within the rod. For element 5, the nodes 5 and 6; the shape function defining the surface at node 6 is constant; $N_5 = 0$; $N_6 = 1$.

Thus, the surface area given by eq. [4], after integration gives

$$\int_S [N]^T q ds = qA \begin{Bmatrix} 0 \\ 1 \end{Bmatrix}$$

$$= -1.8(10^6)\,(0.4)\,(10^{-3}) \begin{Bmatrix} 0 \\ 1 \end{Bmatrix}$$

$$\int_s [N]^T q ds = \begin{Bmatrix} 0 \\ -720 \end{Bmatrix} \qquad \text{...[5]}$$

(The flux term is –ve, ∵ heat is being removed from the system).

Now, we add eq. [5] to the RHS of eq. [3]. Since the flux is applied at node 6, only that term is modified as:

$$[24\ 48\ 48\ 48\ 48\ -696]^T \qquad \text{...[6]}$$

We write eq. [3] as:

$$\begin{bmatrix} 1 & 0 & 0 & 0 & 0 & 0 \\ 0 & 120 & -60 & 0 & 0 & 0 \\ 0 & -60 & 120 & -60 & 0 & 0 \\ 0 & 0 & -60 & 120 & -60 & 0 \\ 0 & 0 & 0 & -60 & 120 & -60 \\ 0 & 0 & 0 & 0 & -60 & 60 \end{bmatrix} \begin{Bmatrix} T_1 \\ T_2 \\ T_3 \\ T_4 \\ T_5 \\ T_6 \end{Bmatrix} = \begin{Bmatrix} 100 \\ 48 + (60)\,(100) \\ 48 \\ 48 \\ 48 \\ -696 \end{Bmatrix} \qquad \text{...[7]}$$

In eq. (7) we have incorporated the flux b.c. given by eq. [6] and the temperature b.c.

$$\text{Node } \frac{Ak}{L} = 60 \text{ W/K},\ T_0 = 100$$

$$\therefore\quad 48 + \frac{Ak}{L} T_0 = 48 + (60)\,(100)$$

The solution of equation [7] gives the temperature distribution along the rod.

The exact solution is:

$$T = \frac{3.0(10^6)}{6000}[0.2x - 0.5x^2] - \frac{1.8(10^6)x}{6000} + 100 \qquad \text{...[8]}$$

EXERCISES

1. Express the shape functions of quadratic element in terms of line coordinates.
2. For the pin-jointed configuration shown in Fig. P3.2, determine the stiffness values K_{11}, K_{12}, K_{22} of the global stiffness matrix.
3. An axial element has a varying cross-section as shown in Fig. P3.3. Derive the stiffness matrix.

[Ans: $[k] = \dfrac{EA_0\alpha}{L \ln(1+\alpha)} \begin{bmatrix} 1 & -1 \\ -1 & 1 \end{bmatrix}$ where α is the variation factor]

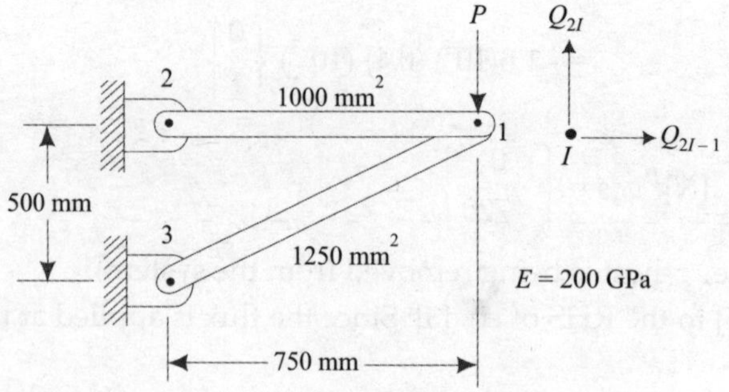

Fig. P3.2

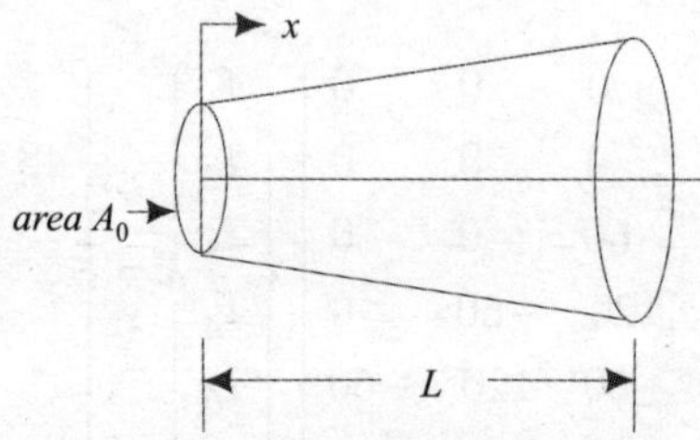

Fig. P3.3

4. For a bar subjected to a pure torque (Fig. P3.4) derive the stiffness matrix.

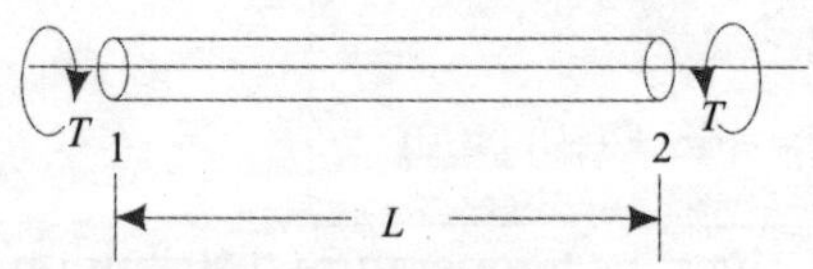

Fig. P3.4

5. For an axial element with nodes 1 and 2 the displacement field is given as $u = ax + bx^2$. Find the stiffness matrix and comment on the choice of displacement function (Fig. P3.5) .

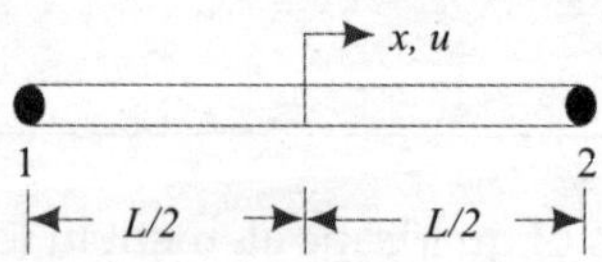

Fig. P3.5

$$\left[\text{Ans: } k = \frac{EA}{3L}\begin{bmatrix} 7 & 1 \\ 1 & 7 \end{bmatrix}\right]$$

4

Trusses

4.1 TWO-DIMENSIONAL TRUSSES (PLANE TRUSSES)

The present chapter is concerned with the finite element analysis of trusses (or truss structures) in two dimensions. The treatment can be readily generated to handle three-dimensional trusses. Earlier we saw an axial element with two degrees of freedom. The truss elements are part of a truss structure linked together by pinned joints that transmit only axial force to the elements. Unlike an axial element, a truss element is two-dimensional, i.e. it can displace both in x and y directions. Thus, we have two degrees of freedom at each node and totally four degrees of freedom for a truss element (Fig. 4.1).

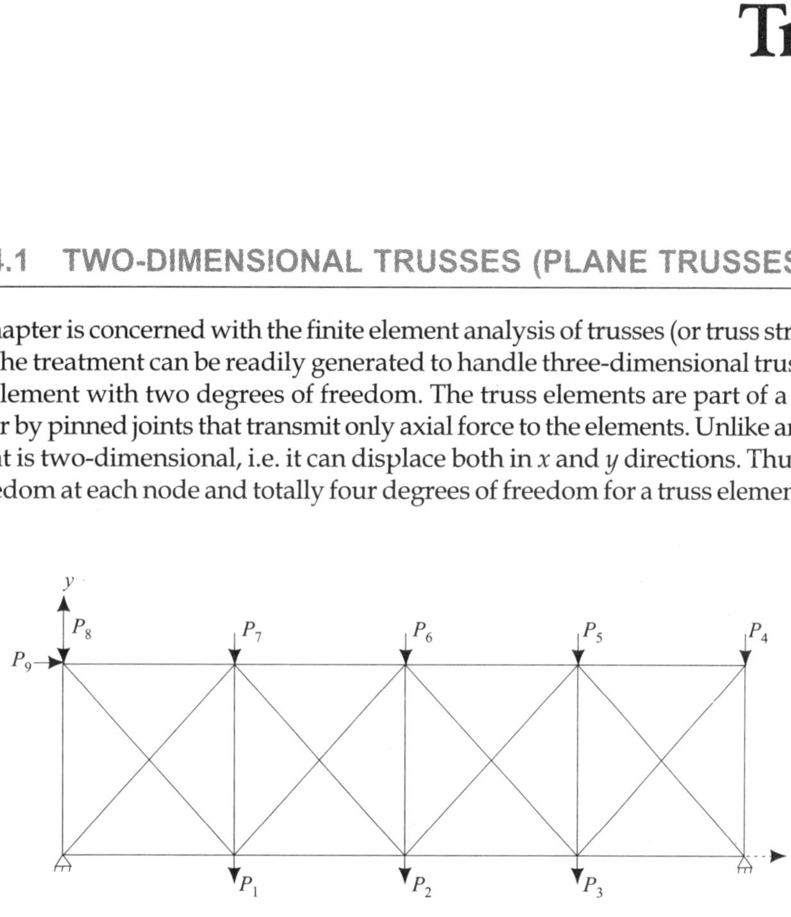

Fig. 4.1 A typical plane truss

Now, we are interested to see the *FEA* procedure for such trusses. Regarding field variables and elements, joint displacements are selected as basic field variables. Since there is no bending of the members we have to ensure only displacement continuity (C°-Continuity) and there is no need to worry about slope continuity (c′-Continuity). Hence we select two noded bar elements for the analysis of trusses.

At node 1, q_1 and q_2 are two displacements along x and y directions. Similarly at node 2, q_3 and q_4 are the two displacements. Thus, we have four degrees of freedom for a truss element. F_1, F_2 are the force components at node 1 and F_3, F_4 at node 2. For convenience let the odd subscript refer to x direction and even to y-direction. Also note that $q^1{}_1$ and $q^2{}_2$ are the displacements of nodes 1 and 2 in the local coordinate direction, i.e., along the element.

(a) Relationship between the Displacements in the Local and Global Coordinates

In the local coordinates, the displacement vector is :

$$q' = \begin{Bmatrix} q_1' \\ q_2' \end{Bmatrix} = [q'_1\ q'_2]^T \qquad ...[1]$$

While in global coordinates the displacement vector is:

$$q = \begin{Bmatrix} q_1 \\ q_2 \\ q_3 \\ q_4 \end{Bmatrix} [q_1\ q_2\ q_3\ q_4]^T \qquad ...[2]$$

From the Fig. 4.1, we find that

$$q'_1 = q_1 \cos\alpha + q_2 \sin\alpha \qquad ...[3]$$

$$q'_2 = q_3 \cos\alpha + q_4 \sin\alpha \qquad ...[4]$$

Now, let l and m are the direction cosines (D.C's) given by $l = \cos\alpha$ and $m = \sin\alpha$, then equations [3] are [4] can be written as $q' = Lq$, where L is the transformation matrix given by

$$L = \begin{bmatrix} \cos\alpha & \sin\alpha & 0 & 0 \\ 0 & 0 & \cos\alpha & \sin\alpha \end{bmatrix} = \begin{bmatrix} l & m & 0 & 0 \\ 0 & 0 & l & m \end{bmatrix} \qquad ...[5]$$

The values of l and m can be determined from global coordinates of the end nodes. Thus,

$$l = \frac{x_2 - x_1}{L_e} \text{ and } m = \frac{y_2 - y_1}{L_e} \qquad ...[6]$$

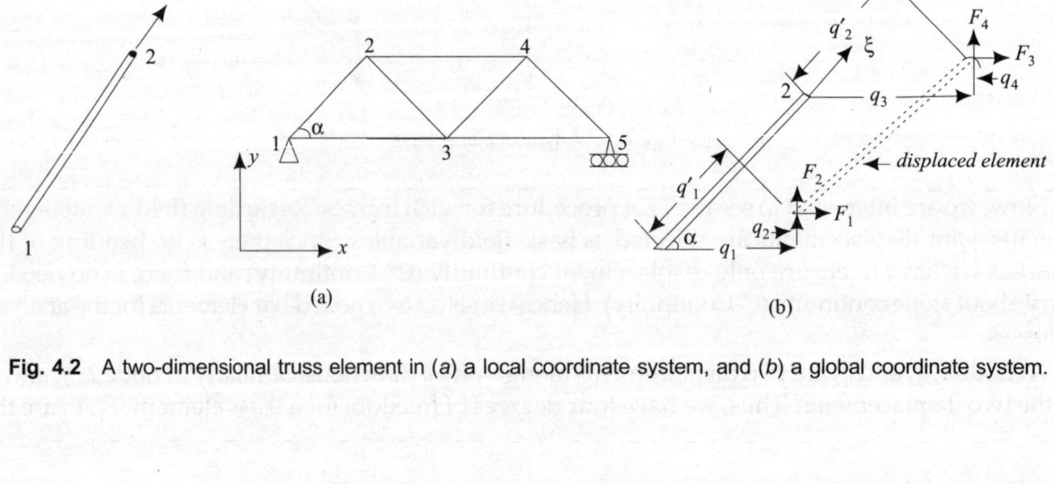

Fig. 4.2 A two-dimensional truss element in (*a*) a local coordinate system, and (*b*) a global coordinate system.

where the length of the element L_e is given by

$$L_e = \sqrt{(x_2 - x_1)^2 + (y_2 - y_1)^2} \quad ...[7]$$

(b) (i) Stiffness Matrix for the Truss Element in Local Coordinates

The element stiffness matrix of bars and columns is given by

$$[k]_e = \frac{EA_e}{L_e}\begin{bmatrix} 1 & -1 \\ -1 & 1 \end{bmatrix} \quad ...[1]$$

when viewed in local coordinate system, the truss is also a one-dimensional two-noded bar element.

Thus, from the local coordinate point, the truss element is no different from the bar element. The only two degrees of freedom are q'_1 and q'_2. Hence, the stiffness matrix in local coordinate is given by

$$[k']_e = \frac{EA_e}{L_e}\begin{bmatrix} 1 & -1 \\ -1 & 1 \end{bmatrix} \quad ...[2]$$

The strain energy in the element is given by

$$U_e = \frac{1}{2}\{q'\}^{\mathrm{T}}[k']\{q'\} \quad ...[3]$$

Now, substituting for q' from equation $q' = Lq$,

we get
$$U_e = \frac{1}{2}([L]\{q\})^{\mathrm{T}}[k'][L]\{q\}$$

$$= \frac{1}{2}\{q\}^{\mathrm{T}}[L]^{\mathrm{T}}[k'][L]\{q\}$$

$$= \frac{1}{2}\{q^T\}[L^T k' L]\{q\} \quad ...[4]$$

The quantity within the bracket is the stiffness matrix of the truss element in the global coordinates and is given by

$$[k]_e = [L^T k' L]$$

$$[k]_e = \begin{bmatrix} l & 0 \\ m & 0 \\ 0 & l \\ 0 & m \end{bmatrix} \frac{E_e A_e}{L_e}\begin{bmatrix} 1 & -1 \\ -1 & 1 \end{bmatrix}\begin{bmatrix} l & m & 0 & 0 \\ 0 & 0 & l & m \end{bmatrix}$$

$$= \frac{E_e A_e}{L_e}\begin{bmatrix} l & 0 \\ m & 0 \\ 0 & l \\ 0 & m \end{bmatrix}\begin{bmatrix} l & m & -l & -m \\ -l & -m & l & m \end{bmatrix}$$

$$= \frac{E_e A_e}{L_e}\begin{bmatrix} l^2 & lm & -l^2 & -lm \\ lm & m^2 & -lm & -m^2 \\ -l^2 & -lm & l^2 & lm \\ -lm & -m^2 & lm & m^2 \end{bmatrix}$$

Useful formulas for calculating *l* and *m*:

Referring to Fig. 4.3.

Let (x_1, y_1) and (x_2, y_2) be the coordinates of nodes 1 and 2, respectively, we then have:

$$l = \frac{x_2 - x_1}{L_e}; \text{ and } m = \frac{y_2 - y_1}{L_e} \quad \text{...[1]}$$

$$L_e = \sqrt{(x_2 - x_1)^2 + (y_2 - y_1)^2} \quad \text{...[2]}$$

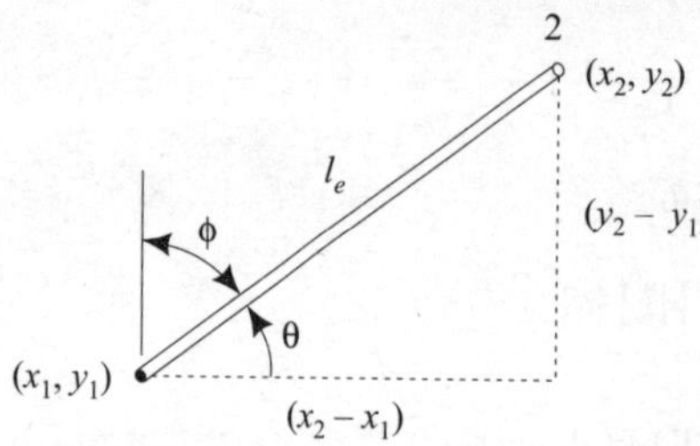

$$l = \cos\theta = \frac{x_2 - x_1}{l_e}$$

$$m = \cos\phi = \frac{y_2 - y_1}{l_e} \; (= \sin\theta)$$

$$l_e = \sqrt{(x_2 - x_1)^2 + (y_2 - y_1)^2}$$

Fig. 4.3 Direction cosines

The expressions in Eqs. [1] and [2] are obtained from nodal coordinate data, and can readily be implemented in a computer program.

(ii) Formulation of the Stiffness Matrix of the Truss-element in Global Coordinates

Consider any arbitrary truss member of length L_m, cross-sectional area A_m, modulus of elasticity E_m joining the node (*i*) to node (*j*) as shown in Fig. 4.4.

It has got 4 dof, U_i, V_i, U_j and V_j, associated with these dof are forces (or loads) F_{x_i}, F_{y_i}, F_{x_j} and F_{y_j} respectively.

From the Fig. 4.4 we can obtain the first column of $[k]$, that is k_{11} by defining the compressive force at *i* due to unit displacement at *j*, that is $u_i = 1$ along *AB*, keeping other dof (they are df_{y_i}, df_{x_j} and df_{y_j}) arrested, and the member *ij* is compressed by 'δ' along *AC*

where $$AC = \delta = (-1)\cos\theta = -C \quad \text{...[1]}$$

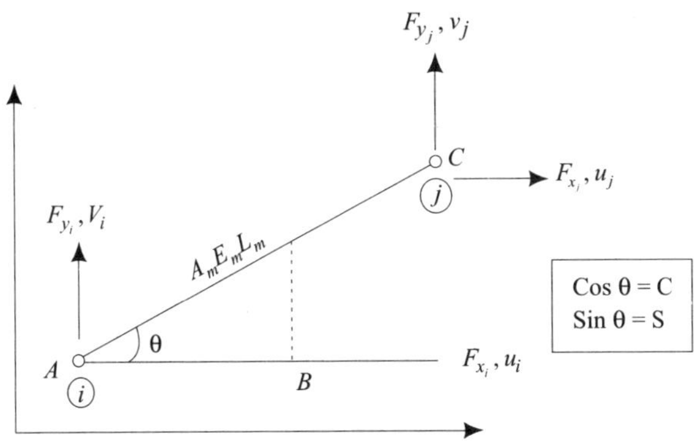

Fig. 4.4 Pin jointed truss element in global coordinates and sign convention

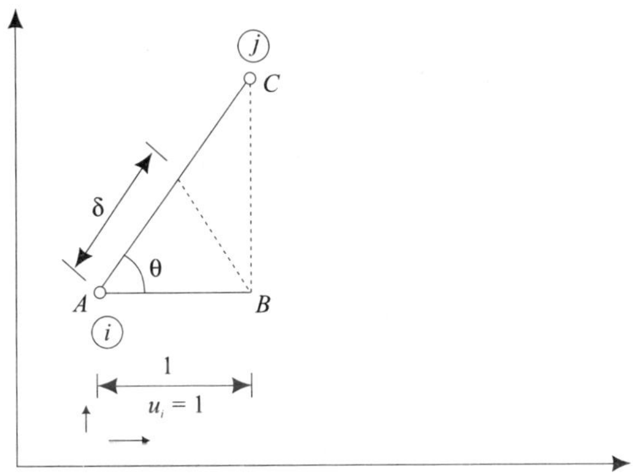

Fig. 4.5

(–ve sign denotes compressive force).

Thus, the force developed along the member AC, denoted by P_m is given by

$$P_m = \frac{A_m E_m}{L_m}(-\cos\theta) = -\frac{A_m E_m}{L_m} C \qquad \text{(Using [1])}$$

i.e.

$$\boxed{P_m = -\frac{E_m A_m}{L_m} C} \qquad ...[2]$$

Resolving the force P_m along x- and y-directions:

Along x-direction:

$$F_{x_i} = P_m \cos\theta = P_m C = -\frac{A_m E_m}{L_m} C^2 = K_{11}$$

i.e.,
$$F_{x_i} = -\frac{A_m E_m}{L_m} C^2 = k_{11} \qquad ...[3]$$

Along y-direction:

$$F_{y_i} = P_m \sin\theta = P_m S = -\frac{A_m E_m}{L_m} C.S$$

i.e.,
$$F_{y_i} = -\frac{A_m E_m}{L_m} C\,S = k_{21} \qquad ...[4]$$

For the member *ij* to be in equilibrium; the remaining reactions are given by

$$F_{x_j} = k_{31} = -k_{11} = \frac{A_m E_m}{L_m} C^2$$

i.e.
$$F_{x_j} = k_{31} = \frac{A_m E_m}{L_m} C^2 \qquad ...[5]$$

and
$$F_{y_j} = k_{41} = -k_{21} = \frac{A_m E_m}{L_m} CS$$

or
$$F_{y_j} = k_{41} = \frac{A_m E_m}{L_m} CS \qquad ...[6]$$

For equilibrium of the joints: (shown in Fig. 4.6)

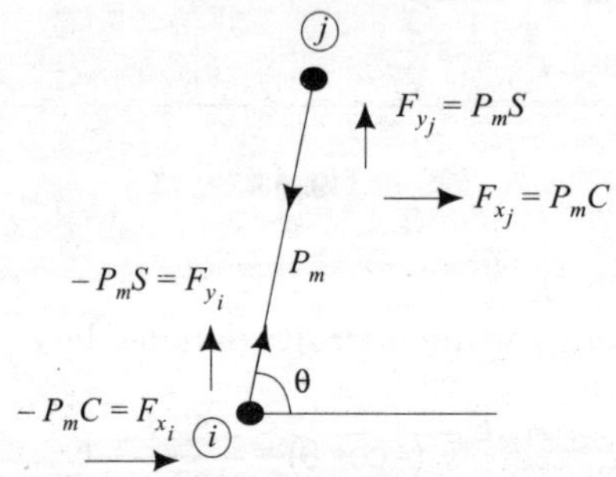

Fig. 4.6

$$\begin{aligned} F_{x_i} + P_m C &= 0 \\ F_{y_i} + P_m S &= 0 \\ F_{x_j} - P_m C &= 0 \\ F_{y_j} - P_m S &= 0 \end{aligned} \qquad ...[7]$$

where $P_m = -\frac{E_m A_m}{L_m} C$, from equation [2].

Substituting for P_m, we obtain

$$\text{Matrix form of [7] as: } \begin{Bmatrix} F_{x_i} \\ F_{y_i} \\ F_{x_j} \\ F_{y_j} \end{Bmatrix} = \frac{E_m A_m}{L_m} \begin{Bmatrix} C^2 \\ CS \\ -C^2 \\ -CS \end{Bmatrix}$$

This forms the first column of $[K]$.

The same procedure is used to find the second, third and fourth column coefficients by defining forces at the nodes i and j (say dF_{y_i}, dF_{x_j} and dF_{y_j}) due to unit displacements ($V_i = 1$, $U_j = 1$ and $V_j = 1$) one at a time, keeping the other 3 locked and the remaining. 3 columns are generated.

Thus, the element Stiffness Matrix in Global coordinates is given by

Global number → 1 2 3 4

$$[K]_e = \frac{A_m E_m}{L_m} \begin{bmatrix} C^2 & CS & -C^2 & -CS \\ CS & S^2 & -CS & -S^2 \\ -C^2 & -CS & C^2 & CS \\ -CS & -S^2 & CS & S^2 \end{bmatrix}$$

Example: Find the member forces of the truss shown in Fig. 4.7 by direct stiffness method. $\frac{AE}{L}$ values are shown hinged beside the member.

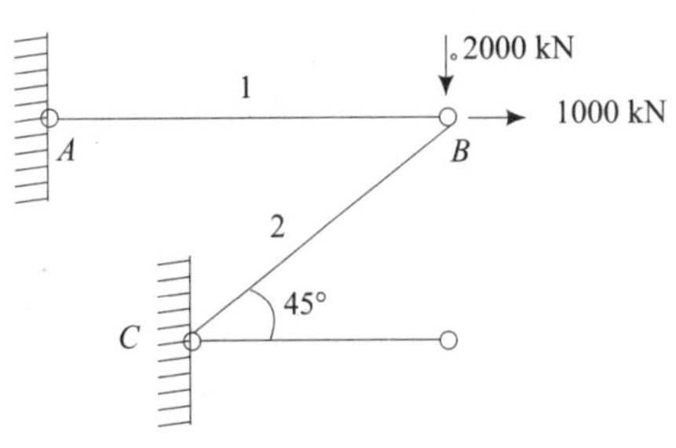

Fig. 4.7 Given truss

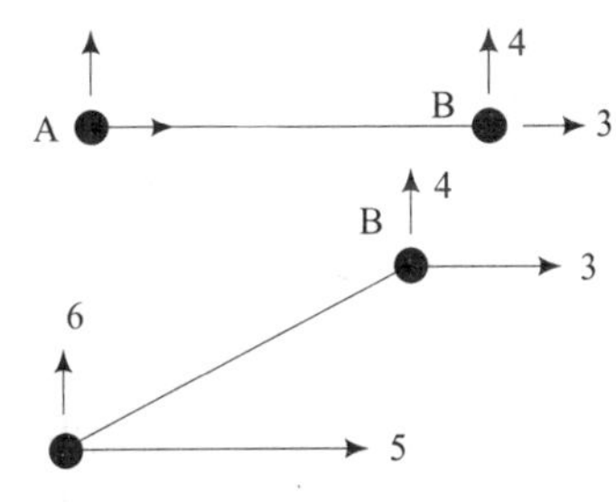

Fig. 4.8 Discretized elements

Solution: The truss is a natural system. It has got two elements. The degrees of freedom w.r.t. global axes are shown in Fig. 4.8.

Totally, there are six degrees of unrestrained freedom.

For member 1, the *DF*5 and *DF*6 are absent. So let us put zeros in the respective columns and row of the expanded matrix.

Member 1:

$$K^a = 1 \times \begin{array}{c} \begin{matrix} 1 & 2 & 3 & 4 & 5 & 6 \\ \downarrow & \downarrow & \downarrow & \downarrow & \downarrow & \downarrow \end{matrix} \\ \begin{Bmatrix} 1 & 0 & -1 & 0 & 0 & 0 \\ 0 & 0 & 0 & 0 & 0 & 0 \\ -1 & 0 & 1 & 0 & 0 & 0 \\ 0 & 0 & 0 & 0 & 0 & 0 \\ 0 & 0 & 0 & 0 & 0 & 0 \\ 0 & 0 & 0 & 0 & 0 & 0 \end{Bmatrix} \end{array}$$

Note: $\cos\theta = 1$, $\sin\theta = 0$ and $\frac{AE}{L} = 1$.

Member 2:

$$\cos\theta = \frac{1}{\sqrt{2}},\ \sin\theta = \frac{1}{\sqrt{2}}$$

$$\frac{AE}{L} = 2.$$

Note: Df_1 and Df_2 are absent

$$K^b = \begin{array}{c} \begin{matrix} 1 & 2 & 3 & 4 & 5 & 6 \\ \downarrow & \downarrow & \downarrow & \downarrow & \downarrow & \downarrow \end{matrix} \\ \begin{bmatrix} 0 & 0 & 0 & 0 & 0 & 0 \\ 0 & 0 & 0 & 0 & 0 & 0 \\ 0 & 0 & 1 & 1 & -1 & -1 \\ 0 & 0 & 1 & 1 & -1 & -1 \\ 0 & 0 & -1 & -1 & 1 & 1 \\ 0 & 0 & -1 & -1 & 1 & 1 \end{bmatrix} \end{array} \begin{matrix} \leftarrow 1 \\ \leftarrow 2 \\ \leftarrow 3 \\ \leftarrow 4 \\ \leftarrow 5 \\ \leftarrow 6 \end{matrix}$$

$$K = K^a + K^b$$

$$K = \begin{array}{c} \begin{array}{cccccc} 1 & 2 & 3 & 4 & 5 & 6 \\ \downarrow & \downarrow & \downarrow & \downarrow & \downarrow & \downarrow \end{array} \\ \begin{bmatrix} 1 & 0 & -1 & 0 & 0 & 0 \\ 0 & 0 & 0 & 0 & 0 & 0 \\ -1 & 0 & 2 & 1 & -1 & -1 \\ 0 & 0 & 1 & 1 & -1 & -1 \\ 0 & 0 & -1 & -1 & 1 & 1 \\ 0 & 0 & -1 & -1 & 1 & 1 \end{bmatrix} \end{array} \begin{array}{c} \\ \\ \leftarrow 1 \\ \leftarrow 2 \\ \leftarrow 3 \\ \leftarrow 4 \\ \leftarrow 5 \\ \leftarrow 6 \end{array}$$

Now let us impose the boundary conditions.

Columns 1, 2, 5 and 6 have to be deleted. Similarly, rows 1, 2, 5 and 6 also.

The K matrix corresponding to the Unrestrained degrees of freedom

$$K = \begin{bmatrix} 2 & 1 \\ 1 & 1 \end{bmatrix}$$

Now $r = KR$

The load 2000 kN acts in the opposite direction to the global y-axis. Hence –ve sign will have to be inserted.

$$V = K^{-1} R$$

$$\begin{bmatrix} r_3 \\ V_3 \end{bmatrix} = \frac{1}{1} \begin{bmatrix} 1 & -1 \\ -1 & 2 \end{bmatrix} \begin{bmatrix} 1000 \\ -2000 \end{bmatrix}$$

$$V_3 = 3000,\ V_4 = -5000.$$

The displacements are shown in Fig. 4.9

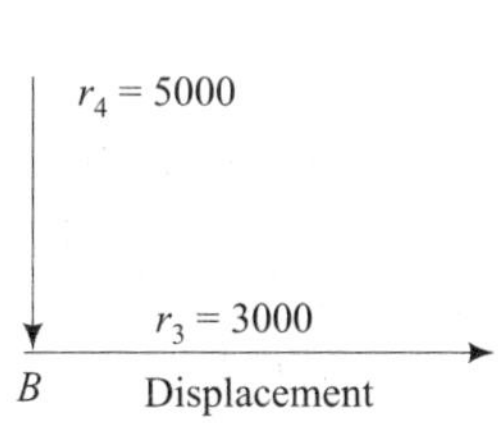

Fig. 4.9

Displacement in the inclined member

$$= \frac{5000}{\sqrt{2}} - \frac{3000}{\sqrt{2}} = \frac{2000}{\sqrt{2}} \text{ (compression)}$$

Force in the inclined member

$$= \frac{AE}{L} \times \text{displacement}$$

$$= 2 \times \frac{2000}{\sqrt{2}} = 2000\ \sqrt{2}\ \text{kN (compression)}$$

Displacement in the Horizontal member

$$= 3000.$$

Force in the member$= \frac{AE}{L} \times$ displacement

$$= 1 \times 3000 = 3000\ \text{kN (Tension)}$$

When the member becomes horizontal or vertical sin θ or cos θ becomes zero.

The element stiffness matrix

$$K_i = \frac{AE}{L}\begin{bmatrix} 1 & -1 \\ -1 & 1 \end{bmatrix}$$

Expression for Stress in the Truss Element

We know that stress is given by

$$\sigma = E_e \varepsilon \qquad \text{...[1]}$$

Now, the strain in the element is

$$\varepsilon = \frac{\text{change in length}}{\text{original length}} = \frac{q_2' - q_1'}{L_e} \qquad \text{...[2]}$$

Hence,

$$\sigma = E_e \frac{q_2' - q_1'}{L_e} = \frac{E_e}{L_e}[-1\ 1]\begin{Bmatrix} q_1' \\ q_2' \end{Bmatrix} \qquad \text{...[3]}$$

Now, substituting for q' from equation: $q' = Lq$ into equation [3], we get

$$\sigma = \frac{E_e}{L_e}[-1\ 1]\begin{Bmatrix} q_1' \\ q_2' \end{Bmatrix}$$

$$= \frac{E_e}{L_e}[-1\ 1]\, Lq$$

$$= \frac{E_e}{L_e}[-1\ 1]\begin{bmatrix} l & m & 0 & 0 \\ 0 & 0 & l & m \end{bmatrix} q$$

$$= \frac{E_e}{L_e}[-1 \quad -m \quad l \quad m]\begin{Bmatrix} q_1 \\ q_2 \\ q_3 \\ q_4 \end{Bmatrix} \qquad ...[4]$$

Equation [4] gives the stress in the truss element which can be calculated once the nodal displacements are known.

Example: Analyse the truss shown in Fig. 4.10 and find the stresses in all the members. Given $E = 200 \times 10^9\ \mathrm{N/mm^2}$ and area of cross-section of all the members equal to 200 mm^2.

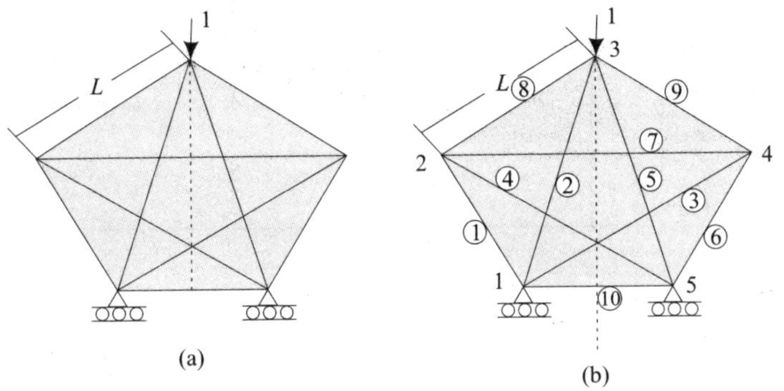

Fig. 4.10

The nodes are numbered consecutively along the shorter dimension (Fig. 4.11). This way of numbering is usually the most effective way of achieving the smallest band width. Taking node 1 as origin of global coordinates of the other nodes are

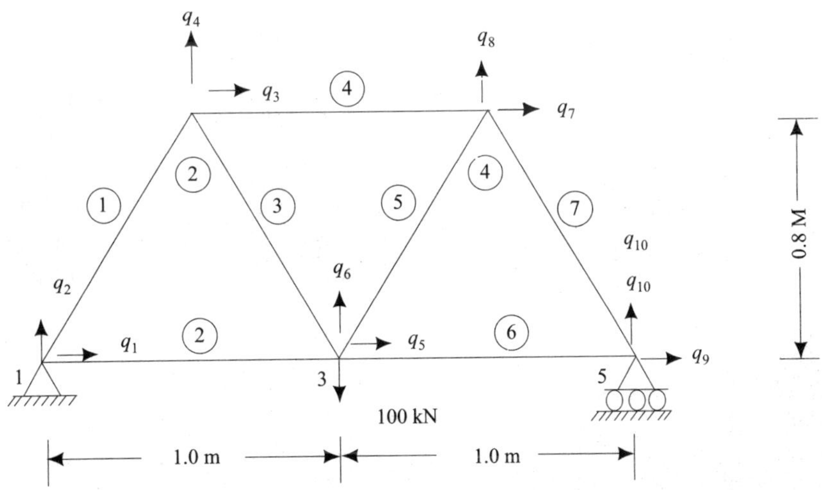

Fig. 4.11

node	x-coord. (m)	y-coord. (m)
1	0.0	0.0
2	0.5	0.8
3	1.0	0.0
4	1.5	0.8
5	2.0	0.0

Following is the element connectivity table.

element	1	2 ← local
1	1	2
2	1	3
3	2	3
4	2	4 ← global
5	3	4
6	3	5
7	4	5

The global degrees of freedom are $q_1, q_2 \ldots q_{10}$.

The lengths of various elements are calculated as

$$L_1 = \sqrt{(x_2 - x_1)^2 + (y_2 - y_1)^2} = \sqrt{(0.5 - 0)^2 + (0.8 - 0)^2} = 0.9434 \text{ m}$$

$$L_2 = \sqrt{(x_2 - x_1)^2 + (y_2 - y_1)^2} = \sqrt{(1.0 - 0)^2 + (0.0 - 0)^2} = 1.0 \text{ m}$$

By symmetry we find that $L_3 = L_5 = L_7 = L_1$.

Let us now find the direction cosines for each element

element	element length L_e (m)	$l = \frac{x_2 - x_1}{L_e}$	$m = \frac{y_2 - y_1}{L_e}$
1	0.9434	$\frac{0.5 - 0.0}{0.9434} = 0.523$	$\frac{0.8 - 0.0}{0.9434} = 0.848$
2	1.0	$\frac{1.0 - 0.0}{1.0} = 1.0$	$\frac{0.0 - 0.0}{1.0} = 0.0$
3	0.9434	$\frac{0.5 - 1.0}{0.9434} = -0.523$	$\frac{0.8 - 0.0}{0.9434} = 0.848$
4	1.0	1.0	0.0
5	0.9434	0.523	0.848
6	1.0	1.0	0.0
7	0.9434	–0.523	0.848

Next, calculate the element stiffness matrices. For element 1,

$$[k]^1 = \frac{EA_e}{L_e}\begin{bmatrix} l^2 & lm & -l^2 & -lm \\ lm & m^2 & -lm & -m^2 \\ -l^2 & -lm & l^2 & lm \\ -lm & -m^2 & lm & m^2 \end{bmatrix}$$

Substituting the numerical values we obtain

$$[k]^1 = \frac{EA}{0.9434}\begin{bmatrix} 0.523^2 & 0.523\times 0.848 & -0.523^2 & -0.523\times 0.848 \\ 0.523\times 0.848 & 0.848^2 & -0.523\times 0.848 & -0.848^2 \\ -0.523^2 & -0.523\times 0.848 & 0.523^2 & 0.523\times 0.848 \\ -0.523\times 0.848 & -0.848^2 & 0.523\times 0.848 & 0.848^2 \end{bmatrix}$$

$$\begin{matrix} 1 & 2 & 3 & 4 & \leftarrow \text{global degree of freedom} \end{matrix}$$

$$= EA\begin{bmatrix} 0.2899 & 0.4701 & -0.2899 & -0.4701 \\ 0.4701 & 0.7622 & -0.4701 & -0.7622 \\ -0.2899 & -0.4701 & 0.2899 & 0.4701 \\ -0.4701 & -0.7622 & 0.4701 & 0.7622 \end{bmatrix}$$

The superscript above k represents element number. Working in a similar way, we find for other elements the stiffness matrices as

$$\begin{matrix} 1 & 2 & 5 & 6 & \leftarrow \text{global degree of freedom} \end{matrix}$$

$$[k]^2 = EA\begin{bmatrix} 1 & 0 & -1 & 0 \\ 0 & 0 & 0 & 0 \\ -1 & 0 & 1 & 0 \\ 0 & 0 & 0 & 0 \end{bmatrix}$$

$$\begin{matrix} 3 & 4 & 5 & 6 & \leftarrow \text{global degree of freedom} \end{matrix}$$

$$[k]^3 = EA\begin{bmatrix} .2899 & -.4701 & -.2899 & .4701 \\ -.4701 & .7622 & .4701 & -.7622 \\ -.2899 & .4701 & .2899 & -.4701 \\ .4701 & -.7622 & -.4701 & .7622 \end{bmatrix}$$

$$\begin{matrix} 3 & 4 & 7 & 8 & \leftarrow \text{global degree of freedom} \end{matrix}$$

$$[k]^4 = EA\begin{bmatrix} 1 & 0 & -1 & 0 \\ 0 & 0 & 0 & 0 \\ -1 & 0 & 1 & 0 \\ 0 & 0 & 0 & 0 \end{bmatrix}$$

$$\begin{array}{c} \\ [k]^5 = EA \end{array} \begin{array}{c} \begin{matrix} 5 & \quad 6 & \quad 7 & \quad 8 \end{matrix} \leftarrow \text{global degree of freedom} \\ \begin{bmatrix} .2899 & .4701 & -.2899 & -.4701 \\ .4701 & .7622 & -.4701 & -.7622 \\ -.2899 & -.4701 & .2899 & .4701 \\ -.4701 & -.7622 & .4701 & .7622 \end{bmatrix} \end{array}$$

$$\begin{array}{c} \\ [k]^6 = EA \end{array} \begin{array}{c} \begin{matrix} 5 & 6 & 9 & 10 \end{matrix} \leftarrow \text{global degree of freedom} \\ \begin{bmatrix} 1 & 0 & -1 & 0 \\ 0 & 0 & 0 & 0 \\ -1 & 0 & 1 & 0 \\ 0 & 0 & 0 & 0 \end{bmatrix} \end{array}$$

$$\begin{array}{c} \\ [k]^7 = EA \end{array} \begin{array}{c} \begin{matrix} 7 & \quad 8 & \quad 9 & \quad 10 \end{matrix} \leftarrow \text{global degree of freedom} \\ \begin{bmatrix} .2899 & -.4701 & -.2899 & .4701 \\ -.4701 & .7622 & .4701 & -.7622 \\ -.2899 & .4701 & .2899 & -.4701 \\ .4701 & -.7622 & -.4701 & .7622 \end{bmatrix} \end{array}$$

Hence, the global stiffness matrix is

$$\begin{array}{cccccccccc} 1 & 2 & 3 & 4 & 5 & 6 & 7 & 8 & 9 & 10 \end{array}$$

$$[K] = EA \begin{bmatrix} 1.2899 & .4701 & -.2899 & -.4701 & -1 & 0 & 0 & 0 & 0 & 0 \\ .4701 & .7622 & -.4701 & -.7622 & 0 & 0 & 0 & 0 & 0 & 0 \\ -.2899 & -.4701 & 1.5798 & 0 & -.2899 & .4701 & -1 & 0 & 0 & 0 \\ -.4701 & -.7622 & 0 & 1.5244 & .4701 & -.7622 & 0 & 0 & 0 & 0 \\ -1 & 0 & -.2899 & .4701 & 2.5798 & 0 & -.2899 & -.4701 & -1 & 0 \\ 0 & 0 & .4701 & -.7622 & 0 & 1.5244 & -.4701 & -.7622 & 0 & 0 \\ 0 & 0 & -1 & 0 & -.2899 & -.4701 & 1.5798 & 0 & -.2899 & .4701 \\ 0 & 0 & 0 & 0 & -.4701 & -.7622 & 0 & 1.524 & .4701 & -.7622 \\ 0 & 0 & 0 & 0 & -1 & 0 & -.2899 & .4701 & 1.2899 & -.4701 \\ 0 & 0 & 0 & 0 & 0 & 0 & .4701 & -.7622 & -.4701 & .7622 \end{bmatrix}$$

The global vector F is given by

$$\boldsymbol{F} = [0 \;\; 0 \;\; 0 \;\; 0 \;\; 0 \;\; -100 \;\; 0 \;\; 0 \;\; 0 \;\; 0]^T$$

The boundary conditions for this case are

$$q_1 = q_2 = q_9 = q_{10} = 0$$

Hence, deleting the rows and columns containing these nodes as shown below,

$$\begin{bmatrix} 1.2899 & .4701 & -.2899 & -.4701 & -1 & 0 & 0 & 0 & 0 & 0 \\ .4701 & .7622 & -.4701 & -.7622 & 0 & 0 & 0 & 0 & 0 & 0 \\ -.2899 & -.4701 & 1.5798 & 0 & -.2899 & .4701 & -1 & 0 & 0 & 0 \\ -.4701 & -.7622 & 0 & 1.5244 & .4701 & -.7622 & 0 & 0 & 0 & 0 \\ -1 & 0 & -.2899 & .4701 & 2.5798 & 0 & -.2899 & -.4701 & -1 & 0 \\ 0 & 0 & .4701 & -.7622 & 0 & 1.5244 & -.4701 & -.7622 & 0 & 0 \\ 0 & 0 & -1 & 0 & -.2899 & -.4701 & 1.5798 & 0 & -.2899 & .4701 \\ 0 & 0 & 0 & 0 & -.4701 & -.7622 & 0 & 1.524 & .4701 & -.7622 \\ 0 & 0 & 0 & 0 & -1 & 0 & -.2899 & .4701 & 1.2899 & -.4701 \\ 0 & 0 & 0 & 0 & 0 & 0 & .4701 & -.7622 & -.4701 & .7622 \end{bmatrix}$$

Following finite element equations are obtained in the six unknowns, viz, $q_3, q_4, q_5, q_6, q_7, q_8$:

$$\begin{bmatrix} 1.5798 & 0 & -.2899 & .4701 & -1 & 0 \\ 0 & 1.5244 & .4701 & -.7622 & 0 & 0 \\ -.2899 & .4701 & 2.5798 & 0 & -.2899 & -.4701 \\ .4701 & -.7622 & 0 & 1.5244 & -.4701 & -.7622 \\ -1 & 0 & -.2899 & -.4701 & 1.5798 & 0 \\ 0 & 0 & .4701 & -.7622 & 0 & 1.5244 \end{bmatrix} \begin{Bmatrix} q_3 \\ q_4 \\ q_5 \\ q_6 \\ q_7 \\ q_8 \end{Bmatrix} = \begin{Bmatrix} 0 \\ 0 \\ 0 \\ -100 \\ 0 \\ 0 \end{Bmatrix}$$

It is difficult to solve the above equation manually. The Gaussian scheme is normally used to solve such equations. Using the method, it is found that:

$$q_3 = 0.678 \text{ mm} = -q_7;\ q_4 = 1.838 \text{ mm} = q_8;\ q_5 = 0;\ q_6 = -3.676 \text{ mm}$$

The negative sign indicates that the direction of the displacement is opposite to the direction assumed. Next, the stress in the element 1 is given by

$$\sigma^1 = \frac{E_e}{L_e}[-l\ -m\ l\ m]\begin{Bmatrix} q_1 \\ q_2 \\ q_3 \\ q_4 \end{Bmatrix} = \frac{200\times 10^9}{0.9434}[-.523\ -.848\ .523\ -.848]\begin{Bmatrix} 0 \\ 0 \\ .678\times 10^{-3} \\ -1.838\times 10^{-3} \end{Bmatrix}$$

$$= -2.9525\times 10^4 \text{ N/cm}^2 = \sigma^7$$

The negative sign indicates that the member is in compression. Similarly, the stresses in all other members can be found out. They are $\sigma^2 = \sigma^6 = 1.5625\times 10^4$ N/cm^2 (tension); $\sigma^3 = \sigma^5 = 2.9525\times 10^4$ N/cm^2 (tension); $\sigma^4 = 0$.

4.2 THREE-DIMENSIONAL TRUSSES (SPACE TRUSSES)

The plane truss element can be extended straightforward to space truss element. Typical two-noded three-dimensional truss element in local and global coordinate systems is shown in Fig. 4.12.

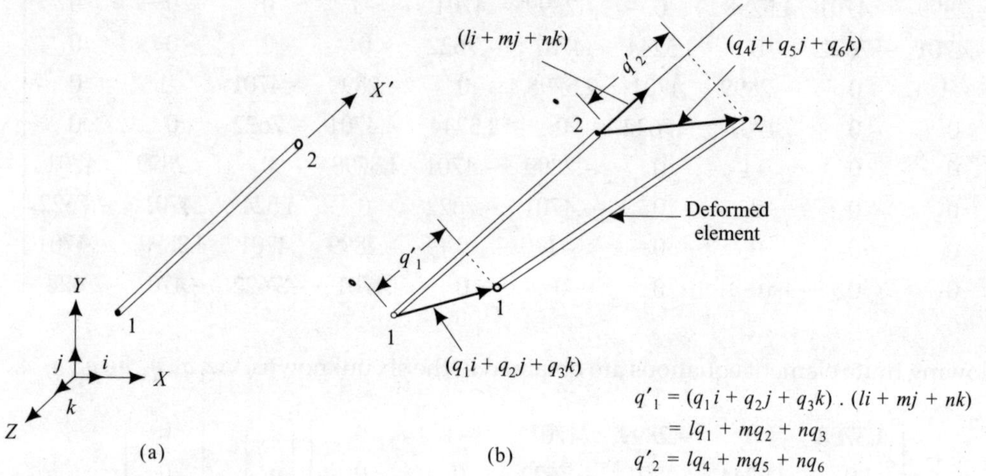

$$q'_1 = (q_1 i + q_2 j + q_3 k) \cdot (li + mj + nk)$$
$$= lq_1 + mq_2 + nq_3$$
$$q'_2 = lq_4 + mq_5 + nq_6$$
$$l = \cos(x', X) \quad m = \cos(x', Y) \quad n = \cos(x', Z)$$
$$= \frac{x_2 - x_1}{l_e} \quad = \frac{y_2 - y_1}{l_e} \quad = \frac{z_2 - z_1}{l_e}$$

Fig. 4.12 A three-dimensional truss element in local and global coordinate systems.

In local coordinate system, the displacement vector

$$\{q'\} = \begin{Bmatrix} q'_1 \\ q'_2 \end{Bmatrix} = [q'_1 \ q'_2]^T \qquad \text{...[1]}$$

In global coordinate system, the displacement vector is

$$\{q\}^T = [q_1 \ q_2 \ q_3 \ q_4 \ q_5 \ q_6]; \qquad \text{...[2]}$$

Stiffness Matrix of the Element

We find the transformation between local and global coordinates is

$$q' = Lq \qquad \text{...[3]}$$

where the transformation matrix L is given by

$$L = \begin{bmatrix} l & m & n & 0 & 0 & 0 \\ 0 & 0 & 0 & l & m & n \end{bmatrix} \qquad \text{...[4]}$$

where l, m and n are the direction cosines of the local x' axis w.r.t. to the global x, y and z axes, respectively.

Since l, m and n are direction cosines, we know

$$l = \frac{x_2 - x_1}{L_e},\ m = \frac{y_2 - y_1}{L_e} \text{ and } n = \frac{z_2 - z_1}{L_e}$$

The length of the member

$$L_e = \sqrt{(x_2 - x_1)^2 + (y_2 - y_1)^2 + (z_2 - z_1)^2}$$

we know, $\quad q'_1 = lq_1 + mq_2 + nq_3$

and $\quad q'_2 = lq_4 + mq_5 + nq_6$

In matrix form

$$\begin{Bmatrix} q'_1 \\ q'_2 \end{Bmatrix} = \begin{bmatrix} l & m & n & 0 & 0 & 0 \\ 0 & 0 & 0 & l & m & n \end{bmatrix} \begin{Bmatrix} q_1 \\ q_2 \\ \vdots \\ q_6 \end{Bmatrix}$$

$$\{q'\} = [L]\{q\} \qquad \text{...[5]}$$

From equation,

$$\left([K]_e = [L]^T [K'][L]\right)$$

we know the relationship between global and local stiffness matrices is,

$$[K]_e = [L]^T [K'] [L] = \begin{bmatrix} l & 0 \\ m & 0 \\ n & 0 \\ 0 & l \\ 0 & m \\ 0 & n \end{bmatrix} \frac{E_e A_e}{L_e} \begin{bmatrix} 1 & -1 \\ -1 & 1 \end{bmatrix} \begin{bmatrix} l & m & n & 0 & 0 & 0 \\ 0 & 0 & 0 & l & m & n \end{bmatrix}$$

$$[K]_e = \frac{E_e A_e}{L_e} \begin{bmatrix} l & 0 \\ m & 0 \\ n & 0 \\ 0 & l \\ 0 & m \\ 0 & n \end{bmatrix} \begin{bmatrix} l & m & n & -l & -m & -n \\ -l & -m & -n & l & m & n \end{bmatrix}$$

$$[K]_e = \frac{E_e A_e}{L_e} \begin{bmatrix} l^2 & lm & ln & -l^2 & -lm & -ln \\ lm & m^2 & mn & -lm & -m^2 & -mn \\ ln & mn & n^2 & -ln & -mn & -n^2 \\ -l^2 & -lm & -ln & l^2 & lm & ln \\ -lm & -m^2 & -mn & lm & m^2 & mn \\ -ln & -mn & -n^2 & ln & mn & n^2 \end{bmatrix} \quad ...[6]$$

Example: The tripod shown in Fig. 4.13 carries a vertically downward load of 10 kN at joint 4. If Young's modulus of the material of tripod stand is 200 kN/mm^2 and the cross-sectional area of each leg is 2000 mm^2, determine the forces developed in the legs of the tripod.

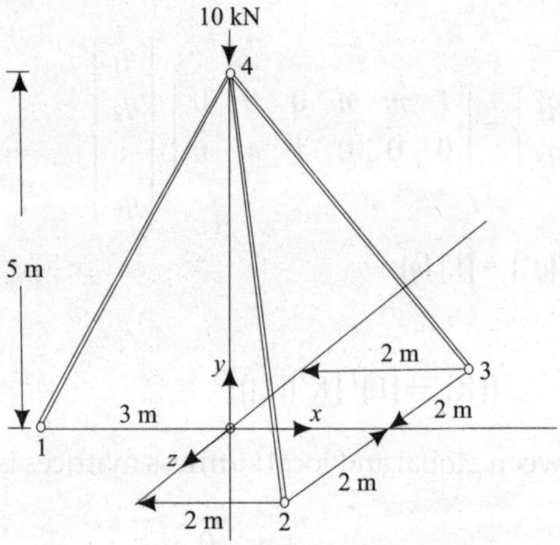

Fig. 4.13

Solution: If each member is taken as a bar element in 3-*D* space, then the coordinates of various joints (points) are:

Joints	*Coordinates*
1	(–3, 0, 0)
2	(2, 0, 2)
3	(2, 0, –2)
4	(0, 5, 0)

Length of each element is given by

$$L_e = \sqrt{(x_2 - x_1)^2 + (y_2 - y_1)^2 + (z_2 - z_1)^2}$$

and D. C's are given by

$$l = \frac{x_2 - x_1}{L_e},\ m = \frac{y_2 - y_1}{L_e} \text{ and } n = \frac{z_2 - z_1}{L_e}$$

The following Table 4.1 provides the details of three elements.

Table 4.1

Element No.	Node1	Node 2	l_e in mm	l	m	n
1	1	4	5831	0.514	0.857	0
2	2	4	5745	–0.348	0.870	– 0.348
3	3	4	5745	–0.348	0.870	– 0.348

$$\begin{matrix} & 1 & 2 & 3 & 10 & 11 & 12 & \leftarrow \text{Global} \\ & & & & & & & \downarrow \end{matrix}$$

$$k_1 = \begin{bmatrix} 18.11 & 30.184 & 0 & -18.11 & -30.184 & 0 \\ 30.184 & 50.350 & 0 & -30.184 & -50.350 & 0 \\ 0 & 0 & 0 & 0 & 0 & 0 \\ -18.11 & -30.184 & 0 & 18.11 & 30.184 & 0 \\ -30.184 & -50.350 & 0 & 30.184 & 50.350 & 0 \\ 0 & 0 & 0 & 0 & 0 & 0 \end{bmatrix} \begin{matrix} 1 \\ 2 \\ 3 \\ 10 \\ 11 \\ 12 \end{matrix}$$

Using equation:

$$\mathbf{k} = \frac{E_e A_e}{l_e} \begin{bmatrix} l^2 & lm & ln & -l^2 & -lm & -ln \\ lm & m^2 & mn & -lm & -m^2 & -mn \\ ln & mn & n^2 & -ln & -mn & -n^2 \\ -l^2 & -lm & -ln & l^2 & lm & ln \\ -lm & -m^2 & -mn & lm & m^2 & mn \\ -ln & -mn & -n^2 & ln & mn & n^2 \end{bmatrix}$$

$$\begin{matrix} & 4 & 5 & 6 & 10 & 11 & 12 & \leftarrow \text{Global} \\ & & & & & & & \downarrow \end{matrix}$$

$$[K_2] = \begin{bmatrix} 8.425 & -21.097 & 8.425 & -8.425 & 21.097 & -8.425 \\ -21.097 & 52.707 & -21.037 & 21.09 & -52.707 & 21.097 \\ 8.425 & -21.097 & 8.425 & -8.425 & 21.097 & -8.425 \\ -8.425 & 21.097 & -8.425 & 8.425 & -21.097 & 8.425 \\ 21.097 & -52.707 & 21.097 & -21.097 & 52.707 & -21.097 \\ -8.425 & 21.097 & -8.425 & 8.425 & -21.097 & 8.425 \end{bmatrix} \begin{matrix} 4 \\ 5 \\ 6 \\ 10 \\ 11 \\ 12 \end{matrix}$$

$$\begin{matrix} & 7 & 8 & 9 & 10 & 11 & 12 & \leftarrow \text{Global} \\ & & & & & & & \downarrow \end{matrix}$$

$$[K_3] = \begin{bmatrix} 8.425 & -21.097 & -8.425 & -8.425 & 21.097 & 8.425 \\ -21.097 & 52.707 & 21.097 & 21.097 & -52.707 & -21.097 \\ -8.425 & 21.097 & 8.425 & 8.425 & -21.097 & -8.425 \\ -8.425 & 21.097 & 8.425 & 8.425 & -21.097 & -8.425 \\ 21.097 & -52.707 & -21.097 & -21.097 & 52.707 & 21.097 \\ 8.245 & -21.097 & -8.425 & -8.425 & 21.097 & 8.425 \end{bmatrix} \begin{matrix} 7 \\ 8 \\ 9 \\ 10 \\ 11 \\ 12 \end{matrix}$$

Global matrix of size 12×12 can be assembled.

The load vector is

$$[F]^T = [0\ 0\ 0\ 0\ 0\ 0\ 0\ 0\ 0\ 0\ -10\ 0]$$

Since $$q_1 = q_2 = \ldots q_9 = 0,$$

this implies that the reduced matrix will be of size 3×3, the elements being from rows and columns of 10, 11, 12.

Thus, the reduced equation of equilibrium is:

$$\begin{array}{|c|c|c|}\hline \begin{matrix}18.110\\ 8.425\\ 8.425\end{matrix} & \begin{matrix}30.148\\ -21.097\\ -21.097\end{matrix} & \begin{matrix}0\\ 8.425\\ -8.425\end{matrix} \\ \hline \begin{matrix}30.148\\ -21.097\\ -21.097\end{matrix} & \begin{matrix}50.35\\ 52.707\\ 52.707\end{matrix} & \begin{matrix}0\\ -21.097\\ 21.097\end{matrix} \\ \hline \begin{matrix}0\\ 8.425\\ -8.425\end{matrix} & \begin{matrix}0\\ -21.097\\ 21.097\end{matrix} & \begin{matrix}0\\ 8.425\\ 8.425\end{matrix} \\ \hline \end{array} \begin{Bmatrix} q_{10}\\ \\ q_{11}\\ \\ q_{12}\end{Bmatrix} = \begin{Bmatrix} 0\\ \\ -10\\ \\ 0\end{Bmatrix}$$

$$\begin{bmatrix} 34.96 & -12.046 & 0\\ -12.046 & 155.764 & 0\\ 0 & 0 & 16.950\end{bmatrix} \begin{Bmatrix} q_{10}\\ q_{11}\\ q_{12}\end{Bmatrix} = \begin{Bmatrix} 0\\ -10\\ 0\end{Bmatrix}$$

$$\begin{bmatrix} 34.96 & -12.046 & 0\\ 0 & 151.589 & 0\\ 0 & 0 & 16.950\end{bmatrix} \begin{Bmatrix} q_{10}\\ q_{11}\\ q_{12}\end{Bmatrix} = \begin{Bmatrix} 0\\ -10\\ 0\end{Bmatrix}$$

$\therefore$ $$\Delta_{12} = 0$$

$\Rightarrow$ $$151.\ 584\ q_{12} = -10$$

Or $$q_{11} = -0.065697$$

$\therefore$ $$q_{10} = -0.02273$$

$$F_1 = \frac{200 \times 2000}{5831}\ [-0.514\ \ -0.857\ \ 0\ \ 0.514\ \ 0.857\ \ 0] \begin{Bmatrix} 0\\ 0\\ 0\\ -0.02273\\ 0.06597\\ 0\end{Bmatrix}$$

$$= -4.680 \text{ kN}$$

$$F_2 = \frac{200 \times 2000}{5745} [0 - 0 + 0 - 0.348(-0.02273) + 0.87 (-0.06597) + 0]$$

$$= -3.445 \text{ kN}$$

$$F_3 = \frac{200 \times 2000}{5745} [0 - 0 + 0 - 0.348(-0.02273) + 0.87 (-0.06597) + 0]$$

$$= -3.445 \text{ kN}$$

$$\{F\}^T = [- 4.680 - 3.445 - 3.445]$$

EXERCISES

1. Differentiate between a bar element and a truss element.
2. Using variational approach derive element stiffness matrix of
   (*a*) bar element
   (*b*) plane truss element.
3. Derive the expressions for nodal load vector in a two-noded bar element due to
   (*a*) Body force
   (*b*) Surface load.

   Use variational approach.
4. Obtain the forces in the plane truss shown in Fig. P4.4 and determine the support reactions also. Use finite element method. Take $E = 200$ GPa and $A = 2000 \text{ mm}^2$.

   [Ans: $[F]^T = [10 \quad -37.25 \quad 18.75]$ in kN
   $R_1 = -11.25$ kN $\quad R_2 = -15$ kN $\quad R_4 = 31.25$ kN]

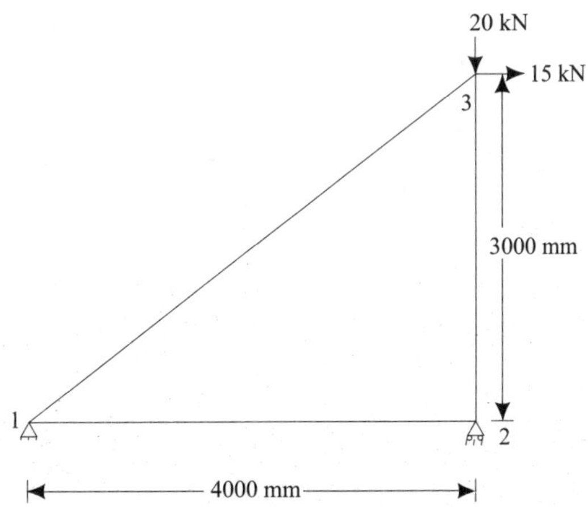

Fig. P4.4

5. Determine the forces in the members of the truss shown in Fig. P4.5. Take E = 200 GPa, A = 2000 mm^2.

[Ans: $\{F\}^T$=[20 –26.25 4.961 0 –6.2752] in kN]

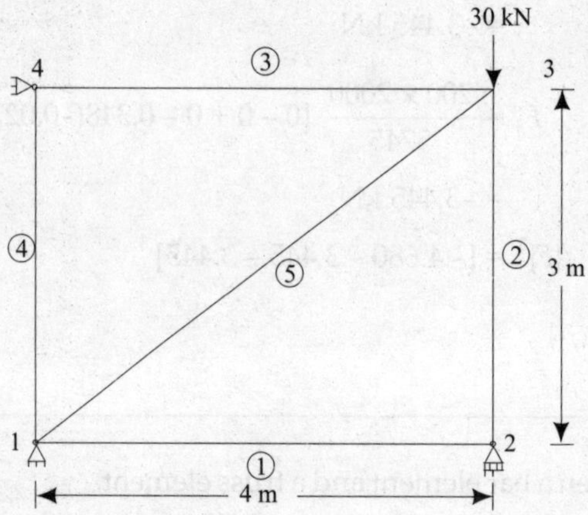

Fig. P4.5

6. Determine the displacements at the joint C of the truss loaded as shown in Fig. P4.6. E = 200 GPa; A = 19.35 cm^2.

[Ans: u = –0.5257 mm; v = –2.0142 mm]

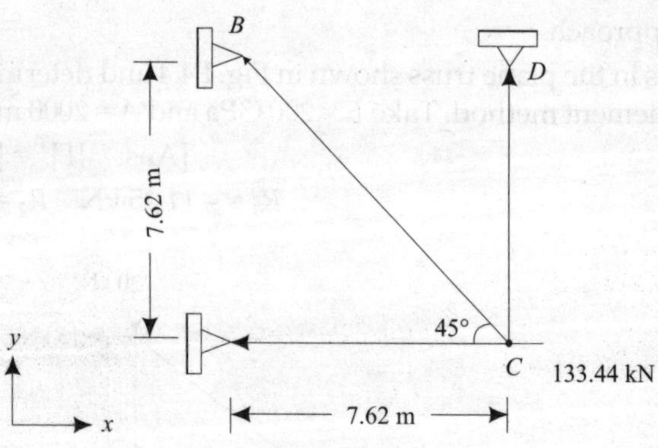

Fig. P4.6

7. Find the displacements at the joint C of the truss loaded as shown in Fig. P4.7. Here E and A are all same for all the members.

[Ans: $\begin{Bmatrix} u \\ v \end{Bmatrix} = \dfrac{L}{AE}\begin{bmatrix} 1+\dfrac{1}{\sqrt{2}} & -\dfrac{1}{2\sqrt{2}} \\ -\dfrac{1}{2\sqrt{2}} & -\dfrac{1}{2\sqrt{2}} \end{bmatrix}^{-1} \begin{Bmatrix} H \\ -V \end{Bmatrix}$]

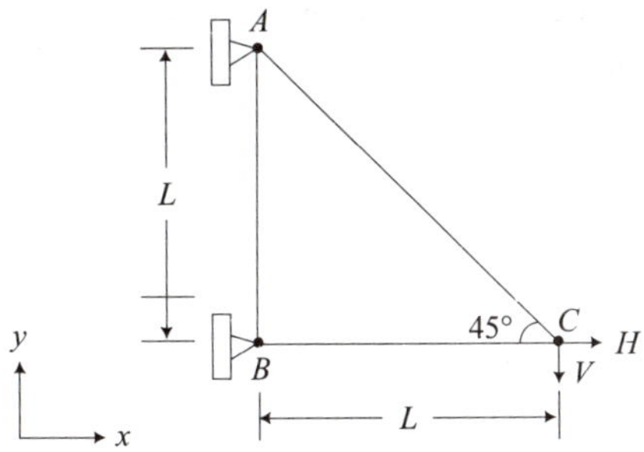

Fig. P4.7

8. Determine the displacements at the joint C of the truss loaded as shown in Fig. P4.8.
   $EA = 38.7 \times 10^7$ N.

   [Ans: $u = 31 \times 10^{-5}$ mm; $v = 11.63 \times 10^{-5}$ mm]

9. Find the displacement at the joint C of the truss loaded as shown in Fig. P4.9. Take $E = 200$

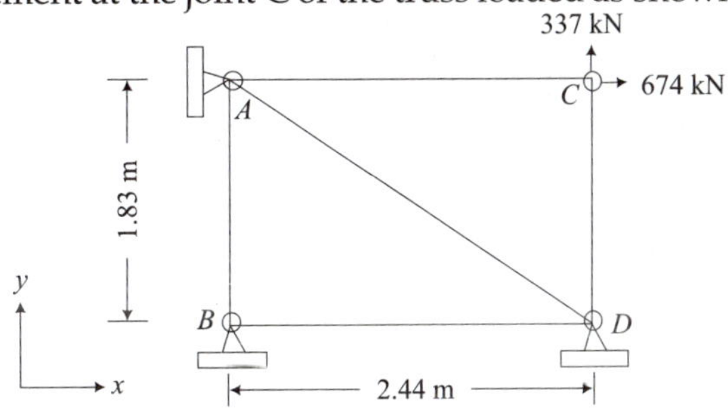

Fig. P4.8

GPa, area A for all the members = 4.84 cm^2. What are the equivalent loads at the nodes?

[Ans: $u = 0.496$ mm; $v = 1.749$ mm]

10. Consider the truss element shown in Fig. P4.10. The x, y coordinates of the two nodes are

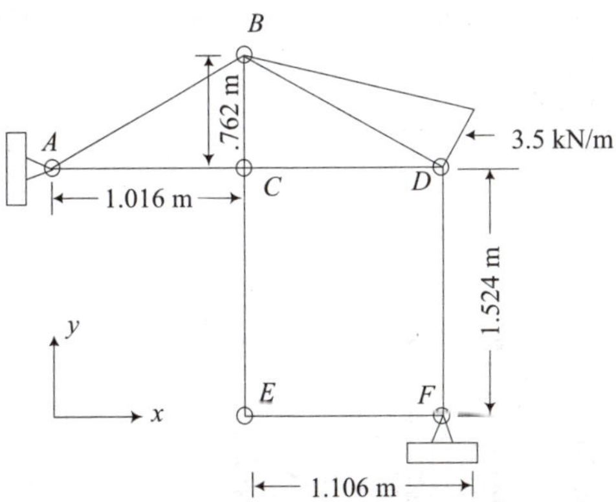

Fig. P4.9

indicated in the figure. If $\mathbf{q} = [1.5, 1.0, 2.1, 4.3]^T \times 10^2$ in.

(a) Determine the vector $\mathbf{q}'$.

(b) Determine the stress in the element.

(c) Determine the $\mathbf{k}$ matrix.

(d) Determine the strain energy in the element.

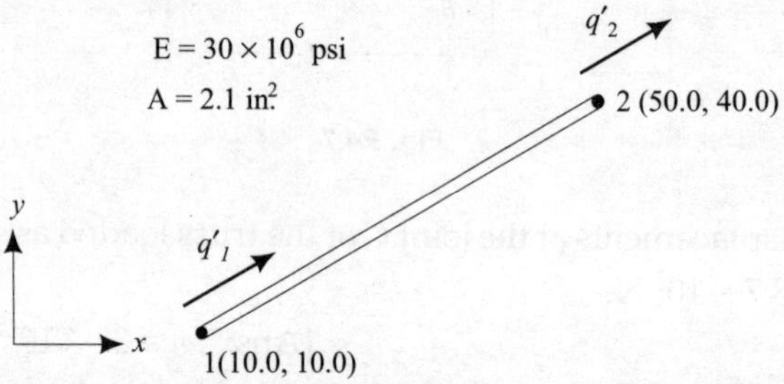

Fig. P4.10

11. For the two-bar truss shown in Fig. P4.11 determine the displacements of node 1 and the stress in element 2.

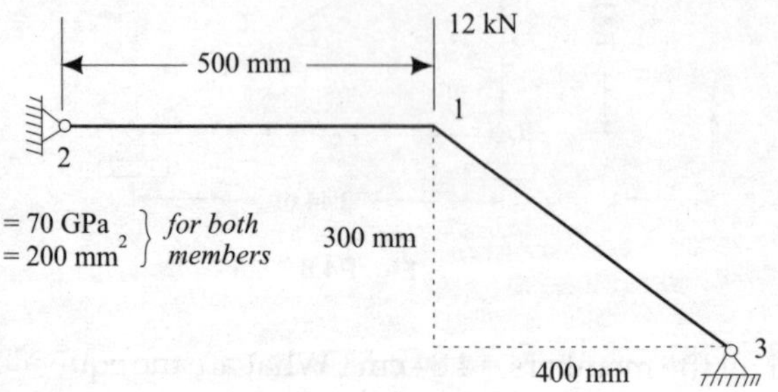

Fig. P4.11

12. For the two-dimensional truss configuration shown in Fig. P4.12 determine the bandwidth for stiffness storage in a banded form. Choose an alternative numbering scheme and determine the corresponding bandwidth. Comment on the strategy that you use for decreasing the bandwidth.

13. For the three-bar truss shown in Fig. P4.13, determine the displacements of node 1 and the stress in element 3.

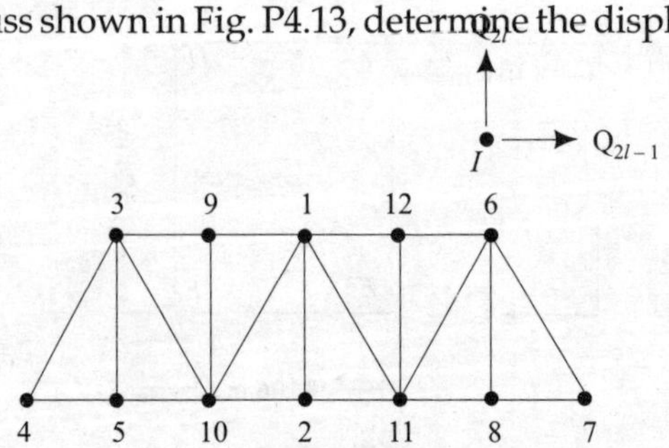

Fig. P4.12

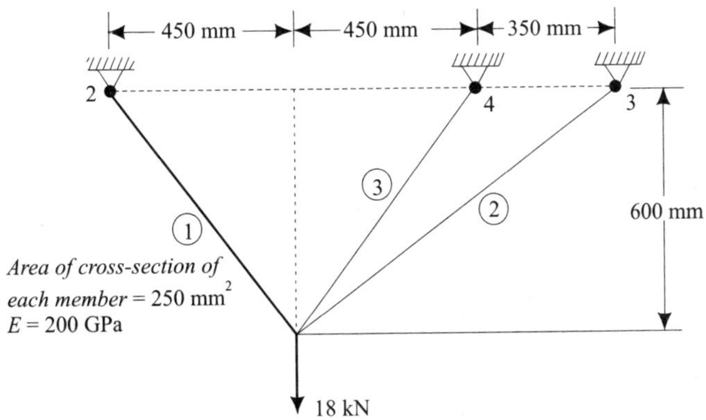
450 mm
450 mm
350 mm
2
4
3
3
2
1
600 mm
Area of cross-section of
each member = 250 mm2
E = 200 GPa
18 kN

Fig. P4.13

5

Two-Dimensional Problems Using Constant Strain Triangles

5.1 INTRODUCTION

The one-dimensional concept will be extended to two dimensions in two-dimensional finite element problems. A two-dimensional problem will be either a plane stress or a plane strain approximation to a real three-dimensional situation. The difference arises only in the elasticity matrix D that relates the stress and strain.

In two dimensions, the elements can have different shapes, *viz.* the 3-noded triangular element, the four-noded rectangular or quadrilateral elements, in addition. In case of symmetry, we can have axi-symmetric elements too.

Brief summary of the basic results in elasticity and structured mechanics has been given below.

Displacements

Two-dimensional displacements along x and y directions will be represented by

$$u = \begin{Bmatrix} u \\ v \end{Bmatrix} = [u \;\; v]^T \qquad \text{...[1]}$$

Stress and Strain

The two-dimensional stress and strain states are given by

$$\sigma = \begin{Bmatrix} \sigma_x \\ \sigma_y \\ \tau_{xy} \end{Bmatrix} = [\sigma_x \;\; \sigma_y \;\; \tau_{xy}]^T \qquad \text{...[2]}$$

$$\in = \begin{Bmatrix} \varepsilon_x \\ \varepsilon_y \\ \gamma_{xy} \end{Bmatrix} = [\varepsilon_x \ \varepsilon_y \ \gamma_{xy}]^T \quad ...[3]$$

Strain–Displacement

The strain-displacement relations are represented by

$$\varepsilon = \begin{Bmatrix} \dfrac{\partial u}{\partial x} \\ \dfrac{\partial v}{\partial y} \\ \dfrac{\partial u}{\partial y} + \dfrac{\partial v}{\partial x} \end{Bmatrix} = \left[\dfrac{\partial u}{\partial x} \ \dfrac{\partial v}{\partial y} \ \dfrac{\partial u}{\partial y} + \dfrac{\partial v}{\partial x} \right]^T \quad ...[4]$$

The body force vector f is given by (the force distributed over the volume of the body is called the body force)

$$f = \begin{Bmatrix} f_x \\ f_y \end{Bmatrix} = [f_x \ f_y]^T \quad ...[5]$$

where f_x, f_y are complements in the x-direction.

Traction Vector or Surface Force

The force distributed over the surface is called surface force or surface traction or traction vector denoted by T and is defined as:

$$T = \begin{Bmatrix} T_x \\ T_y \end{Bmatrix} = [T_x \ T_y]^T \quad ...[6]$$

where T_x, T_y are the traction components in the x and y directions respectively.

Hooke's Law

It states that the deformation (ε) (strain) is proportional to the force (σ) (stress).

Thus, we have $\boxed{\sigma = D\varepsilon}$ ...[7]

D → Elasticity matrix alone has different forms depending upon plane stress–plane strain problems.

The differential volume is defined as

$$\boxed{dV = tdA} \quad ...[8]$$

where t is the thickness of the element.

In two-dimensional case the problems are modelled as plane stress and plane strain. If u and v be the displacements along x and y directions, then

$$\left.\begin{aligned} \varepsilon_x &= \frac{\partial u}{\partial x}; \\ \varepsilon_y &= \frac{\partial v}{\partial y}; \\ \gamma_{xy} &= \frac{\partial u}{\partial y} + \frac{\partial v}{\partial x} \end{aligned}\right\} \text{(strain–displacement relationship)}$$

Plane stress: It is an analysis of stress conditions in one plane only, i.e. the stresses in the third direction are neglected. In fact, the stresses in the thickness directions for such cases are quite small to fit for neglection.

If a rectangular plate is loaded in such a way that the stress components acting on the plane (parallel to the base of Fig. 5.1(i) throughout the thickness are zero, then the plate is said to be in plane stress.

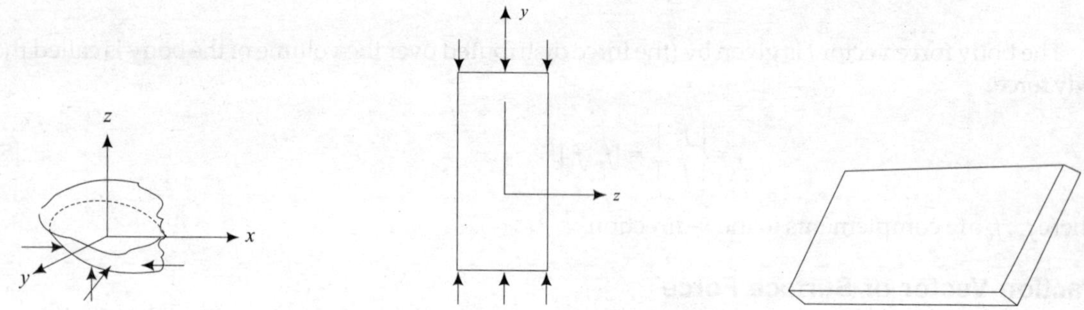

(i) Thin plate under plane stress (ii) Thin plate within plane loading (iii) A case of plane stress

Fig. 5.1

For these two plates [Fig. 5.1(i) & (ii)] the stress components σ_x, σ_y and τ_{xy} represent the state of stress. The stress components

$$\sigma_z = \tau_{xz} = \tau_{yz} = 0 \qquad \text{...[1]}$$

Beams, buttress of dams, web frame of tankers, super structure deck interaction problems of ships are some of the examples of plane stress problems. The stress-strain relations for the plane stress problems then become

$$\begin{bmatrix} \varepsilon_x \\ \varepsilon_y \\ \gamma_{xy} \end{bmatrix} = \frac{1}{E}\begin{bmatrix} 1 & -\nu & 0 \\ -\nu & 1 & 0 \\ 0 & 0 & 2(1+\nu) \end{bmatrix}\begin{bmatrix} \sigma_x \\ \sigma_y \\ \tau_{xy} \end{bmatrix} \qquad \text{...[2]}$$

$$\downarrow \qquad\qquad \downarrow \qquad\qquad \downarrow$$

$$\varepsilon \qquad\qquad S \qquad\qquad \sigma$$

The equation [2] has resulted after making necessary substitution of plane stress conditions into the following stress-strain relation

$$\left\{\begin{aligned} \epsilon_x &= \frac{1}{E}[\sigma_x - \nu(\sigma_y + \sigma_z)] \\ \epsilon_y &= \frac{1}{E}[\sigma_y - \nu(\sigma_z + \sigma_x)] \\ \epsilon_z &= \frac{1}{E}[\sigma_z - \nu(\sigma_x + \sigma_y)] \\ \gamma_{xy} &= \frac{\tau_{xy}}{G};\ \gamma_{yz} = \frac{\tau_{yz}}{G};\ \gamma_{zx} = \frac{\tau_{zx}}{G} \end{aligned}\right\} \quad ...[3]$$

and replacing G in terms of E from equation $G = \dfrac{E}{2(1+\nu)}$.

Plane Strain

As the name says, it is an analysis of strain conditions in one plane only i.e., the strain in the third direction (direction of thickness) are neglected/considered to be zero.

The dimension (geometry) of the body in the Z-direction is very much larger in comparison to other two dimensions, the section at any point of the body perpendicular to the Z-axis is constant and the necessary forces applied on the body do not vary along the length, it is assumed that deformation of the plane which is perpendicular to Z-axis becomes independent of its distance from Z-axis, i.e., all (cross-sections are assumed to be in the same condition. For this situation, naturally the displacement (longitudinal) ω is zero.

Examples are a culvert or tunnel, a dam, a retaining wall, a very long cylindrical tube filled with pressure, etc.

Strain in the third direction (of thickness) is considered to be zero,

i.e., $$\varepsilon_z = 0;\ \gamma_{zx} = \gamma_{zy} = 0$$

For a two-dimensional case, we have to eliminate, to the extent possible, the variables of the third dimension.

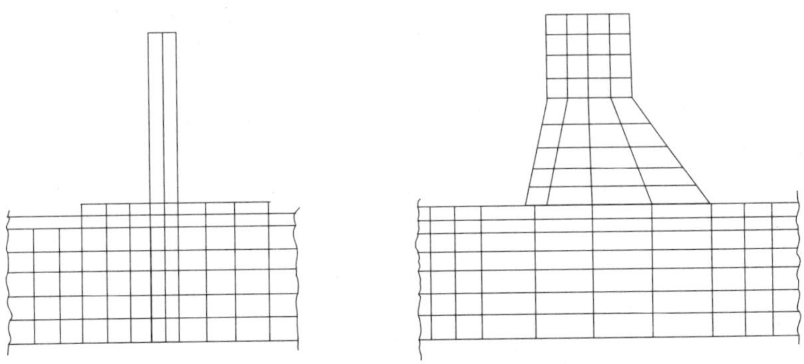

Fig. 5.2 Cases of plane strain

This condition $\varepsilon_z = 0$ leads to

$$\sigma_z - \nu(\sigma_x + \sigma_y) = 0$$

or,

$$\sigma_z = \nu(\sigma_x + \sigma_y) \qquad ...(4)$$

Substituting this in the stress-strain relationships, (given by [3]), we get

$$\varepsilon_x = \frac{(1-\nu^2)}{E}\sigma_x - \frac{\nu(1+\nu)}{E}\sigma_y \qquad ...[5]$$

$$\varepsilon_y = \frac{(1-\nu^2)}{E}\sigma_y - \frac{\nu(1+\nu)}{E}\sigma_x$$

$\therefore$ The stress-strain relation of the plane strain problem is

$$\underset{\downarrow\ \varepsilon}{\begin{Bmatrix} \varepsilon_x \\ \varepsilon_y \\ \gamma_{xy} \end{Bmatrix}} = \underset{\downarrow\ S}{\frac{1-\nu^2}{E}\begin{bmatrix} 1 & -\frac{\nu}{1-\nu} & 0 \\ -\frac{\nu}{1-\nu} & 1 & 0 \\ 0 & 0 & \frac{2}{1-\nu} \end{bmatrix}} \underset{\downarrow\ \sigma}{\begin{Bmatrix} \sigma_x \\ \sigma_y \\ \tau_{xy} \end{Bmatrix}} \qquad ...[6]$$

or,

$$\{\varepsilon\} = [S]\{\sigma\}$$

Inverting the same, $\{\sigma\} = [S]^{-1}\varepsilon = D\varepsilon$

where D is known as material stiffness matrix.

$$D = \frac{E(1-\nu)}{(1-2\nu)(1+\nu)}\begin{bmatrix} 1 & \frac{\nu}{1-\nu} & 0 \\ \frac{\nu}{1-\nu} & 1 & 0 \\ 0 & 0 & \frac{(1-2\nu)}{2(1-\nu)} \end{bmatrix}$$

Similarly, In the case of beams, the moment–curvature relationship becomes

$$\underset{\downarrow\ \sigma}{M} = \underset{\downarrow\ D}{(-EI)}\ \underset{\downarrow\ \varepsilon}{\frac{d^2y}{dx^2}}$$

In compact form, equation [2] can be written as

$$\{\varepsilon\} = [S]\{\sigma\};\ \{\sigma\} = \{S\}^{-1}\{\varepsilon\}$$

Or, $\{\sigma\} = [D]\{\varepsilon\}$

where $$[D] = [S]^{-1} = \frac{E}{(1-\nu^2)}\begin{bmatrix} 1 & \nu & 0 \\ \nu & 1 & 0 \\ 0 & 0 & \frac{1-\nu}{2} \end{bmatrix}$$

is known as material stiffness matrix. It is a square matrix :

$$D = \frac{E}{(1-\nu^2)}\begin{bmatrix} 1 & \nu & 0 \\ \nu & 1 & 0 \\ 0 & 0 & \frac{1-\nu}{2} \end{bmatrix}$$

5.2 FINITE ELEMENT MODELING

We discretize the two-dimensional region R into elements. Figure. 5.3 shows a typical discretization which is a triangulation.

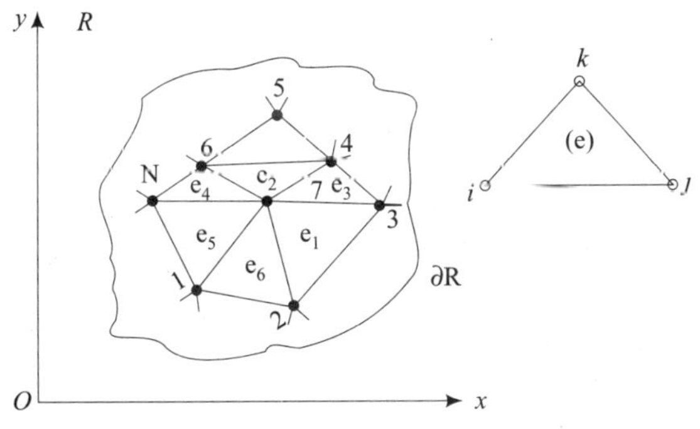

Fig. 5.3 Discretization of two-dimensional region into triangular elements.

Here, each element (e) is a triangle with nodes i, j and k numbered anticlockwise and straight edged. The idea of the *FEM* is to solve the continuous problem approximately. The elements fill the entire region except a small region at the boundary. This unfilled region exists for curved boundaries and it can be reduced by choosing smaller elements or elements with curved boundaries. This unfilled region contributes to some part of the approximate solution.

In two dimensions, each node is permitted to displace in x and y directions. Thus, each node has two dof's.

We denote the global displacement vector as $Q = [Q_1, Q_2, \ldots Q_N]^T$ ...[1]

where N is the number of degrees of freedom.

5.3 CONSTANT STRAIN TRIANGLE (CST)

Since there is a provision for constant strain within the element, so we assume that the strain is constant within a triangular element, then it is called constant strain triangle (usually CST). This can be seen from the displacement function. This assumption ensures continuity between the elements.

Stiffness Matrix

There are many ways of evaluating the stiffness matrix. We may consider 3 different procedures to get the element stiffness and consistent load vector for CST.

Let us consider a consistent strain triangular (CST) element as shown in Fig. 5.4, with orientation in global system.

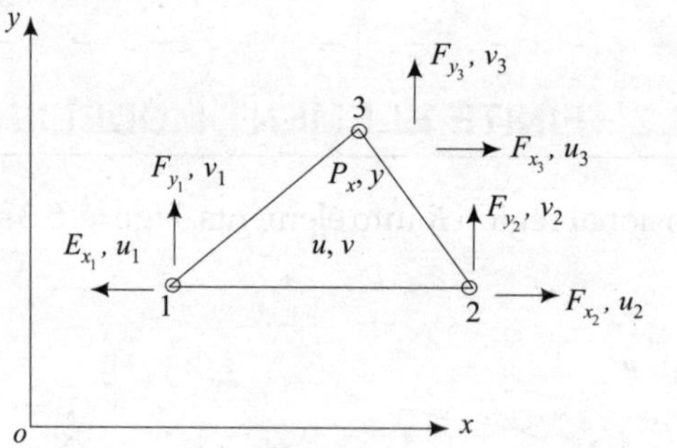

Fig. 5.4 Constant strain triangle

Each node is free to move in *x* and *y* directions so that there are two degrees of freedom per node. Hence, for 3 nodes there are in all 6 dof. The element deformations each one is associated with ε_x, ε_y and γ_{xy}.

The triangular element considered in this section is a constant strain triangular element, normally called as *CST element*. This assumption ensures continuity between elements. However, this will allow only straight line or linear displacements.

Shape Functions

Figure 5.5 shows a triangular element with three nodes marked 1, 2 and 3.

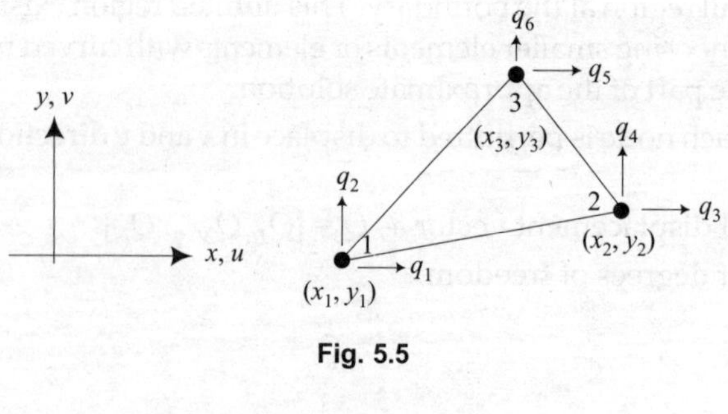

Fig. 5.5

Note that the numbering is in anti-clockwise direction to ensure that the area calculated will be positive. At each node we have two degrees of freedom marked by q_i's where once again the odd and even subscripts refer to displacements in the x and y directions respectively. As in the one-dimensional case, here also we define a pair of natural coordinates ξ and η. As explained earlier, we must have three shape functions associated with each node. Since by definition the shape functions N_i has value 1 at the node i and zero elsewhere, then

$$N_1 = 1 \text{ at node 1 and } N_1 = 0 \text{ at nodes 2 and 3}$$

$$N_2 = 1 \text{ at node 2 and } N_2 = 0 \text{ at nodes 1 and 3}$$

$$N_3 = 1 \text{ at node 3 and } N_3 = 0 \text{ at nodes 2 and 1}$$

This can be represented by defining

$$N_1 = \xi; \qquad N_2 = \eta; \qquad N_3 = 1 - \xi - \eta \qquad \text{...[1]}$$

The graphical representation of the above is shown in Fig. 5.6. Since the shape functions are linear, they represent a plane surface as shown. Also, note that the sum $N_1 + N_2 + N_3 = 1$ is also a plane parallel to the triangle but at height 1 above it. This clarifies that N_1, N_2, N_3 are not independent, only two are independent.

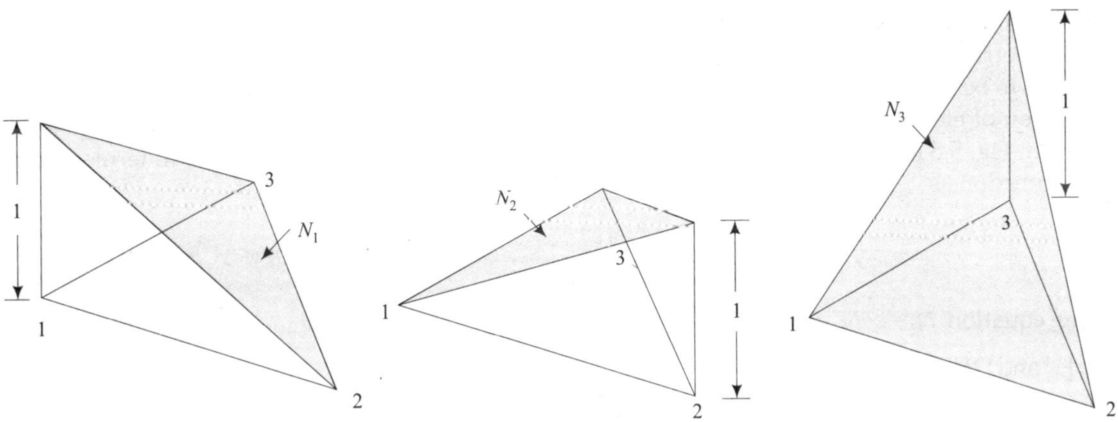

Fig. 5.6

Area Coordinates

Area coordinates are natural coordinates for a triangular element with a value between 0 and 1which is equal to the ratio of the area of triangular portion of the element to the total area of the element. From Fig. 5.7, a point (x, y) within the element divides the triangle into three areas A_1, A_2, A_3. The shape functions defined in the earlier problem can now be redefined as Fig. 5.7 where the total area $A = A_1 + A_2 + A_3$. It may also be noted that $N_1 = 0$ on side 1; $N_2 = 0$ on side 2 and $N_3 = 0$ on side 3.

$$N_1 = \frac{A_1}{A}; N_2 = \frac{A_2}{A}; N_3 - \frac{A_3}{A} \qquad \text{...[1]}$$

where A is the area of the element.

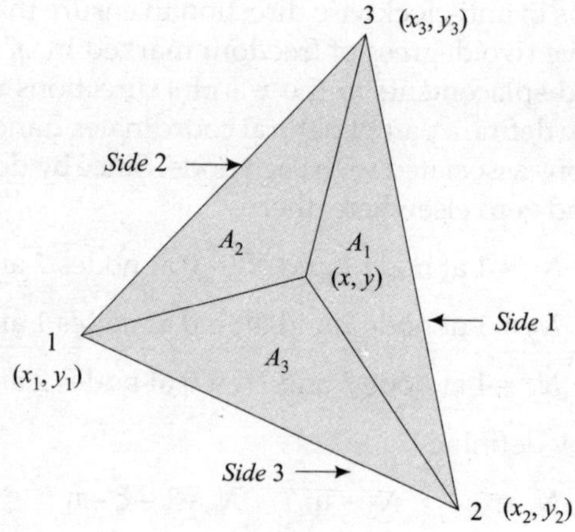

Fig. 5.7

Clearly, $N_1 + N_2 + N_3 = 1$ at every point inside the triangle.

Isoparametric Representation

In isoparametric element the geometry and the field interpolation functions are of the same order. Isoparametric coordinates are another type of natural coordinates. Their use enables the basic elements to be mapped into distorted shapes which is of considerable importance for it allows the generation of elements having curved sides or curved faces. Considering the triangular element shown in Fig. 5.8 the displacements u and v at any point (x, y) can be represented in terms of the shape function as

$$u = N_1q_1 + N_2q_3 + N_3q_5 \quad ...[1]$$

$$v = N_1q_2 + N_2q_4 + N_3q_6 \quad ...[2]$$

(Using equation, $N_1 = \xi$, $N_2 = \eta$; $N_3 = 1 - \xi - \eta$)

Eqs. [1] and [2] become

$$u = \xi q_1 + \eta q_3 + (1 - \xi - \eta)q_5 = (q_1 - q_5)\xi + (q_3 - q_5)\eta + q_5 \quad ...[3]$$

$$v = \xi q_2 + \eta q_4 + (1 - \xi - \eta)q_6 = (q_2 - q_6)\xi + (q_4 - q_6)\eta + q_6 \quad ...[4]$$

The coordinates (x, y) can also be expressed in terms of the nodal coordinates as

$$x = N_1x_1 + N_2x_3 + N_3x_5 \quad ...[5]$$

$$y = N_1y_1 + N_2y_3 + N_3y_5 \quad ...[6]$$

Note that the same shape functions are used to define both variables and the coordinates. Such a representation is called *isoparametric* and the element itself the isoparametric. Eqs. [5] and [6] can also be expressed as

$$x = (x_1 - x_3)\,\xi + (x_2 - x_3)\eta + x_3 \quad ...[7]$$

$$y = (y_1 - y_3)\,\xi + (y_2 - y_3)\eta + y_3 \quad ...[8]$$

$$x = x_{13}\,\xi + x_{23}\eta + x_3 \quad ...[9]$$

$$y = y_{13}\xi + y_{23}\eta + y_3 \qquad ...[10]$$

$$(x_{ij} = x_i - x_j; \quad y_{ij} = y_i - y_j)$$

This equation relates x and y coordinates to the ξ and η coordinates. Equations [3] and [4] express u and v as functions of ξ, η.

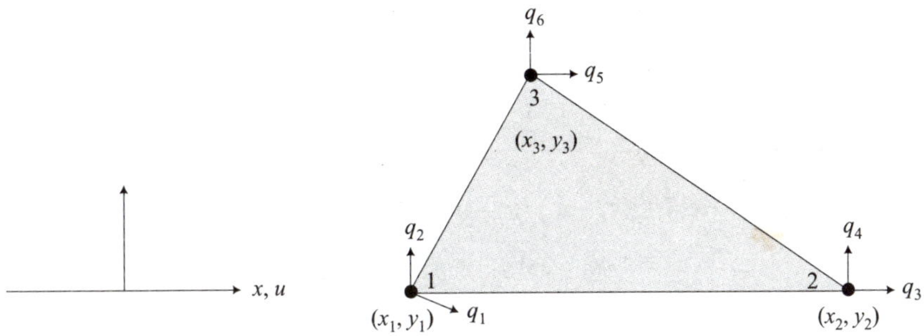

Fig. 5.8

Jacobian of the Element

Jacobian (J) is the short form for the determinant of the Jacobian matrix which contains the derivatives of the global coordinates with respect to the natural coordinates of a particular element. The magnitude of the det **J** is twice the area of the element.

Jacobian for the triangular element.

The field variables (u, v) and (x, y) are all functions of (ξ, η) i.e.,

$$u = u\,(x, (\xi, \eta), y(\xi, \eta) \text{ and } v = (x\,(\xi, \eta), y\,(\xi, \eta).$$

Hence, the derivatives of u and v are

$$\frac{\partial u}{\partial \xi} = \frac{\partial u}{\partial x} \cdot \frac{\partial x}{\partial \xi} + \frac{\partial u}{\partial y} \cdot \frac{\partial y}{\partial \xi} \qquad ...[1]$$

$$\frac{\partial u}{\partial \eta} = \frac{\partial u}{\partial x} \cdot \frac{\partial x}{\partial \eta} + \frac{\partial u}{\partial y} \cdot \frac{\partial y}{\partial \eta} \qquad ...[2]$$

This can be put in Matrix form as:

$$\begin{Bmatrix} \dfrac{\partial u}{\partial \xi} \\ \dfrac{\partial u}{\partial \eta} \end{Bmatrix} = \begin{bmatrix} \dfrac{\partial x}{\partial \xi} & \dfrac{\partial y}{\partial \xi} \\ \dfrac{\partial x}{\partial \eta} & \dfrac{\partial y}{\partial \eta} \end{bmatrix} \begin{Bmatrix} \dfrac{\partial u}{\partial x} \\ \dfrac{\partial u}{\partial y} \end{Bmatrix} \qquad ...[3]$$

The 2×2 matrix in equation [3] is called the Jacobian of the transformation denoted by

$$J = \begin{bmatrix} \dfrac{\partial x}{\partial \xi} & \dfrac{\partial y}{\partial \xi} \\ \dfrac{\partial x}{\partial \eta} & \dfrac{\partial y}{\partial \eta} \end{bmatrix} \qquad ...[4]$$

In view of equations [1] and [2], eq. [4] takes the form

$$J = \begin{bmatrix} x_1 - x_3 & y_1 - y_3 \\ x_2 - x_3 & y_2 - y_3 \end{bmatrix} \qquad ...[5]$$

The determinant of J is given by

$$\begin{aligned} \det J &= (x_1 - x_3)(y_2 - y_3) - (x_2 - x_3)(y_1 - y_3) \\ &= x_2y_3 - x_3y_2 + x_1y_2 - x_2y_1 + x_3y_1 - x_1y_3 \end{aligned} \qquad ...[6]$$

Comparing eq. [6] with

$$2A = \det \begin{vmatrix} 1 & x_1 & y_1 \\ 1 & x_2 & y_2 \\ 1 & x_3 & y_3 \end{vmatrix} = x_2y_3 - x_3y_2 + x_1y_2 - x_2y_1 + x_3y_1 - x_1y_3$$

we find that det J = twice the area of the triangle. Hence $A = \frac{1}{2}\,|\det J|$.

Note that if the numbering of the node is done anticlockwise as done here, we always get positive value for det *J*. Otherwise we will get only negative value.

Now, inverting equation [3] we get

$$\begin{Bmatrix} \dfrac{\partial u}{\partial x} \\ \dfrac{\partial u}{\partial y} \end{Bmatrix} = J^{-1} \begin{Bmatrix} \dfrac{\partial u}{\partial \xi} \\ \dfrac{\partial u}{\partial \eta} \end{Bmatrix} \qquad ...[7]$$

where J^{-1} is the inverse of the Jacobian given by

$$J^{-1} = \frac{1}{\det J} \begin{bmatrix} y_2 - y_3 & -(y_1 - y_3) \\ -(x_2 - x_3) & x_1 - x_3 \end{bmatrix} \qquad ...[8]$$

Strain–displacement matrix* B *for the triangular element

We know that the strains are given by

$$\varepsilon = \begin{Bmatrix} \dfrac{\partial u}{\partial x} \\ \dfrac{\partial v}{\partial y} \\ \dfrac{\partial u}{\partial y} + \dfrac{\partial v}{\partial x} \end{Bmatrix} \qquad ...[1]$$

Now, from equation

$$\begin{Bmatrix} \dfrac{\partial u}{\partial x} \\ \dfrac{\partial u}{\partial y} \end{Bmatrix} = J^{-1} \begin{Bmatrix} \dfrac{\partial u}{\partial \xi} \\ \dfrac{\partial u}{\partial \eta} \end{Bmatrix}.$$

we have

$$\begin{Bmatrix} \dfrac{\partial u}{\partial x} \\ \dfrac{\partial u}{\partial y} \end{Bmatrix} = J^{-1} \begin{Bmatrix} \dfrac{\partial u}{\partial \xi} \\ \dfrac{\partial u}{\partial \eta} \end{Bmatrix} = \frac{1}{\det J} \begin{bmatrix} y_2 - y_3 & -(y_1 - y_3) \\ -(x_2 - x_3) & x_1 - x_3 \end{bmatrix} \begin{Bmatrix} \dfrac{\partial u}{\partial \xi} \\ \dfrac{\partial u}{\partial \eta} \end{Bmatrix}$$

$$= \frac{1}{\det J} \begin{Bmatrix} (y_2 - y_3)\dfrac{\partial \mu}{\partial \xi} - (y_1 - y_3)\dfrac{\partial \mu}{\partial \eta} \\ -(x_2 - x_3)\dfrac{\partial \mu}{\partial \xi} + (x_1 - x_3)\dfrac{\partial \mu}{\partial \eta} \end{Bmatrix} \qquad \text{...[2]}$$

Now, replacing u by v it is obtained

$$\begin{Bmatrix} \dfrac{\partial v}{\partial x} \\ \dfrac{\partial v}{\partial y} \end{Bmatrix} = \frac{1}{\det J} \begin{Bmatrix} (y_2 - y_3)\dfrac{\partial v}{\partial \xi} - (y_1 - y_3)\dfrac{\partial v}{\partial \eta} \\ -(x_2 - x_3)\dfrac{\partial v}{\partial \xi} + (x_1 - x_3)\dfrac{\partial v}{\partial \eta} \end{Bmatrix} \qquad \text{...[3]}$$

Substituting Eqs. [2] and [3] into [1] we get

$$\varepsilon = \begin{Bmatrix} \dfrac{\partial u}{\partial x} \\ \dfrac{\partial v}{\partial y} \\ \dfrac{\partial u}{\partial y} + \dfrac{\partial v}{\partial x} \end{Bmatrix} = \frac{1}{\det J} \begin{Bmatrix} (y_2 - y_3)\dfrac{\partial u}{\partial \xi} - (y_1 - y_3)\dfrac{\partial u}{\partial \eta} \\ -(x_2 - x_3)\dfrac{\partial v}{\partial \xi} + (x_1 - x_3)\dfrac{\partial v}{\partial \eta} \\ -(x_2 - x_3)\dfrac{\partial u}{\partial \xi} + (x_1 - x_3)\dfrac{\partial u}{\partial \eta} + (y_2 - y_3)\dfrac{\partial v}{\partial \xi} - (y_1 - y_3)\dfrac{\partial v}{\partial \eta} \end{Bmatrix} \qquad \text{...[4]}$$

Now, using equations:

$$u = \xi q_1 + \eta q_3 + (1 - \xi - \eta)\, q_5 = (q_1 - q_5)\, \xi + (q_3 - q_5)\eta + q_5$$

$$v = \xi q_2 + \eta q_4 + (1 - \xi - \eta)\, q_6 = (q_2 - q_6)\, \xi + (q_4 - q_6)\eta + q_6$$

and taking the derivatives these w. r. t. ξ and η and substituting in eq. [4], the expression for the strain become:

$$\varepsilon = \frac{1}{\det J} \begin{Bmatrix} (y_2 - y_3)q_1 + (y_3 - y_1)q_3 + (y_1 - y_2)q_5 \\ (x_3 - x_2)q_2 + (x_1 - x_3)q_4 + (x_2 - x_1)q_6 \\ (x_3 - x_2)q_1 + (y_2 - y_3)q_2 + (x_1 - x_3)q_3 + (y_3 - y_1)q_4 + (x_2 - x_1)q_5 + (y_1 - y_2)q_6 \end{Bmatrix} \qquad \text{...[5]}$$

Eq. [5] can be put in the form

$$\varepsilon = \frac{1}{\det J}\begin{bmatrix} y_2-y_3 & 0 & y_3-y_1 & 0 & y_1-y_2 & 0 \\ 0 & x_3-x_2 & 0 & x_1-x_3 & 0 & x_2-x_1 \\ x_3-x_2 & y_2-y_3 & x_1-x_3 & y_3-y_1 & x_2-x_1 & y_1-y_2 \end{bmatrix}\begin{Bmatrix} q_1 \\ q_2 \\ q_3 \\ q_4 \\ q_5 \\ q_6 \end{Bmatrix} \quad ...[6]$$

In simple notation [6] takes the form

$$\varepsilon = \mathbf{Bq}. \quad ...[7]$$

where the strain-displacement matrix **B** is given by

$$\mathbf{B} = \frac{1}{\det J}\begin{bmatrix} y_2-y_3 & 0 & y_3-y_1 & 0 & y_1-y_2 & 0 \\ 0 & x_3-x_2 & 0 & x_1-x_3 & 0 & x_2-x_1 \\ x_3-x_2 & y_2-y_3 & x_1-x_3 & y_3-y_1 & x_2-x_1 & y_1-y_2 \end{bmatrix} \quad ...[8]$$

Shape Functions in Global Coordinates

A 3-noded triangular element is shown in Fig. 5.9, in an x, y coordinate system with nodes 1, 2 and 3 located at (x_1, y_1), (x_2, y_2), and (x_3, y_3), respectively, in the global system.

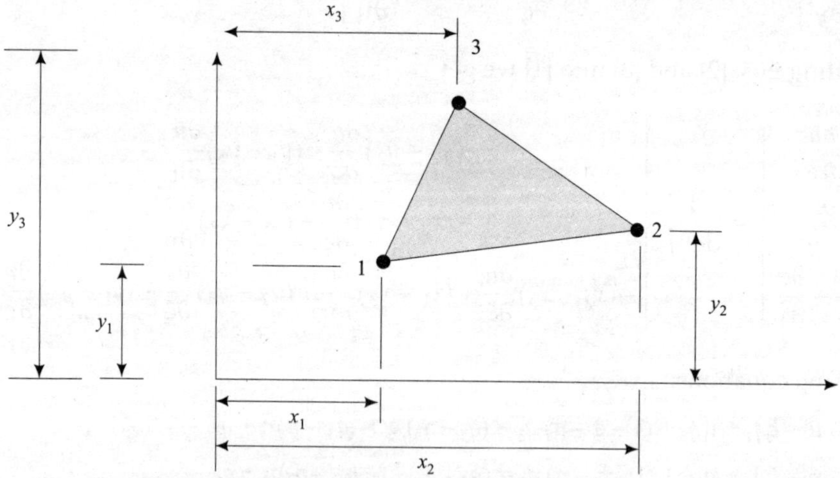

Fig. 5.9 A (triangle) global system

Let us assume $\phi(x, y) = \phi$ an interpolation function to represent the variation of the unknown quantity,

$$\phi = c_1 + c_2x + c_3y \quad ...[1]$$

The matrix equation of [1] can be written as:

$$\phi = [1 \ x \ y] \begin{Bmatrix} c_1 \\ c_2 \\ c_3 \end{Bmatrix}$$

or in a matrix format

$$\phi = [\alpha]\{c\} \qquad ...[2]$$

The boundary conditions in terms of nodal coordinates of ϕ;

$$\left.\begin{aligned} \phi(x_1, y_1) &= \phi_1 \\ \phi(x_2, y_2) &= \phi_2 \\ \phi(x_3, y_3) &= \phi_3 \end{aligned}\right] \qquad ...[3]$$

Substitute eqn. [3] into equation [1] to obtain 3 equations that can be solved for C_1, C_2 and C_3:

$$\phi_1 = c_1 + c_2x_1 + c_3y_1$$

$$\phi_2 = c_1 + c_2x_2 + c_3y_2$$

$$\phi_3 = c_1 + c_2x_3 + c_3y_3$$

Or,

$$\begin{Bmatrix} \phi_1 \\ \phi_2 \\ \phi_3 \end{Bmatrix} = \begin{bmatrix} 1 & x_1 & y_1 \\ 1 & x_2 & y_2 \\ 1 & x_3 & y_3 \end{bmatrix} \begin{Bmatrix} c_1 \\ c_2 \\ c_3 \end{Bmatrix} \qquad ...[4]$$

The matrix form of this is:

$$\{\phi\} = [X]\{C\} \qquad ...[5]$$

$$\Rightarrow \qquad \{C\} = [X]^{-1}\{\phi\} \qquad ...[6]$$

Substituting from eqn [6] into equation [2], we get

$$\phi = \{\alpha\}[X]^{-1}\{\phi\}$$

$$= [N]\{\phi\} \qquad ...[7]$$

The shape functions are the product of the first two matrices on the *RHS* of equation [7]

$$[N] = [\alpha][X]^{-1}$$

Or,

$$[N_1 \, N_2 \, N_3] = [1 \ x \ y][X]^{-1} \qquad ...[8]$$

In order to solve for N_1, N_2 and N_3 we proceed as follows, where

$$\det[X] = 2A$$

and [C] is the cofactor, also

$$[X]^{-1} = \frac{[C]^T}{\det[X]} \qquad ...[9]$$

See note for computation of $[X]^{-1}$

Hence, equation [8] gives

$$\left.\begin{aligned} N_1 &= \frac{[(x_2y_3 - x_3y_2) + x(y_2 - y_3) + y(x_3 - x_2)]}{2A} \\ N_2 &= \frac{[(x_3y_1 - x_1y_3) + x(y_3 - y_1) + y(x_1 - x_3)]}{2A} \\ N_3 &= \frac{[(x_1y_2 - x_2y_1) + x(y_1 - y_2) + y(x_2 - x_1)]}{2A} \end{aligned}\right\} \qquad ...[10]$$

where $$A = \frac{1}{2}\det\begin{bmatrix} 1 & x_1 & y_1 \\ 1 & x_2 & y_2 \\ 1 & x_3 & y_3 \end{bmatrix} \qquad ...[11]$$

A is the area of the triangular element.

Note: Computation of $[X]^{-1}$

Since $$X = \begin{bmatrix} 1 & x_1 & y_1 \\ 1 & x_2 & y_2 \\ 1 & x_3 & y_3 \end{bmatrix}$$

To find X^{-1}: also det $[X] = 2A$ and $[C]$ is the cofactor:

$$c_{11} = (-1)^{1+1}\begin{vmatrix} x_2 & y_2 \\ x_3 & y_3 \end{vmatrix} = x_2y_3 - x_3y_2;$$

$$c_{12} = (-1)^{1+2}\begin{vmatrix} 1 & y_2 \\ 1 & y_3 \end{vmatrix} = y_2 - y_3;$$

$$c_{13} = (-1)^{1+3}\begin{vmatrix} 1 & x_2 \\ 1 & x_3 \end{vmatrix} = x_3 - x_2;$$

$$c_{21} = (-1)^{2+1}\begin{vmatrix} x_1 & y_1 \\ x_3 & y_3 \end{vmatrix} = x_3y_1 - x_1y_3;$$

$$c_{22} = (-1)^{2+2}\begin{vmatrix} 1 & y_1 \\ 1 & y_3 \end{vmatrix} = y_3 - y_1;$$

$$c_{23} = (-1)^{2+3}\begin{vmatrix} 1 & x_1 \\ 1 & x_3 \end{vmatrix} = x_1 - x_3;$$

$$c_{31} = (-1)^{3+1}\begin{vmatrix} x_1 & y_1 \\ x_2 & y_2 \end{vmatrix} = x_1y_2 - x_2y_1;$$

$$c_{32} = (-1)^{3+2} \begin{vmatrix} 1 & y_1 \\ 1 & y_2 \end{vmatrix} = y_1 - y_2$$

$$c_{33} = (-1)^{3+3} \begin{vmatrix} 1 & x_1 \\ 1 & x_2 \end{vmatrix} = x_2 - x_1$$

Thus,
$$[C]^T = \begin{bmatrix} c_{11} & c_{21} & c_{31} \\ c_{12} & c_{22} & c_{32} \\ c_{13} & c_{23} & c_{33} \end{bmatrix}$$

$\therefore$
$$[X]^{-1} = \frac{1}{2A}[C]^T$$

$$= \frac{1}{2A}\begin{bmatrix} c_{11} & c_{21} & c_{31} \\ c_{12} & c_{22} & c_{32} \\ c_{13} & c_{23} & c_{33} \end{bmatrix}$$

$$= \frac{1}{2A}\begin{bmatrix} x_2y_3 - x_3y_2 & x_3y_1 - x_1y_3 & x_1y_2 - x_2y_1 \\ y_2 - y_3 & y_3 - y_1 & y_1 - y_2 \\ x_3 - x_2 & x_1 - x_3 & x_2 - x_1 \end{bmatrix}$$

Area coordinates are related to shape functions for a three-noded triangular finite element.

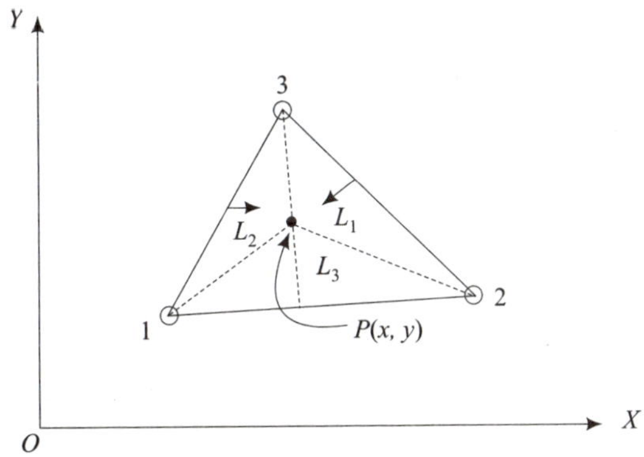

Fig. 5.10

Consider a triangular element as shown in Fig. 5.10. Here the point $P(x, y)$, which is in the interior to the element, is not a node but an arbitrary reference point.

Area coordinates are defined as:

$$L_1 = \frac{\text{area }(23P)}{\text{area }123},\ L_2 = \frac{\text{area }(31P)}{\text{area }123},\ L_3 = \frac{\text{area }(12P)}{\text{area }123}. \qquad \text{...[1]}$$

The coordinates L_1, L_2 and L_3 are defined as area coordinates because each represents a fractional part of the total triangular area and their sum is equal to the total triangular area [Fig. 5.10].

Area 23P + area 31P + area 12P

$$\Rightarrow \qquad L_1 + L_2 + L_3 = 1, L_i > 0, i = 1, 2, 3 \ldots$$

The area coordinates system is called natural or local coordinates system. The coordinate L_1 emanates from side 2 – 3 and corresponds to the area defined by side 2 – 3 and the point P. Note that if L_1 is allowed to increase and reach node point 1, its magnitude will be unity and $L_2 = L_3 = 0$.

It appears that L_1 behaves similarly to N_1, the shape function for Node 1.

From Fig. 5.10, the following relations are deduced:

$$L_1 = \begin{cases} 1 & \text{at node 1} \\ 0 & \text{at nodes 2 and 3} \end{cases}$$

$$L_2 = \begin{cases} 1 & \text{at node 2} \\ 0 & \text{at nodes 1 and 3} \end{cases}$$

$$L_3 = \begin{cases} 1 & \text{at node 3} \\ 0 & \text{at nodes 1 and 2} \end{cases}$$

Figure 5.10 clearly shows the coordinate location of each node. The ratio of area coordinate L_3 relative to the total area is (x, y is the location of point, P)

$$L_3 = \frac{\text{area } (12P)}{\text{area } 123} = \frac{1}{2} \det \begin{bmatrix} 1 & 1 & 1 \\ x_1 & x_2 & x \\ y_1 & y_2 & y \end{bmatrix} \div A_{\text{total}}$$

or,

$$L_3 = \frac{[(x_1y_2 - x_2y_1) + x(y_1 - y_2) + y(x_2 - x_1)]}{2A_{\text{total}}} \qquad \ldots[2]$$

By definition of the area of a triangular element, eqn. [2] gives the area and it is concluded that

$$L_3 = N_3;$$

Similary,

$$L_1 = N_1 = \frac{[(x_2y_3 - x_3y_2) + x(y_2 - y_3) + y(x_3 - x_2)]}{2A_{\text{total}}}$$

and

$$L_2 = N_2 = \frac{[(x_3y_1 - x_1y_3) + x(y_3 - y_1) + y(x_1 - x_3)]}{2A_{\text{total}}}$$

Example (1): Express the shape functions in terms of the global coordinates (x_1, y_1), (x_2, y_2), and (x_3, y_3).

We know that the equation for area of Δ gives the relation between the shape functions and the area coordinates. Now, if we express the area in terms of coordinates we will get the required relations. We know that the area of the triangle 123 is given by

$$2A = \det \begin{vmatrix} 1 & x_1 & y_1 \\ 1 & x_2 & y_2 \\ 1 & x_3 & y_3 \end{vmatrix} = x_2y_3 - x_3y_2 + x_1y_2 - x_2y_1 + x_3y_1 - x_1y_3 \qquad [1]$$

Similarly,

$$2A_1 = \det\begin{vmatrix}1 & x & y\\ 1 & x_2 & y_2\\ 1 & x_3 & y_3\end{vmatrix};\quad 2A_2 = \det\begin{vmatrix}1 & x & y\\ 1 & x_3 & y_3\\ 1 & x_1 & y_1\end{vmatrix};\quad 2A_3 = \det\begin{vmatrix}1 & x & y\\ 1 & x_1 & y_1\\ 1 & x_2 & y_2\end{vmatrix} \qquad ...[2]$$

Hence,

$$N_1 = \frac{A_1}{A} = \frac{(x_2y_3 - x_3y_2) - x(y_3 - y_2) + y(x_3 - x_2)}{A} \qquad ...[3a]$$

$$N_2 = \frac{A_2}{A} = \frac{(x_3y_1 - x_1y_3) - x(y_1 - y_3) + y(x_1 - x_3)}{A} \qquad ...[3b]$$

$$N_3 = \frac{A_3}{A} = \frac{(x_1y_2 - x_2y_1) - x(y_2 - y_1) + y(x_2 - x_1)}{A} \qquad ...[3c]$$

Equations [3a], [3b] and [3c] are the required expressions.

Example (2): Develop the **B** matrix for the triangular element as shown in Fig. 5.11.

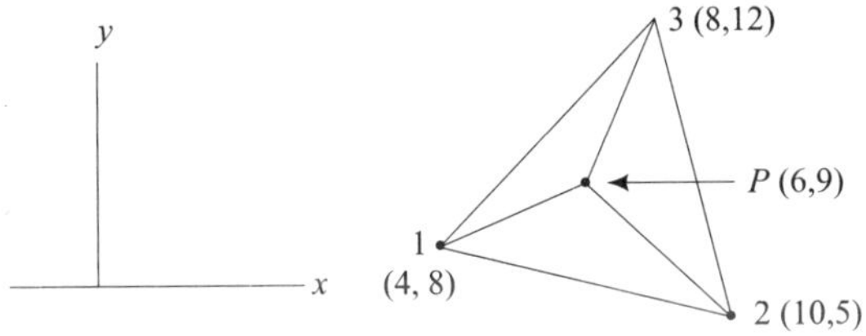

Fig. 5.11

$$\mathbf{B} = \frac{1}{\det J}\begin{bmatrix} 5-12 & 0 & 12-8 & 0 & 8-5 & 0\\ 0 & 8-10 & 0 & 4-8 & 0 & 10-4\\ 8-10 & 5-12 & 4-8 & 12-8 & 10-4 & 8-5 \end{bmatrix}$$

$$= \frac{1}{\det J}\begin{bmatrix} -7 & 0 & 4 & 0 & 3 & 0\\ 0 & -2 & 0 & -4 & 0 & 6\\ -2 & -7 & -4 & 4 & 6 & 3 \end{bmatrix}$$

det $\qquad J = 10 \times 12 - 8 \times 5 + 4 \times 5 - 4 \times 12 + 8 \times 8 - 8 \times 10 = 36$

Hence $\qquad \mathbf{B} = \dfrac{1}{36}\begin{bmatrix} -7 & 0 & 4 & 0 & 3 & 0\\ 0 & -2 & 0 & -4 & 0 & 6\\ -2 & -7 & -4 & 4 & 6 & 3 \end{bmatrix}$

Stiffness Matrix of a Constant Strain Three-noded Simple Triangular Element

Natural Coordinate Method

A three-noded constant strain triangle (*CST*) is as shown in Fig. 5.12(*a*). The interior point *P* is not a node but merely an arbitrary reference point.

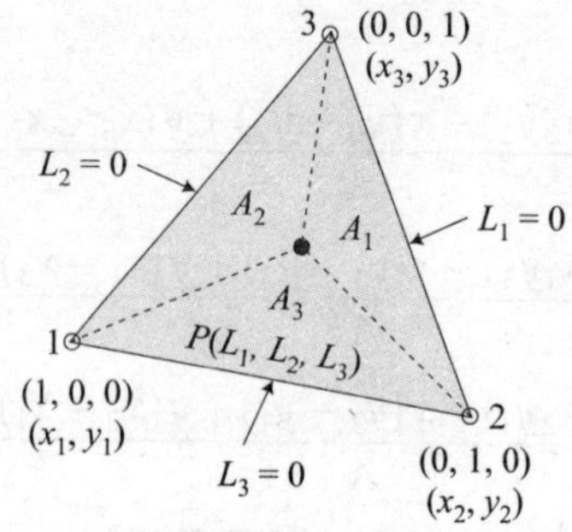

Fig. 5.12(a) A CST element is natural coordinates

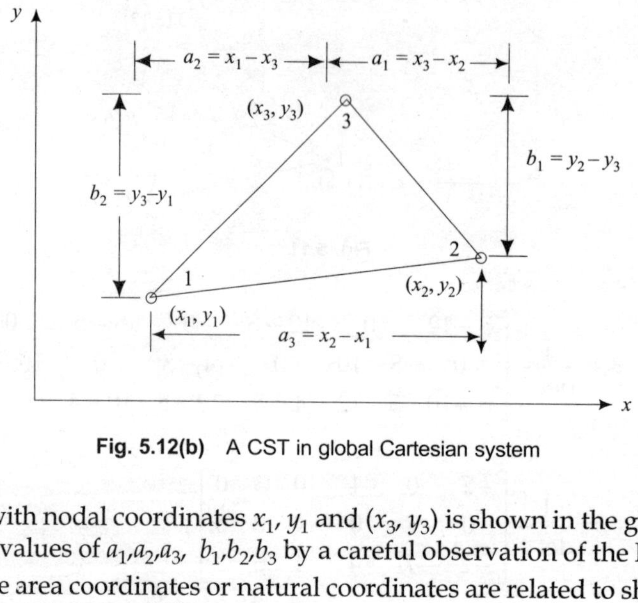

Fig. 5.12(b) A CST in global Cartesian system

A CST element with nodal coordinates x_1, y_1 and (x_3, y_3) is shown in the global system. We can easily remember the values of a_1, a_2, a_3, b_1, b_2, b_3 by a careful observation of the Fig. 5.12[*b*].

We know that the area coordinates or natural coordinates are related to shape functions (Refer already proved result in the same topic).

i.e.,

$$L_1 = N_1 = \frac{A_1}{A}$$

$$L_2 = N_2 = \frac{A_2}{A}$$

$$L_3 = N_3 = \frac{A_3}{A}$$

By the definition of Isoparametric elements, the geometry can be defined by the shape functions and nodal parameters, and the given element is an isoparametric element.

We know

$$L_1 + L_2 + L_3 = 1$$
$$L_1x_1 + L_2x_2 + L_3x_3 = x$$
$$L_1y_1 + L_2y_2 + L_3y_3 = y$$

The matrix equation of this is:

$$\begin{bmatrix} 1 & 1 & 1 \\ x_1 & x_2 & x_3 \\ y_1 & y_2 & y_3 \end{bmatrix} \begin{bmatrix} L_1 \\ L_2 \\ L_3 \end{bmatrix} = \begin{bmatrix} 1 \\ x \\ y \end{bmatrix}$$

Inverting the above matrix, we get

$$\begin{bmatrix} L_1 \\ L_2 \\ L_3 \end{bmatrix} = \frac{1}{2A} \begin{bmatrix} c_1 & b_1 & a_1 \\ c_2 & b_2 & a_2 \\ c_3 & b_3 & a_3 \end{bmatrix} \begin{bmatrix} 1 \\ x \\ y \end{bmatrix}$$

The above matrix gives the shape functions L_1, L_2, L_3 in terms of Cartesian coordinates. A is the total area of the triangle in terms of nodal parameters

i.e., here,

$$c_1 = x_2y_3 - y_2x_3$$
$$c_2 = x_3y_1 - y_2x_1$$
$$c_3 = x_1y_2 - y_1x_2$$

One can easily remember by recalling cyclic order. We now have

$$\left.\begin{aligned} L_1 = N_1 &= \frac{1}{2A}\,[c_1 + b_1x + a_1y] \\ L_2 = N_2 &= \frac{1}{2A}\,[c_2 + b_2x + a_2y] \\ L_3 = N_3 &= \frac{1}{2A}\,[c_3 + b_3x + a_3y] \end{aligned}\right\} \quad \text{...[1]}$$

The shape functions N_1, N_2 and N_3 can be found from $N = PC^{-1}$ but for the sake of convenience, we can apply isoparametric concept which requires

$$\left.\begin{aligned} u &= N_1u_1 + N_2u_2 + N_3u_3 \\ v &= N_1v_1 + N_2v_2 + N_3v_3 \end{aligned}\right\} \quad \text{...[2]}$$

Substituting eqn. [1] into eqn. [2], we obtain:

$$U = \frac{1}{2A}\,\{[c_1 + b_1x + a_1y]\,u_1 + [c_2 + b_2x + a_2y]\,u_2 + [c_3 + b_3x + a_3y]\,u_3\} \quad \text{...[3]}$$

$$\therefore \quad \frac{\partial u}{\partial x} = \frac{1}{2A}[b_1u_1 + b_2u_2 + b_3u_3] = \varepsilon_x \quad ...[3]$$

$$v = \frac{1}{2A}\{[c_1 + b_1x + a_1y]v_1 + \{[c_2 + b_2x + a_2y]v_2 + [c_3 + b_3x + a_3y]v_3\}$$

$$\therefore \quad \frac{\partial v}{\partial y} = \varepsilon_y = \frac{1}{2A}[a_1v_1 + a_2v_2 + a_3v_3] \quad ...[4]$$

Along similar lines

$$\gamma_{xy} = \frac{\partial u}{\partial y} + \frac{\partial v}{\partial x}$$

$$= \frac{1}{2A}[a_1u_1 + b_1v_1 + a_2u_2 + b_2v_2 + a_3u_3 + b_3v_3] \quad ...[5]$$

The matrix form of eqns. [3], [4] and [5], is

$$\underset{\varepsilon}{\begin{bmatrix} \varepsilon_x \\ \varepsilon_y \\ \gamma_{xy} \end{bmatrix}} = \frac{1}{2A} \underset{B}{\begin{bmatrix} b_1 & 0 & b_2 & 0 & b_3 & 0 \\ 0 & a_1 & 0 & a_2 & 0 & a_3 \\ a_1 & b_1 & a_2 & b_2 & a_3 & b_3 \end{bmatrix}} \underset{q}{\begin{Bmatrix} u_1 \\ v_1 \\ u_2 \\ v_2 \\ u_3 \\ v_3 \end{Bmatrix}}$$

$$\downarrow \qquad\qquad \downarrow \qquad\qquad \downarrow$$

$$\varepsilon \qquad\qquad B \qquad\qquad q$$

Now $dv = t\,dA$

t = thickness of the element

A = Area

$$k = \int B^T d B \, dv$$

where d = elasticity matrix

$$d = \begin{bmatrix} c_1 & c_1c_2 & 0 \\ c_1c_2 & c_1 & 0 \\ 0 & 0 & c_{12} \end{bmatrix}$$

Polynomials Method for Stiffness Matrix K

The displacement functions at any point $P(x, y)$ for u and v are represented by linear polynomials (or displacement field).

$$u = \alpha_1 + \alpha_2 x + \alpha_3 y \qquad ...[1]$$

$$v = \alpha_4 + \alpha_5 x + \alpha_6 y \qquad ...[2]$$

Or,

$$\begin{bmatrix} u \\ v \end{bmatrix} = \begin{bmatrix} 1 & x & y & 0 & 0 & 0 \\ 0 & 0 & 0 & 1 & x & y \end{bmatrix} \begin{Bmatrix} \alpha_1 \\ \alpha_2 \\ \alpha_3 \\ \alpha_4 \\ \alpha_5 \\ \alpha_6 \end{Bmatrix}$$

where

$$\{\phi\} = [1 \;\; x \;\; y] \qquad ...[3]$$

and

$$[Q] = \begin{vmatrix} \{\phi\} & \{0\} \\ \{0\} & \{\phi\} \end{vmatrix} \qquad ...[4]$$

Or,

$$\begin{bmatrix} u \\ v \end{bmatrix} = [Q] \begin{Bmatrix} \alpha_x \\ \alpha_y \end{Bmatrix} \qquad ...[5]$$

Substituting for x, y the coordinates of points 1, 2, 3, we get

$$\begin{Bmatrix} u_1 \\ u_2 \\ u_3 \\ v_1 \\ v_2 \\ v_3 \end{Bmatrix} = \begin{bmatrix} 1 & x_1 & y_1 & 0 & 0 & 0 \\ 1 & x_2 & y_2 & 0 & 0 & 0 \\ 1 & x_3 & y_3 & 0 & 0 & 0 \\ 0 & 0 & 0 & 1 & x_1 & y_1 \\ 0 & 0 & 0 & 1 & x_2 & y_2 \\ 0 & 0 & 0 & 1 & x_3 & y_3 \end{bmatrix} \begin{Bmatrix} \alpha_1 \\ \alpha_2 \\ \alpha_3 \\ \alpha_4 \\ \alpha_5 \\ \alpha_6 \end{Bmatrix} \qquad ...[6]$$

Or,

$$\begin{Bmatrix} \underline{U} \\ \underline{V} \end{Bmatrix} = \begin{vmatrix} [A_s] & [0] \\ [0] & [A_s] \end{vmatrix} \begin{Bmatrix} \underline{\alpha_x} \\ \underline{\alpha_y} \end{Bmatrix} \qquad ...[7]$$

where

$$\{\underline{\alpha_x}\} = \begin{Bmatrix} \alpha_1 \\ \alpha_2 \\ \alpha_3 \end{Bmatrix}; \{\underline{\alpha_y}\} = \begin{Bmatrix} \alpha_4 \\ \alpha_5 \\ \alpha_6 \end{Bmatrix}$$

Or,

$$\begin{Bmatrix} \underline{\alpha_x} \\ \underline{\alpha_y} \end{Bmatrix} = \begin{vmatrix} A_s^{-1} & 0 \\ 0 & A_s^{-1} \end{vmatrix} \begin{Bmatrix} \underline{U} \\ \underline{V} \end{Bmatrix}; \qquad ...[8]$$

where

$$[A_s]^{-1} = \frac{1}{2\Delta}\begin{bmatrix} t_1 & t_2 & t_3 \\ y_{23} & y_{31} & y_{12} \\ x_{32} & x_{13} & x_{21} \end{bmatrix} \qquad ...[9]$$

where Δ = area of Δ 123,

$$t_1 = x_2y_3 - y_2x_3$$
$$t_2 = x_3y_1 - x_1y_3$$
$$t_3 = x_1y_2 - x_2y_1$$

and

$$y_{23} = y_2 - y_3$$
$$x_{31} = x_3 - x_1$$
$$y_{31} = y_3 - y_1 \qquad \text{etc.}$$
$$y_{12} = y_1 - y_2$$
$$x_{32} = x_3 - x_2$$
$$x_{21} = x_2 - x_1$$

$$x_{ij} = x_i - x_j, \quad i, j = 1, 2, 3$$
$$y_{ij} = y_i - y_j, \quad i, j = 1, 2, 3$$

The strain at any point $P(x, y)$ within the element may be expressed as:

$$\{\varepsilon\} = \begin{Bmatrix} \varepsilon \\ \varepsilon_y \\ \gamma_{xy} \end{Bmatrix}$$

$$= \begin{Bmatrix} \dfrac{\partial u}{\partial x} \\ \dfrac{\partial u}{\gamma y} \\ \dfrac{\partial u}{\partial y} + \dfrac{\partial v}{\partial x} \end{Bmatrix}$$

$$= \begin{bmatrix} 0 & 1 & 0 & 0 & 0 & 0 \\ 0 & 0 & 0 & 0 & 0 & 1 \\ 0 & 0 & 1 & 0 & 1 & 0 \end{bmatrix} \begin{Bmatrix} \underline{\alpha_x} \\ \underline{\alpha_y} \end{Bmatrix} \qquad ...[10]$$

and substitute for $\{\alpha_x, \alpha_y\}$,

we get

$$\{\varepsilon\} = \{B\}\begin{Bmatrix}\underline{U}\\ \underline{V}\end{Bmatrix} \text{ where}$$

$$\{B\} = \begin{bmatrix}0 & 1 & 0 & 0 & 0 & 0\\ 0 & 0 & 0 & 0 & 0 & 1\\ 0 & 0 & 1 & 0 & 1 & 0\end{bmatrix}\begin{vmatrix}[A_s]^{-1} & [0]\\ [0] & [A_s]^{-1}\end{vmatrix}\begin{Bmatrix}\underline{U}\\ \underline{V}\end{Bmatrix} \quad \text{...[11]}$$

Then
$$[k] = \int_v [B]^T [C]\,[B]\,dv$$
$$= t\Delta\,[B]^T [C]\,[B] \quad \text{...[12]}$$

where Δ is the area of the triangle and C is the constitute law for plane stress–strain.

Consistent Load Vector

Using the principle of virtual work we have,

$$\{R\} = \int_V [N]^T \begin{Bmatrix}F_x\\ F_y\end{Bmatrix} dv + \int_s [N]^T \begin{Bmatrix}T_x\\ T_y\end{Bmatrix} ds \quad \text{...[1]}$$

Note: 1. The shape function that we select to define the displacement must provide constant strain state.

2. For CST, all matrices are constant matrices and hence integration is handy.

[K}: *Direct Shape Function Method*

The shape functions ϕ_i for u_i which will give unity at (x_i, y_i)

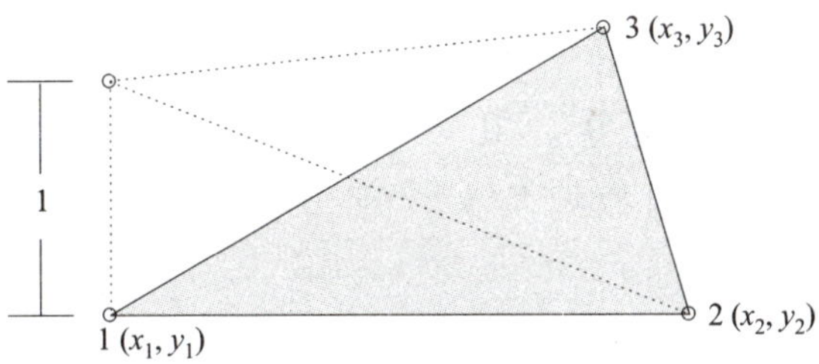

Fig. 5.13 Shape function ϕ

$$\begin{Bmatrix}u\\ v\end{Bmatrix} = \begin{bmatrix}\phi_1 & \phi_2 & \phi_3 & 0 & 0 & 0\\ 0 & 0 & 0 & \phi_1 & \phi_2 & \phi_3\end{bmatrix}\begin{Bmatrix}u_1\\ u_2\\ u_3\\ v_1\\ v_2\\ v_3\end{Bmatrix} \quad \text{...[1]}$$

Equation of the line joining (2-3) is given by

$$\frac{(x - x_2)}{(x_2 - x_3)} = \frac{(y - y_2)}{(y_2 - y_3)} \quad ...[2]$$

Or,
$$(x - x_2)y_{23} - (y - y_2)x_{23} = 0 \quad ...[3]$$

The displacement function $\phi_1 = 1$ at *i*.

= 0 on the line joining (2 – 3)

$$\phi_1 = Kf \quad ...[4]$$

$$\phi_1 = K[(x - x_2)y_{23} - (y - y_2)x_{23}] \quad ...[5]$$

$$\phi_1 = 1 \text{ at } x = x_1, y = y_1$$

Equation [5] implies

$$1 = K[(x_1 - x_2)y_{23} - (y_1 - y_2)x_{23}] \quad ...[6]$$

Or,
$$K = \frac{1}{2}\Delta;$$

and
$$\phi_1 = \frac{1}{2\Delta}[(x_1 - x_2)y_{23} - (y - y_2)x_{23}] \quad ...[7]$$

Similarly,
$$\phi_2 = \frac{1}{2\Delta}[(x - x_3)y_{31} - (y - y_3)x_{13}]$$

and
$$\phi_3 = \frac{1}{2\Delta}[(x - x_1)y_{12} - (y - y_1)x_{21}]$$

We can now write:

$$\{\varepsilon\} = \begin{Bmatrix} \dfrac{\partial u}{\partial x} \\ \dfrac{\partial u}{\partial y} \\ \dfrac{\partial u}{\partial y} + \dfrac{\partial v}{\partial x} \end{Bmatrix} = \begin{bmatrix} \phi_{1,x} & \phi_{2,x} & \phi_{3,x} & 0 & 0 & 0 \\ 0 & 0 & 0 & \phi_{1,y} & \phi_{2,y} & \phi_{3,y} \\ \phi_{1,y} & \phi_{2,y} & \phi_{3,y} & \phi_{1,x} & \phi_{2,x} & \phi_{3,x} \end{bmatrix} \left[\frac{U}{V}\right] \quad ...[8]$$

$$\{\varepsilon\} = [B]\left[\frac{U}{V}\right];$$

$$[K] = \int_V [B]^T [C] [B]\, dv \quad ...[9]$$

Consistent load vector

$$\{R\} = \int_V [N]^T \begin{Bmatrix} F_x \\ F_y \end{Bmatrix} dv + \int_s [N]^T \begin{Bmatrix} T_x \\ T_y \end{Bmatrix} ds \quad ...[10]$$

$$\{R\} = \int_V [N]^T \{F\} dv + \int_S [N]^T \{T\} ds \qquad ...[1]$$

where

$$[N] = \begin{bmatrix} N_1 & N_2 & N_3 & 0 & 0 & 0 \\ \hline 0 & 0 & 0 & N_1 & N_2 & N_3 \end{bmatrix} \qquad ...[2]$$

and

$$\{F\} = \begin{Bmatrix} F_x \\ F_y \end{Bmatrix} \text{ Body force/unit volume}$$

$$\{T\} = \begin{Bmatrix} T_x \\ T_y \end{Bmatrix} \text{ Surface force/unit area}$$

For finding stiffness matrix or consistent load vector, we get the Integral of the form

$$I = \int_A L_1^m L_2^n L_3^p \, dA \qquad ...[3]$$

Example 1 Calculate the consistent load vector when self weight of L/unit volume is acting in a CST (L is acting in –ve y-direction).

Solution: We know that, the formula for consistent load vector is given by

$$\begin{Bmatrix} Q_x \\ Q_y \end{Bmatrix} = \int_V [N]^T \{F\} dv + \int_S [N]^T \{T\} ds \qquad ...[1]$$

Since surface forces are zero

$$\begin{Bmatrix} Q_x \\ Q_y \end{Bmatrix} = t \int \begin{bmatrix} L_1 & 0 \\ L_2 & 0 \\ L_3 & 0 \\ 0 & L_1 \\ 0 & L_2 \\ 0 & L_3 \end{bmatrix} \begin{Bmatrix} 0 \\ L \end{Bmatrix} dA \qquad ...[2]$$

Integrating, we get

$$\begin{Bmatrix} Q_{x_i} \\ Q_{x_j} \\ Q_{x_k} \\ Q_{y_i} \\ Q_{y_j} \\ Q_{y_k} \end{Bmatrix} = \frac{tL\Delta}{3} \begin{Bmatrix} 0 \\ 0 \\ 0 \\ 1 \\ 1 \\ 1 \end{Bmatrix} \qquad ...[3]$$

Example 2 Find the shape function at point *P* shown in the Fig. 5.14.

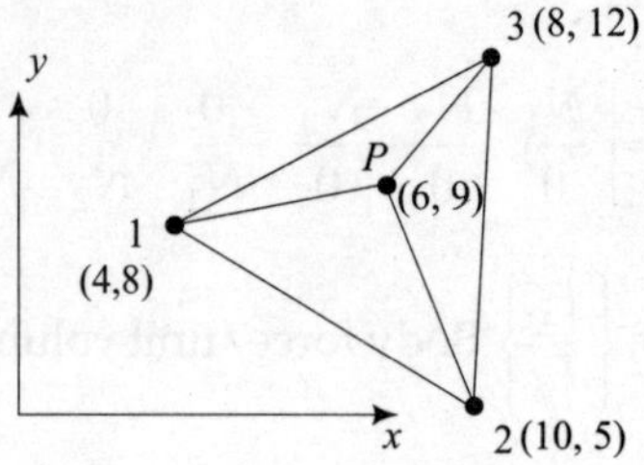

Fig. 5.14

Solution: The isomorphic representation can be expressed as

$$\left.\begin{aligned} x &= (x_1 - x_3)\xi + (x_2 - x_1)\eta + x_3 \\ y &= (y_1 - y_3)\xi + (y_2 - y_3)\eta + y_3 \end{aligned}\right] \quad ...[1]$$

$$\therefore \quad 6 = (4-8)\xi + (10-8)\eta + 8$$

$$= -4\xi + 2\eta + 8$$

$$9 = (8-12)\xi + (5-12)\eta + 12 = -4\xi + 7\eta + 12$$

On solving, we get

$$\xi = \frac{5}{9}; \eta = \frac{1}{8},$$

Hence,

$$N_1 = \frac{5}{9}$$

$$N_2 = \frac{1}{9}$$

$$N_3 = 1 - \frac{5}{9} - \frac{1}{9} = \frac{3}{9}$$

5.4 POTENTIAL ENERGY APPROACH

The potential energy Π of the system is given by

$$\Pi = \frac{1}{2}\int_{Area} \epsilon^T D \epsilon t\,dA - \int_{Area} u^T f t\,dA - \int_{Line} u^T T t\,dl - \sum_i u_i^T P_i \quad ...[1]$$

where *f*, *T* and P_i refer to body force, surface traction and point load applied at point *i*, respectively.

Here, $P_i = [P_x\ P_y]_i^T$. The summation on i gives the total potential energy due to all point loads. The first term on r.h.s of equation [1] represents the total strain energy of the system. Since this total strain energy is the sum of energy of all the elements then the above can be written as

$$\Pi = \sum_{NE} U_e - \sum_{NE} \int_{Area} u^T f t dA - \int_{Line} u^T T t dl - \sum_i u_i^T P_i \qquad ...[2]$$

where NE is the number of elements and $U_e = \frac{1}{2} \int_{Area} \varepsilon^T D E t dA$ ...[3]

is the element strain energy.

Element Stiffness

We now substitute for the strain-displacement relationship $\varepsilon = Bq$ into [3], the latter becomes

$$U_e = \frac{1}{2} \int_{Area} q^T B^T D B q \ t dA \qquad ...[4]$$

If we assume that the element has a constant thickness throughout and since all terms in B and D matrices are constant then the equation [4] will reduce to

$$U_e = \frac{1}{2} q^T B^T D B t_e \left(\int_{Area} dA \right) dA q \qquad ...[5]$$

$$= \frac{1}{2} t_e A_e q^T B^T D B \qquad ...[6]$$

which can be put in the form

$$U_e = \frac{1}{2} q^T k q \qquad ...[7]$$

where the element stiffness k is given by

$$k = t_e A_e \ B^T D B \qquad ...[8]$$

Expression for the Total Potential

If we now assemble all the elements then we get for the total strain energy

$$U = \sum_{NE} U_e = \sum_{NE} \frac{1}{2} q^T k q = \frac{1}{2} Q^T K Q \qquad ...[1]$$

where K is the global stiffness matrix. The global stiffness matrix will be symmetric, banded and sparse. Similarly, the total work done by the external loads can be put as

$$W = Q^T F, \quad \text{where} \quad F = \sum_{NE} (f^e + T^e) + P \qquad ...[2]$$

Hence, the total potential of the system is

$$\Pi = \frac{1}{2} Q^T K Q - Q^T F \qquad \text{...[3]}$$

minimizing equation [3] leads to the system equation

$$KQ = F \qquad \text{...[4]}$$

Equation [4] is to be solved after substituting for the boundary conditions.

Example 3: Assuming a plane stress condition find the stiffness matrix for the triangular element shown in Fig. 5.15. Assume constant thickness t equal to 1 unit and the Poisson's ratio = 0.3.

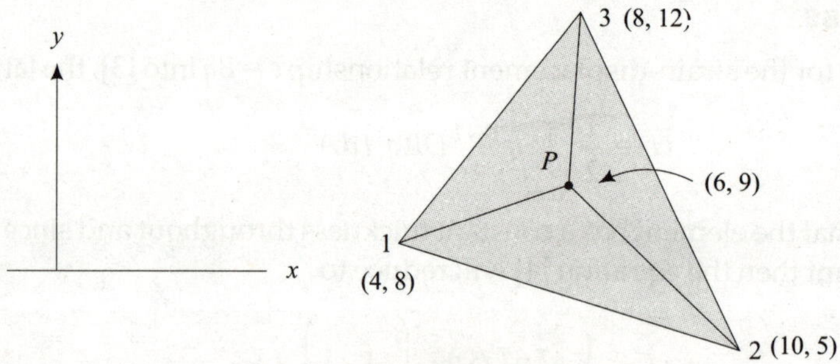

Fig. 5.15

We have already developed **B** matrix for this element and the area of the element was also found to be 18 units. Hence,

$$k = (1 \times 18)\frac{1}{36}\begin{bmatrix} -7 & 0 & -2 \\ 0 & -2 & -7 \\ 4 & 0 & -4 \\ 0 & -4 & 4 \\ 3 & 0 & 6 \\ 0 & 6 & 3 \end{bmatrix} \frac{E}{1-0.3^2}\begin{bmatrix} 1 & .3 & 0 \\ .3 & 1 & 0 \\ 0 & 0 & \frac{1-0.3}{2} \end{bmatrix} \frac{1}{36}\begin{bmatrix} -7 & 0 & 4 & 0 & 3 & 0 \\ 0 & -2 & 0 & -4 & 0 & 6 \\ -2 & -7 & -4 & 4 & 6 & 3 \end{bmatrix}$$

$$k = \frac{E}{65.52}\begin{bmatrix} 50.4 & 9.1 & -25.2 & 5.6 & -25.2 & -14.7 \\ 9.1 & 21.15 & 7.4 & -1.8 & -16.5 & -19.35 \\ -25.2 & 7.4 & 21.6 & -10.4 & 3.6 & 3.0 \\ 5.6 & -1.8 & -10.4 & 21.6 & 4.8 & -19.8 \\ -25.2 & -16.5 & 3.6 & 4.8 & 21.6 & 11.7 \\ -14.7 & -19.35 & 3.0 & -19.8 & 11.7 & 39.15 \end{bmatrix}$$

is the element stiffness matrix.

Force Terms

(i) Body force term in equation:

$$\Pi = \sum_{NE} U_e - \sum_{NE} \int_{Area} u^T f t dA - \int_{Line} u^T T t dl - \sum_i u_i^T P_i$$

We know that the work done by the body force is given by $\int_{Area} u^T f t dA$. This can be written as

$$\int_{Area} u^T f t dA = t_e \int_{Area} (u f_x + v f_y) dA \qquad ...[1]$$

Using the interpolation for u and v as given by equation:

$$\left.\begin{aligned} u &= N_1 q_1 + N_2 q_3 + N_3 q_5 \\ v &= N_1 q_2 + N_2 q_4 + N_3 q_6 \end{aligned}\right\}$$

into equation [1] it is obtained

$$\int_{Area} u^T f.t\, dA = q_1\left(t_e f_x \int_e N_1 dA\right) + q_2\left(t_e f_y \int_e N_1 dA\right) + q_3\left(t_e f_x \int_e N_2 dA\right) + q_4\left(t_e f_y \int_e N_2 dA\right)$$

$$+ q_5\left(t_e f_x \int_e N_3 dA\right) + q_6\left(t_e f_y \int_e N_3 dA\right) \qquad ...[2]$$

We know from problem 1 that by definition N_i's represent linear measure and hence the quantity $\int_e N_i dA$ represents the volume of a tetrahedron with base area that of the element A_e and height of corner equal to 1. This is shown in Fig. 5.16.

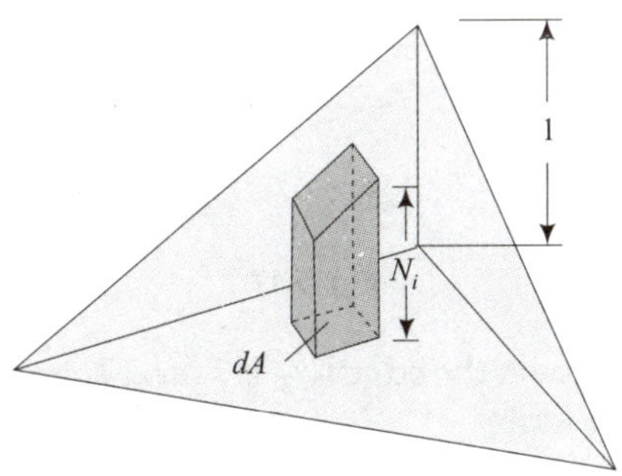

Fig. 5.16

Now, the volume of tetrahedron $= \frac{1}{3} \times \text{height} \times \text{base area} = \frac{1}{3} A_e$. Hence,

$$\int_e N_i \, dA = \frac{1}{3} A_e \quad (i = 1, 2, 3)$$

Eq. [2] can therefore be written as

$$\int_{\text{Area}} u^T f t \, dA = q^T f^e$$

where $f^e = \dfrac{t_e A_e}{3} \begin{Bmatrix} f_x \\ f_y \\ f_x \\ f_y \\ f_x \\ f_y \end{Bmatrix} = \dfrac{t_e A_e}{3} [f_x \, f_y \, f_x \, f_y \, f_x \, f_y]^T$ [3]

Eq. [3] gives the body force term.

Surface Traction Force Term

We know that the surface traction force term is given by $\int_{\text{Line}} u^T T t \, dl$. The surface force (or surface traction) acts over part or whole of the surface area of the body. On the element, the surface force acts on edges connecting boundary nodes as shown in Fig. 5.17.

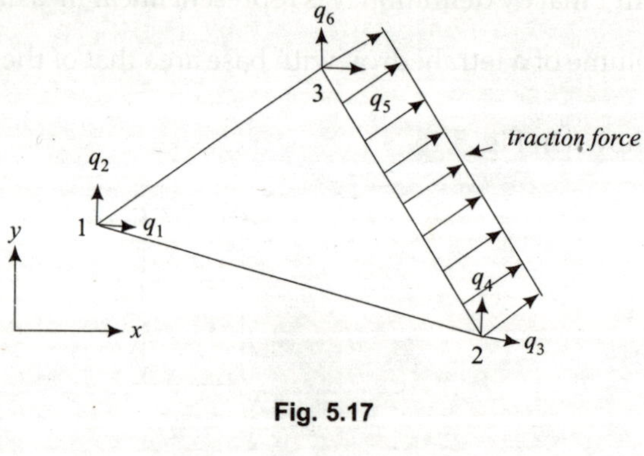

Fig. 5.17

As shown, the traction is acting over the edge l_{2-3}. If T_x and T_y are the components of traction along the x and y directions then we can write

$$\int_{\text{Line}} u^T T t \, dl = \int_{l_{2-3}} (uT_x + vT_y) t \, dl \qquad ...[1]$$

We substitute for u and v from equations:

$$\begin{Bmatrix} u = N_1 q_1 + N_2 q_3 + N_3 q_5 \\ v = N_1 q_1 + N_2 q_4 + N_3 q_6 \end{Bmatrix}$$

into equation $\int_{Area} u^T T t dA = t_e \int_{Area} (uf_x + vf_y)\, dA$

We obtain

$$\int_{Line} u^T T t dl = q_3\left(t_e f_x \int_e N_2 dA\right) + q_4\left(t_e f_y \int_e N_2 dA\right) + q_5\left(t_e f_x \int_e N_3 dA\right) + q_6\left(t_e f_y \int_e N_3 dA\right) \quad ...[2]$$

It should be noted that along the edge 2–3, the shape function $N_1 = 0$ and

$$\int_{l_{2-3}} N_i\, dl = \frac{1}{2} l_{2\text{-}3} \quad ...[3]$$

where the length $l_{2\text{-}3}$ is given by

$$l_{2\text{-}3} = \sqrt{(x_2 - x_3)^2 + (y_2 - y_3)^2}$$

Hence, the traction force term becomes

$$\int_{Line} \mathbf{u}^T \mathbf{T}\, t\, dl = \bar{q}^T \mathbf{T}^e \quad ...[4]$$

where $\bar{q}$ is the actual nodal degrees of freedom that have connection to the surface traction and is

$$\bar{q} = [q_3\ q_4\ q_5\ q_6]^T \quad ...[5]$$

and $$\mathbf{T}^e = \frac{t_e l_{2-3}}{2}\, [T_x\ T_y\ T_x\ T_y]^T \quad ...[6]$$

Eq. [6] is the surface traction force term. It is, however, better to express the traction term in a different way.

$$\int_{Line} \mathbf{u}^T \mathbf{T}\, t\, dl = \boldsymbol{q}^T \boldsymbol{T}^e \quad ...[7]$$

where $\boldsymbol{q}^T$ is the usual nodal displacements and $\boldsymbol{T}^e$ for this case is

$$\boldsymbol{T}^e = \frac{t_e}{2}\begin{Bmatrix} 0 \\ 0 \\ l_{23}T_x \\ l_{23}T_y \\ l_{23}T_x \\ l_{23}T_y \end{Bmatrix} \quad ...[8]$$

Example 4: Determine the surface traction vector of the element loaded as shown in Fig. 5.18. The thickness of the element is 2 cm.

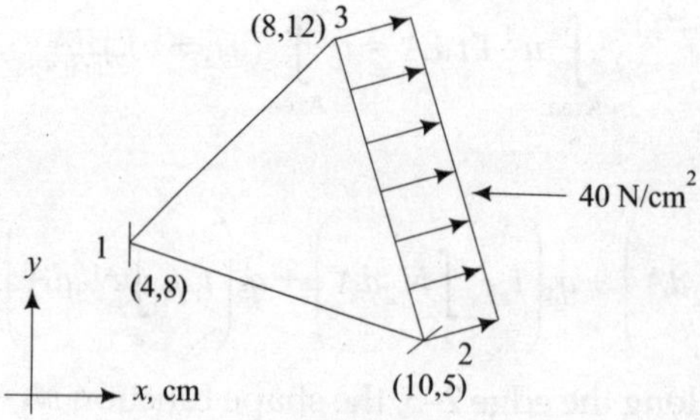

Fig. 5.18

Solution: The surface traction is acting over the side 2–3. Hence,

$$l_{2\text{-}3} = \sqrt{(2-8)^2 + (6-12)^2} = 6.48 \text{ cm.}$$

The components of the surface traction parallel to the x and y axes are given by

$$T_x = 40 \cos\theta = 40(6/6.48) = 37.03 \text{ N/cm}^2.$$

$$T_y = 40 \sin\theta = 40(6/6.48) = 24.70 \text{ N/cm}^2.$$

Substituting the above into eq. [8] we obtain

$$T^e = \frac{2}{2}[0 \;\; 0 \;\; 239.95 \;\; 160.05 \;\; 239.95 \;\; 160.05]^T$$

Example 5: For the element loaded as shown in Fig. 5.19, find the stresses if in addition to the surface traction the element is subjected to a temperature change of 10 °C. Take $t = 2$ cm; $E = 200$ GPa: $\mu = 0.3$; $\alpha = 7 \times 10^{-6}$ cm/cm/°C.

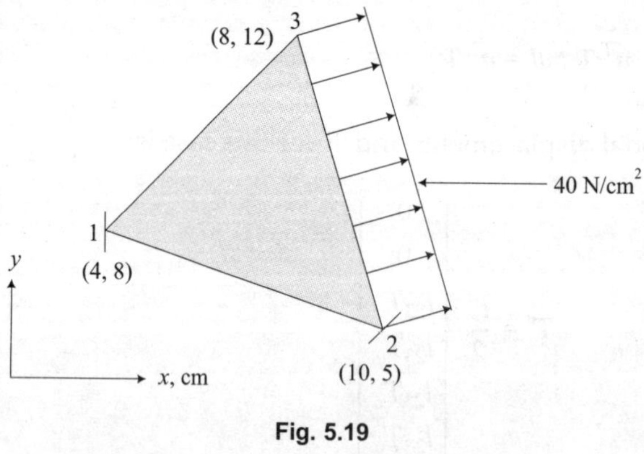

Fig. 5.19

Solution: The stiffness matrix for the element was obtained in the solved example with thickness $t = 1$ cm. Since in this problem $t = 2$ cm, we get for the element stiffness

$$k = \frac{2E}{65.52}\begin{bmatrix} 50.4 & 9.1 & -25.2 & 5.6 & -25.2 & -14.7 \\ 9.1 & 21.15 & 7.4 & -1.8 & -16.5 & -19.35 \\ -25.2 & 7.4 & 21.6 & -10.4 & 3.6 & 3.0 \\ 5.6 & -1.8 & -10.4 & 21.6 & 4.8 & -19.8 \\ -25.2 & -16.5 & 3.6 & 4.8 & 21.6 & 11.7 \\ -14.7 & -19.35 & 3 & -19.8 & 11.7 & 39.15 \end{bmatrix}$$

The traction force is acting over the side 2 – 3. Hence,

$$l_{2-3} = (10-8)^2 + (5-12)^2 = 7.28 \text{ cm.}$$

The x and y components of traction is

$$T_x = 40 \cos\theta = 40(7/7.28) = 38.46$$

$$T_y = 40 \sin\theta = 40(2/7.28) = 10.99$$

Hence, the surface traction vector is

$$T = \frac{t_e}{2}\begin{Bmatrix} 0 \\ 0 \\ l_{23}T_x \\ l_{23}T_y \\ l_{23}T_x \\ l_{23}T_y \end{Bmatrix} = \frac{2}{2}\begin{Bmatrix} 0 \\ 0 \\ 280 \\ 80 \\ 280 \\ 80 \end{Bmatrix}$$

Next, let us calculate the thermal load vector. We have already obtained the **B** matrix for this triangle

$$\mathbf{B} = \frac{1}{36}\begin{bmatrix} -7 & 0 & 4 & 0 & 3 & 0 \\ 0 & -2 & 0 & -4 & 0 & 6 \\ -2 & -7 & -4 & 4 & 6 & 3 \end{bmatrix}$$

and the elasticity matrix **D** for this element is

$$\mathbf{D} = \frac{2\times 10^7}{0.91}\begin{bmatrix} 1 & .3 & 0 \\ .3 & 1 & 0 \\ 0 & 0 & 0.35 \end{bmatrix}$$

Hence,

$$\mathbf{B^T D} = \frac{1}{36}\begin{bmatrix} -7 & 0 & -2 \\ 0 & -2 & -7 \\ 4 & 0 & -4 \\ 0 & -4 & 4 \\ 3 & 0 & 6 \\ 0 & 6 & 3 \end{bmatrix} \frac{2\times 10^7}{0.91}\begin{bmatrix} 1 & .3 & 0 \\ .3 & 1 & 0 \\ 0 & 0 & 0.35 \end{bmatrix}$$

Therefore,

$$F_T^{\,e} = \begin{Bmatrix} T_1 \\ T_2 \\ T_3 \\ T_4 \\ T_5 \\ T_6 \end{Bmatrix} = t_e A_e\,(\mathbf{B}^T \mathbf{D}\,\varepsilon_T) = \begin{Bmatrix} -389 \\ -111.11 \\ 222.22 \\ -222.22 \\ 166.67 \\ 333.33 \end{Bmatrix}$$

The total load vector for this element

$$F = \begin{Bmatrix} -389 \\ -111.11 \\ 502.22 \\ -142.22 \\ 446.67 \\ 413.33 \end{Bmatrix}$$

Hence, the system equation is

$$\frac{2E}{65.52}\begin{bmatrix} 50.4 & 9.1 & -25.2 & 5.6 & -25.2 & -14.7 \\ 9.1 & 21.15 & 7.4 & -1.8 & -16.5 & -19.35 \\ -25.2 & 7.4 & 21.6 & -10.4 & 3.6 & 3.0 \\ 5.6 & -1.8 & -10.4 & 21.6 & 4.8 & -19.8 \\ -25.2 & -16.5 & 3.6 & 4.8 & 21.6 & 11.7 \\ -14.7 & -19.35 & 3 & -19.8 & 11.7 & 39.15 \end{bmatrix}\begin{bmatrix} q_1 \\ q_2 \\ q_3 \\ q_4 \\ q_5 \\ q_6 \end{bmatrix} = \begin{Bmatrix} -389 \\ -111.11 \\ 502.22 \\ -142.22 \\ 446.67 \\ 413.33 \end{Bmatrix}$$

Note that we have only one element which happens to represent the system. Now, that the nodes 1 and 2 are restrained to move, then $q_1 = q_2 = q_3 = q_4 = 0$. Hence, neglecting the rows and columns containing these nodal displacements the above equation reduces to

$$\frac{2E}{65.52}\begin{bmatrix}21.6 & 11.7\\ 11.7 & 39.15\end{bmatrix}\begin{Bmatrix}q_5\\ q_6\end{Bmatrix}=\begin{Bmatrix}446.67\\ 413.33\end{Bmatrix}$$

Solving the above yields

$$q_5 = 292.305 \times 10^{-7}\text{ cm}$$

$$q_6 = 85.691 \times 10^{-7}\text{ cm}$$

Next, the stresses are obtained from using stress-strain relations or by Hooke's law;

$$\begin{Bmatrix}\sigma_x\\ \sigma_y\\ \tau_{xy}\end{Bmatrix}=\frac{2\times 10^7}{0.91}\begin{bmatrix}1 & .3 & 0\\ .3 & 1 & 0\\ 0 & 0 & .35\end{bmatrix}\frac{1}{36}\begin{bmatrix}-7 & 0 & 4 & 0 & 3 & 0\\ 0 & -2 & 0 & -4 & 0 & 6\\ -2 & -7 & -4 & 4 & 6 & 3\end{bmatrix}\begin{Bmatrix}0\\ 0\\ 0\\ 0\\ 292.3\times 10^{-7}\\ 85.61\times 10^{-7}\end{Bmatrix}$$

$$-\frac{2\times 10^7}{0.91}\begin{bmatrix}1 & 0.3 & 0\\ 0.3 & 1 & 0\\ 0 & 0 & 0.35\end{bmatrix}\begin{bmatrix}7\times 10^{-6}\times 10\\ 7\times 10^{-6}\times 10\\ 0\end{bmatrix}$$

$$=\begin{Bmatrix}62.94\\ 47.42\\ 42.96\end{Bmatrix}-\begin{Bmatrix}2000\\ 2000\\ 0\end{Bmatrix}=\begin{Bmatrix}-1937.06\\ -1952.58\\ 42.96\end{Bmatrix}\text{N/cm}^2$$

5.5 GALERKIN'S APPROACH

Let us introduce a virtual displacement field represented by $\phi = [\phi_x, \phi_y]^T$ consistent with the boundary conditions

$$\varepsilon(\phi) = \left[\frac{\partial\phi_x}{\partial x}, \frac{\partial\phi_y}{\partial y}, \frac{\partial\phi_x}{\partial y}+\frac{\partial\phi_y}{\partial x}\right]^T \qquad ...[1]$$

The Galerkin's variation form given in will become for this two-dimensional case

$$\int_A \sigma^T \varepsilon(\phi)\, tdA - \left(\int_A \phi^T f\, tdA + \int_l \phi^T T\, tdl + \sum_i \phi_i^T P_i\right) = 0 \qquad ...[2]$$

The first term on left hand side represents the strain energy or the internal virtual work, while the other terms represent the external virtual work by various loads.

If we discretize the region, the equation [2] becomes

$$\sum_e \int_e \varepsilon^T D\varepsilon(\phi)\, t\, dA - \left(\sum_e \int_e \phi^T f\, tdA + \int_l \phi^T T\, tdl + \sum_i \phi_i^T P_i\right) = 0 \qquad ...[3]$$

where ε is the actual strain and $\varepsilon(\phi)$ is the virtual strain. Now, representing the above in terms of shape function expression and strain–displacement matrix,

i.e.,
$$\phi = N\psi \qquad \text{...[4]}$$
$$\varepsilon(\phi) = B\psi \qquad \text{...[5]}$$
where
$$\psi = [\psi_1, \psi_2, \psi_3, \psi_4, \psi_5, \psi_6]^T \qquad \text{...[6]}$$
is the nodal displacement vector.

Hence, the first term in equation [3] will take the form:

$$\int_e \varepsilon^T D\varepsilon(\phi)\, t\, dA = \int_e q^T B^T DB\, \psi\, t\, dA$$

Noting that all the terms of B and D are constant, and denoting t_e and A_e as thickness and area of element, respectively,
we find

$$\int_e \varepsilon^T D\varepsilon(\phi)\, t\, dA = q^T B^T DB\, t_e \int_e dA\psi$$
$$= q^T t_e A_e B^T DB\, \psi \qquad \text{...[7]}$$
$$= q^T k^e \psi$$

where k^e is the element stiffness matrix given by,

$$\boxed{k^e = t_e A_e B^T DB} \qquad \text{...[8]}$$

The material property matrix D is symmetric, and, hence, the element stiffness matrix is also symmetric.

The Global stiffness is

$$\sum_e \int_e \varepsilon^T D\varepsilon(\phi)\, t\, dA = \sum_e q^T k^e \psi$$
$$= \sum_e \psi^T k^e q$$
$$= \psi^T kQ \qquad \text{...[9]}$$

The global stiffness matrix k is symmetric and banded.

Force Terms

$$\int_e \phi^T f\, t\, dA = \psi^T f^e \qquad \text{...[10]}$$

$$\text{Internal virtual work} = \psi^T kQ \qquad \text{...[11]}$$
$$\text{External virtual work} = \psi^T F \qquad \text{...[12]}$$

From the Galerkin's form (Eq. [3]), the arbitrariness of ψ gives

$$KQ = F \qquad ...[13]$$

where K and F are modified to account for boundary conditions.

5.6 TEMPERATURE EFFECTS

The strains {ε} that we considered earlier was primarily due to displacements. However, there are strains arising due to, for example, temperature variations, $\{\varepsilon_T\}$ and initial strains $\{\varepsilon_0\}$ etc. We will consider here the temperature variations. The strains due to a change in temperature $T(x, y)$ is given as

$$\{\varepsilon_T\} = \begin{Bmatrix} \alpha T \\ \alpha T \\ 0 \end{Bmatrix} = [\alpha T \ \ \alpha T \ \ 0]^T \qquad ...[1]$$

where, α is the coefficient of thermal expansion. Eq. (i) is for plane stress case. For, plane strain the expression for strains will be

$$\{\varepsilon_T\} = (1+\mu)\begin{Bmatrix} \alpha T \\ \alpha T \\ 0 \end{Bmatrix} = (1+\mu)\,[\alpha_T \ \ \alpha_T \ \ 0]^T \qquad ...[2]$$

The stress-strain relation is, now, given by

$$\sigma = D(\varepsilon + \varepsilon_T) \qquad ...[3]$$

which can also be written as

$$\sigma = \mathbf{D}(\mathbf{B}\, q + \varepsilon_T) \qquad ...[4]$$

Now, the strain energy can be expressed as

$$U = \frac{1}{2}\int (\varepsilon + \varepsilon_T)^T D(\varepsilon + \varepsilon_T)t\, dA$$

$$= \frac{1}{2}\int (\varepsilon^T D\varepsilon + 2\varepsilon^T D\varepsilon_T + \varepsilon^T{}_T D\varepsilon_T)t\, dA \qquad ...[5]$$

The first term on the right-hand side of eq. (5) is the known familiar stiffness matrix expression. The third term is a constant quantity which will be eliminated during minimization process. The only additional contribution will be due to the second term which can be written as, using the relation $\varepsilon = \mathbf{B}\,\mathbf{q}$

$$\int \varepsilon^T \mathbf{D}\, \varepsilon_T\, tdA = \sum_{NE} q^T (\mathbf{B}^T \mathbf{D}\, \varepsilon_T) t_e A_e = \sum_{NE} q^T F_T^e \qquad ...[6]$$

In the above, F_T^e represents the thermal load where

$$F_T^e = (\mathbf{B}^T \mathbf{D}\, \varepsilon_T) t_e A_e = [T_1 \ \ T_2 \ \ T_3 \ \ T_4 \ \ T_5 \ \ T_6]^T \qquad ...[7]$$

Example 5: The element shown in Fig. 5.20 is subjected to a temperature change 10°C. Its coefficient of thermal expansion, $\alpha = 7 \times 10^{-6}$ cm/(cm)(°C). The thickness of the element is 2 cm; Poisson's ratio =

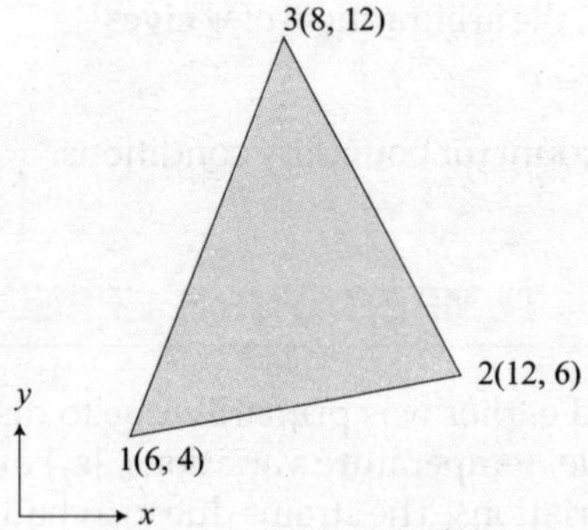

Fig. 5.20

0.3 and Young's modulus $E = 200$ GPa. Find the load due to temperature change. Assume plane stress condition.

Solution: Area, A of the element is

$$A = \frac{1}{2}[12\times12-8\times6+6\times6-6\times12+8\times4-4\times12] = 22\ \text{cm}^2$$

The **B** matrix for the element is obtained as

$$\mathbf{B} = \frac{1}{\det J}\begin{bmatrix} -6 & 0 & 8 & 0 & -2 & 0 \\ 0 & -4 & 0 & -2 & 0 & 6 \\ -4 & -6 & -2 & 8 & 6 & -2 \end{bmatrix} = \frac{1}{44}\begin{bmatrix} -6 & 0 & 8 & 0 & -2 & 0 \\ 0 & -4 & 0 & -2 & 0 & 6 \\ -4 & -6 & -2 & 8 & 6 & -2 \end{bmatrix}$$

The **D** matrix for the plane stress condition from the equation shown is obtained as

$$\left\{\because D = \frac{E}{1-\mu^2}\begin{bmatrix} 1 & \mu & 0 \\ \mu & 1 & 0 \\ 0 & 0 & \frac{1-\mu}{2} \end{bmatrix}\right\}$$

$$\mathbf{D} = \frac{2\times10^7}{(1-.3^2)}\begin{bmatrix} 1 & .3 & 0 \\ .3 & 1 & 0 \\ 0 & 0 & \frac{1-.3}{2} \end{bmatrix} = \frac{2\times10^7}{0.91}\begin{bmatrix} 1 & .3 & 0 \\ .3 & 1 & 0 \\ 0 & 0 & 0.35 \end{bmatrix}$$

Next, let us calculate $\mathbf{B}^T\mathbf{D}$

$$\mathbf{B}^T\mathbf{D} = \frac{1}{44}\begin{bmatrix} -6 & 0 & -4 \\ 0 & -4 & -6 \\ 8 & 0 & -2 \\ 0 & -2 & 8 \\ -2 & 0 & 6 \\ 0 & 6 & -2 \end{bmatrix}\frac{2\times10^7}{0.91}\begin{bmatrix} 1 & .3 & 0 \\ .3 & 1 & 0 \\ 0 & 0 & .35 \end{bmatrix} = \frac{10^7}{20.02}\begin{bmatrix} -6 & -1.8 & -1.4 \\ -1.2 & -4 & -2.1 \\ 8 & 2.4 & -0.7 \\ -0.6 & -2 & 2.8 \\ -2 & -0.6 & 2.1 \\ 1.8 & 6 & -0.7 \end{bmatrix}$$

Hence, the thermal load vector is given by

$$F_T^e = \begin{Bmatrix} T_1 \\ T_2 \\ T_3 \\ T_4 \\ T_5 \\ T_6 \end{Bmatrix} = t_e A_e B^T D\, \varepsilon_T = \frac{2 \times 22 \times 10^7}{20.02} \begin{bmatrix} -6 & -1.8 & -1.4 \\ -1.2 & -4 & -2.1 \\ 8 & 2.4 & -0.7 \\ -0.6 & -2 & 2.8 \\ -2 & -0.6 & 2.1 \\ 1.8 & 6 & -0.7 \end{bmatrix} \begin{Bmatrix} 7 \times 10^{-6} \times 10 \\ 7 \times 10^{-6} \times 10 \\ 0 \end{Bmatrix} = \begin{Bmatrix} -36923 \\ -24615 \\ 16000 \\ -4143 \\ -4143 \\ 10836 \end{Bmatrix}$$

EXERCISES

1. What are area coordinates ?
2. What is an isoparametric element ? Give an example.
3. Express the shape functions in terms of the global coordinates (x_1, y_1) etc.
4. What is Jacobian ?
5. Derive the strain-displacement matrix **B** for the triangular element.
6. Define the **D** matrix for the plane stress and plane strain cases.
7. Evaluate the surface traction force term in equation:

   $$\Pi = \sum_{NE} U_e - \sum_{NE} \int_{Area} u^T f\, t\, dA - \int_{Line} u^T T dl - \sum_i u_i^T P_i$$

8. Evaluate the body force term in equation

   $$\Pi = \sum_{NE} U_e - \sum_{NE} \int_{Area} u^T f\, t\, dA - \int_{Line} u^T T dl - \sum_i u_i^T P_i$$

9. What are line coordinates?
10. Find the shape function at point P as shown in Fig. P5.10.

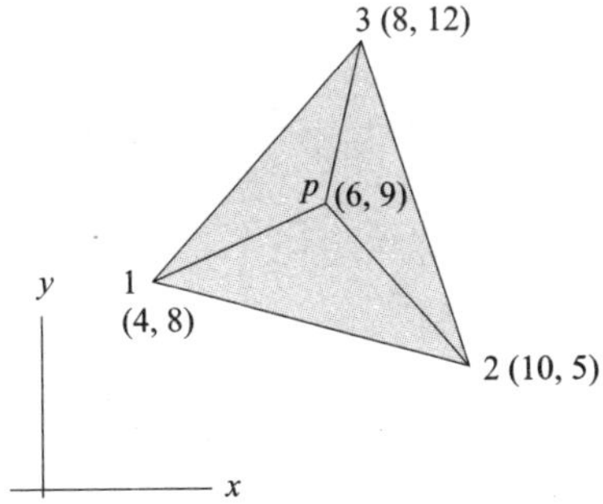

Fig. P5.10

11. Explain how you would incorporate the effect due to temperature change.

6

Three-dimensional Finite Elements

6.1 INTRODUCTION

A three-dimensional analysis is an ultimate by its nature but the definition and description of the system is quite voluminous. Each of the elements of 3*D* system that is idealised has a large number of nodes. Volume, centre of mass, moment of inertia and other geometric properties of rigid homogeneous solids are defined by triple integrals over domains of three-dimensional Euclidean space. Most of the engineering problems are three-dimensional.

The displacements along x, y and z directions will be represented by

$$u = [u, v, w]^T \qquad ...[1]$$

where u, v and w are displacements along x, y and z directions, respectively. The three-dimensional stress and strain states are represented by

$$\sigma = [\sigma_x, \sigma_y, \sigma_z, \tau_{yz}, \tau_{xz}, \tau_{xy}]^T \qquad ...[2]$$

$$\varepsilon = [\varepsilon_x, \varepsilon_y, \varepsilon_z, \gamma_{yz}, \gamma_{xz}, \gamma_{xy}]^T \qquad ...[3]$$

The strain–displacement relations are given by

$$\varepsilon = \left[\frac{\partial u}{\partial x}, \frac{\partial v}{\partial y}, \frac{\partial w}{\partial z}, \frac{\partial v}{\partial z} + \frac{\partial w}{\partial y}, \frac{\partial u}{\partial z} + \frac{\partial w}{\partial x}, \frac{\partial u}{\partial y} + \frac{\partial v}{\partial x}\right]^T \qquad ...[4]$$

The body force and traction vector are given by

$$\mathbf{f} = [f_x, f_y, f_z]^T \qquad ...[5]$$

$$\mathbf{T} = [T_x, T_y, T_z]^T \qquad ...[6]$$

The stress-strain relations are given by

$$\sigma = \mathbf{D}\varepsilon \qquad ...[7]$$

where **D** is a (6×6) symmetric matrix. For isotropic materials **L** is given by the equation:

$$\mathbf{D} = \frac{E}{(1+\nu)(1-2\nu)}\begin{bmatrix} 1-\nu & \nu & \nu & 0 & 0 & 0 \\ \nu & 1-\nu & \nu & 0 & 0 & 0 \\ \nu & \nu & 1-\nu & 0 & 0 & 0 \\ 0 & 0 & 0 & 0.5-\nu & 0 & 0 \\ 0 & 0 & 0 & 0 & 0.5-\nu & 0 \\ 0 & 0 & 0 & 0 & 0 & 0.5-\nu \end{bmatrix} \quad ...[8]$$

Note: The total potential and the Galerkin virtual work form for three dimensions are given in the chapter "Fundamental Concepts".

Though there are a number of three-dimensional elements but the most prime type are based on the shapes of a rectangular parallelopiped, tetrahedran and triangular prisms. All others are basically of same type differing only in the order of parametric variation/number and distribution of nodes taken, as shown in Fig. 6.1.

6.2 FINITE ELEMENT FORMULATION

In three-dimensional case, the tetrahedral family (Fig. 6.2) exhibits the properties similar to those of the triangular family of elements in 2D.

The first formation of a simple tetrahedral element was by Gallagher, earlier elaborations were by Melosh, Argyris and Clough among others. Considering the tetrahedral element (master element) shown in Fig. 6.3, the displacements u, v and w at any point (x, y, z) can be represented in terms of the shape functions as:

$$N_1 = \xi \quad N_2 = \eta \quad N_3 = \zeta \quad N_4 = 1 - \xi - \eta - \zeta \quad ...[1]$$

The displacements u, v, w at $\mathbf{x}$ can be written in terms of the unknown nodal values as

$$\mathbf{u} = \mathbf{Nq} \quad ...[2]$$

where

$$\mathbf{N} = \begin{bmatrix} N_1 & 0 & 0 & N_2 & 0 & 0 & N_3 & 0 & 0 & N_4 & 0 & 0 \\ 0 & N_1 & 0 & 0 & N_2 & 0 & 0 & N_3 & 0 & 0 & N_4 & 0 \\ 0 & 0 & N_1 & 0 & 0 & N_2 & 0 & 0 & N_3 & 0 & 0 & N_4 \end{bmatrix} \quad ...[3]$$

Now, by using equation [1], we can define the coordinates x, y, z of the point at which the displacements u, v and w are interpolated.

The isoparametric transformation is given by

$$\begin{aligned} x &= N_1x_1 + N_2x_2 + N_3x_3 + N_4x_4 \\ y &= N_1y_1 + N_2y_2 + N_3y_3 + N_4y_4 \\ z &= N_1z_1 + N_2z_2 + N_3z_3 + N_4z_4 \end{aligned} \quad ...[4]$$

which, on substituting for N_i from Eq. [1] and using the notation $x_{ij} = x_i - x_j$, $y_{ij} = y_i - y_j$, $z_{ij} = z_i - z_j$, yields

$$x = x_4 + x_{14}\xi + x_{24}\eta + x_{34}\zeta$$
$$y = y_4 + y_{14}\xi + y_{24}\eta + y_{34}\zeta \qquad ...[5]$$
$$z = z_4 + z_{14}\xi + z_{24}\eta + z_{34}\zeta$$

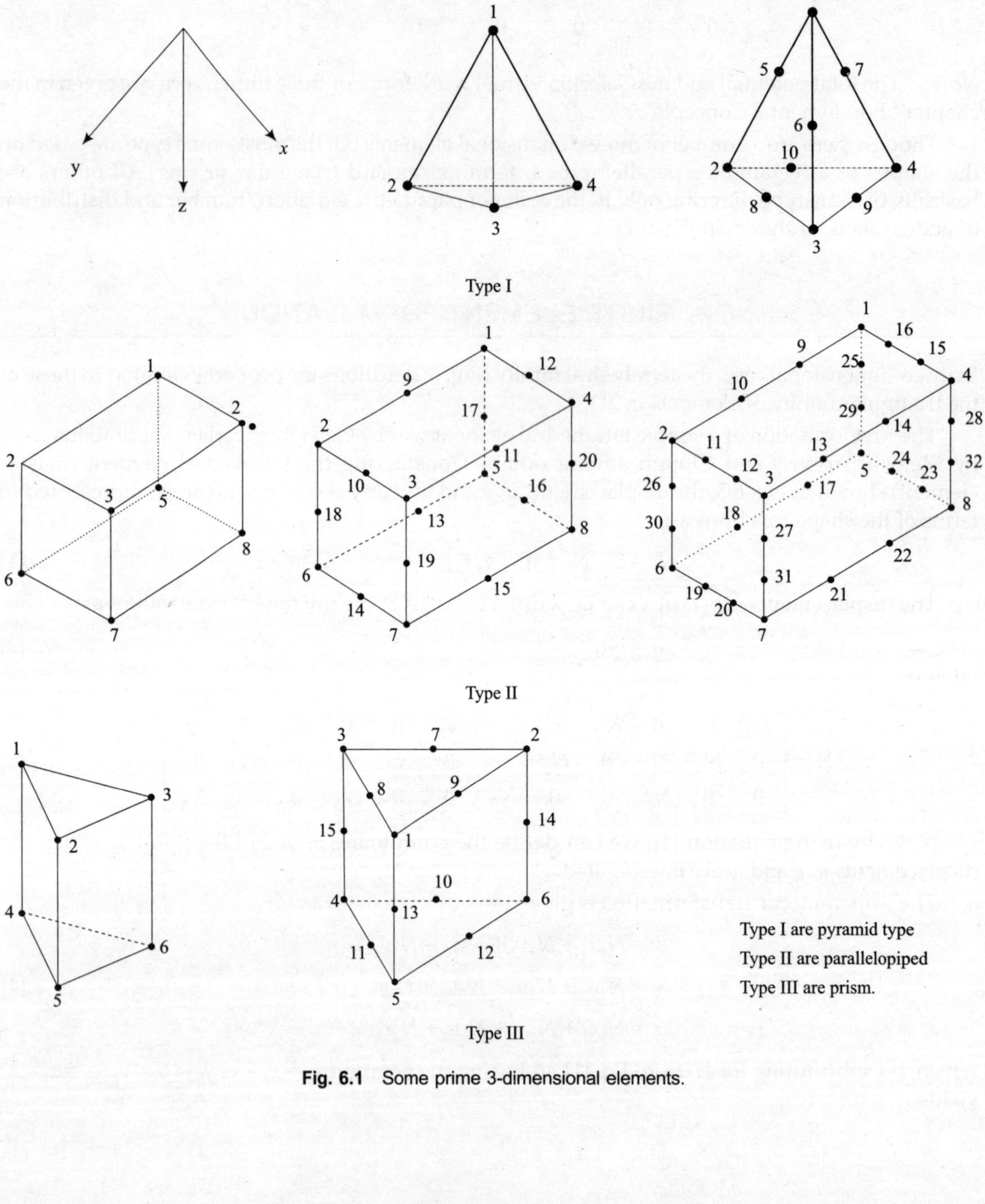

Fig. 6.1 Some prime 3-dimensional elements.

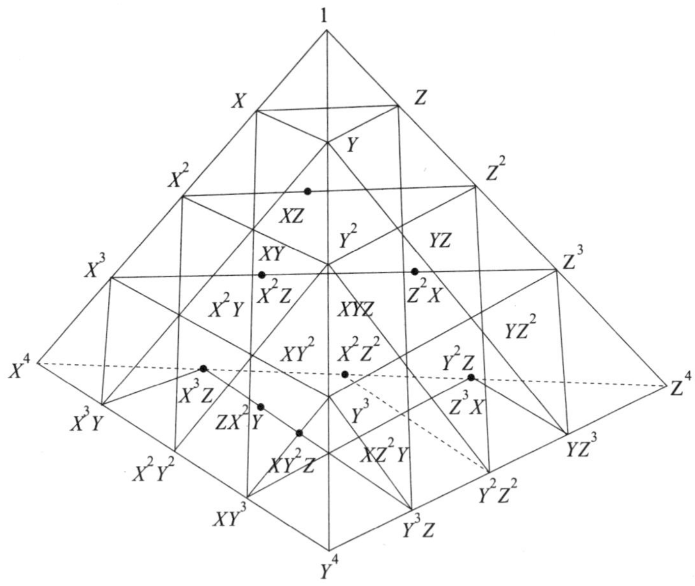

Fig. 6.2 Tetrahedron element

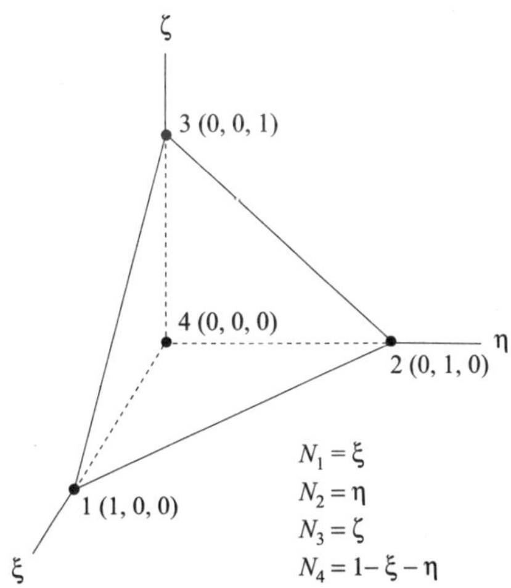

Fig. 6.3 Master element for shape functions

Jacobian for the Tetrahedron Element

We know that Jacobian gives the relation between the derivatives in the global and natural coordinate systems. To this end we need the derivatives.

Let $u = u(x, y, z)$. Then, derivative of this function w. r. t. natural coordinates yields,

$$\frac{\partial u}{\partial \xi} = \frac{\partial u}{\partial x} \cdot \frac{\partial x}{\partial \xi} + \frac{\partial u}{\partial y} \cdot \frac{\partial y}{\partial \xi} + \frac{\partial u}{\partial z} \cdot \frac{\partial z}{\partial \xi} \qquad \text{...[1a]}$$

$$\frac{\partial u}{\partial \eta} = \frac{\partial u}{\partial x} \cdot \frac{\partial x}{\partial \eta} + \frac{\partial u}{\partial y} \cdot \frac{\partial y}{\partial \eta} + \frac{\partial u}{\partial z} \cdot \frac{\partial z}{\partial \eta} \qquad \text{...[1b]}$$

$$\frac{\partial u}{\partial \zeta} = \frac{\partial u}{\partial x} \cdot \frac{\partial x}{\partial \zeta} + \frac{\partial u}{\partial y} \cdot \frac{\partial y}{\partial \zeta} + \frac{\partial u}{\partial z} \cdot \frac{\partial z}{\partial \zeta} \qquad \text{...[1c]}$$

The above equations are called chain rule on partial differentiation.

Using the chain rule for partial derivatives, say of u, we have

$$\begin{Bmatrix} \dfrac{\partial u}{\partial \xi} \\ \dfrac{\partial u}{\partial \eta} \\ \dfrac{\partial u}{\partial \zeta} \end{Bmatrix} = \mathbf{J} \begin{Bmatrix} \dfrac{\partial u}{\partial z} \\ \dfrac{\partial u}{\partial y} \\ \dfrac{\partial u}{\partial z} \end{Bmatrix} \qquad \text{...[2]}$$

Thus, the partial derivatives with respect to ξ, η, ζ are related to x, y, z derivatives by the above relationship. The Jacobian **J** of the transformation is given by

$$\mathbf{J} = \begin{Bmatrix} \dfrac{\partial x}{\partial \xi} & \dfrac{\partial y}{\partial \xi} & \dfrac{\partial z}{\partial \xi} \\ \dfrac{\partial x}{\partial \eta} & \dfrac{\partial y}{\partial \eta} & \dfrac{\partial z}{\partial \eta} \\ \dfrac{\partial x}{\partial \zeta} & \dfrac{\partial y}{\partial \zeta} & \dfrac{\partial z}{\partial \zeta} \end{Bmatrix} = \begin{bmatrix} x_{14} & y_{14} & z_{14} \\ x_{24} & y_{24} & z_{24} \\ x_{34} & y_{34} & z_{34} \end{bmatrix} \qquad \text{...[3]}$$

We note here that

$$\det \mathbf{J} = x_{14}(y_{24}z_{34} - y_{34}z_{24}) + y_{14}(z_{24}x_{34} - z_{34}x_{24}) + z_{14}(x_{24}y_{34} - x_{34}y_{24})$$

Volume of the Element

The volume of the element is given by

$$V_e = \left| \int_0^1 \int_0^{1-\xi} \int_0^{1-\xi-\eta} \det \mathbf{J}\, d\xi\, d\eta\, d\zeta \right| \qquad \text{...[1]}$$

Since det **J** is constant,

$$V_e = |\det \mathbf{J}| \int_0^1 \int_0^{1-\xi} \int_0^{1-\xi-\eta} d\xi\, d\eta\, d\zeta \qquad \text{...[2]}$$

Using the polynomial integral formula

$$\int_0^1 \int_0^{1-\xi} \int_0^{1-\xi-\eta} \xi^m \eta^n \zeta^p\, d\xi\, d\eta\, d\zeta = \frac{m!\, n!\, p!}{(m+n+p+3)!} \qquad \text{...[3]}$$

we get

$$V_e = \frac{1}{6}\, |\det \mathbf{J}| \qquad \text{...[4]}$$

The inverse relation corresponding to the equation:

$$\begin{Bmatrix} \dfrac{\partial u}{\partial \xi} \\ \dfrac{\partial u}{\partial \eta} \\ \dfrac{\partial u}{\partial \zeta} \end{Bmatrix} = \mathbf{J} \begin{Bmatrix} \dfrac{\partial u}{\partial x} \\ \dfrac{\partial u}{\partial y} \\ \dfrac{\partial u}{\partial z} \end{Bmatrix} \quad \text{is given by}$$

$$\begin{Bmatrix} \dfrac{\partial u}{\partial x} \\ \dfrac{\partial u}{\partial y} \\ \dfrac{\partial u}{\partial z} \end{Bmatrix} = \mathbf{A} \begin{Bmatrix} \dfrac{\partial u}{\partial \xi} \\ \dfrac{\partial u}{\partial \eta} \\ \dfrac{\partial u}{\partial \zeta} \end{Bmatrix} \qquad \text{...[5]}$$

where **A** is the inverse of the Jacobian **J** given in equation:

$$\mathbf{J} = \begin{bmatrix} \dfrac{\partial x}{\partial \xi} & \dfrac{\partial y}{\partial \xi} & \dfrac{\partial z}{\partial \xi} \\ \dfrac{\partial x}{\partial \eta} & \dfrac{\partial y}{\partial \eta} & \dfrac{\partial z}{\partial \eta} \\ \dfrac{\partial x}{\partial \zeta} & \dfrac{\partial y}{\partial \zeta} & \dfrac{\partial z}{\partial \zeta} \end{bmatrix} = \begin{bmatrix} x_{14} & y_{14} & z_{14} \\ x_{24} & y_{24} & z_{24} \\ x_{34} & y_{34} & z_{34} \end{bmatrix} \qquad \text{...[6]}$$

$$\mathbf{A} = \mathbf{J}^{-1} = \frac{1}{\det \mathbf{J}} \begin{bmatrix} y_{24}z_{34} - y_{34}z_{24} & y_{34}z_{14} - y_{14}z_{34} & y_{14}z_{24} - y_{24}z_{14} \\ z_{24}x_{34} - z_{34}x_{24} & z_{34}x_{14} - z_{14}x_{34} & z_{14}x_{24} - z_{24}x_{14} \\ x_{24}y_{34} - x_{34}y_{24} & x_{34}y_{14} - x_{14}y_{34} & x_{14}y_{24} - x_{24}y_{14} \end{bmatrix} \qquad \text{...[7]}$$

Now, using the strain–displacement relations in equation,

$$\varepsilon = \left[\frac{\partial u}{\partial x}, \frac{\partial v}{\partial y}, \frac{\partial w}{\partial z}, \frac{\partial v}{\partial z} + \frac{\partial w}{\partial y}, \frac{\partial u}{\partial z} + \frac{\partial w}{\partial x}, \frac{\partial u}{\partial y} + \frac{\partial v}{\partial x} \right]^T \qquad \text{...[8]}$$

the relation between derivatives in x, y, z and ξ, η, ζ in equation [5] above and the assumed displacement field in equation $u = Nq$, we get

$$\varepsilon = Bq \qquad \text{...[9]}$$

where **B** is a (6×12)matrix given by

$$\mathbf{B} = \begin{bmatrix} A_{11} & 0 & 0 & A_{12} & 0 & 0 & A_{13} & 0 & 0 & -\tilde{A}_1 & 0 & 0 \\ 0 & A_{21} & 0 & 0 & A_{22} & 0 & 0 & A_{23} & 0 & 0 & -\tilde{A}_2 & 0 \\ 0 & 0 & A_{31} & 0 & 0 & A_{32} & 0 & 0 & A_{33} & 0 & 0 & -\tilde{A}_3 \\ 0 & A_{31} & A_{21} & 0 & A_{32} & A_{22} & 0 & A_{33} & A_{23} & 0 & -\tilde{A}_3 & -\tilde{A}_2 \\ A_{31} & 0 & A_{11} & A_{32} & 0 & A_{12} & A_{33} & 0 & A_{13} & -\tilde{A}_3 & 0 & -\tilde{A}_1 \\ A_{21} & A_{11} & 0 & A_{22} & A_{12} & 0 & A_{23} & A_{13} & 0 & -\tilde{A}_2 & -\tilde{A}_1 & 0 \end{bmatrix} \qquad \text{...[10]}$$

where $\tilde{A}_1 = A_{11} + A_{12} + A_{13}$, $\tilde{A}_2 = A_{21} + A_{22} + A_{23}$, and $\tilde{A}_3 = A_{31} + A_{32} + A_{33}$. All the terms of **B** are constants. Thus, equation $\varepsilon = Bq$ gives constant strains after the nodal displacements are calculated.

Element Stiffness

Since $U_e = \frac{1}{2}\int_e \varepsilon^T D\varepsilon\, dV$ is the element strain energy in the total potential energy.

Or
$$U_e = \frac{1}{2} q^T B^T DBq \int_e dV \quad \text{...[1]}$$

(where $\varepsilon = Bq$)

$$= \frac{1}{2} q^T V_e B^T DB\, q$$

$$= \frac{1}{2} q^T k^e q$$

where the element stiffness k_e is given by

$$k^e = V_e B^T DB \quad \text{...[2]}$$

where V_e is the volume of the element given by $\frac{1}{6}\,|\det \mathbf{J}|$.

In the Galerkin's approach, we have

Internal virtual work is:
$$\int_e \sigma^T \varepsilon(\phi)\, dV = \psi^T V_e B^T DBq$$

which gives the element stiffness in equation:

$$k^e = V_e B^T DB$$

Force Terms

(i) *Body force*

$$\int_e u^T f dV = q^T \iiint \mathbf{N}^T f \det \mathbf{J}\, d\xi\, d\eta\, d\zeta \quad \text{...[1]}$$

$$= q^T f^e$$

Now, using

$$\int_0^1\int_0^{1-\xi}\int_0^{1-\xi-\eta} \xi^n \eta^n \zeta^n\, d\xi\, d\eta\, d\zeta = \frac{m!\,n!\,p!}{(m+n+p+3)!} \quad \text{...[2]}$$

we have

$$f^e = \frac{V_e}{4}\,[f_x, f_y, f_z, f_x, f_y, f_z, \ldots f_z]^T \quad \text{...[3]}$$

f^e is of dimension 12×1.

$V_e f_x$ is the x component of the body force which is distributed to the degrees of freedom q_1, q_4, q_7 and q_{10}.

(ii) *Surface traction force*

$$\int_{A_e} u^T T\,dA = q^T \int_{A_e} N^T T dA = q^T T^e \qquad \text{...[4]}$$

The element traction force vector T^e is given by

$$T^e = \frac{A_e}{3}[T_x, T_y, T_z, \ldots T_z, 0, 0, 0] \qquad \text{...[5]}$$

The energy and Galerkin's approaches yield the set of equations

$$KQ = F \qquad \text{...[6]}$$

6.3 STRESS CALCULATIONS

It may be observed from equation [6] that the energy and Galerkin's approaches yield the set of equations $KQ = F$

On solving equation [1], the nodal displacements (q) can be obtained.

Also, we have $\sigma = D\varepsilon$

and $\varepsilon = Bq$

$\Rightarrow$ $\sigma = DBq$ (the element stresses) ...[1]

Principal stresses are computed using the equation given below:

The 3 invariants of the (3×3) stress tensor are

$$\left\{\begin{aligned} I_1 &= \sigma_x + \sigma_y + \sigma_z \\ I_2 &= \sigma_x\sigma_y + \sigma_y\sigma_z + \sigma_x\sigma_z - \tau^2_{yz} - \tau^2_{xz} - \tau^2_{xy} \\ I_3 &= \sigma_x\sigma_y\sigma_z + 2\tau_{yz}\,\tau_{xz}\,\tau_{xy} - \sigma_x\tau^2_{yz} - \sigma_y\tau^2_{xz} - \sigma_z\tau^2_{xy} \end{aligned}\right\} \qquad \text{...[2]}$$

Define:

$$\left\{\begin{aligned} a &= \frac{I_1^2}{3} - I_2 \\ b &= -2\left(\frac{I_1}{3}\right)^3 + \frac{I_1 I_2}{3} - I_3 \\ c &= 2\sqrt{\frac{a}{3}} \\ \theta &= \frac{1}{3}\cos^{-1}\left(-\frac{3b}{ac}\right) \end{aligned}\right\} \qquad \text{...[3]}$$

The principal stresses are given by

$$\sigma_1 = \frac{I_1}{3} + c\cos\theta$$

$$\sigma_2 = \frac{I_1}{3} + c\cos\left(\theta + \frac{2\pi}{3}\right)$$

$$\sigma_3 = \frac{I_1}{3} + c\cos\left(\theta + \frac{4\pi}{3}\right) \qquad ...[4]$$

TETRAHEDRAL FAMILY

We wish to discretize the 3-dimensional region R using tetrahedron elements. We consider an arbitrary tetrahedron element (e) with four nodes labelled 1, 2, 3 and 4 as shown in Fig. 6.4(a). The nodes 1, 2 and 3 are labelled in a counter clockwise sequence when viewed from the node 4.

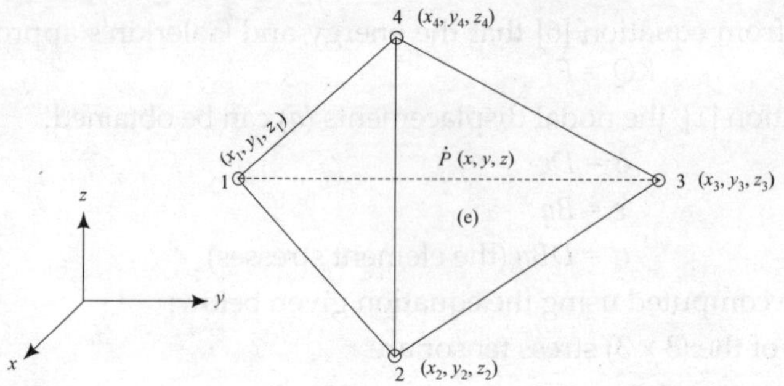

Fig. 6.4(a) Tetrahedron in Cartesian coordinates

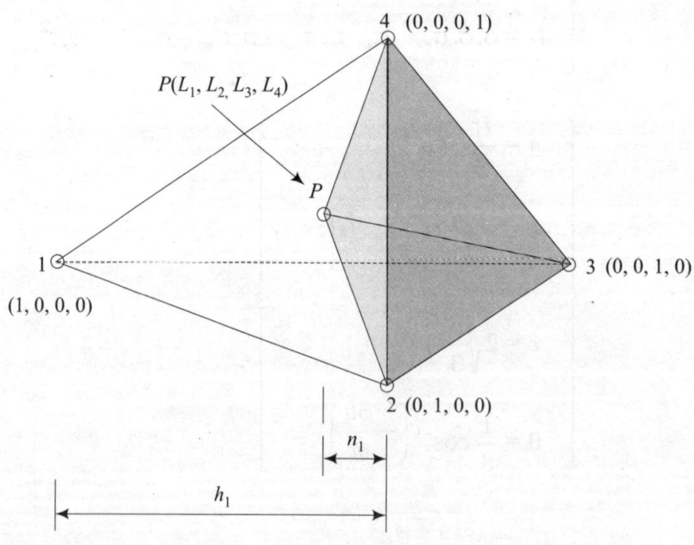

Fig. 6.4(b) Tetrahedron in natural coordinates or volume coordinate.

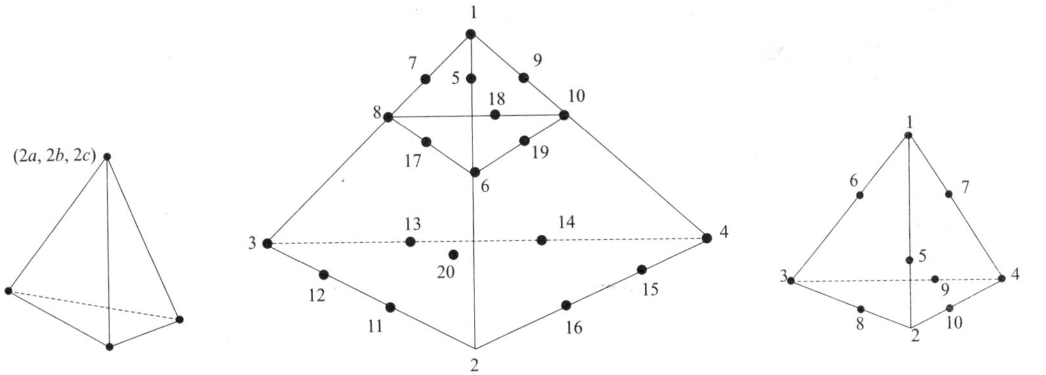

(a) 4 noded linear tetrahedron (b) 10 noded quadratic tetrahedron (c) 20 noded cubic tetrahedron

Fig. 6.5

From the Figs. 6.4(a) and (b) the volume of the tetrahedron is given by

$$V = \frac{1}{6}\begin{Vmatrix} 1 & 1 & 1 & 1 \\ x_1 & x_2 & x_3 & x_4 \\ y_1 & y_2 & y_3 & y_4 \\ z_1 & z_2 & z_3 & z_4 \end{Vmatrix} \qquad ...[1]$$

The location of any point $P(x, y, z) \in (e)$ can be defined in terms of the volume coordinates (L_1, L_2, L_3, L_4) given by

$$L_1 = \frac{Vol\ P\ 234}{V},\ L_2 = \frac{Vol\ P\ 341}{V},\ L_3 = \frac{Vol\ P\ 412}{V},\ L_4 = \frac{Vol\ P\ 123}{V} \qquad ...[2]$$

The volume coordinate system is also known as local or natural coordinate system.

(i) Relation between Cartesian and tetrahedron coordinates :

$$\begin{Bmatrix} 1 \\ x \\ y \\ z \end{Bmatrix} = \begin{bmatrix} 1 & 1 & 1 & 1 \\ x_1 & x_2 & x_3 & x_4 \\ y_1 & y_2 & y_3 & y_4 \\ z_1 & z_2 & z_3 & z_4 \end{bmatrix} \begin{Bmatrix} L_1 \\ L_2 \\ L_3 \\ L_4 \end{Bmatrix} \qquad ...[3]$$

$$\Rightarrow \quad \begin{Bmatrix} L_1 \\ L_2 \\ L_3 \\ L_4 \end{Bmatrix} = \frac{1}{6V}\begin{bmatrix} V_1 & a_1 & b_1 & c_1 \\ V_2 & a_2 & b_2 & c_2 \\ V_3 & a_3 & b_3 & c_3 \\ V_4 & a_4 & b_4 & c_4 \end{bmatrix} \begin{Bmatrix} 1 \\ x \\ y \\ z \end{Bmatrix} \qquad ...[4]$$

where V is given by eq. [1]. V_1, V_2, V_3, V_4 are the sub-volume subtended from face i and the terms a_i, b_i, c_i represent the projected area of face i on the x, y, z coordinate planes respectively.

The cofactors of x, y, z (Cartesian coordinates) in equation [4] are given by:
The cofactor of x_1 in eq. [1] is a_1

$$a_1 = -\begin{vmatrix} 1 & 1 & 1 \\ y_2 & y_3 & y_4 \\ z_2 & z_3 & z_4 \end{vmatrix} \quad \text{(along column)}$$

The cofactor of x_2 in eq. [1] is a_2

$$a_2 = -\begin{vmatrix} 1 & 1 & 1 \\ y_1 & y_3 & y_4 \\ z_1 & z_3 & z_4 \end{vmatrix}$$

The cofactor of x_3 in eq. [1] is a_3

$$a_3 = -\begin{vmatrix} 1 & 1 & 1 \\ y_1 & y_2 & y_4 \\ z_1 & z_2 & z_4 \end{vmatrix} \text{ and so on.} \qquad ...[5]$$

Obviously, we can find the partial derivatives of the tetrahedron coordinate w.r.t. Cartesian coordinates, that is, they are

$$\frac{\partial L_1}{\partial x} = \frac{a_1}{6V}; \frac{\partial L_2}{\partial x} = \frac{a_2}{6V}; \frac{\partial L_3}{\partial x} = \frac{a_3}{6V}; \frac{\partial L_4}{\partial x} = \frac{a_4}{6V} \qquad ...[6]$$

Similarly, $$\frac{\partial L_i}{\partial y} = \frac{b_i}{6V}; \text{ and } \frac{\partial L_i}{\partial z} = \frac{c_i}{6V}$$

Note: (1) We can easily evaluate the integrals of product tetrahedron coordinates, taken over the volume of the tetrahedron element, which is one of the advantages of the tetrahedron coordinates.

Note: (2) Interpolation function defines a linear variation from one node to every other node, in the constant strain tetrahedron.

$$\begin{Bmatrix} \bar{u} \\ \bar{v} \\ \bar{w} \end{Bmatrix} = \begin{bmatrix} \{N\} & \{0\} & \{0\} \\ \{0\} & \{N\} & \{0\} \\ \{0\} & \{0\} & \{N\} \end{bmatrix} \begin{Bmatrix} u \\ v \\ w \end{Bmatrix}$$

$$= [N] \begin{Bmatrix} \overline{U} \\ \overline{V} \\ \overline{W} \end{Bmatrix} \qquad ...[1]$$

where $$\{N\} = [L_1 \; L_2 \; L_3 \; L_4]$$

$$\{u\} = [u_1 \; u_2 \; u_3 \; u_4]$$

$$\{v\} = [v_1\ v_2\ v_3\ v_4]$$

$$\{w\} = [w_1\ w_2\ w_3\ w_4]$$

Element Stiffness *K*: for Tetrahedron Element

The element stiffness for a tetrahedron element represents the nodal force vectors in the element resulting from unit nodal displacements.

For a three-dimensional solid object, the strain-displacement may be written as

$$\begin{Bmatrix} \varepsilon_x \\ \varepsilon_y \\ \varepsilon_z \\ \nu_{xy} \\ \nu_{yz} \\ \nu_{zx} \end{Bmatrix} = \begin{Bmatrix} \dfrac{\partial u}{\partial x} \\ \dfrac{\partial v}{\partial y} \\ \dfrac{\partial w}{\partial z} \\ \dfrac{\partial u}{\partial y} + \dfrac{\partial v}{\partial x} \\ \dfrac{\partial v}{\partial z} + \dfrac{\partial w}{\partial y} \\ \dfrac{\partial u}{\partial z} + \dfrac{\partial w}{\partial x} \end{Bmatrix} = \frac{1}{6V} \begin{bmatrix} \{a_i\} & \{0\} & \{0\} \\ \{0\} & \{b_i\} & \{0\} \\ \{0\} & \{0\} & \{c_i\} \\ \{b_i\} & \{a_i\} & \{0\} \\ \{0\} & \{c_i\} & \{b_i\} \\ \{c_i\} & \{0\} & \{a_i\} \end{bmatrix} \begin{Bmatrix} U \\ V \\ W \end{Bmatrix} \quad ...[1]$$

in which

$$\{a_i\} = \{\, a_1\ a_2\ a_3\ a_4 \,\}$$

$$\{b_i\} = \{\, b_1\ b_2\ b_3\ b_4 \,\} \quad ...[2]$$

$$\{c_i\} = \{\, c_1\ c_2\ c_3\ c_4 \,\}$$

The stress-strain relationship for the elastic isotropic may be written as :

$$\{\sigma\} = [D]\,\{\varepsilon\} \quad ...[3]$$

where

$$[D] = \frac{E}{(1+\nu)(1-2\nu)} \begin{bmatrix} (1-\nu) & \nu & \nu & 0 & 0 & 0 \\ \nu & (1-\nu) & \nu & 0 & 0 & 0 \\ \nu & \nu & (1-\nu) & 0 & 0 & 0 \\ 0 & 0 & 0 & \dfrac{(1-2\nu)}{2} & 0 & 0 \\ 0 & 0 & 0 & 0 & \dfrac{(1-2\nu)}{2} & 0 \\ 0 & 0 & 0 & 0 & 0 & \dfrac{(1-2\nu)}{2} \end{bmatrix} \quad ...[4]$$

The stiffness matrix is given by

$$k = \int_V B^T DB\, dV \qquad ...[5]$$

and in the case of CS tetrahedron, we have $K = B^T DBV$ ...[6]

and
$$\{Q\} = \int_S [N]^T \begin{Bmatrix} T_x \\ T_y \end{Bmatrix} ds + \int_V [N]^T \begin{bmatrix} F_x \\ F_y \\ F_z \end{bmatrix} dV \qquad ...[7]$$

Shape function for a 4 noded Tetrahedron element in terms of Cartesian coordinates

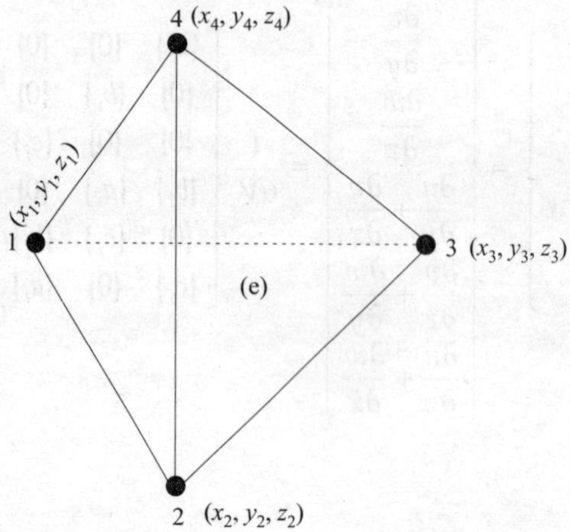

Fig. 6.6 Tetrahedron element in Cartesian coordinates

Method: *Linear polynomial method*

Let us consider, the linear (piecewise) polynomial in the form

$$u = a + bx + cy + dz \qquad ...[1]$$

Since $u(x_i, y_i, z_i) = u_i, \quad i = 1(1)\,4$

This

$$\left.\begin{aligned} u_1 &= a + bx_1 + cy_1 + dz_1 \\ u_2 &= a + bx_2 + cy_2 + dz_2 \\ u_3 &= a + bx_3 + cy_3 + dz_3 \\ u_4 &= a + bx_4 + cy_4 + dz_4 \end{aligned}\right\} \qquad ...[2]$$

Solving for a, b, c and d, we get

$$a = \frac{1}{6V}\,(a_1u_1 + a_2u_2 + a_3u_3 + a_4u_4)$$

$$\left.\begin{aligned} b &= \frac{1}{6V}(b_1u_1 + b_2u_2 + b_3u_3 + b_4u_4) \\ c &= \frac{1}{6V}(c_1u_1 + c_2u_2 + c_3u_3 + c_4u_4) \\ d &= \frac{1}{6V}(d_1u_1 + d_2u_2 + d_3u_3 + d_4u_4) \end{aligned}\right\} \quad \text{...[3]}$$

The matrix form of this is given as:

$$\begin{Bmatrix} a \\ b \\ c \\ d \end{Bmatrix} = \frac{1}{6V}\begin{bmatrix} a_1 & a_2 & a_3 & a_4 \\ b_1 & b_2 & b_3 & b_4 \\ c_1 & c_2 & c_3 & c_4 \\ d_1 & d_2 & d_3 & d_4 \end{bmatrix}\begin{Bmatrix} u_1 \\ u_2 \\ u_3 \\ u_4 \end{Bmatrix} \quad \text{...[4]}$$

where V is the volume of the tetrahedron element (e) given by

$$V = \frac{1}{6}\begin{vmatrix} 1 & x_1 & y_1 & z_1 \\ 1 & x_2 & y_2 & z_2 \\ 1 & x_3 & y_3 & z_3 \\ 1 & x_4 & y_4 & z_4 \end{vmatrix}$$

and

$$a_1 = \begin{vmatrix} x_2 & y_2 & z_2 \\ x_3 & y_3 & z_3 \\ x_4 & y_4 & z_4 \end{vmatrix}$$

$$b_1 = -\begin{vmatrix} 1 & y_2 & z_2 \\ 1 & y_3 & z_3 \\ 1 & y_4 & z_4 \end{vmatrix}$$

$$c_1 = +\begin{vmatrix} x_2 & 1 & z_2 \\ x_3 & 1 & z_3 \\ x_4 & 1 & z_4 \end{vmatrix}$$

$$d_1 = -\begin{vmatrix} x_2 & y_2 & 1 \\ x_3 & y_3 & 1 \\ x_4 & y_4 & 1 \end{vmatrix}$$

The other constants are obtained by the interchange of the subscripts in the order 1 2 3 4. Substituting from [3] into eq. [1],

we get

$$u = \phi_1 u_1 + \phi_2 u_2 + \phi_3 u_3 + \phi_4 u_4 \qquad ...[5]$$

The matrix form of equation [5] is:

$$u = \phi q$$

where

$$\phi = [\phi_1 \ \phi_2 \ \phi_3 \ \phi_4]$$

$$q = [u_1 \ u_2 \ u_3 \ u_4]^T$$

$$\phi_i = \frac{1}{6V} (a_i + b_i x + c_i y + d_i z) \ i = 1 \ (1) \ 4.$$

We find that the shape functions satisfy

$$\phi_i (x_j, y_j, z_j) = \begin{vmatrix} 1, i = j \\ 0, i \neq j \end{vmatrix}$$

Example: Using natural coordinates, find the shape function for a tetrahedron element.

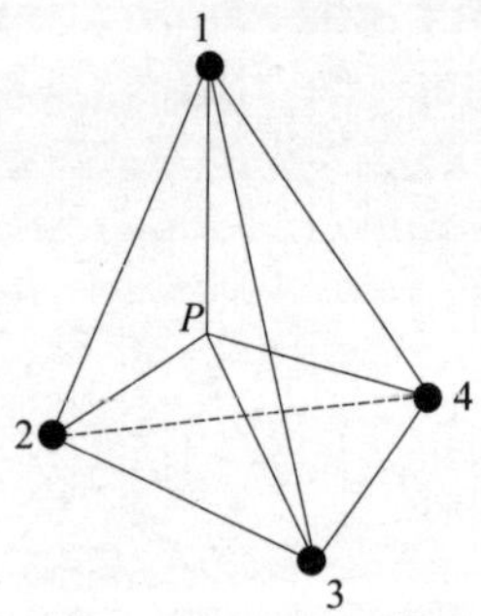

(a) Noded linear tetrahedral element

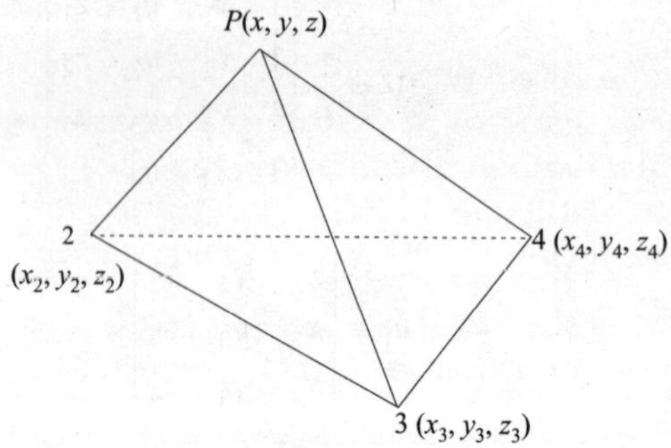

(b) Enlarged view of P 234

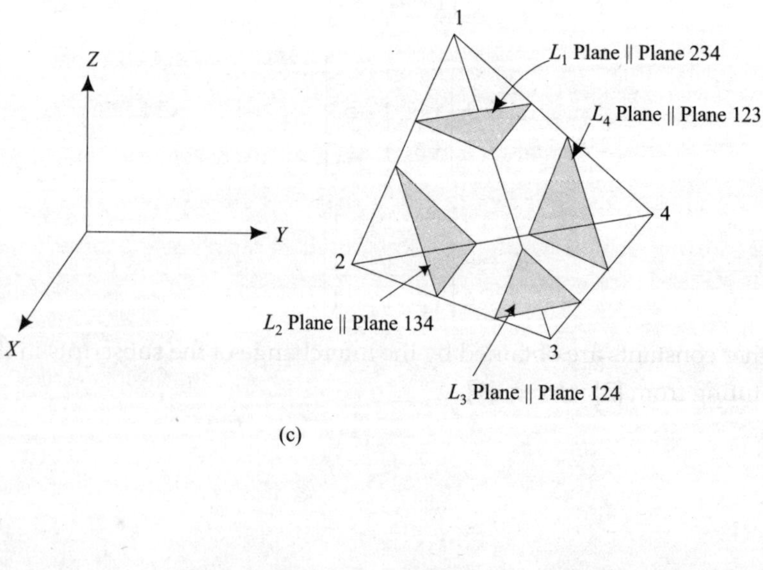

(c)

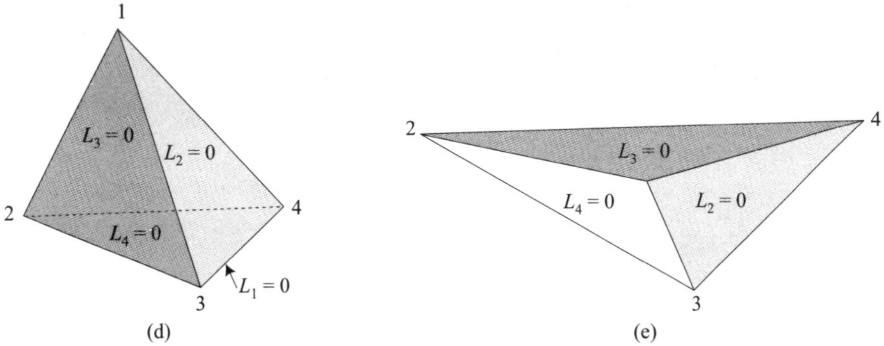

$$\begin{Bmatrix} \text{At } PL\ 123 & L_4 = 0 \\ \text{At } PL\ 234 & L_1 = 0 \\ \text{At } PL\ 124 & L_3 = 0 \\ \text{At } PL\ 134 & L_2 = 0 \end{Bmatrix}$$

Fig. 6.7 Definition of tetrahedral co-ordinates.

Solution: Corresponding to the triangle in two-dimension the tetrahedron is the basic finite element for three-dimensional problems. A tetrahedron has four primary external nodes as shown in Fig. 6.7. There can be a tetrahedral and volume coordinate system for 3-dimensional triangular i.e., tetrahedral elements. Such elements are convenient in defining in natural coordinates.

The location of any point $P(x, y, z) \in (e)$ can be defined in terms of the volume coordinates (L_1, L_2, L_3, L_4).

Definition: We define L_1, L_2, L_3 and L_4 as :

Volume co-ordinates

$$\left\{\begin{array}{l} \text{1. Plane } L_1 = \text{Plane parallel to face 2, 3, 4} \\ \text{2. Plane } L_2 = \text{Plane parallel to face 1, 3, 4} \\ \text{3. Plane } L_3 = \text{Plane parallel to face 1, 2, 4} \\ \text{4. Plane } L_4 = \text{Plane parallel to face 1, 2, 3} \\ \qquad\qquad L_i = 1 \text{ at node } i \text{ and 0 at others.} \end{array}\right\}$$

i.e., these volume co-ordinates L_i of a point within the element is defined as the ratio of the volume enclosed between the point and the plane normal to the node i to the total volume of the tetrahedral. That is,

$$L_1 = \frac{\text{Vol. P234}}{\text{Vol. 1234}} \qquad ...[1]$$

$$L_2 = \frac{\text{Vol. P134}}{\text{Vol. 1234}} \qquad ...[2]$$

$$L_3 = \frac{\text{Vol. P124}}{\text{Vol. 1234}} \qquad ...[3]$$

$$L_4 = \frac{\text{Vol. P123}}{\text{Vol. 1234}} \qquad ...[4]$$

Adding all the above

$$L_1 + L_2 + L_3 + L_4 = \frac{1}{\text{Vol. }1234}\,[\text{P234} + \text{P134} + \text{P123} + \text{P124}]$$

or

$$= \frac{\text{Vol. }1234}{\text{Vol. }1234} \qquad ...[5]$$

Therefore, $L_1 + L_2 + L_3 + L_4 = 1$

Further, we have $L_i = N_i;\quad i = 1(1)4,$ ...[7]

and the coordinates of the vertices 1, 2, 3 and 4 in terms of the natural coordinates become (1, 0, 0,0), (0, 1, 0, 0), (0, 0, 1, 0), (0, 0, 0, 1) respectively.

Let us assume the displacement in x direction at any point in the element be

$$u = \alpha_1 L_1 + \alpha_2 L_2 + \alpha_3 L_3 + \alpha_4 L_4$$

$$\therefore \qquad \{u\}_e = \begin{bmatrix} 1 & 0 & 0 & 0 \\ 0 & 1 & 0 & 0 \\ 0 & 0 & 1 & 0 \\ 0 & 0 & 0 & 1 \end{bmatrix} \begin{Bmatrix} \alpha_1 \\ \alpha_2 \\ \alpha_3 \\ \alpha_4 \end{Bmatrix}$$

$$\Rightarrow \qquad \begin{Bmatrix} \alpha_1 \\ \alpha_2 \\ \alpha_3 \\ \alpha_4 \end{Bmatrix} = \begin{bmatrix} 1 & 0 & 0 & 0 \\ 0 & 1 & 0 & 0 \\ 0 & 0 & 1 & 0 \\ 0 & 0 & 0 & 1 \end{bmatrix}^{-1} \{u\}_e$$

$$= \begin{bmatrix} 1 & 0 & 0 & 0 \\ 0 & 1 & 0 & 0 \\ 0 & 0 & 1 & 0 \\ 0 & 0 & 0 & 1 \end{bmatrix} \{u\}_e$$

$$\therefore \qquad u = \alpha_1 L_1 + \alpha_2 L_2 + \alpha_3 L_3 + \alpha_4 L_4$$

$$= [L_1\ L_2\ L_3\ L_4] \begin{Bmatrix} \alpha_1 \\ \alpha_2 \\ \alpha_3 \\ \alpha_4 \end{Bmatrix}$$

$$= [L_1\ L_2\ L_3\ L_4]\begin{bmatrix} 1 & 0 & 0 & 0 \\ 0 & 1 & 0 & 0 \\ 0 & 0 & 1 & 0 \\ 0 & 0 & 0 & 1 \end{bmatrix}^{-1} \{u\}_e$$

$$= [L_1\ L_2\ L_3\ L_4]\ \{u\}_e$$

$$= [N_1\ N_2\ N_3\ N_4]\ \{u\}_e$$

$$= [N]\ \{u\}_e$$

where $N_1 = L_1,\ N_2 = L_2,\ N_3 = L_3, \text{ and } N_4 = L_4$

Similarly, for displacement v and w, we get

$$v = [N]\ \{V\}_e \text{ and } w = [N]\ \{w\}_e$$

$$\begin{Bmatrix} u \\ v \\ w \end{Bmatrix} = \begin{bmatrix} \underset{1\times 4}{N} & \underset{1\times 4}{0} & \underset{1\times 4}{0} \\ \underset{1\times 4}{0} & \underset{1\times 4}{N} & \underset{1\times 4}{0} \\ \underset{1\times 4}{0} & \underset{1\times 4}{0} & \underset{1\times 4}{N} \end{bmatrix} \{\delta\}_e$$

where

$$\{\delta\}_e^T = [\ u_1\ u_2\ u_3\ u_4\ v_1\ v_2\ v_3\ v_4\ w_1\ w_2\ w_3\ w_4]$$

Note: $\left\{ N_i = \begin{vmatrix} 1, i = j \\ 0, i \neq j \end{vmatrix} \right\}$

Variation is linear.

Let us suppose we wish to develop a wedge shaped finite element as shown in Fig. 6.7 with a linear variation of displacement in three coordinate directions (the triangle lies in the local xy plane and the axis of the element in z-direction).

Example: Use non-dimensional coordinate to

(i) write the vector of interpolating functions

(ii) express the vectors

(iii) indicate how the vectors of (i) and (ii) would be used to evaluate the element stiffness and statically equivalent nodal forces for distributed forces.

Solution: Choose the origins at 3, ξ_1, ξ_2, ξ_3, are the usual triangular coordinates and $\xi_4 = z/2c$

$$\begin{Bmatrix} \phi_1 = \xi_1(1 - \xi_4) & \phi_4 = \xi_1\xi_4 \\ \phi_2 = \xi_2(1 - \xi_4) & \phi_5 = \xi_2\xi_4 \\ \phi_3 = \xi_3(1 - \xi_4) & \phi_6 = \xi_3\xi_4 \end{Bmatrix}$$

ξ_3 is dependent on ξ_1 and ξ_2; $\xi_3 = 1 - \xi_1 - \xi_2$

$$\begin{Bmatrix} \phi, x \\ \phi, y \\ \phi, z \end{Bmatrix} = \begin{bmatrix} \frac{y_{23}}{2\Delta}(1-\xi_4), & \frac{y_{31}}{2\Delta}, (1-\xi_4), & \frac{y_{12}}{2\Delta}(1-\xi_4), & \frac{y_{23}}{2\Delta}\xi_4; & \frac{y_{31}}{2\Delta}\xi_4 & \frac{y_{12}}{2\Delta}\xi_4 \\ \frac{x_{32}}{2\Delta}(1-\xi_4), & \frac{x_{13}}{2\Delta}(1-\xi_4), & \frac{x_{21}}{2\Delta}(1-\xi_4), & \frac{x_{32}}{2\Delta}\xi_4, & \frac{x_{13}}{2\Delta}\xi_4, & \frac{x_{21}}{2\Delta}\xi_4 \\ \frac{-\xi_1}{2c} & \frac{-\xi_2}{2c} & \frac{-\xi_3}{2c} & \frac{\xi_1}{2c} & \frac{\xi_2}{2c} & \frac{\xi_3}{2c} \end{bmatrix}$$

The above two can be used as shown below.

$$\begin{Bmatrix} U \\ V \\ W \end{Bmatrix} = \begin{bmatrix} \{\phi\} & \{0\} & \{0\} \\ \{0\} & \{\phi\} & \{0\} \\ \{0\} & \{0\} & \{\phi\} \end{bmatrix} \begin{Bmatrix} \underline{U} \\ \underline{V} \\ \underline{W} \end{Bmatrix} = [N] \begin{Bmatrix} \underline{U} \\ \underline{V} \\ \underline{W} \end{Bmatrix}$$

$$\{\varepsilon\} = \begin{Bmatrix} U, x \\ V, y \\ W, z \\ U, y + V, x \\ V, z + W, y \\ U, z + W, x \end{Bmatrix} = \begin{Bmatrix} \{\phi, x\} & \{0\} & \{0\} \\ \{0\} & \{\phi, y\} & \{0\} \\ \{0\} & \{0\} & \{\phi, z\} \\ \{\phi, y\} & \{\phi, x\} & \{0\} \\ \{0\} & \{\phi, z\} & \{\phi, y\} \\ \{\phi, z\} & \{0\} & \{\phi, x\} \end{Bmatrix} \begin{Bmatrix} \underline{U} \\ \underline{V} \\ \underline{W} \end{Bmatrix}$$

$$= [B] \begin{Bmatrix} \underline{U} \\ \underline{V} \\ \underline{W} \end{Bmatrix}$$

$$[K] = \int B^T CB \, dV$$

$$= 2C \int_0^1 d\,\xi_4 \int B^T CB \, dA$$

$$\{Q\} = \int [N]^T \begin{Bmatrix} F_x \\ F_y \\ F_z \end{Bmatrix} dV$$

$$= 2C \int_0^1 d\,\xi_4 \int [N]^T \begin{Bmatrix} F_x \\ F_y \\ F_z \end{Bmatrix} dA$$

6.4 HEXAHEDRON ELEMENTS

Suppose we wish to discretize 3-dimensional region into hexahedron elements with four quadrilateral faces.

We choose an arbitrary hexahedron element (e) with eight nodes (x_i, y_i, z_i) i = 1(1) 8 at the corners. The displacements at the node i is represented by u_i. We take the origin of the local coordinates (ξ, η, ζ) at the point of intersection of the lines joining the midpoints of the opposite faces of the hexahedron and define the sides by $\xi = \pm 1$, $\eta = \pm 1$, $\zeta = \pm 1$ as shown in Fig. 6.8.

ZIBB–Zienkiewicz–Irons Brick with 8 Nodes

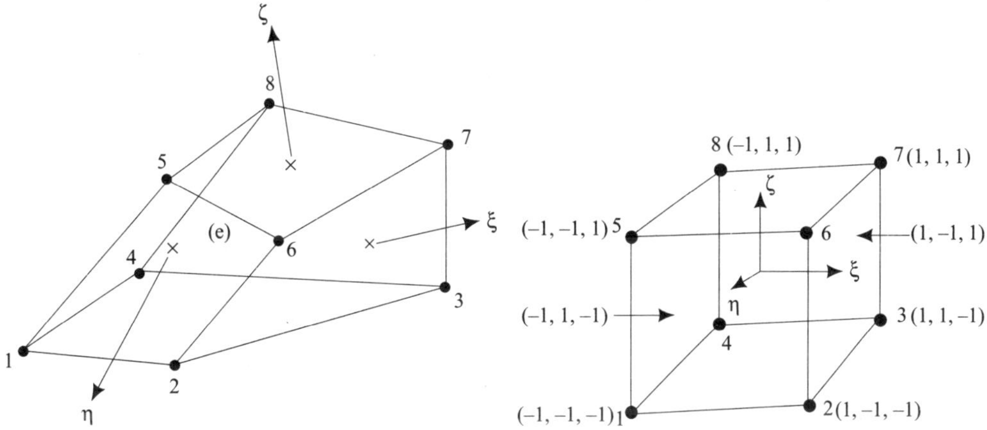

(a) Hexahedron element in ξ, η ζ spaces

(b) Mapping into cube

Fig. 6.8 Zienkiewicz–Irons brick with 8 nodes

The isoparametric transformations are :

$$x = N_1x_1 + N_2x_2 + N_3x_3 + N_4x_4 + N_5x_5 + N_6x_6 + N_7x_7 + N_8x_8 \text{ etc.}$$

or,

$$\left.\begin{aligned} x &= \sum_{i=1}^{8} N_i x_i \\ y &= \sum_{i=1}^{8} N_i y_i \\ z &= \sum_{i=1}^{8} N_i y_i \end{aligned}\right\} \quad ...[1]$$

where $$N_i = \frac{1}{8}(1 + \xi_i\xi)\,(1 + \eta_i\eta)\,(1 + \zeta_i\zeta) \quad ...[2]$$

$$[\xi_1\ \xi_2\ \xi_3\ \xi_4\ \xi_5\ \xi_6\ \xi_7\ \xi_8]^T = [-1\ 1\ 1\ -1\ -1\ 1\ 1\ -1]^T$$
$$[\eta_1\ \eta_2\ \eta_3\ \eta_4\ \eta_5\ \eta_6\ \eta_7\ \eta_8]^T = [-1\ -1\ 1\ 1\ -1\ -1\ 1\ 1]^T \qquad ...[3]$$
$$[\zeta_1\ \zeta_2\ \zeta_3\ \zeta_4\ \zeta_5\ \zeta_6\ \zeta_7\ \zeta_8]^T = [-1\ -1\ -1\ -1\ 1\ 1\ 1\ 1]^T$$

transforms the Hexahedron element (*e*) into a cube $|\xi| = 1$, $|\eta| = 1$, $|\zeta| = 1$ as shown in Fig. 6.8(b).

The deformation of the element can be expressed in terms of displacement components

$$\{u\} = \begin{Bmatrix} u_1 \\ u_2 \\ u_3 \end{Bmatrix} \qquad ...[4]$$

Three-dimensional elements with eight primary nodes are either in the form of a general hexahedron (Fig. 6.9(a)), or a rectangular prism (Fig. 6.9(b)).

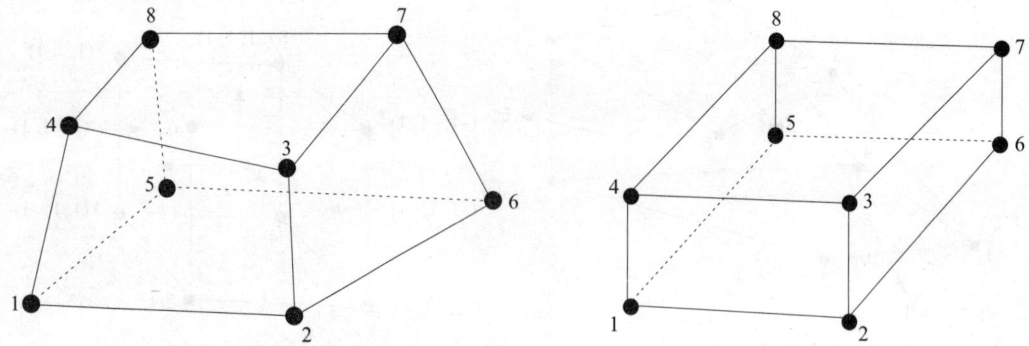

(a) An aribitrary hexahedron (b) A rectangular prism

Fig. 6.9

Stiffness Matrix *K*: For Hexahedron Element

Consider a hexahedron element as shown in Fig. 6.10(a). This element in the natural coordinate system is a cube as shown in Fig. 6.10(b).

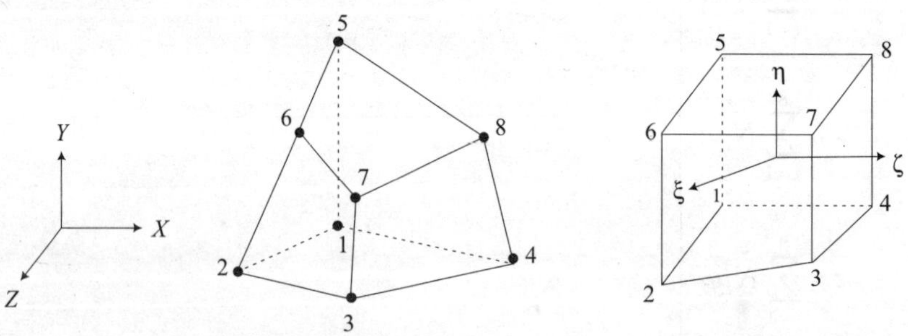

(a) Real (arbitrary) element. (b) Parent element

Fig. 6.10 Linear Hexahedron (ZIB8) element

(1) The Cartesian coordinates can be expressed in terms of natural coordinates.

$$\begin{Bmatrix} x \\ y \\ z \end{Bmatrix} = \begin{bmatrix} x_1 & x_2 & x_3 & \dots & x_8 \\ y_1 & y_2 & y_3 & \dots & y_8 \\ z_1 & z_2 & z_3 & \dots & z_8 \end{bmatrix} \begin{bmatrix} N_1 \\ N_2 \\ N_3 \\ \vdots \\ N_8 \end{bmatrix} \quad \dots[1]$$

where

$$N_i = \frac{1}{8}(1+\xi_0)(1+\eta_0)(1+\zeta_0)$$

where $\xi_0 = \xi\xi_i,\ \eta_0 = \eta\eta_i,\ \zeta_0 = \zeta\zeta_i$ ...[2]

(2) Jacobian can be determined as:

$$[J] = \begin{bmatrix} N_{1,\xi} & \dots & N_{8,\xi} \\ N_{1,\eta} & \dots & N_{8,\eta} \\ N_{1,\zeta} & \dots & N_{8,\zeta} \end{bmatrix} \begin{bmatrix} x_1 & y_1 & z_1 \\ \dots & \dots & \dots \\ x_8 & y_8 & z_8 \end{bmatrix} \quad \dots[3]$$

(3) Strain-Displacement function: For isoparametric element, the shape function used for the displacement is the same as the shape function employed for the relationship between Cartesian and natural coordinates.

$$\{\varepsilon\} = \begin{Bmatrix} \dfrac{\partial u}{\partial x} \\ \dfrac{\partial v}{\partial y} \\ \dfrac{\partial w}{\partial z} \\ \dfrac{\partial u}{\partial y} + \dfrac{\partial v}{\partial x} \\ \dfrac{\partial v}{\partial z} + \dfrac{\partial w}{\partial y} \\ \dfrac{\partial u}{\partial z} + \dfrac{\partial w}{\partial x} \end{Bmatrix}$$

$$= \begin{bmatrix} 1 & 0 & 0 & 0 & 0 & 0 & 0 & 0 & 0 \\ 0 & 0 & 0 & 0 & 1 & 0 & 0 & 0 & 0 \\ 0 & 0 & 0 & 0 & 0 & 0 & 0 & 0 & 1 \\ 0 & 1 & 0 & 1 & 0 & 0 & 0 & 0 & 0 \\ 0 & 0 & 0 & 0 & 0 & 1 & 0 & 1 & 0 \\ 0 & 0 & 1 & 0 & 0 & 0 & 1 & 0 & 0 \end{bmatrix} \begin{Bmatrix} \dfrac{\partial \bar{U}}{\partial x} \\ \vdots \\ \vdots \\ \vdots \\ \dfrac{\partial \bar{W}}{\partial z} \end{Bmatrix} \quad \dots[4]$$

$$= [T]\begin{bmatrix} [J]^{-1} & [0] & [0] \\ [0] & [J]^{-1} & [0] \\ [0] & [0] & [J]^{-1} \end{bmatrix} \begin{Bmatrix} \dfrac{\partial \bar{u}}{\partial \xi} \\ \dfrac{\partial \bar{u}}{\partial \eta} \\ \dfrac{\partial \bar{u}}{\partial \zeta} \\ \vdots \\ \vdots \\ \dfrac{\partial \bar{w}}{\partial \zeta} \end{Bmatrix} \qquad \text{...[5]}$$

$$= [T]\begin{bmatrix} [J]^{-1} & [0] & [0] \\ [0] & [J]^{-1} & [0] \\ [0] & [0] & [J]^{-1} \end{bmatrix} \begin{bmatrix} B_\eta & & \\ & B_\eta & \\ & & B_\eta \end{bmatrix} \begin{Bmatrix} \underline{u} \\ \underline{v} \\ \underline{w} \end{Bmatrix} \qquad \text{...[6]}$$

where

$$[B_\eta] = \begin{bmatrix} N_{1,\xi} & \dots & N_{8,\xi} \\ N_{1,\eta} & \dots & N_{8,\eta} \\ N_{1,\zeta} & \dots & N_{8,\zeta} \end{bmatrix}$$

$$\{\underline{u}\} = \begin{Bmatrix} u_1 \\ u_2 \\ \vdots \\ u_8 \end{Bmatrix}$$

$$\{\underline{v}\} = \begin{Bmatrix} v_1 \\ v_2 \\ \vdots \\ v_8 \end{Bmatrix}$$

$$\{\underline{w}\} = \begin{Bmatrix} w_1 \\ w_2 \\ \vdots \\ w_8 \end{Bmatrix} \qquad \text{...[7]}$$

$$\{\varepsilon\} = [B]\begin{Bmatrix} \underline{u} \\ \underline{v} \\ \underline{w} \end{Bmatrix} \qquad \text{...[8]}$$

(4) The stiffness matrix is given by

$$[K] = \int_V [B]^T [C][B]dv$$

$$= \int_{-1}^{1}\int_{-1}^{1}\int_{-1}^{1} [B]^T [C]\, B \det J \, d\xi \, d\eta \, d\zeta \qquad ...[9]$$

where C is the constitutive matrix and is given by

$$[D] = \frac{E}{(1+v)(1-2v)} \begin{bmatrix} (1-v) & v & v & 0 & 0 & 0 \\ & (1-v) & v & 0 & 0 & 0 \\ & & (1-v) & 0 & 0 & 0 \\ & & & \frac{(1-2v)}{2} & 0 & 0 \\ & & & & \frac{(1-2v)}{2} & 0 \\ & \text{SYMMETRIC} & & & & \frac{(1-2v)}{2} \end{bmatrix}$$

Stiffness Matrix K: For a twenty noded Hexahedron Isoparametric element. (*ZIB* 20)

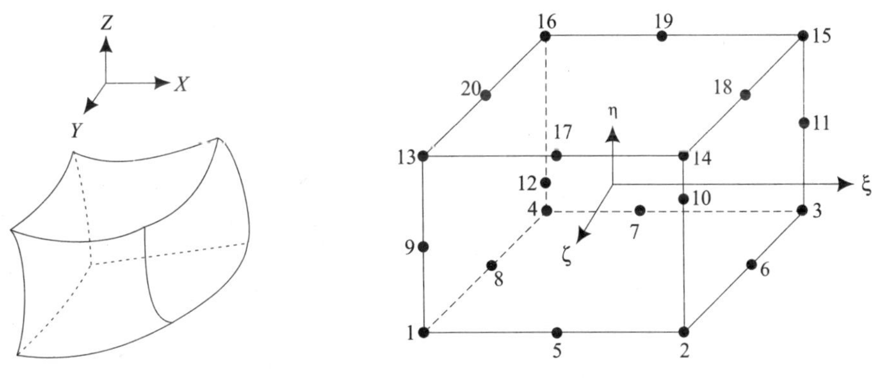

(a) Arbitrary element (b) Parent element

Fig. 6.11 ZIB20 element (Zienkiewicz and Irons brick with 20 nodes)

Formulation of Stiffness Matrix

(*a*) The local coordinate system is related to global Cartesian coordinates as :

$$\begin{Bmatrix} x \\ y \\ z \end{Bmatrix} = \begin{bmatrix} \{N\} & \{0\} & \{0\} \\ \{0\} & \{N\} & \{0\} \\ \{0\} & \{0\} & \{N\} \end{bmatrix} \begin{Bmatrix} \underline{x}_n \\ \underline{y}_n \\ \underline{z}_n \end{Bmatrix}$$

or,

$$\begin{Bmatrix} x \\ y \\ z \end{Bmatrix} = N \begin{Bmatrix} \underline{x}_n \\ \underline{y}_n \\ \underline{z}_n \end{Bmatrix} \qquad \text{...[1]}$$

where

$$\underline{x_n} = \begin{Bmatrix} x_1 \\ \vdots \\ x_{20} \end{Bmatrix}; \ \underline{y_n} = \begin{Bmatrix} y_1 \\ \vdots \\ y_{20} \end{Bmatrix}; \ \underline{z_n} = \begin{Bmatrix} z_1 \\ \vdots \\ z_{20} \end{Bmatrix} \qquad \text{...[2]}$$

$\{N\} \rightarrow$ Shape functions, (a row vector).

(*b*) The displacement model : isoparametric representation.

$$\begin{Bmatrix} u \\ v \\ w \end{Bmatrix} = \begin{Bmatrix} \{N\} & \{0\} & \{0\} \\ \{0\} & \{N\} & \{0\} \\ \{0\} & \{0\} & \{N\} \end{Bmatrix} \begin{Bmatrix} \underline{u_n} \\ \underline{v_n} \\ \underline{w_n} \end{Bmatrix} \qquad \text{...[3]}$$

where

$$\underline{u_n} = \begin{Bmatrix} u_1 \\ u_2 \\ \vdots \\ u_{20} \end{Bmatrix};$$

$$\underline{v_n} = \begin{Bmatrix} v_1 \\ v_2 \\ \vdots \\ v_{20} \end{Bmatrix}; \qquad \text{...[4]}$$

$$\underline{w_n} = \begin{Bmatrix} w_1 \\ w_2 \\ \vdots \\ w_{20} \end{Bmatrix}$$

(*c*) Strain–displacement equations:

$$\{\varepsilon\} = \begin{Bmatrix} \varepsilon_x \\ \varepsilon_y \\ \varepsilon_z \\ \gamma_{xy} \\ \gamma_{yz} \\ \gamma_{zx} \end{Bmatrix} = \begin{bmatrix} \{N,x\} & \{0\} & \{0\} \\ \{0\} & \{N,y\} & \{0\} \\ \{0\} & \{0\} & \{N,z\} \\ \{N,y\} & \{N,x\} & \{0\} \\ \{0\} & \{N,z\} & \{N,y\} \\ \{N,z\} & 0 & \{N,x\} \end{bmatrix} \begin{Bmatrix} \underline{u_n} \\ \underline{v_n} \\ \underline{w_n} \end{Bmatrix}$$

i.e., $$\{\varepsilon\} = [B]\begin{Bmatrix} \underline{u_n} \\ \underline{v_n} \\ \underline{w_n} \end{Bmatrix} \quad ...[5]$$

where $N = N(\xi, \eta, \zeta)$; and N,x denotes the P_1 differentiation $\frac{\partial N}{\partial x}$ and so on. To obtain necessary differentiation formulae, we write:

$$\begin{Bmatrix} \frac{\partial N}{\partial x} \\ \frac{\partial N}{\partial y} \\ \frac{\partial N}{\partial z} \end{Bmatrix} = [J]^{-1}\begin{Bmatrix} \frac{\partial N}{\partial \xi} \\ \frac{\partial N}{\partial \eta} \\ \frac{\partial N}{\partial \zeta} \end{Bmatrix} \quad ...[6]$$

where [J] is (3 × 3) Jacobian matrix and is evaluated as:

(*d*) $$[J]_{3\times 3} = \begin{Bmatrix} \frac{\partial \{N\}}{\partial \xi} \\ \frac{\partial \{N\}}{\partial \eta} \\ \frac{\partial \{N\}}{\partial \zeta} \end{Bmatrix}_{3\times 8} [x_n \;\; y_n \;\; z_n]_{8\times 3} \quad ...[7]$$

and we have from eqn. [6]:

$$\begin{Bmatrix} \frac{\partial u}{\partial x} \\ \frac{\partial u}{\partial y} \\ \frac{\partial u}{\partial z} \end{Bmatrix} = [J]^{-1}\begin{Bmatrix} \frac{\partial u}{\partial \xi} \\ \frac{\partial u}{\partial \eta} \\ \frac{\partial u}{\partial \zeta} \end{Bmatrix}; \quad ...[8a]$$

$$\begin{Bmatrix} \frac{\partial v}{\partial x} \\ \frac{\partial v}{\partial y} \\ \frac{\partial v}{\partial z} \end{Bmatrix} = [J]^{-1}\begin{Bmatrix} \frac{\partial v}{\partial \xi} \\ \frac{\partial v}{\partial \eta} \\ \frac{\partial v}{\partial \zeta} \end{Bmatrix} \quad ...[8b]$$

and
$$\begin{Bmatrix} \dfrac{\partial w}{\partial x} \\ \dfrac{\partial w}{\partial y} \\ \dfrac{\partial w}{\partial z} \end{Bmatrix} = [J]^{-1} \begin{Bmatrix} \dfrac{\partial w}{\partial \xi} \\ \dfrac{\partial w}{\partial \eta} \\ \dfrac{\partial w}{\partial \zeta} \end{Bmatrix} \qquad \text{...[8c]}$$

It follows from the equation [3],

$$\begin{Bmatrix} \dfrac{\partial u}{\partial \xi} \\ \dfrac{\partial u}{\partial \eta} \\ \dfrac{\partial u}{\partial \zeta} \end{Bmatrix} = \begin{Bmatrix} \dfrac{\partial N}{\partial \xi} \\ \dfrac{\partial N}{\partial \eta} \\ \dfrac{\partial N}{\partial \zeta} \end{Bmatrix} \{\underline{u}_n\} = [B_i]\,\{\underline{u}_n\} \qquad \text{...[9]}$$

(*e*) Strain-displacement matrix $[B]$ can be determined as:

$$\{\varepsilon\} = [T] = \begin{Bmatrix} \dfrac{\partial u}{\partial x} \\ \vdots \\ \dfrac{\partial w}{\partial z} \end{Bmatrix}$$

$$= [T] \begin{bmatrix} [J]^{-1} & & \\ & [J]^{-1} & \\ & & [J]^{-1} \end{bmatrix} \begin{Bmatrix} \dfrac{\partial u}{\partial \xi} \\ \vdots \\ \dfrac{\partial w}{\partial \zeta} \end{Bmatrix}$$

$$= [T] \begin{bmatrix} [J]^{-1} & & \\ & [J]^{-1} & \\ & & [J]^{-1} \end{bmatrix} \begin{bmatrix} B_i & & \\ & B_i & \\ & & B_i \end{bmatrix} \begin{Bmatrix} \underline{u}_n \\ \underline{v}_n \\ \underline{w}_n \end{Bmatrix} \qquad \text{...[10]}$$

$$\{\varepsilon\}_{60 \times 1} = [B]_{6 \times 60} \begin{Bmatrix} \underline{u}_n \\ \underline{v}_n \\ \underline{w}_n \end{Bmatrix}_{60 \times 1}$$

(*f*) The element stiffness is evaluated by numerical integration of the standard formulae. Thus, the Stiffness Matrix is given by

$$[k] = \int_V [B]^T [C] [B] \, dV$$

$$= \int_{-1}^{1} \int_{-1}^{1} \int_{-1}^{1} [B]^T [C] [B] \det([J]) \, . \, d\xi \, d\eta \, d\zeta \quad ...[11]$$

Shape Functions for ZIB20 Element Hexahedron with 20 Nodes

In order to determine the shape function for node 1 (Fig. 6.11) *ZIB* 20), we can multiply the equation of plane $\zeta = 1$, $\eta = 1$ and $\xi = -1$ and a plane passing through the points 9, 8, 5 the equation of which is

$$\zeta = a + b\xi + c\eta \quad ...[1]$$

Substituting the coordinates of 9, 8, 5 we can find *a*, *b* and *c*.

$$N_1 = c(\xi + 1)(\eta - 1)(\zeta - 1)(\xi - \eta - \zeta - 2) \quad ...[2]$$

Substituting the coordinate at 1, we get

$$N_1 = \frac{1}{8}(1 + \xi)(1 - \eta)(1 - \zeta)(\xi - \eta - \zeta - 2) \quad ...[3]$$

or N_1 for corner nodes :

$$N_1 = \frac{1}{8}(1 + \xi_0)(1 + \eta_0)(1 + \zeta_0)(\xi + \eta_0 + \zeta_0 - 2) \quad ...[4]$$

Example: Describe the method of finding shape function for a hexahedral element.

Solution: A hexahedral element with the natural coordinates of various nodal points is as shown in Fig. 6.12.

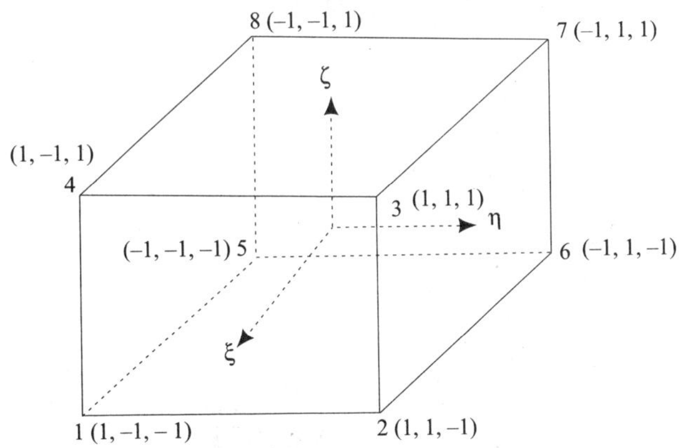

Fig. 6.12 Typical hexahedron element
1(1, −1, −1) ; 2(1, 1, −1) ; 3(1, 1, 1) ; 4(1, −1, 1);
5(−1, −1, −1) ; 6(−1, 1, −1) ; 7(−1, 1, 1) ; 8(−1, −1, 1).

It may be seen from Fig. 6.12 that there are only eight nodal values for defining displacement inside the element.

So, we assume a polynomial with 8 constants (as in equation [1]), keeping in view that geometric isotropy is maintained.

$$u = \alpha_1 + \alpha_2\xi + \alpha_3\eta + \alpha_4\zeta + \alpha_5\xi\eta + \alpha_6\eta\zeta + \alpha_7\zeta\xi + \alpha_8\xi\eta\zeta \quad ...[1]$$

$$= [1\ \xi\ \eta\zeta\ \xi\eta\ \eta\zeta\ \zeta\xi\ \xi\eta\zeta \begin{Bmatrix} \alpha_1 \\ \vdots \\ \alpha_8 \end{Bmatrix}$$

$$\therefore \quad \{u\}_e = \begin{bmatrix} 1 & 1 & -1 & -1 & -1 & 1 & -1 & 1 \\ 1 & 1 & 1 & -1 & 1 & -1 & -1 & -1 \\ 1 & 1 & 1 & 1 & 1 & 1 & 1 & 1 \\ 1 & 1 & -1 & 1 & -1 & -1 & 1 & -1 \\ 1 & -1 & -1 & -1 & 1 & 1 & 1 & -1 \\ 1 & -1 & 1 & -1 & -1 & -1 & 1 & 1 \\ 1 & -1 & 1 & -1 & 1 & -1 & 1 & -1 \\ 1 & -1 & -1 & 1 & 1 & -1 & -1 & 1 \end{bmatrix} \begin{Bmatrix} \alpha_1 \\ \alpha_2 \\ \alpha_3 \\ \alpha_4 \\ \alpha_5 \\ \alpha_6 \\ \alpha_7 \\ \alpha_8 \end{Bmatrix} = [A]\{\alpha\}$$

$$\Rightarrow \quad \{\alpha\} = [A]^{-1}\{u\}_e \quad (A^{-1} \text{ can be computed by elementary method})$$

$$u = [1\ \xi\ \eta\ \zeta\ \xi\eta\ \eta\zeta\ \zeta\xi\ \xi\eta\zeta]\{\alpha\}$$

$$= [1\ \xi\ \eta\ \zeta\ \xi\eta\ \eta\zeta\ \zeta\xi\ \xi\eta\zeta][A]^{-1}\{u\}_e = [N]\{u\}_e\,;\ (u = [N]\{u\}_e)$$

where

$$[N] = [1\ \xi\ \eta\ \zeta\ \xi\eta\ \eta\zeta\ \zeta\eta\ \xi\eta\zeta]\ [A]^{-1}$$

$$= [N_1\ N_2\ N_3\ N_4\ N_5\ N_6\ N_7\ N_8]$$

$$\Rightarrow \quad N_i = \frac{1}{8}(1 + \xi\xi_i)(1 + \eta\eta_i)(1 + \zeta\zeta_i) \quad ...[2]$$

$$\begin{Bmatrix} u \\ v \\ w \end{Bmatrix} = \begin{bmatrix} \underset{1\times 8}{N} & \underset{1\times 8}{0} & \underset{1\times 8}{0} \\ \underset{1\times 8}{0} & \underset{1\times 8}{N} & \underset{1\times 8}{0} \\ \underset{1\times 8}{0} & \underset{1\times 8}{0} & \underset{1\times 8}{N} \end{bmatrix} \{\delta\}_e$$

where

$$\{\delta\}_e^T = [u_1 \ldots u_8, v_1 \ldots v_8, w_1 \ldots w_8]$$

Note: $\left\{ N_i = \begin{Bmatrix} 1, i = j \\ 0, i \neq j \end{Bmatrix} \right.$

Variation is linear.

Lagrangian Polynomial

It is always preferable to express directly the assumed displacement function in terms of its nodal degrees of freedom.

Lagrangian interpolation provides an easy means of generating shape functions.

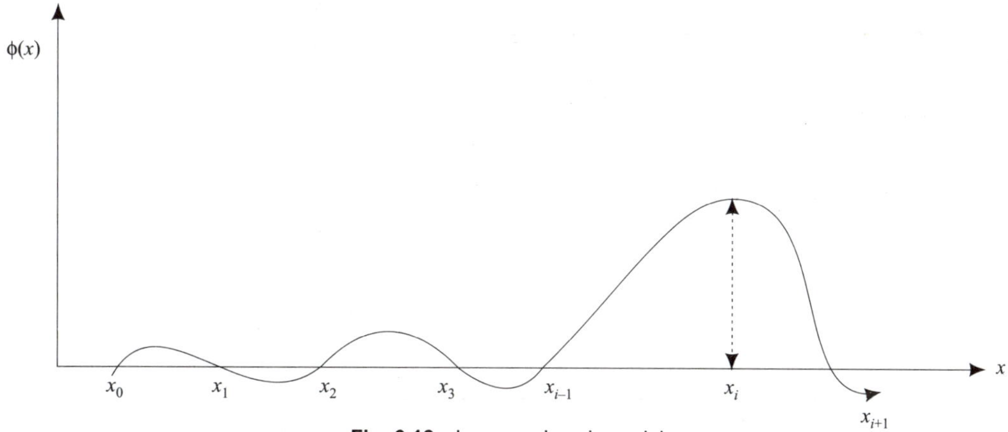

Fig. 6.13 Lagrange's polynomial

The characteristic polynomial $\phi_i(x)$ is that it has got $\phi_i(x_i) = 1$ and at all other points (nodes) zero as seen from Fig. 6.13.

$$\phi_i(x) = C(x - x_0)(x - x_1) \dots (x - x_{i-1})(x - x_{i+1})$$

when $x = x_i$

$$\phi_i(x_i) = 1 = C(x - x_0)(x - x_1) \dots (x - x_{i-1})(x - x_{i+1})$$

Substituting the value of C in equation [1], we get,

$$\phi_i(x) = \frac{(x - x_0)(x - x_1) \dots (x - x_{i-1})(x - x_{i+1})}{(x_i - x_0)(x_i - x_1) \dots (x_i - x_{i-1})(x_i - x_{i+1})}$$

Note the absence of $(x - x_i)$ term in the numerator. The particular point under consideration must be left out.

The straight and simple multiplication of 2 or 3 such polynomials, one for each direction, gives shape functions in 2 or 3 dimensions respectively.

Using natural coordinates, ξ, we can write the Lagrangian interpolation formula as:

$$N_k = \frac{(\xi - \xi_1)(\xi - \xi_2) \dots (\xi - \xi_{k-1})(\xi - \xi_{k+1}) \dots (\xi - \xi_n)}{(\xi_k - \xi_1)(\xi_k - \xi_2) \dots (\xi_k - \xi_{k-1})(\xi_k - \xi_{k+1}) \dots (\xi_k - \xi_n)} \quad \text{...[1]}$$

For different nodes, we have,

$$N_1 = \frac{(\xi - \xi_2)(\xi - \xi_3) \dots (\xi - \xi_n)}{(\xi_1 - \xi_2)(\xi_1 - \xi_3) \dots (\xi_1 - \xi_n)}$$

Example: Using Lagrange functions, derive shape function for hexahedron (brick) element. Typical hexahedron element is shown in Fig. 6.14.

The nodal coordinates are:

$\xi_1 = 1;$	$\xi_2 = 1;$	$\xi_3 = 1;$	$\xi_4 = 1;$	$\xi_5 = -1;$
$\eta_1 = -1;$	$\eta_2 = 1;$	$\eta_3 = 1;$	$\eta_4 = -1;$	$\eta_5 = -1;$
$\zeta_1 = -1;$	$\zeta_2 = -1;$	$\zeta_3 = 1;$	$\zeta_4 = 1;$	$\zeta_5 = -1;$
$\xi_6 = -1;$	$\xi_7 = -1;$	$\xi_8 = -1;$		
$\eta_6 = 1;$	$\eta_7 = 1;$	$\eta_8 = -1;$		
$\zeta_6 = -1;$	$\zeta_7 = 1;$	$\zeta_8 = 1$		

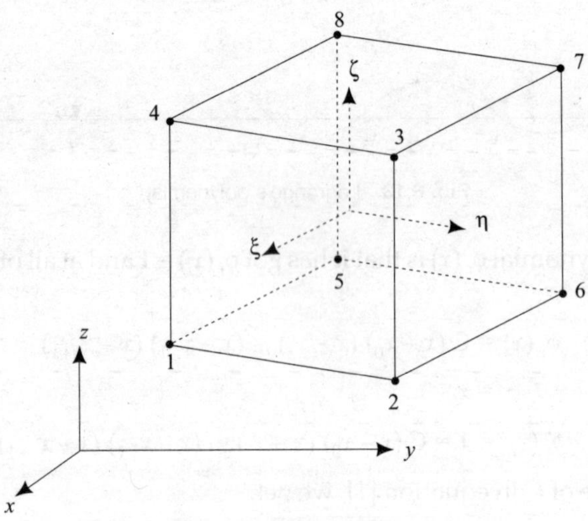

Fig. 6.14 Typical hexahedron element

In general, shape function by Lagrange function for three-dimensional case is given by

$$N_i = L_i\{\xi\}\, L_i(\eta)\, L_i(\zeta)$$

$$N_1 = \frac{(\xi - \xi_5)}{\xi_1 - \xi_5} \frac{\eta - \eta_2}{\eta_1 - \eta_2} \frac{\zeta - \zeta_4}{\zeta_1 - \zeta_4}$$

$$= \frac{(\xi + 1)(\eta - 1)(\zeta - 1)}{(1+1)(-1-1)(-1-1)}$$

$$= \frac{(\xi + 1)(\eta - 1)(\zeta - 1)}{8}$$

$$N_2 = L_2\{\xi\}\, L_2(\eta)\, L_2(\zeta)$$

$$= \frac{(\xi - \xi_6)}{(\xi_2 - \xi_6)} \frac{(\eta - \eta_1)}{(\eta_2 - \eta_1)} \frac{(\zeta - \zeta_3)}{(\zeta_2 - \zeta_3)}$$

$$= \frac{(\xi + 1)(\eta + 1)(\zeta - 1)}{(1 + 1)(1 + 1)(-1 - 1)} = \frac{(\xi + 1)(\eta + 1)(\zeta - 1)}{-8}$$

$$N_3 = L_3\{\xi\} L_3(\eta) L_3(\zeta)$$

$$= \frac{(\xi - \xi_7)}{(\xi_3 - \xi_7)} \frac{\eta - \eta_4}{\eta_3 - \eta_4} \frac{\zeta - \zeta_2}{\zeta_3 - \zeta_2}$$

$$= \frac{\xi + 1}{1 + 1} \frac{\eta + 1}{1 + 1} \frac{\zeta + 1}{1 + 1} = \frac{(\xi + 1)(\eta + 1)(\zeta + 1)}{8}$$

$$N_4 = L_4\{\xi\} L_4(\eta) L_4(\zeta) = \frac{\xi - \xi_8}{\xi_4 - \xi_8} \frac{\eta - \eta_3}{\eta_4 - \eta_3} \frac{\zeta - \zeta_1}{\zeta_4 - \zeta_1}$$

$$= \frac{\xi + 1}{1 + 1} \frac{\eta - 1}{-1 - 1} \frac{\zeta + 1}{1 + 1} = \frac{(\xi + 1)(\eta - 1)(\zeta + 1)}{-8}$$

$$N_5 = L_5\{\xi\} L_5(\eta) L_5(\zeta)$$

$$= \frac{(\xi - \xi_1)}{\xi_5 - \xi_1} \frac{\eta - \eta_6}{\eta_5 - \eta_6} \frac{\zeta - \zeta_8}{\zeta_5 - \zeta_8} = \frac{(\xi - 1)}{-1 - 1} \frac{\eta - 1}{-1 - 1} \frac{\zeta - 1}{-1 - 1} = \frac{(\xi - 1)(\eta - 1)(\zeta - 1)}{-8}$$

$$N_6 = L_6\{\xi\} L_6(\eta) L_6(\zeta)$$

$$= \frac{\xi - \xi_2}{\xi_6 - \xi_2} \frac{\eta - \eta_5}{\eta_6 - \eta_5} \frac{\zeta - \zeta_7}{\zeta_6 - \zeta_7} = \frac{\xi - 1}{-1 - 1} \frac{\eta + 1}{1 + 1} \frac{\zeta - 1}{-1 - 1} = \frac{(\xi - 1)(\eta + 1)(\zeta - 1)}{8}$$

$$N_7 = L_7\{\xi\} L_7(\eta) L_7(\zeta)$$

$$= \frac{\xi - \xi_3}{\xi_7 - \xi_3} \frac{\eta - \eta_8}{\eta_7 - \eta_8} \frac{\zeta - \zeta_6}{\zeta_8 - \zeta_6} = \frac{\xi - 1}{-1 - 1} \frac{\eta + 1}{1 + 1} \frac{\zeta + 1}{1 + 1} = \frac{(\xi - 1)(\eta + 1)(\zeta + 1)}{-8}$$

$$N_8 = L_8\{\xi\} L_8(\eta) L_8(\zeta)$$

$$= \frac{\xi - \xi_4}{\xi_8 - \xi_4} \frac{\eta - \eta_7}{\eta_8 - \eta_7} \frac{\zeta - \zeta_5}{\zeta_8 - \zeta_5} = \frac{(\xi - 1)}{-1 - 1} \frac{(\eta - 1)}{-1 - 1} \frac{(\zeta + 1)}{1 + 1} = \frac{(\xi - 1)(\eta - 1)(\zeta + 1)}{8}$$

In general, it may be noted that,

$$N_i = \frac{1}{8}(1 + \xi\xi_i)(1 + \eta\eta_i)(1 + \zeta\zeta_i)$$

Parallelopiped Elements

Higher order or refinements of three-dimensional elements by introduction of internal displacement nodes similar to the plane stress problems have not led to significant improvement. It is difficult manually to model a three-dimensional solid with tetrahedra without having mix up. If, however, tetrahedra are assembled to form hexahedron the modeling of 3*D* solid becomes more practical.

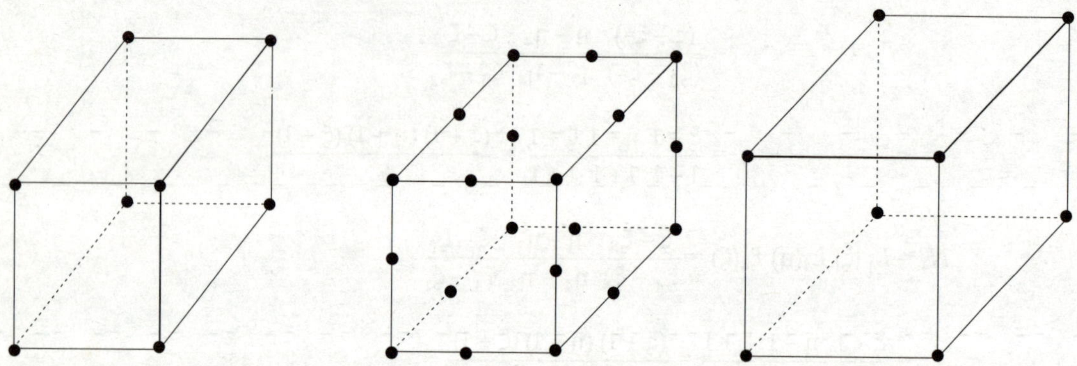

H24 obtained by combining two T12   H60 combining two T30   H96 combining two T48

Fig. 6.15 Parallelopiped elements combining tetrahedron elements

Derivation of shape function for the 3*D* isoparametric finite element with 20 edge nodes.
We have choosen twenty nodes on the element as shown in Fig. 6.16. The approximate function within the hexahedron element is given by

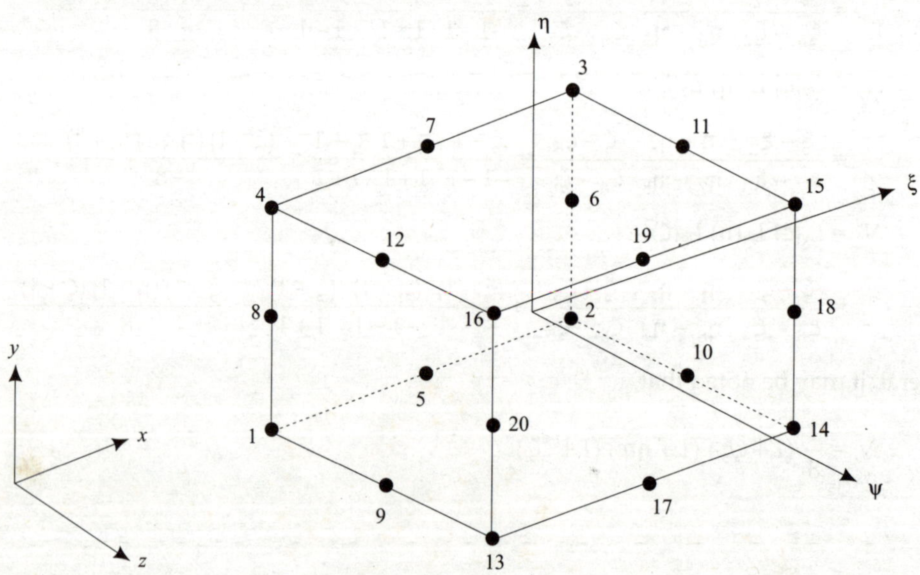

Fig. 6.16 20-node element.

$$u^{(e)}(x, y, z) = \sum_{i=1}^{20} N_i u_i$$

The local axes ξ, η and ψ are located at the centroid of the element (see Fig. 6.16)
Now, on basis of Fig. 6.17,

$$\boxed{N_i = \frac{1}{8}(1 + \xi)(1 + \eta)(1 + \psi) - \frac{1}{2}N_9 - \frac{1}{2}N_{12} - \frac{1}{2}N_{17}} \qquad \text{...[1]}$$

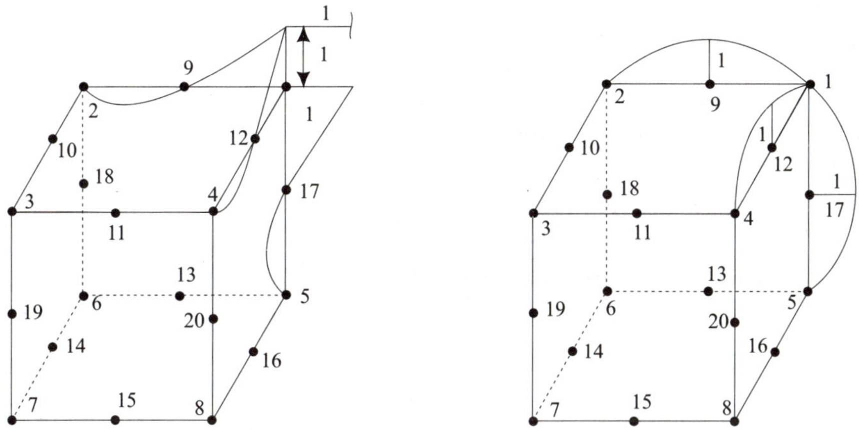

Fig. 6.17 Derivation of shape function for 20-noded three-*D* element

In two-dimensional quadratic variation, the shape functions for midside nodes were

$$N_i = \frac{1}{2}(1 - \xi^2)(1 + \eta) \quad \text{for } \xi = 0$$

and $\eta = +/-1.$

Therefore, the functions for midside nodes with effect of the third dimension can be given as :

$$N_9 = \frac{1}{2}(1 - \xi^2)(1 + \eta)(1 + \psi)/_2 \qquad \text{...[2a]}$$

$$N_{12} = \frac{1}{2}(1 - \eta^2)(1 + \xi)(1 + \psi)/_2 \qquad \text{...[2b]}$$

$$N_{17} = \frac{1}{2}(1 + \psi^2)(1 + \xi)(1 + \eta)/_2 \qquad \text{...[2c]}$$

$$\therefore \quad N_i = \frac{1}{8}(1 + \xi)(1 + \eta)(1 + \psi)[1 - (1 - \xi) - (1 - \eta) - (1 - \psi)]$$

$$\Rightarrow \quad \boxed{N_i = \frac{1}{8}(1 + \xi)(1 + \eta)(1 + \psi)(\xi + \eta + \psi - 2)} \qquad \text{...[3]}$$

That is, the shape function for corner nodes can be given as

$$N_i = \frac{1}{8}(1 + \xi_0\xi_i)(1 + \eta_0\eta_i)(1 + \psi_0\psi_i)(\xi_0\xi_i + \eta_0\eta_i + \psi_0\psi_i - 2)$$

where $\xi_0 = +/-1$, $\eta_0 = +/-1$, $\psi_0 = +/-1$ and for midside nodes they can be given as:

(i) for $\xi = 0, \eta_0 = +/-1\ \psi_0;\ N_i = \frac{1}{4}(1 - \xi_i^2)(1 + \eta_0\eta_i)(1 + \psi_0\psi_i)$

(ii) for $\eta = 0, \xi_0 = \psi_0 = +/-1;\ N_i = \frac{1}{4}(1 - \eta_i^2)(1 + \xi_0\xi_i)(1 + \psi_0\psi_i)$

(iii) for $\psi = 0, \xi_0 = \eta_0 = +/-1;\ N_i = \frac{1}{4}(1 - \psi_i^2)(1 + \xi_0\xi_i)(1 + \eta_0\eta_i)$

The node numbering scheme of Fig. 6.16 shows the face that corresponds to $\psi = -1$ having nodes numbered 1– 8, with corner nodes numbered 1–4 and midside nodes 5–8.

$N_i = \frac{1}{8}(1 + \xi)(1 + \eta)(1 + \psi)(\xi + \eta + \psi - 2)$	
$N_1 = (1 - \xi)(1 - \eta)(1 - \psi)(-\xi - \eta - \psi - 2)/8$	$N_5 = (1 + \xi^2)(1 - \eta)(1 - \psi)/4$
$N_2 = (1 + \xi)(1 - \eta)(1 - \psi)(\xi - \eta - \psi - 2)/8$	$N_6 = (1 + \xi)(1 - \eta^2)(1 - \psi)/4$
$N_3 = (1 + \xi)(1 + \eta)(1 - \psi)(\xi + \eta - \psi - 2)/8$	$N_7 = (1 - \xi^2)(1 + \eta)(1 - \psi)/4$
$N_4 = (1 - \xi)(1 + \eta)(1 - \psi)(-\xi + \eta - \psi - 2)/8$	$N_8 = (1 - \xi)(1 - \eta^2)(1 - \psi)/4$
$N_9 = (1 - \xi)(1 - \eta)(1 - \psi^2)/4$	$N_{10} = (1 + \xi)(1 - \eta)(1 - \psi^2)/4$
$N_{11} = (1 + \xi)(1 + \eta)(1 - \psi^2)/4$	$N_{12} = (1 - \xi)(1 + \eta)(1 - \psi^2)/4$
$N_{13} = (1 - \xi)(1 - \eta)(1 + \psi)(-\xi - \eta + \psi - 2)/8$	$N_{17} = (1 - \xi^2)(1 - \eta)(1 + \psi)/4$
$N_{14} = (1 + \xi)(1 - \eta)(1 + \psi)(\xi - \eta + \psi - 2)/8$	$N_{18} = (1 + \xi)(1 - \eta^2)(1 + \psi)/4$
$N_{15} = (1 + \xi)(1 + \eta)(1 + \psi)(\xi + \eta + \psi - 2)/8$	$N_{19} = (1 - \xi^2)(1 + \eta)(1 + \psi)/4$
$N_{16} = (1 - \xi)(1 + \eta)(1 + \psi)(-\xi + \eta + \psi - 2)/8$	$N_{20} = (1 - \xi)(1 - \eta^2)(1 + \psi)/4$

Derivation of Lagrange shape function for 27-node Lagrange element

See the diagram (Fig. 6.18) on the next page.

Solution: The plane that corresponds to $\psi = -1$ contains the first nine nodes.

The shape functions are as follows.

Corner nodes:

$N_1 = -(1 - \xi)(1 - \eta)(1 - \psi)\xi\eta\psi/8$ $\quad$ $N_3 = (1 + \xi)(1 - \eta)(1 - \psi)\xi\eta\psi/8$

$N_5 = -(1 + \xi)(1 + \eta)(1 - \psi)\xi\eta\psi/8$ $\quad$ $N_7 = (1 - \xi)(1 + \eta)(1 - \psi)\xi\eta\psi/8$

$N_{19} = (1 - \xi)(1 - \eta)(1 + \psi)\xi\eta\psi/8$ $\quad$ $N_{21} = -(1 + \xi)(1 - \eta)(1 + \psi)\xi\eta\psi/8$

$N_{23} = (1 + \xi)(1 + \eta)(1 + \psi)\xi\eta\psi/8$ $\quad$ $N_{25} = -(1 - \xi)(1 + \eta)(1 + \psi)\xi\eta\psi/8$

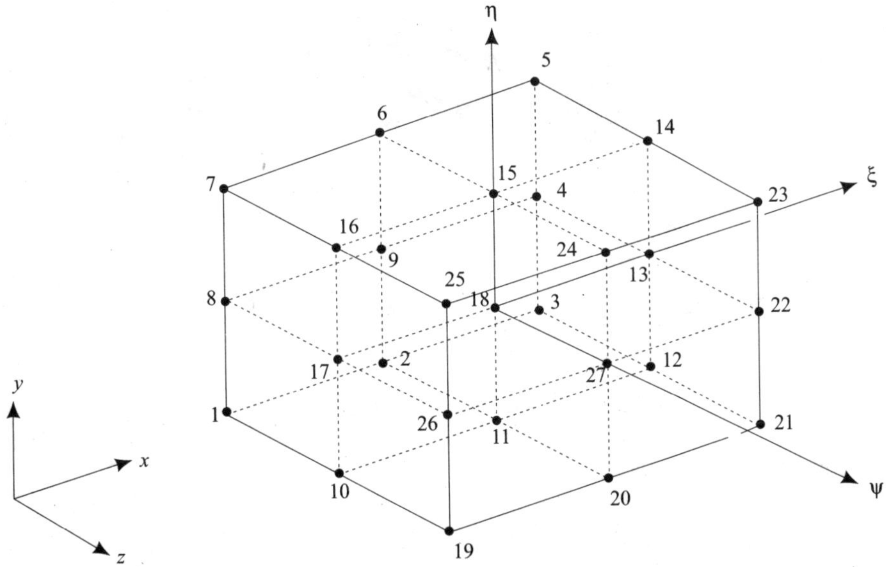

Fig. 6.18 27-node Lagrange element.

Midside nodes:

$$N_2 = (1 - \xi^2)(1 - \eta)(1 - \psi)\,\eta\psi/4 \qquad N_4 = -(1 + \xi)(1 - \eta^2)(1 - \psi)\,\xi\psi/4$$

$$N_6 = -(1 - \xi^2)(1 + \eta)(1 - \psi)\,\eta\psi/4 \qquad N_8 = (1 - \xi)(1 - \eta^2)(1 - \psi)\,\xi\psi/4$$

$$N_{10} = (1 - \xi)(1 - \eta)(1 - \psi^2)\,\xi\psi/4 \qquad N_{12} = -(1 + \xi)(1 - \eta)(1 - \psi^2)\,\xi\eta/4$$

$$N_{14} = (1 + \xi)(1 + \eta)(1 - \psi^2)\,\xi\eta/4 \qquad N_{16} = -(1 - \xi)(1 + \eta)(1 - \psi^2)\,\xi\eta/4$$

$$N_{20} = -(1 - \xi^2)(1 - \eta)(1 + \psi)\,\eta\psi/4 \qquad N_{22} = (1 + \xi)(1 - \eta^2)(1 + \psi)\,\xi\psi/4$$

$$N_{24} = (1 - \xi^2)(1 + \eta)(1 + \psi)\,\eta\psi/4 \qquad N_{26} = -(1 - \xi^2)(1 - \eta^2)(1 + \psi)\,\xi\psi/4$$

Face centre nodes:

$$N_9 = -(1 - \xi^2)(1 - \eta^2)(1 - \psi)\,\psi/2 \qquad N_{11} = -(1 - \xi^2)(1 - \eta)(1 - \psi^2)\eta/2$$

$$N_{13} = (1 + \xi)(1 - \eta^2)(1 - \psi^2)\,\xi/2 \qquad N_{15} = (1 - \xi^2)(1 + \eta)(1 - \psi^2)\eta/2$$

$$N_{17} = -(1 - \xi)(1 - \eta^2)(1 - \psi^2)\,\xi/2 \qquad N_{27} = (1 - \xi^2)(1 - \eta^2)(1 - \psi)\,\psi/2$$

Centre node:

$$N_{18} = (1 - \xi^2)(1 - \eta^2)(1 - \psi^2)$$

Cubic Element with 32 Nodes

The hexahedral element shown in Fig. 6.19 has 32 nodes. The piecewise cubic polynomial is given by

$$u^{(e)} = \sum_{i=1}^{32} N_i u_i \qquad \text{...[1]}$$

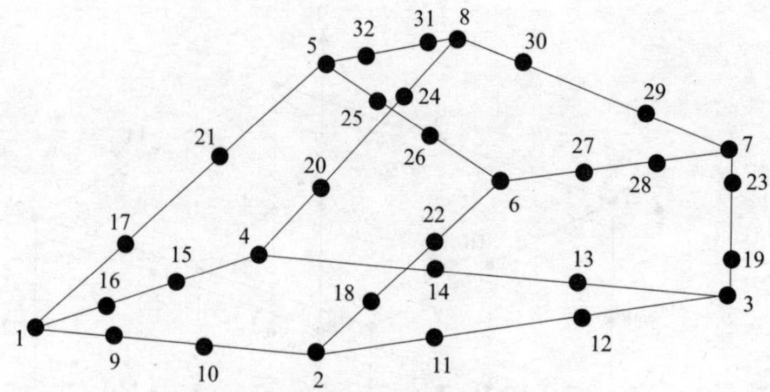

Fig. 6.19 Cubic element with thirty-two nodes

where for corner nodes, $i = 1\ (1)\ 8$

$$N_i = \frac{1}{64}(1 + \xi\xi_i)\,(1 + \eta\eta_i)\,(1 + \zeta\zeta_i)\,(9\xi^2 + 9\eta^2 + 9\zeta^2 - 19) \quad \text{...[2]}$$

for one-third side nodes, $i = 9, 10, 13, 14, 25, 26\ 29, 30$

$$N_i = \frac{9}{64}(1 - \xi^2)(1 + 9\,\xi\xi_i)\,(1 + \eta\eta_i)(1 + \zeta\zeta_i)$$

and for one-third side nodes, i = 11, 12, 15, 16, 27, 28 31, 32

$$N_i = \frac{9}{64}(1 - \eta^2)\,(1 + 9\,\eta\eta_i)\,(1 + \xi\xi_i)\,(1 + \zeta\zeta_i) \quad \text{...[2]}$$

and for one-third side nodes, $i = 17, 18, 19, 20, 21, 22, 23, 24$

$$N_i = \frac{9}{64}(1 - \xi^2)\,(1 + 9\,\zeta\zeta_i)\,(1 + \xi\xi_i)\,(1 + \eta\eta_i)$$

6.5 MESH PREPARATION

Hexahedron as an Assemblage of Tetrahedra

The division of a space volume into individual tetrahedra sometimes presents difficulties of visualisation and could easily lead to errors in nodal numbering etc. A more convenient subdivision of space is into eight cornered brick elements. Such elements could be assembled automatically from several tetrahedra and process of creating these tetrahedra left to a simple logical program. In Fig. 6.20(a) and (b), an eight cornered element of a brick shape which could conveniently be used in computer aided design and stress analysis etc. is shown. It will be readily appreciated from the "exploded" view that an element of this shape (hexahedron) could be built in two, and only two, distinct ways from five tetrahedral shape elements. Both possible divisions into tetrahedral elements are illustrated.

Hexahedral Element and its Subdivision

A hexahedral (eight cornered brick) element can be systematically divided into five tetrahedra. This is illustrated in Fig. 6.20(a) and (b) and the nodal connectivity is shown in Tables 6.1 and 6.2.

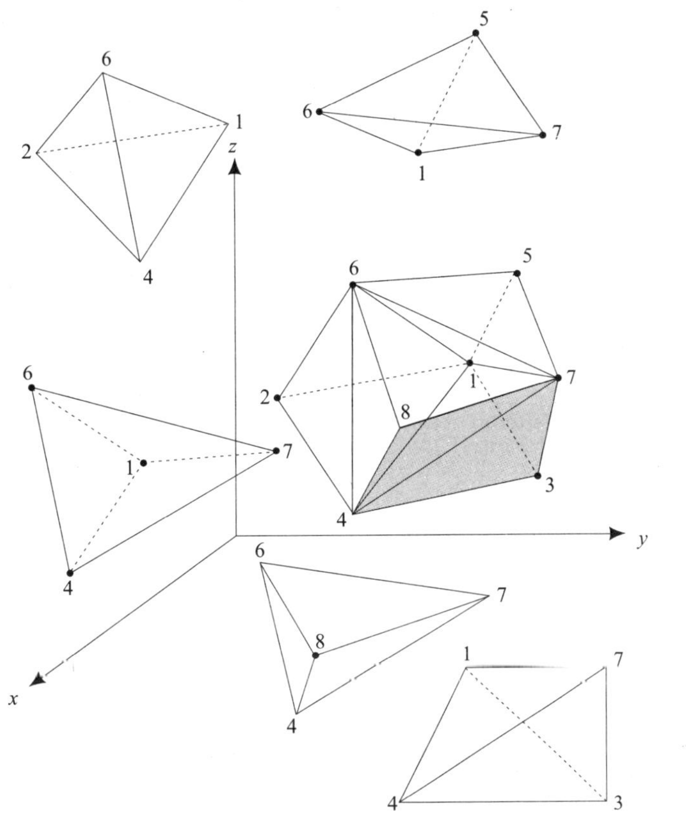

Fig. 6.20(a) Systematic way of splitting an eight cornered hexahedron shaped brick into five tetrahedra

Table 6.1 Nodal connectivity of Hexahedron of Fig. 6.20(a) as a subdivision into five tetrahedra

Element No.	*Nodes*			
	i	*j*	*k*	*l*
1	1	4	2	6
2	1	4	3	7
3	6	7	5	1
4	6	7	8	4
5	1	4	6	7

Table 6.2 Nodal connectivity of Hexahedron of Fig. 6.20(b) as a subdivision into five tetrahedra

Element No.	*Nodes*			
	i	*j*	*k*	*l*
1	1	2	3	5
2	2	6	5	8
3	2	3	4	8
4	3	5	7	8
5	2	3	5	8

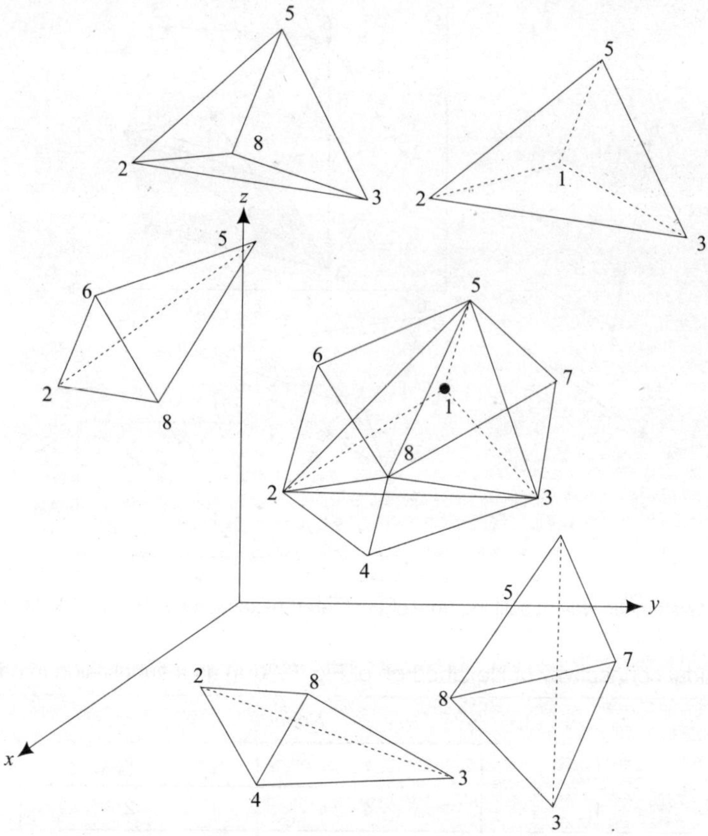

Fig. 6.20(b) An alternative systematic way of splitting an eight cornered hexahedron shaped brick into five tetrahedra

EXERCISES

1. What is an isoparametric element ? Give an example.
2. What is Jacobian ? Develop the Jacobian for the tetrahedral element.
3. Derive the shape function for 3*D* isoparametric element with 20 nodes
4. Use the Lagrange polynomial to derive the shape function for the quadratic 3*D* Lagrange finite element with 27 nodes.
5. Define cubic element with 32 nodes and find the shape functions.

7

Isoparametric Elements, Numerical Integration and Higher Order Elements

7.1 INTRODUCTION

The previous chapters have been concerned with the derivation of local finite elements for various physical problems, and applications have centered about the rectangular finite element. For higher order elements the nodes may also occur on the faces of the element as well as inside it. The use of additional nodes increases the accuracy of the process, so that in general fewer elements need be used in such cases than for simple elements. But using comparatively larger elements then poses a problem on curved and irregular boundaries.

For this reason it would appear advisable to be able to form elements with curved boundaries from those with straight boundaries such as triangles and rectangles. The geometry of the element and the relation between interpolation functions and shape functions will be emphasized in this chapter.

Concept of mapping in isoparametric elements (Fig. 7.1).

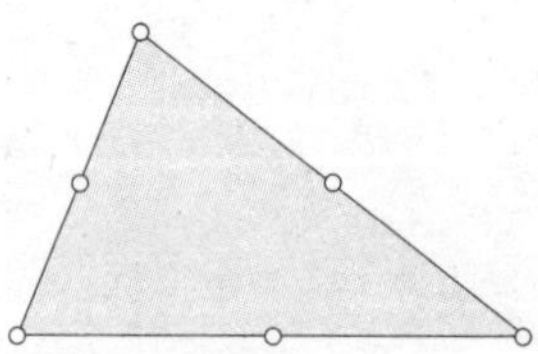

Natural coordinate system (Parent element)

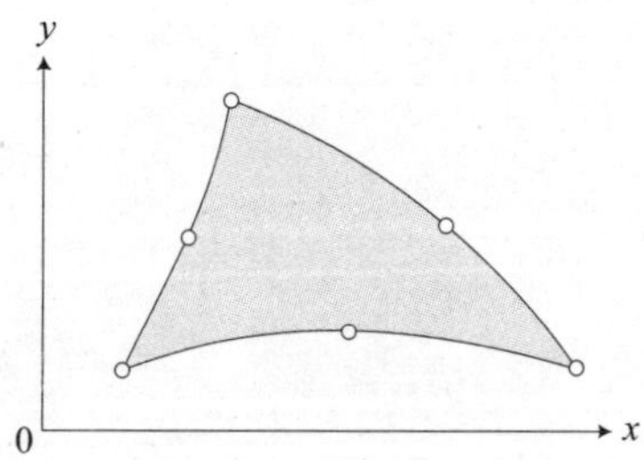

(Global system) Mapped element

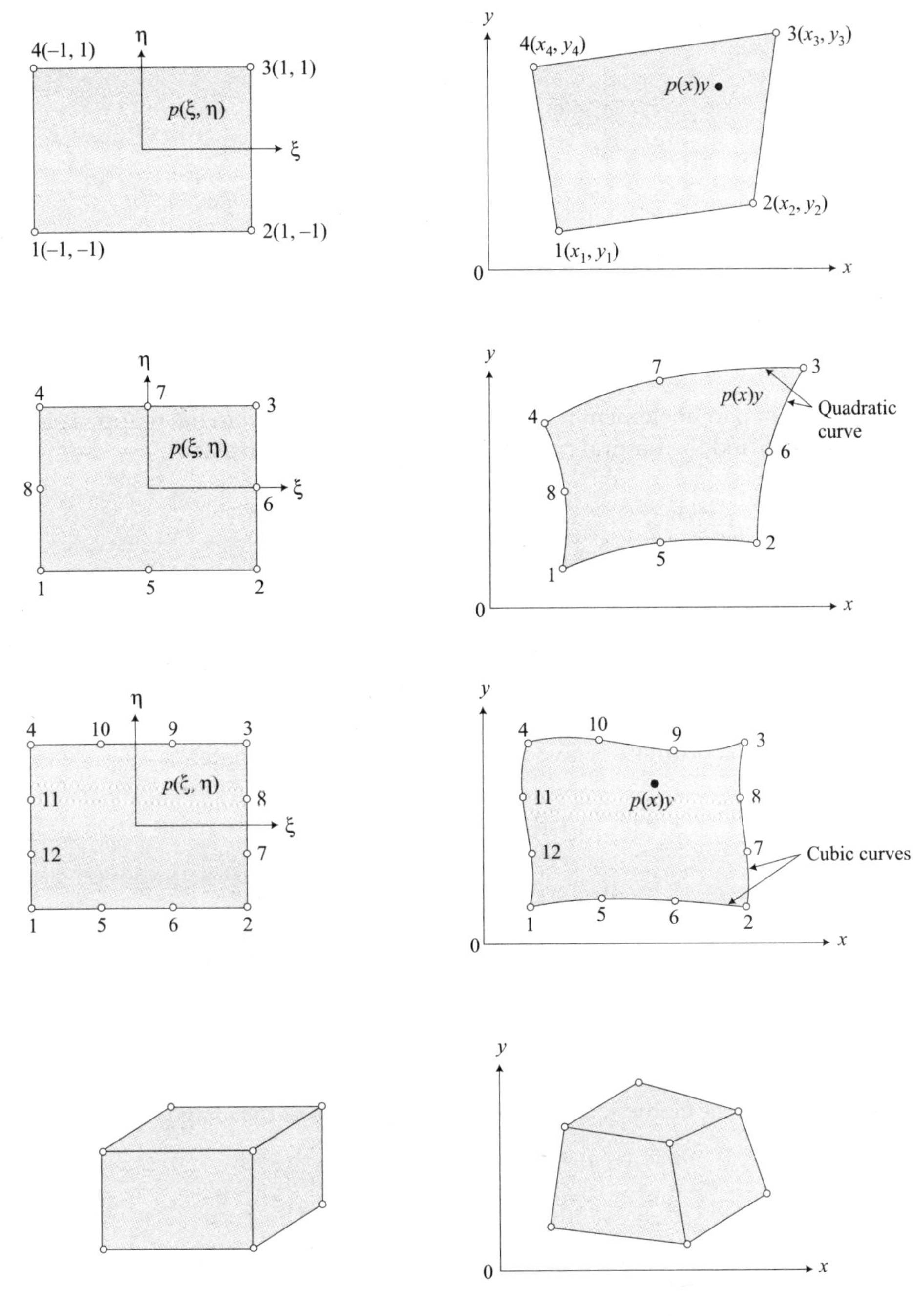

Natural coordinate system (Parent element)

Global system (Mapped element)

(Contd.)

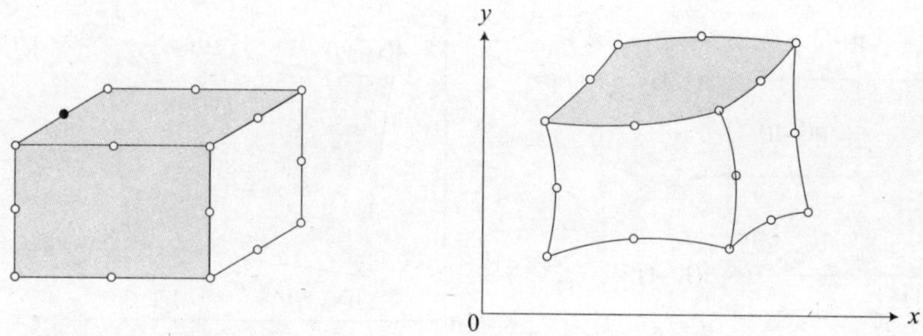

(Natural Coordinate system) (Global system)

Fig. 7.1

For every point *P* in parent element there is a corresponding point in the mapped element and the relation between (x) and the natural coordinate (ξ) system is given as

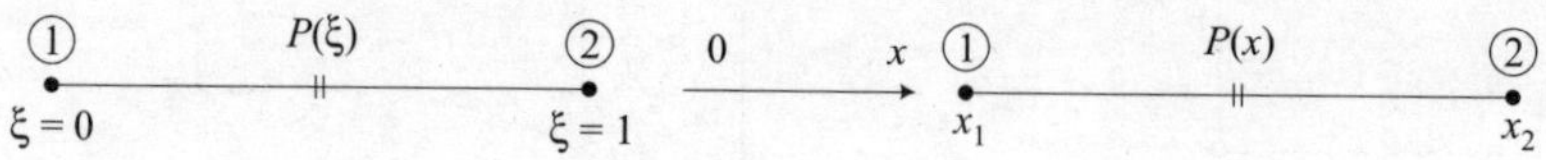

(Parent element) (Mapped element)

$$x = [\,x_{21}\; x_1] \begin{Bmatrix} u \\ 1 \end{Bmatrix} \quad \text{where } x_{21} = x_2 - x_1$$

and the (local or global) x coordinate is given in terms of nodal coordinates as

$$x = [(1-u)\; u] \begin{Bmatrix} x_1 \\ x_2 \end{Bmatrix} \qquad ...[2]$$

or $$x = [N^1]\,\{\underline{x}\} \qquad ...[3]$$

where $[N^1]$ is known as the shape function matrix. ...[4]

7.2 ISOPARAMETRIC FORMULATION

The displacements within the element can be expressed in terms of the shape functions as

$$\left.\begin{aligned} u &= N_1 q_1 + N_2 q_3 + N_3 q_5 + N_5 q_7 \\ v &= N_1 q_2 + N_2 q_4 + N_3 q_6 + N_4 q_8 \end{aligned}\right\}$$

The above can be put in matrix notation as

$$\boldsymbol{u} = \begin{Bmatrix} u \\ v \end{Bmatrix} = \boldsymbol{N}\,\boldsymbol{q}$$

where

$$N = \begin{bmatrix} N_1 & 0 & N_2 & 0 & N_3 & 0 & N_4 & 0 \\ 0 & N_1 & 0 & N_2 & 0 & N_3 & 0 & N_4 \end{bmatrix}$$

Now, if the relation between the Cartesian coordinates (x, y) and the natural coordinates (ξ, η) is also expressed in terms of the same shape functions N_1 etc. as

$$\left.\begin{aligned} x &= N_1x_1 + N_2x_2 + N_3x_3 + N_4x_4 \\ y &= N_1y_1 + N_2y_2 + N_3y_3 + N_4y_4 \end{aligned}\right\}$$

and is called *isoparametric formulation*.

Advantages of Isoparametric Element

(1) These elements are useful in modeling structures with curved edges and in grading a mesh from coarse to fine.

(2) Isoparametric elements are versatile and are used in two- and three-dimensional elasticity problems.

Definition: Elements in which the same interpolating functions are used to define the shape of the element as well as the displacement functions are defined as isoparametric.

For example, in a *truss* element, we have

$$x = N_1x_1 + N_2x_2$$

where x defines the geometry and x_1 and x_2 are nodal coordinates. The displacement function is defined by

$$u = N_1u_1 + N_2u_2$$

Coordinate Transformation

The concept of isoparametric element is based on the transformation of the parent element is local or natural coordinate system to an arbitrary shape in the Cartesian coordinate system as shown by examples in Fig. 7.1. A convenient way of expressing the transformation is to make use of the shape function of the rectilinear elements in their natural coordinate system and the nodal values of the coordinates.

Thus, the Cartesian coordinate of a point in an element may be expressed as

$$\left\{\begin{aligned} x &= N_1x_1 + N_2x_2 + \cdots + N_nx_n \\ y &= N_1y_1 + N_2y_2 + \cdots + N_ny_n \\ z &= N_1z_1 + N_2z_2 + \cdots + N_nz_n \end{aligned}\right\} \quad ...[1]$$

or in matrix form

$$\{x\} = [N]\,\{x\}_e \quad ...[2]$$

where N are shape functions and $(x)_e$ are the coordinates of nodal points of the element. The shape functions are to be expressed in natural coordinate system.

Illustration: Consider the mapping of rectangular element in natural local coordinate system to global Cartesian coordinate system as shown in Fig. 7.2 below:

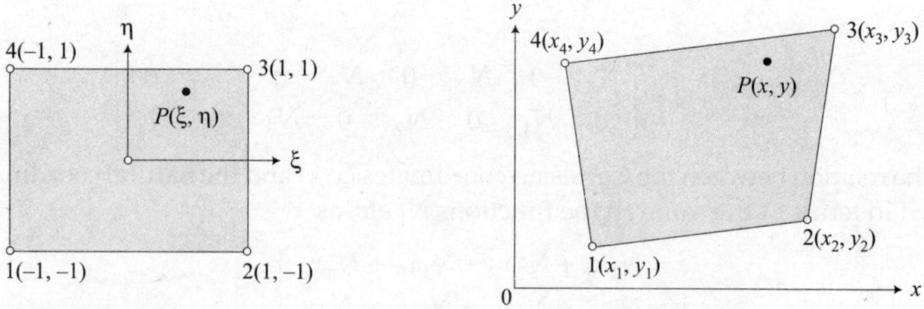

(a) Parent element (b) Mapped element

Fig. 7.2

The parent element has:

Nodes:	1,	2,	3,	4
	↓	↓	↓	↓
Coordinates:	(–1, –1),	(1, –1),	(1, 1),	(–1, 1)

The shape functions of this element are

$$N_1 = \frac{(1-\xi)(1-\eta)}{4}, \quad N_2 = \frac{(1+\xi)(1-\eta)}{4}$$

$$N_3 = \frac{(1+\xi)(1+\eta)}{4} \text{ and } N_4 = \frac{(1-\xi)(1+\eta)}{4}$$

$$P = P(\xi, \eta).$$

In global system the coordinates of the nodal points are

$$(x_1, y_1), (x_2, y_2), (x_3, y_3) \text{ and } (x_4, y_4)$$

To get the mapping we define the coordinate of point P as

$$\begin{Bmatrix} x = N_1x_1 + N_2x_2 + N_3x_3 + N_4x_4 \\ y = N_1y_1 + N_2y_2 + N_3y_3 + N_4y_4 \end{Bmatrix}$$

Example (1): Derive the interpolation functions for a four-node isoparametric quadrilateral element.

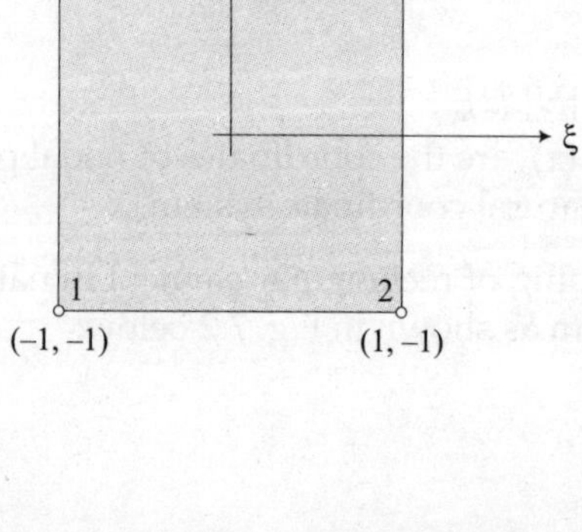

Solution: Rectangular element with a normalized reference $\xi\eta$ at the centroid, so that we have linear variation of the field variable u. We will propose for this case that

$$u = \sum_{i=1}^{4} N_i u_i$$

or $$u = N_1 u_1 + N_2 u_2 + N_3 u_3 + N_4 u_4$$

$$N_1 = [L_1(\xi)]\,[L_1(\eta)] \quad \text{...(1)}$$

$$N_2 = [L_2(\xi)]\,[L_2(\eta)] \quad \text{...(2)}$$

$$N_3 = [L_3(\xi)]\,[L_3(\eta)] \quad \text{...(3)}$$

$$N_4 = [L_4(\xi)]\,[L_4(\eta)] \quad \text{...(4)}$$

Note that the interpolation function (or shape function) at node 1, (N_1) is the product of shape functions along the lines $\xi = -1$ and $\eta = 1$.

$$N_1 = L_1(\xi)\, L_1(\eta) = \frac{\xi - 1}{-1-1} \cdot \frac{\eta - 1}{-1-1}$$

$$= \frac{(1-\xi)\,(1-\eta)}{4}$$

Along the similar lines,
we have,

$$\left\{ \begin{aligned} N_2 &= L_{2\xi} L_{2\eta} = \frac{\xi+1}{1+1}\,\frac{\eta-1}{-1-1} = \frac{(1+\xi)\,(1-\eta)}{4} \\ N_3 &= L_{3\xi} L_{3\eta} = \frac{\xi+1}{1+1}\,\frac{\eta+1}{1+1} = \frac{(1+\xi)\,(1-\eta)}{4} \\ N_4 &= L_{4\xi} L_{4\eta} = \frac{\xi-1}{-1-1}\,\frac{\eta+1}{1+1} = \frac{(1-\xi)\,(1+\eta)}{4} \end{aligned} \right\}$$

Properties of Isoparametric Concept

P_1: If two adjacent elements are generated using shape functions, then there is continuity at the common edge. (See the diagram Fig. 7.3 on the next page).

Proof: From the elements in Fig. 7.3 we notice that for any point on AB, $N_i = 0$, (shape functions). For nodes not on the edge N_i exists for nodes on the edge. This implies that the final function is the same for the common edge. AB in any two adjacent elements, when we give the same coordinate values for the nodes on the common edge.

Hence edge AB is contiguous in the adjacent elements.

P_2: If the shape functions used are such that continuity of displacement is represented in the parent coordinates, then the continuity requirementh will be satisfied in the isoparametric elements also.

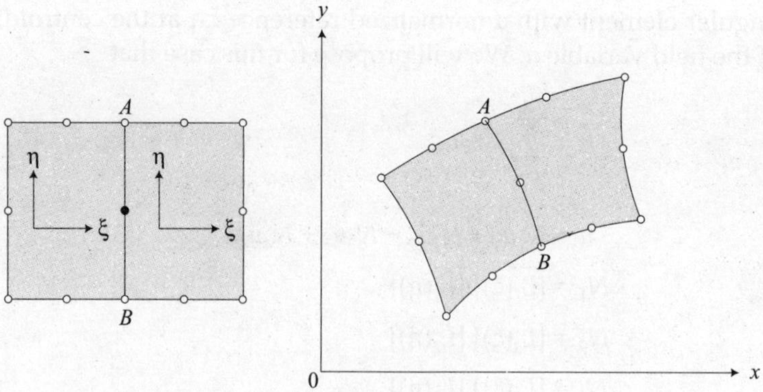

Fig. 7.3

Proceed exactly as in P_1 above for the proof.

P_3: (Convergence criteria for isoparametric element)

The constant derivative conditions and condition for rigid body are satisfied for all isoparametric elements if, $\sum N_i = 1$

Proof: The displacement function over the element may be assumed as

$$u = \alpha_1 + \alpha_2 x + \alpha_3 y + \alpha_4 z \quad \text{...[1]}$$

Hence, displacement function is given by

$$u_i = \alpha_1 + \alpha_2 x_i + \alpha_3 y_i + \alpha_4 z_i$$

We define nodal displacement at any point in the element in terms of nodal displacement as

$$u = \sum N_i u_i$$

$$\therefore \quad u = \sum N_i \ (\alpha_1 + \alpha_2 x_i + \alpha_3 y_i + \alpha_4 z_i)$$

$$= \alpha_1 \sum N_i + \alpha_2 \sum N_i x_i + \alpha_3 \sum N_i y_i + \alpha_4 \sum N_i z_i$$

From the isoparametric concept, we know

$$\sum N_i x_i = x$$

$$\sum N_i y_i = y$$

$$\sum N_i z_i = z$$

$$\therefore \quad u = \alpha_1 \sum N_i + \alpha_2 x + \alpha_3 y + \alpha_4 z \quad \text{...[2]}$$

Hence if equation [2] has to represent equation [1] uniquely, then

$$\sum N_i = 1$$

Uniqueness of Mapping

It is important to note that a point in parent element represents only one point in the isoparametric element. Care should be taken to see that this correspondence is unique. That is, only one set of local coordinates correspond to a particular point in space. With violent distortion, it is possible to obtain the undesirable situation of non-unique correspondence.

For a complete definition of geometry or transformation every set of local co-coordinates will correspond to a set of global co-ordinates and in general to only one such set. Violent distortion will result in non-uniqueness. Violently distorted elements may give singular Jacobian (Transformation) matrix. Some examples of violently distorted elements are shown in Fig. 7.4(a).

Comparing the geometry of these distorted elements with those of Fig. 7.5 we can say that as we go on increasing or decreasing an internal angle beyond or below 90 degree, distortion of element starts increasing. When these angles reach 0 degree or 180 degree the sides or points of the element start coinciding or/and number of sides decrease and as they cross these limits of 0 degree and 180 degree the elements get *fully distorted* with undefinable geometry. Therefore, we can say that for violent distortions 0 degrees or 180 degrees is the critical limit for internal angles to make the elements violently distorted.

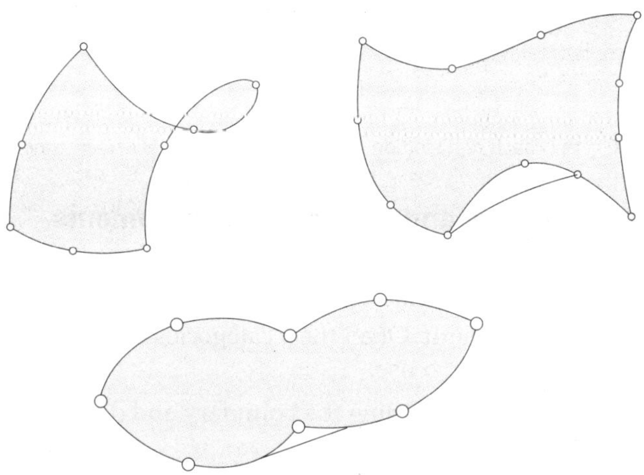

Fig. 7.4(a) Non-uniqueness of mapping

The following figures (Fig. 7.4(b)) explain the distortions.

Another distortion comes through *improper placement of midside nodes*. Regular and symmetrical figures always give good results but if the midside nodes shift to either side of the midpoint by a large distance, a numerically ill-conditioned Jacobian matrix may result. The permissible range for such midside nodes is to keep them within middle third.

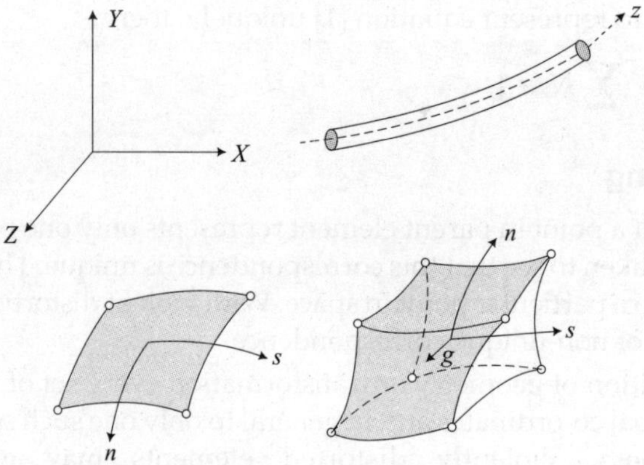

Fig. 7.4(b)

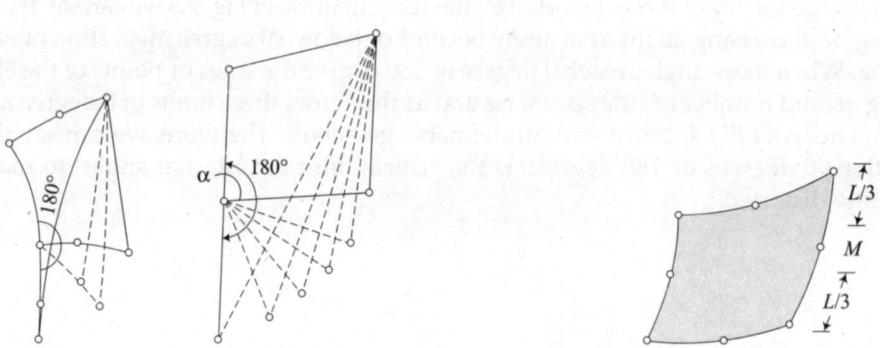

(a) Tolerable distortion 180°—approach of distortion

(b) M is safe zone for midpoints

Fig. 7.5

Isoparametric, Superparametric and Subparametric Elements

While using curved elements it is better to use larger number of small elements than to take up small number of large elements. Yet one rule/base does not fit everywhere.

Situations are to be tackled per merit. Of all the 3 categories of curved elements isoparametric elements are widely used and are easy to use.

Definition (1): If the shape functions defining the boundary and displacements are the same, the element is called as **isoparametric element**.

Example: The following element (Fig. 7.6(a)) is isoparametric as all the eight nodes are used in defining the geometry and displacement.

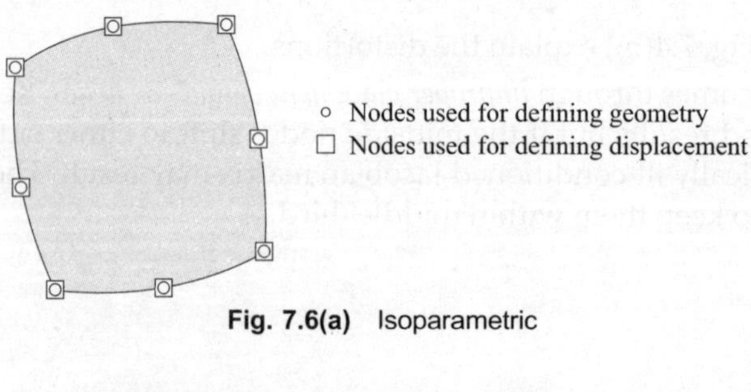

Fig. 7.6(a) Isoparametric

Definition (2): The element in which more than one number of nodes are used to define geometry compared to the number of nodes used to define displacement are known as **superparametric elements**.

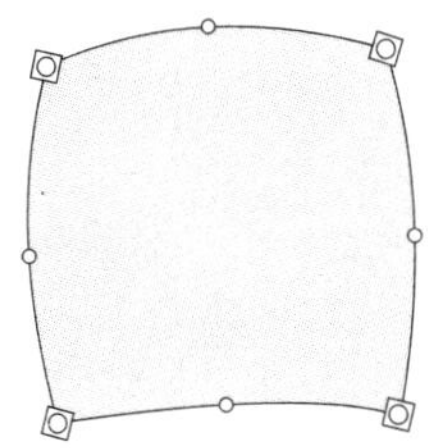

Fig. 7.6(b) Superparametric

Example: Fig. 7.6(b).

Definition (3): Figure 7.6(c) shows a **subparametric element** in which less number of nodes are used to define geometry compared to the number of nodes used for defining the displacements.

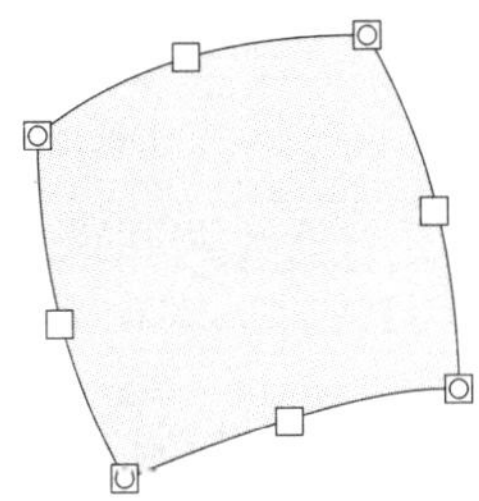

Fig. 7.6(c) Subparametric

Explanation: Consider the following curved elements and parent elements (Fig. 7.7).

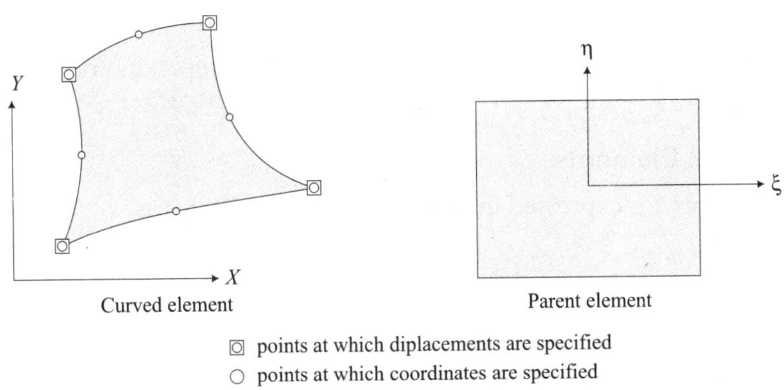

Fig. 7.7(a) Superparametric elements

$$\begin{Bmatrix} X \\ Y \end{Bmatrix} = [N^1] \begin{Bmatrix} x \\ y \end{Bmatrix}; \begin{Bmatrix} U \\ V \end{Bmatrix} = [N] \begin{Bmatrix} u \\ v \end{Bmatrix}$$

Degree of $[N^1]$ > degree of $[N]$

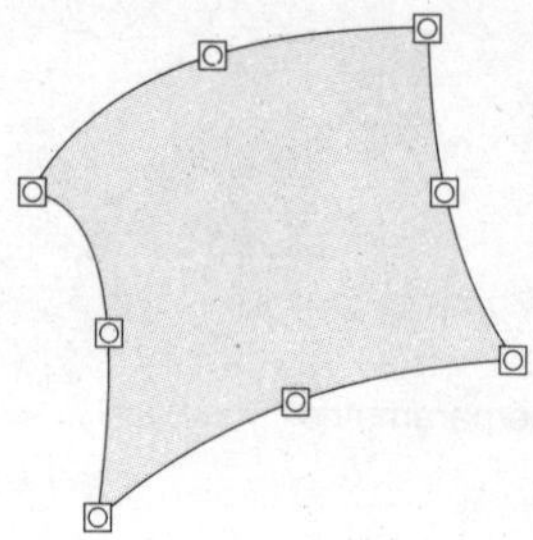

Degree of $[N']$ = Degree of $[N]$

Fig. 7.7(b) Isoparametric element

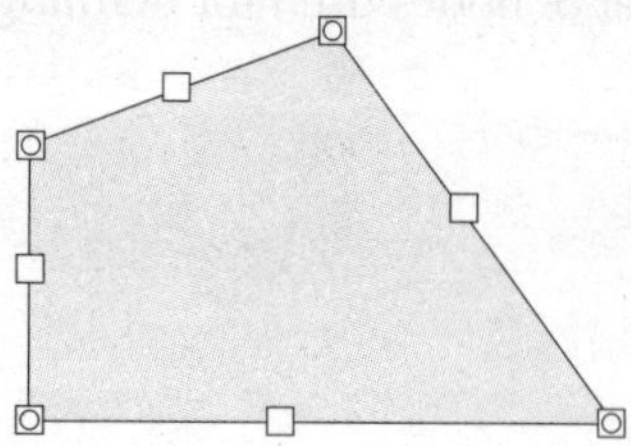

Degree of $[N']$ < Degree of $[N]$

Fig. 7.7(c)

(a) If the degree of N^1 is greater than the degree of N then the element is called superparametric element [as shown in Fig. 7.7(a)].

(b) If the shape function chosen to describe the element coordinates are identical to that used to prescribe the function variable, then the element is known as isoparametric element [as shown in Fig. 7.7(b)].

(c) If the degree of $[N^1]$ is less than the degree $[N]$ then it is subparametric element [as shown in Fig. 7.7(c)].

Isoparametric Axial Elements

The axial coordinate z is expressed in terms of nodal coordinates as:

$$z = \frac{1}{2}\left[(1-\xi)\,(1+\xi)\right]\begin{Bmatrix} z_1 \\ z_2 \end{Bmatrix} \qquad ...[1]$$

and

$$dz = \frac{l}{2}\,d\xi \qquad ...[2]$$

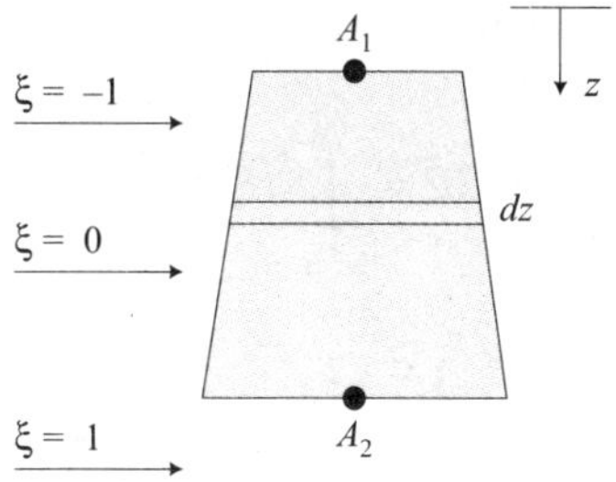

Fig. 7.8 Tapering element

For isoparametric formulation,

$$A = \frac{1}{2}\left[(1-\xi)\,(1+\xi)\right]\begin{Bmatrix} A_1 \\ A_2 \end{Bmatrix} \qquad \text{...[3]}$$

and the axial displacement

$$u = \frac{1}{2}\left[(1-\xi)\,(1+\xi)\right]\begin{Bmatrix} u_1 \\ u_2 \end{Bmatrix} \qquad \text{...[4]}$$

The axial strain ε is given by

$$\varepsilon = \frac{du}{dz} = \frac{du}{d\xi}\cdot\frac{d\xi}{dz} \qquad \text{...[5]}$$

$$= \begin{bmatrix} -\frac{1}{l} & \frac{1}{l} \end{bmatrix}\begin{Bmatrix} u_1 \\ u_2 \end{Bmatrix}$$

$$= [B]\begin{Bmatrix} u_1 \\ u_2 \end{Bmatrix} \qquad \text{...[6]}$$

The stiffness matrix $[k]$ is given by ,

$$[k] = \int_{-l/2}^{l/2} B^T E B dV \qquad \text{...[7]}$$

$$= \frac{lE}{4}\int_{-1}^{1}\begin{Bmatrix} -\frac{1}{l} \\ \frac{1}{l} \end{Bmatrix}\begin{bmatrix} -\frac{1}{l} & \frac{1}{l} \end{bmatrix}\{[(1-\xi)\,(1+\xi)]\}\begin{Bmatrix} A_1 \\ A_2 \end{Bmatrix}\, d\xi \qquad \text{...[8]}$$

$$= \frac{E\,(A_1 + A_2)}{2l}\begin{bmatrix} 1 & -1 \\ -1 & 1 \end{bmatrix}$$

Consistent load vector due to self weight is given by

$$\{Q\} = \int N^T F dV = \int_{-l/2}^{l/2} N^T \xi\, A dz \qquad ...[9]$$

$$= \frac{\xi l}{8} \int_{-1}^{1} \begin{Bmatrix} (1-\xi) \\ (1+\xi) \end{Bmatrix} [(1-\xi)\ (1+\xi)] \begin{Bmatrix} A_1 \\ A_2 \end{Bmatrix} d\xi \qquad ...[10]$$

$$= \frac{\xi l}{8} \begin{bmatrix} 8/3 & 4/3 \\ 4/3 & 8/3 \end{bmatrix} \begin{Bmatrix} A_1 \\ A_2 \end{Bmatrix} \qquad ...[11]$$

$$= \frac{\xi l}{6} \begin{Bmatrix} \frac{A_1}{3} + \frac{A_2}{6} \\ \frac{A_1}{6} + \frac{A_2}{3} \end{Bmatrix} \qquad ...[12]$$

Element stiffness matrix for the 3-noded axial element shown in Fig. 7.9 using isoparametric concept.

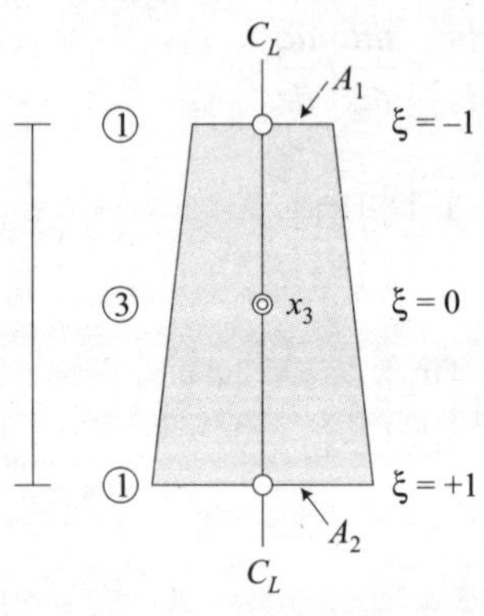

Fig. 7.9 3-noded axial element

Let A be the cross-sectional area of the element, then A can be expressed as

$$A = \frac{(1-\xi)}{2} A_1 + \frac{(1+\xi)}{2} \cdot A_2, \text{ Matrix forms}$$

$$= \begin{bmatrix} \frac{(1-\xi)}{2} & \frac{(1+\xi)}{2} \end{bmatrix} \begin{Bmatrix} A_1 \\ A_2 \end{Bmatrix} \qquad ...[1]$$

From isoparametric concept,

$$x = N_1 x_1 + N_2 x_2 + N_3 x_3$$

The relation between x and ξ for linear or axial element is

$$\xi = \frac{2(x - x_3)}{(x_2 - x_1)} = \frac{x - x_3}{(l/2)}$$

$$\therefore \quad d\xi = \frac{dx}{(l/2)} \quad \text{... [3]}$$

Again, by isoparametric concept,

$$u = N_1 u_1 + N_2 u_2 + N_3 u_3 \quad \text{...[4]}$$

The matrix form of this is ;

$$u = \left[\frac{\xi}{2}(1-\xi)\,\frac{\xi}{2}(1+\xi)\left(1-\xi^2\right)\right]\begin{Bmatrix} u_1 \\ u_2 \\ u_3 \end{Bmatrix} \quad \text{...[5]}$$

$$\varepsilon = \frac{du}{dx} = \frac{du}{d\xi}\cdot\frac{d\xi}{dx} \quad \text{...[6]}$$

$$= \left[(2\xi - 1)\,(2\xi + 1) - 4\xi\right] \times \frac{1}{2} \times \frac{2}{l}\begin{Bmatrix} u_1 \\ u_2 \\ u_3 \end{Bmatrix}$$

(Using [3] and [4] in [6])

$$\therefore \quad \varepsilon = \frac{1}{l}\ \underbrace{[(2\xi - 1)\ (2\xi + 1) - 4\xi]}_{B}\ \underbrace{\begin{Bmatrix} u_1 \\ u_2 \\ u_3 \end{Bmatrix}}_{q}$$

$$\therefore \quad \varepsilon = Bq \quad \text{...[7]}$$

Now $$K = \int_l B^{\mathrm{T}} d\,B\,dV \quad \text{...[8]}$$

where $dV = Adx$ [for axial element]

where $$dV = Adx = \left[\frac{1}{2}(1-\xi)\,A_1 + \frac{1}{2}(1+\xi)\,A_2\right]dx$$

$$dV = \left[\frac{1}{2}(1-\xi)\,A_1 + \frac{1}{2}(1+\xi)\,A_2\right]\frac{l}{2}\,d\xi \quad \text{...[9]}$$

$$K = \int_{-1}^{1} B^T d\, B\left[\frac{1}{2}(1-\xi)A_1 + \frac{1}{2}(1+\xi)A_2\right]\frac{l}{2}d\xi \qquad ...[10]$$

Substituting for B in [10], and also $d = E$, after integrating, we get

$$K = \frac{E}{6l}\left\{A_1\begin{bmatrix} 11 & 1 & -12 \\ 1 & 3 & -4 \\ -12 & -4 & 16 \end{bmatrix} + A_2\begin{bmatrix} 3 & 1 & -4 \\ 1 & 11 & -12 \\ -4 & -12 & 16 \end{bmatrix}\right.$$

Case (i) If $A_2 = 2A_1$, then

$$K = \frac{EA}{6l}\begin{bmatrix} 17 & 3 & -20 \\ 3 & 25 & -28 \\ -20 & -28 & 48 \end{bmatrix}$$

Case (ii) If $= A_2 = A_1 = A$ (for an axial element with 3 nodes)

$$K = \frac{2AE}{6l}\begin{bmatrix} 7 & 1 & -8 \\ 1 & 7 & -8 \\ -8 & -8 & 16 \end{bmatrix}$$

Shape function for the square element in [ξ, η} coordinates:

We know that the shape functions are such that they have the value of unity at a particular node

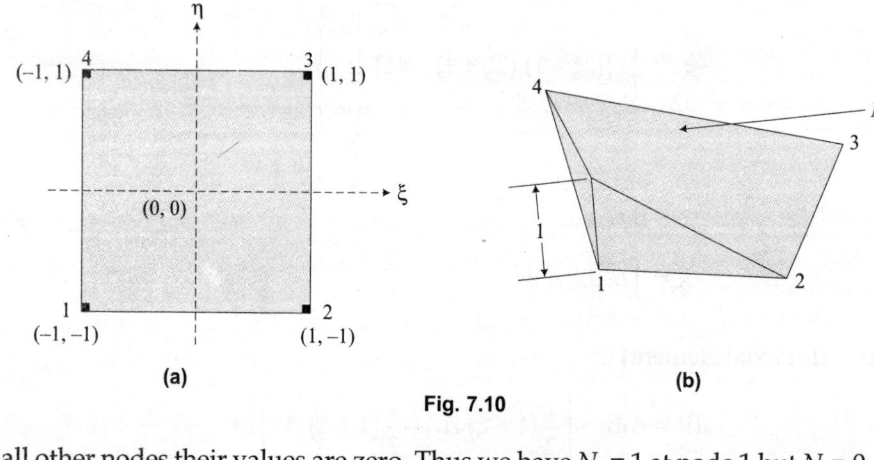

Fig. 7.10

but at all other nodes their values are zero. Thus we have $N_1 = 1$ at node 1 but $N_1 = 0$ at nodes 2, 3, 4. With this in mind we can write a general expression for N_1 as

$$N_1 = K(1-\xi)(1-\eta) \qquad ...[1]$$

where α is a constant to be determined.

Now, @ node 1 where $\xi = -1$, $\eta = -1$, $N_1 = 1$. Hence $1 = K(2)\,(2) \Rightarrow K = 1/4$. Thus,

$$N_1 = \frac{1}{4}(1-\xi)\,(1-\eta) \qquad ...[2]$$

Proceeding in this manner we can obtain the other shape functions and these are

$$N_1 = \frac{1}{4}(1-\xi)(1-\eta)$$

$$N_2 = \frac{1}{4}(1+\xi)(1-\eta)$$

$$N_3 = \frac{1}{4}(1+\xi)(1+\eta) \qquad ...[3]$$

$$N_4 = \frac{1}{4}(1-\xi)(1+\eta)$$

The above can be put in a general form as

$$N_i = \frac{1}{4}(1+\xi\,\xi_i)(1+\eta\eta_i) \qquad ...[4]$$

where (ξ_i, η_i) are the coordinates of node *i*. Note that shape functions are linear. The variation of N_1 is graphically shown in Fig. 7.10(b). This is called the *parent element*.

Jacobian for the Isoparametric Quadrilateral Element

Since the Jacobian gives the relation between the derivatives in the global and natural coordinate systems, to this end, we need the derivatives. Let $\phi = \phi\,(x, y)$. Then, taking derivatives of this function with respect to natural coordinates yields

$$\frac{\partial\phi}{\partial\xi} = \frac{\partial\phi}{\partial x}\frac{\partial x}{\partial\xi} + \frac{\partial\phi}{\partial y}\frac{\partial y}{\partial\xi} \qquad ...[1]$$

$$\frac{\partial\phi}{\partial\eta} = \frac{\partial\phi}{\partial x}\frac{\partial x}{\partial\eta} + \frac{\partial\phi}{\partial y}\frac{\partial y}{\partial\eta} \qquad ...[2]$$

which can be put in the form

$$\begin{Bmatrix} \dfrac{\partial\phi}{\partial\xi} \\ \dfrac{\partial\phi}{\partial\eta} \end{Bmatrix} = \mathbf{J} \begin{Bmatrix} \dfrac{\partial\phi}{\partial x} \\ \dfrac{\partial\phi}{\partial y} \end{Bmatrix} \qquad ...[3]$$

where the Jacobian **J** is given by

$$\mathbf{J} = \begin{Bmatrix} \dfrac{\partial x}{\partial \xi} & \dfrac{\partial y}{\partial \xi} \\ \dfrac{\partial x}{\partial \eta} & \dfrac{\partial y}{\partial \eta} \end{Bmatrix} \qquad \text{...[4]}$$

The Jacobian can also be written as

$$\mathbf{J} = \frac{\partial(x, y)}{\partial(\xi, \eta)} \qquad \text{...[5]}$$

For isoparametric formulation, we have the equations:

$$x = \frac{1}{4}(1-\xi)(1-\eta)\,x_1 + \frac{1}{4}(1+\xi)(1-\eta)\,x_2 + \frac{1}{4}(1+\xi)(1+\eta)\,x_3 + \frac{1}{4}(1-\xi)(1+\eta)\,x_4 \qquad \text{...[6]}$$

$$y = \frac{1}{4}(1-\xi)(1-\eta)\,y_1 + \frac{1}{4}(1+\xi)(1-\eta)\,y_2 + \frac{1}{4}(1+\xi)(1+\eta)\,y_3 + \frac{1}{4}(1-\xi)(1+\eta)\,y_4 \qquad \text{...[7]}$$

Hence,

$$\frac{\partial x}{\partial \xi} = \frac{1}{4}[-(1-\eta)x_1 + (1-\eta)x_2 + (1+\eta)x_3 - (1+\eta)x_4]$$

Similarly, the other derivatives can be obtained. Substituting them into eq. [4] then gives,

$$J = \frac{1}{4} \times$$

$$\begin{bmatrix} -(1-\eta)x_1 + (1-\eta)x_2 + (1+\eta)x_3 - (1+\eta)x_4 & -(1-\eta)y_1 + (1-\eta)y_2 + (1+\eta)y_3 - (1+\eta)y_4 \\ -(1-\xi)x_1 - (1+\xi)x_2 + (1+\xi)x_3 + (1-\xi)x_4 & -(1-\xi)y_1 - (1+\xi)y_2 + (1+\xi)y_3 + (1-\xi)y_4 \end{bmatrix} \qquad \text{...[8]}$$

which can also be put as

$$\mathbf{J} = \frac{1}{4}\begin{bmatrix} -(1-\eta) & 1-\eta & 1+\eta & -(1+\eta) \\ -(1-\xi) & -(1+\xi) & 1+\xi & 1-\xi \end{bmatrix} \begin{Bmatrix} x_1 & y_1 \\ x_2 & y_2 \\ x_3 & y_3 \\ x_4 & y_4 \end{Bmatrix} \qquad \text{...[9]}$$

Eq. [8] can be written simply as

$$\mathbf{J} = \begin{bmatrix} J_{11} & J_{12} \\ J_{21} & J_{22} \end{bmatrix} \qquad \text{...[10]}$$

Please note that $J_{12} \neq J_{21}$. The inverse relation can also be obtained from eq. [3] as

$$\begin{Bmatrix} \dfrac{\partial \phi}{\partial x} \\ \dfrac{\partial \phi}{\partial y} \end{Bmatrix} = \mathbf{J}^{-1} \begin{Bmatrix} \dfrac{\partial \phi}{\partial \xi} \\ \dfrac{\partial \phi}{\partial \eta} \end{Bmatrix} = \frac{1}{\det J}\begin{bmatrix} J_{22} & -J_{12} \\ -J_{21} & J_{11} \end{bmatrix} \begin{Bmatrix} \dfrac{\partial \phi}{\partial \xi} \\ \dfrac{\partial \phi}{\partial \eta} \end{Bmatrix} \qquad \text{....[11]}$$

The Jacobian for the Triangular Element

The field variables (u, v) and (x, y) are all functions of (ξ, η), i.e.,

$$u = u\,(x(\xi, \eta), y(\xi, \eta) \text{ and } v = (x(\xi, \eta), y(\xi, \eta)).$$

Hence, the derivatives of u and v are

$$\frac{\partial u}{\partial \xi} = \frac{\partial u}{\partial x}\frac{\partial x}{\partial \xi} + \frac{\partial u}{\partial y}\frac{\partial y}{\partial \xi} \qquad \text{....[1]}$$

$$\frac{\partial u}{\partial \eta} = \frac{\partial u}{\partial x}\frac{\partial x}{\partial \eta} + \frac{\partial u}{\partial y}\frac{\partial y}{\partial \eta} \qquad \text{....[2]}$$

This can be put in matrix notation as

$$\begin{Bmatrix} \dfrac{\partial u}{\partial \xi} \\ \dfrac{\partial u}{\partial \eta} \end{Bmatrix} = \begin{bmatrix} \dfrac{\partial x}{\partial \xi} & \dfrac{\partial y}{\partial \xi} \\ \dfrac{\partial x}{\partial \eta} & \dfrac{\partial y}{\partial \eta} \end{bmatrix} \begin{Bmatrix} \dfrac{\partial u}{\partial x} \\ \dfrac{\partial u}{\partial y} \end{Bmatrix} \qquad \text{....[3]}$$

The 2×2 square matrix in eq. [3] is called the Jacobian of the transformation denoted by

$$\mathbf{J} = \begin{bmatrix} \dfrac{\partial x}{\partial \xi} & \dfrac{\partial y}{\partial \xi} \\ \dfrac{\partial x}{\partial \eta} & \dfrac{\partial y}{\partial \eta} \end{bmatrix} \qquad \text{....[4]}$$

In view of eq. [1], eq. [4] takes the form

$$\mathbf{J} = \begin{bmatrix} x_1 - x_3 & y_1 - y_3 \\ x_2 - x_3 & y_2 - y_3 \end{bmatrix} \qquad \text{...[5]}$$

The determinant of **J** is given by

$$\begin{aligned} \det \mathrm{J} &= (x_1 - x_3)(y_2 - y_3) - (x_2 - x_3)(y_1 - y_3) \\ &= x_2y_3 - x_3y_2 + x_1y_2 - x_2y_1 + x_3y_1 - x_1y_3 \end{aligned} \qquad \text{....[6]}$$

Comparing eq. [6] with the area of Δ, we find that det **J** = twice the area of the triangle. Hence,

$$A = \frac{1}{2}\,|\det \mathbf{J}| \qquad \text{...[7]}$$

Note that if the numbering of the node is done anticlockwise as done here, we will always get positive value for det **J**. Otherwise we will get only negative value. Now, inverting eq. [3] we get

$$\begin{Bmatrix} \dfrac{\partial u}{\partial x} \\ \dfrac{\partial u}{\partial y} \end{Bmatrix} = J^{-1} \begin{Bmatrix} \dfrac{\partial u}{\partial \xi} \\ \dfrac{\partial u}{\partial \eta} \end{Bmatrix} \qquad \text{....[8]}$$

where J^{-1} is the inverse of the Jacobian given by

$$J^{-1} = \frac{1}{\det J}\begin{bmatrix} y_2 - y_3 & -(y_1 - y_3) \\ -(x_2 - x_3) & x_1 - x_3 \end{bmatrix} \quad \text{....[9]}$$

B Matrix (Strain Displacement Matrix) for the Isoparametric Quadrilateral Element

The strain–displacement relations are given by

$$\varepsilon = \begin{Bmatrix} \varepsilon_x \\ \varepsilon_y \\ \gamma_{xy} \end{Bmatrix} = \begin{Bmatrix} \dfrac{\partial u}{\partial x} \\ \dfrac{\partial v}{\partial y} \\ \dfrac{\partial u}{\partial y} + \dfrac{\partial v}{\partial x} \end{Bmatrix} \quad \text{....[1]}$$

From the Jacobian equations for quadrilateral element, we have,

$$\begin{Bmatrix} \dfrac{\partial u}{\partial x} \\ \dfrac{\partial u}{\partial y} \end{Bmatrix} = \frac{1}{\det J}\begin{bmatrix} J_{22} & -J_{12} \\ -J_{21} & J_{11} \end{bmatrix}\begin{Bmatrix} \dfrac{\partial u}{\partial \xi} \\ \dfrac{\partial u}{\partial \eta} \end{Bmatrix} \quad \text{...[2]}$$

$$\begin{Bmatrix} \dfrac{\partial v}{\partial x} \\ \dfrac{\partial v}{\partial y} \end{Bmatrix} = \frac{1}{\det J}\begin{bmatrix} J_{22} & -J_{12} \\ -J_{21} & J_{11} \end{bmatrix}\begin{Bmatrix} \dfrac{\partial v}{\partial \xi} \\ \dfrac{\partial v}{\partial \eta} \end{Bmatrix} \quad \text{...[3]}$$

Substituting from eqn. [2] into equation [1], we find

$$\varepsilon = \frac{1}{\det J}\begin{bmatrix} J_{22} & -J_{12} & 0 & 0 \\ 0 & 0 & -J_{21} & J_{11} \\ -J_{21} & J_{11} & J_{22} & -J_{12} \end{bmatrix}\begin{Bmatrix} \dfrac{\partial u}{\partial \xi} \\ \dfrac{\partial u}{\partial \eta} \\ \dfrac{\partial v}{\partial \xi} \\ \dfrac{\partial v}{\partial \eta} \end{Bmatrix} \quad \text{....[4]}$$

Now, from the isoparametric formulae the displacements within the element can be expressed in terms of the shape functions as

$$u = N_1 q_1 + N_2 q_3 + N_3 q_5 + N_4 q_7 \quad \text{....[5]}$$

$$v = N_1 q_2 + N_2 q_4 + N_3\, q_6 + N_4 q_8 \qquad \text{....[6]}$$

and
$$N_i = \frac{1}{4}\,(1+\xi\xi_i)\,(1+\eta\eta_i),\; i = 1, 2, 3, 4 \qquad \text{....[7]}$$

We find that from equations [5], [6] and [7]:

Using eq. [2] in terms of u and v successively, we find

$$\begin{Bmatrix} \dfrac{\partial u}{\partial x} \\ \dfrac{\partial u}{\partial y} \end{Bmatrix} = \frac{1}{\det J}\begin{bmatrix} J_{22} & -J_{12} \\ -J_{21} & J_{11} \end{bmatrix}\begin{Bmatrix} \dfrac{\partial u}{\partial \xi} \\ \dfrac{\partial u}{\partial \eta} \end{Bmatrix} \qquad \text{...[8]}$$

$$\begin{Bmatrix} \dfrac{\partial v}{\partial x} \\ \dfrac{\partial v}{\partial y} \end{Bmatrix} = \frac{1}{\det J}\begin{bmatrix} J_{22} & -J_{12} \\ -J_{21} & J_{11} \end{bmatrix}\begin{Bmatrix} \dfrac{\partial v}{\partial \xi} \\ \dfrac{\partial v}{\partial \eta} \end{Bmatrix} \qquad \text{...[9]}$$

Substituting eq. [8] into [9] it is obtained

$$\varepsilon = \frac{1}{\det J}\begin{bmatrix} J_{22} & -J_{12} & 0 & 0 \\ 0 & 0 & -J_{21} & J_{11} \\ -J_{21} & J_{11} & J_{22} & -J_{12} \end{bmatrix}\begin{Bmatrix} \dfrac{\partial u}{\partial \xi} \\ \dfrac{\partial u}{\partial \eta} \\ \dfrac{\partial v}{\partial \xi} \\ \dfrac{\partial v}{\partial \eta} \end{Bmatrix} \qquad \text{...[10]}$$

Now, from eq. [4] we find that

$$\begin{Bmatrix} \dfrac{\partial u}{\partial \xi} \\ \dfrac{\partial u}{\partial \eta} \\ \dfrac{\partial v}{\partial \xi} \\ \dfrac{\partial v}{\partial \eta} \end{Bmatrix} = \frac{1}{4}\begin{bmatrix} -(1-\eta) & 0 & 1-\eta & 0 & 1+\eta & 0 & -(1+\eta) & 0 \\ -(1-\xi) & 0 & -(1+\xi) & 0 & 1+\xi & 0 & 1-\xi & 0 \\ 0 & -(1-\eta) & 0 & 1-\eta & 0 & 1+\eta & 0 & -(1+\eta) \\ 0 & -(1-\xi) & 0 & -(1+\xi) & 0 & 1+\xi & 0 & 1-\xi \end{bmatrix}\begin{Bmatrix} q_1 \\ q_2 \\ q_3 \\ q_4 \\ q_5 \\ q_6 \\ q_7 \\ q_8 \end{Bmatrix} \qquad \text{....[11]}$$

Substituting eq. [11] into [10] we obtain

$$\varepsilon = \mathbf{Bq} \qquad \text{....[12]}$$

where $\mathbf{B} = \mathbf{GH}$ ....[13]

$$G = \frac{1}{\det J}\begin{bmatrix} J_{22} & -J_{12} & 0 & 0 \\ 0 & 0 & -J_{21} & J_{11} \\ -J_{21} & J_{11} & J_{22} & -J_{12} \end{bmatrix} \quad ....[14]$$

$$H = \frac{1}{4}\begin{bmatrix} -(1-\eta) & 0 & 1-\eta & 0 & 1+\eta & 0 & -(1+\eta) & 0 \\ -(1-\xi) & 0 & -(1+\xi) & 0 & 1+\xi & 0 & 1-\xi & 0 \\ 0 & -(1-\eta) & 0 & 1-\eta & 0 & 1+\eta & 0 & -(1+\eta) \\ 0 & -(1-\xi) & 0 & -(1+\xi) & 0 & 1+\xi & 0 & 1-\xi \end{bmatrix} \quad ....[15]$$

Carrying out the operation indicated in eq. [13] the explicit expression for the 3×8 **B** matrix is obtained as

$$B = \frac{1}{4\det J}\begin{bmatrix} -J_{22}(1-\eta)+J_{12}(1-\xi) & 0 & J_{21}(1-\eta)-J_{11}(1-\xi) \\ 0 & J_{21}(1-\eta)-J_{11}(1-\xi) & -J_{22}(1-\eta)+J_{12}(1-\xi) \\ J_{22}(1-\eta)+J_{12}(1+\xi) & 0 & -J_{21}(1-\eta)-J_{11}(1+\xi) \\ 0 & -J_{21}(1-\eta)-J_{11}(1+\xi) & J_{22}(1-\eta)+J_{12}(1+\xi) \\ J_{22}(1+\eta)-J_{12}(1+\xi) & 0 & -J_{21}(1+\eta)+J_{11}(1+\xi) \\ 0 & -J_{21}(1+\eta)+J_{11}(1+\xi) & J_{22}(1+\eta)-J_{12}(1+\xi) \\ -J_{22}(1+\eta)-J_{12}(1-\xi) & 0 & J_{21}(1+\eta)+J_{11}(1-\xi) \\ 0 & J_{21}(1+\eta)+J_{11}(1-\xi) & -J_{22}(1+\eta)-J_{12}(1-\xi) \end{bmatrix}^T \quad ...[16]$$

Now, comparing the **B** matrix of quadrilateral element with that of CST element of last section it may be noticed that the **B** matrix of CST element is a constant, whereas that of quadrilateral element involve the functions ξ and η.

Two-Dimensional Problems Using Constant Strain Triangle (CST)

If the strain is constant within the element (as could be seen from the displacement function), then such a triangular element is called constant strain triangle (CST).

Element Stiffness Matrix of a Constant Strain Triangular Element Using the Isoparametric Concept

A three-noded CST element is shown in natural coordinates in Fig. [7.11].

In Fig. 7.12 the CST element is shown in nodal coordinates (x_1, y_1 etc.) in the global Cartesian system. By carefully observing the figure, the various a_1, a_2 etc., can be remembered.

The natural coordinates are:

$$L_1 = \frac{A_1}{A} = N_1$$

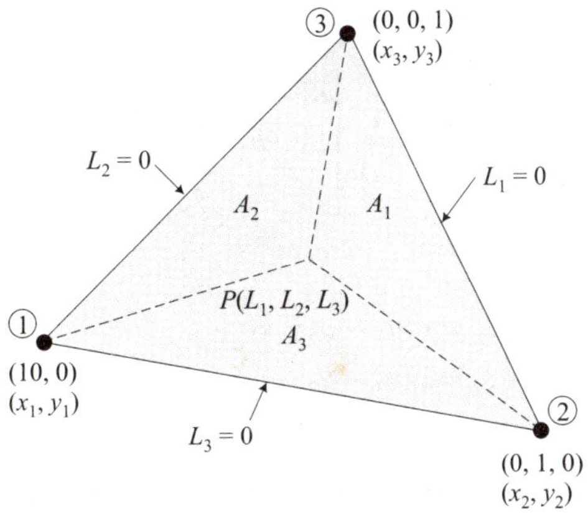

Fig. 7.11 A three-noded CST element in natural coordinates

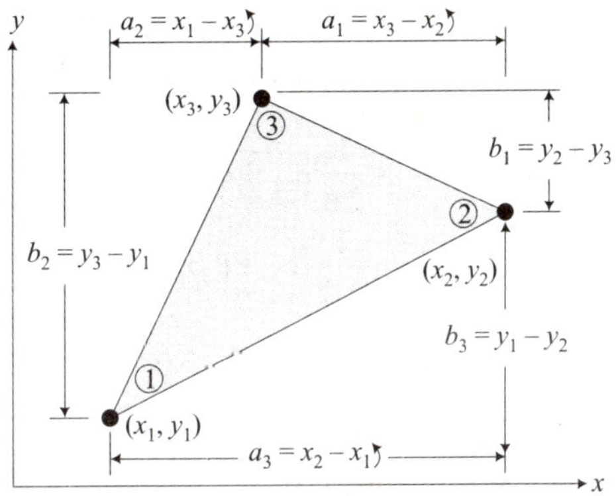

Fig. 7.12 CST in nodal coordinates—Cartesian system.

$$L_2 = \frac{A_2}{A} = N_2$$

$$L_3 = \frac{A_3}{A} = N_3$$

Since it is an isoparametric element, geometry can be defined by the shape functions and nodal parameters. We know

$$L_1 + L_2 + L_3 = 1$$

$$L_1x_1 + L_2x_2 + L_3x_3 = x$$

$$L_1y_1 + L_2y_2 + L_3y_3 = y$$

Rearranging this in matrix form

$$\begin{bmatrix} 1 & 1 & 1 \\ x_1 & x_2 & x_3 \\ y_1 & y_2 & y_3 \end{bmatrix} \begin{bmatrix} L_1 \\ L_2 \\ L_3 \end{bmatrix} = \begin{bmatrix} 1 \\ x \\ y \end{bmatrix}$$

inverting the above matrix

$$\begin{bmatrix} L_1 \\ L_2 \\ L_3 \end{bmatrix} = \frac{1}{2A}\begin{bmatrix} c_1 & b_1 & a_1 \\ c_2 & b_2 & a_2 \\ c_3 & b_3 & a_3 \end{bmatrix}\begin{bmatrix} 1 \\ x \\ y \end{bmatrix}$$

In the above,

$$\begin{Bmatrix} c_1 = x_2y_3 - y_2x_3 \\ c_2 = x_3y_1 - y_3x_1 \\ c_3 = x_1y_2 - y_1x_2 \end{Bmatrix}$$

There is cyclic representation in the above and can be easily remembered. a_1, b_1, etc. are as defined in Fig. 7.11.

Now,

$$\begin{Bmatrix} L_1 = N_1 = \dfrac{1}{2A}[c_1 + b_1x + a_1y] \\ L_2 = N_2 = \dfrac{1}{2A}[c_2 + b_2x + a_2y] \\ L_3 = N_3 = \dfrac{1}{2A}[c_3 + b_3x + a_3y] \end{Bmatrix} \quad ...[1]$$

Now isoparametric concept demands

$$\begin{Bmatrix} u = N_1u_1 + N_2u_2 + N_3u_3 \\ v = N_1v_1 + N_2v_2 + N_3v_3 \end{Bmatrix} \quad ... [2]$$

On substituting Eq. (1) into Eq. (2), we get

$$u = \frac{1}{2A}\{[c_1 + b_1x + a_1y]\, u_1 + [\, c_2 + b_2x + a_2y]\, u_2 + [c_3 + b_3x + a_3y]u_3\}$$

$$\frac{\partial u}{\partial x} = \varepsilon_z = \frac{1}{2A}\,[b_1u_1 + b_2u_2 + b_3u_3] \quad ...(3)$$

$$v = \frac{1}{2A}\{[c_1 + b_1x + a_1y]\, v_1 + [\, c_2 + b_2x + a_2y]\, v_2 + [c_3 + b_3x + a_3y]v_3\}$$

$$\frac{\partial v}{\partial y} = \varepsilon_y = \frac{1}{2A}\,[a_1v_1 + a_2v_2 + a_3v_3] \quad ...(4)$$

Along similar lines

$$\gamma_{xy} = \frac{\partial u}{\partial y} + \frac{\partial v}{\partial x} = \frac{1}{2A}\,[a_1u_1 + b_1\, v_1 + a_2u_2 + b_2v_2 + a_3u_3 + b_3v_3] \quad ...[5]$$

Rearranging Eq. (3), Eq.(4) and Eq. (5) in matrix form, we find

$$\begin{bmatrix} \varepsilon_x \\ \varepsilon_y \\ \gamma_{xy} \end{bmatrix} = \frac{1}{2A} \begin{bmatrix} b_1 & 0 & b_2 & 0 & b_3 & 0 \\ 0 & a_1 & 0 & a_2 & 0 & a_3 \\ a_1 & b_1 & a_2 & b_2 & a_3 & b_3 \end{bmatrix} \begin{Bmatrix} u_1 \\ v_1 \\ u_2 \\ v_2 \\ u_3 \\ v_3 \end{Bmatrix}$$

$$\downarrow \qquad\qquad \downarrow \qquad\qquad \downarrow$$

$$\varepsilon \qquad\qquad \mathbf{B} \qquad\qquad q^*$$

Now $dv = hdA$

where h = thickness of the element

A = area

$$K = \int \mathbf{B}^T d\, \mathbf{B}\, dv$$

where d = elasticity matrix

$$d = \begin{bmatrix} C_1 & C_1C_2 & 0 \\ C_1C_2 & C_1 & 0 \\ 0 & 0 & C_{12} \end{bmatrix}$$

> **Note:**
>
> Plane strain $C_1 = \dfrac{E(1-\mu)}{(1+\mu)(1-2\mu)}$
>
> $C_2 = \dfrac{\mu}{(1-\mu)}$.
>
> Plane stress $C_1 = \dfrac{E}{(1-\mu^2)}$
>
> $C_2 = \mu$.
>
> For both $C_{12} = \dfrac{C_1(1-C_2)}{2}$
>
> Now B^T, B and d are constants.
>
> $$k = \frac{1}{4A^2} B^T d\, B \int_A h\, dA$$

$$= \frac{hA}{4A^2} B^T d B$$

$$k = \frac{h}{4A} [B^T d B]. \text{ (element stiffness)}$$

Note: The product $B^T d B$ is very voluminous. The same can be obtained from standard texts.

7.3 NUMERICAL INTEGRATION

Elementary integration formulas, such as trapezoidal rule, often assume equally spaced data and can become somewhat limited in applicability and accuracy when used in finite element analysis. The basic idea of any of the commonly used numerical integration or quadrature methods as they are normally called is to replace the actual integrand by an approximation in terms of elementary functions. These functions are normally polynomials.

In numerical analysis, numerical integration techniques like trapezoidal rule, Simpson's $\frac{1}{3}$rd rule, Simpson's $\frac{3}{8}$th rule and Gauss quadrature formula are available. In trapezoidal rule, the variation of the function between two sampling points is assumed linear. In Simpson's $\frac{1}{3}$rd rule, 3 sampling points are used and second degree curve is fitted. In Simpson's $\frac{3}{8}$th rule, 4 sampling points are selected and 3rd degree (cubic) curve is fitted. All these methods are based on Newton–Cotes formula in which values at n equally spaced sampling points are required to fit in $n - 1$ degree variation curve.

In finite element analysis, we have seen the element stiffness matrix of isoparametric elements can be formulated using simple natural coordinates whose magnitude varies from –1 to +1. Gauss quadrature formula is preferred. Since in this, values at n points can be used to fit in $(2n - 1)$ degree variation, as the evaluation of functions like $B^T DB dV$ is a time-consuming process. In Gauss quadrature, normally we encounter the problem of evaluation of integral of the form

$$I = \int_{-1}^{1} f(\xi)\, d\xi = \sum_{i=1}^{n} w_i f(\xi_i); \quad \text{(one dimension)} \qquad ...[1]$$

$$= \sum_{i=1}^{n} w_i F_i$$

where w_i is weight function and $f(\xi_i)$ is values of the function at pre-determined sampling points. This type of integral arises in FEM more frequently. The Gaussian quadrature method is most popular even though there are different methods available.

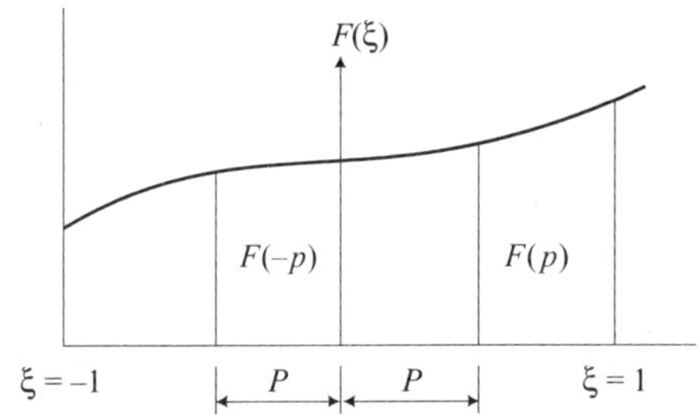

Fig. 7.13 One-dimensional numerical integration

In the Gauss procedure the sampling points are symmetrically located about the centre of the interval of interest (that is about $\xi = 0$) and the magnitudes of the weighting factors are the same at points $\xi = -\xi'$ and $\xi = \xi'$.

Illustrations: (i) If we consider a linear function $F(\xi) = a + b\xi$, then, it is easy to find analytically that $I = \int F(\xi)\, d\xi = 2a$ where a must be $F(0)$, that is, the value of F at $\xi = 0$. Here, we note that only one sampling point is required to evaluate $F(\xi) = a + b\xi$ exactly, it is at $\xi = 0$ and the corresponding weighting factor (value) is 2. Thus, in equation

$$I = \sum_{i=1}^{n} w_i F_i \text{ where } n = 1, \text{ then } \xi_i = 0 \text{ and } w = 2.$$

(*ii*) If we consider a cubic polynomial function

$$F(\xi) = a + b\xi + c\xi^2 d\xi^3$$

analytic evaluation of the definite integral I, then the equation:

$$I = \int_{-1}^{1} F(\xi) d\xi \text{ yields } I = 2a + 2c/3. \qquad \text{[See Fig (7.13)]}$$

It is required to evaluate definite integral numerically be using two sampling (or integrating) points only (*i.e.*, $n = 2$). These points are located at $\xi = p$ and $\xi = -p$.

If we remember that the weighting factors are to have equal value at both points, and we denote this common value as W, the numerical integration method estimates the integral as $I = WF(p) + WF(-p)$ and for the given cubic function of F this becomes $I = 2W(A + Cp^2)$. The difference between the numerical and analytical values of I is $2A(W - 1) + 2C(Wp^2 - 1/3)$. The difference vanishes for any values of A and C if $W = 1$ and $p = \dfrac{1}{\sqrt{3}}$. This result tells us that the exact value of the definite integral of any cubic (or lower-order) function will be obtained numerically by simply summing the two values of the function evaluated at $\xi_1 = +1/\sqrt{3}$ and $= \xi_2 - 1/\sqrt{3}$.

Gauss Propositions

For deriving the Gauss quadrature, Gauss has made certain propositions. They are as follows:

(1) Weights are all positive

(2) Weights are symmetric

(3) With one sampling points we can integrate $(2n - 1) = (2 \times 1 - 1) = 1$, a linear equation

(4) With two sampling points we can integrate $(2 \times 2 - 1) = 3$, we can integrate polynomials of 3rd degree or less than that. The formula used for this purpose is $(2n - 1)$, where n is the number of sampling points.

(5) With three sampling points we can integrate $(2 \times 3 - 1) = 5^{th}$ degree polynomial or less.

Gauss–Legendre Integration Scheme

A general polynomial can be written in the form:

$$P_n(x) = A_0 + A_1 x + A_2 x^2 + A_3 x^3 + \cdots + A_n x^n \quad \text{...[1]}$$

The base points and the coefficients found for this would fit all functions. For convenience in computations the limits in Gauss – Legendre's scheme are kept as from –1 to 1, *i.e.*, in natural coordinates free from any scale. Thus, some transformations are to be carried out for change of limits, before proceeding.

Let the integration to be taken for

$$I = \int_a^b F(x)\, dx \quad \text{...[2]}$$

put $\quad x = p + r\xi$

where, ξ is a new variable and p and r are constants.

For $\xi = a$ and b the limits for Gaussian quadrature for ξ are to be –1 and 1,

therefore, $\quad a = p - r$

and $\quad b = p + r \quad \text{...[3]}$

On solving for p and r which when substituted yields

$$x = \frac{[(b+a) + (b-a)\xi]}{2};\ dx = r d\xi = \frac{(b-a)}{2}\, d\xi.$$

Now, we can check the relationship:

$$x = \frac{(b-a)}{2}\xi + \frac{(b+a)}{2} \quad \text{...[3]}$$

$$\left\{\begin{array}{ll} \text{when} & \xi = -1,\ x = a \\ \text{when} & \xi = 0,\ x = \dfrac{(b+a)}{2} \\ \text{when} & \xi = +1,\ x = b \end{array}\right\}$$

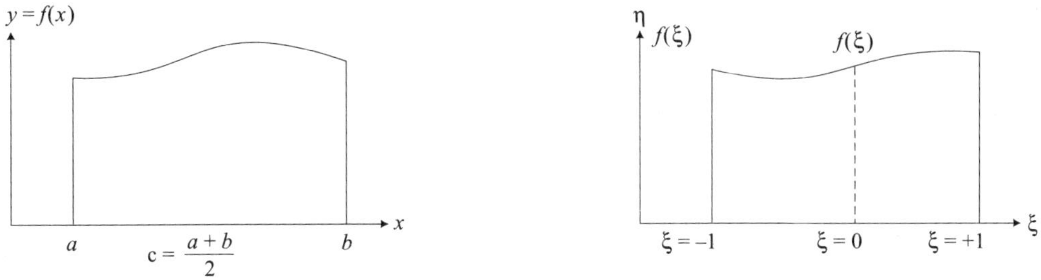

(a) Cartesian coordinate (b) Local coordinate

Therefore, the relationship is proved.

Now
$$\int_a^b f(x)\,dx = \int_{-1}^{1} f\left[\underbrace{\frac{(b-a)}{2}\xi + \frac{(b+a)}{2}}_{x}\right]\underbrace{\frac{(b-a)}{2}d\xi}_{dx} \qquad ...[4]$$

Since
$$dx = \frac{(b-a)}{2}d\xi, \qquad \text{(from [3])}$$

Now, the integral in equation [4] can be rewritten as:

$$\boxed{I = \int_a^b f(x)\,dx = \frac{(b-a)}{2}\int_{-1}^{1} f(\xi)d\xi} \quad \text{(Imp. formula)} \qquad ...[5]$$

We make use of equation [5] more frequently in our problems.

Working Rule

How to evaluate the integral: $\int_{-1}^{1} f\{\xi)d\xi$?

The value of the above integral has been evaluated for one sample point, two, three etc. and it is approximated by the function as:

$$\boxed{I = \int_{-1}^{1} f(\xi)\,d\xi \simeq W_i f(\xi_i) + W_2 f(\xi_2) + W_3 f(\xi_3) + \cdots + W_n f(\xi_n) = \sum_{i=1}^{n} f(\xi_i)\,W_i} \qquad ...[6]$$

The concept underlying in Gaussian quadrature is to select the n-Gauss points and n-weights such that the equation [6] provides an exact answer for polynomials $f(\xi)$ as large a degree as possible. Also, the method will work well if $f(\xi)$ is not a polynomial.

(a) One sampling point formula:

By Gauss quadrature formula, we can integrate $(2n - 1) = (2 \times 1 - 1) = 1$, a linear equation. A linear equation will consist of one constant term and a linear term.

$$\int_{-1}^{1} f(\xi)\,d\xi = w_1 f(\xi_1) \qquad ...[1]$$

We know that $f(\xi) = C$

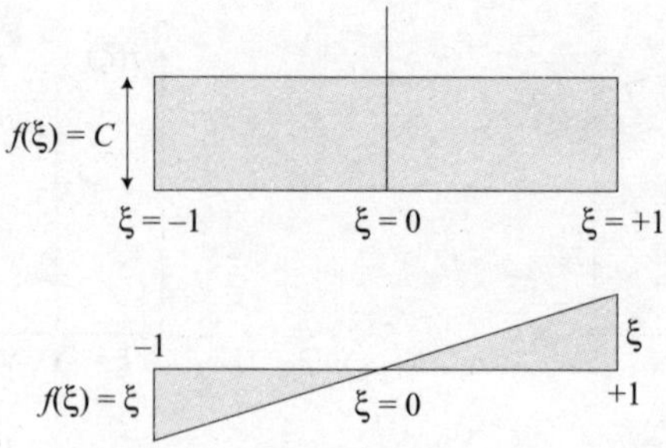

Fig. 7.14 One sampling point

$$\therefore \qquad I = \int_{-1}^{1} C d\xi = 2C = w_1 C$$

Hence, $\qquad w_1 = 2$

Now, $\qquad \int_{-1}^{1} f(\xi) d\xi = \int_{-1}^{1} \xi d\xi = 0 = w_1 f(\xi_1) = w_1 \xi_1$

Since w_1 is not equal to zero, it follows $\xi_1 = 0$.

$$\therefore \qquad I = \int_{-1}^{1} f(\xi) d\xi = 2f(0);$$

Hence, this is the equation, for one sampling point,

$$\text{yields area} = \left[\frac{(b-a)}{2}\right] 2f(0)$$

(b) Two sampling points formula: Degree of curve to be integrated is equal to

$$(2n - 1) = (2 \times 2 - 1) = 3.$$

The polynomial will consist of four terms.

(*i*) $f(\xi) = C; \qquad \int_{-1}^{1} f(\xi) d\xi = \int_{-1}^{1} 2C d\xi = 2C$

$$= w_1 f(\xi_1) + w_2 f(\xi_2)$$

$$= w_1 C + w_2 C = 2C; \qquad ...[1]$$

(*ii*) $f(\xi) = \xi; \qquad \int_{-1}^{1} f(\xi) d\xi = \int_{-1}^{1} \xi d\xi = 0;$

$$= w_1 f(\xi_1) + w_2 f(\xi_2)$$

$$= w_1 \xi_1 + w_2 \xi_2 = 0 \qquad ...[2]$$

(*iii*) $f(\xi) = \xi^2; \qquad \int_{-1}^{1} f(\xi)\, d\xi = \int_{-1}^{1} \xi^2 d\xi = \frac{2}{3}$

$$= w_1 f(\xi_1) + w_2 f(\xi_2)$$

$$= w_1 \xi_1^2 + w_2 \xi_2^2 = \frac{2}{3} \qquad ...[3]$$

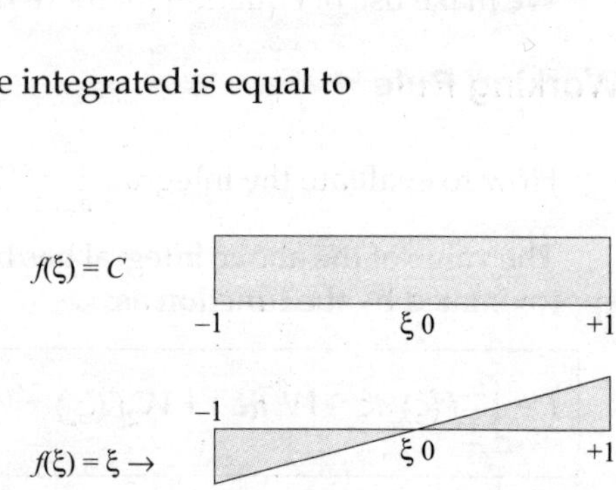

Fig. 7.15 Two sampling points

(*iv*) $f(\xi) = \xi^3$; $\int_{-1}^{1} f(\xi)d\xi = \int_{-1}^{1} \xi^3 d\xi = 0 = w_1 f(\xi_1) + w_2 f(\xi_2) = w_1\xi_1^3 + w_2\xi_2^3 = 0$ ...[1]

According to Gauss, $w_1 = w_2$ (Symmetry)

From equation [1], $2w_1 = 2$

$\Rightarrow$ $w_1 = 1 \quad \therefore w_2 = 1$

From equation [2] $\xi_1 = -\xi_2$

From equation [3], $2\xi_1^2 = \frac{2}{3}$

$\Rightarrow$ $\xi_1 = \frac{1}{\sqrt{3}} = +0.577$

$\therefore$ $\xi_2 = -0.577$

Equation [4] is identically satisfied.

$$\therefore \int_{-1}^{1} f(\xi)\,d\xi = 1 \times f(0.577) + 1 \times f(-0.577)$$

$$= f\left(\frac{1}{\sqrt{3}}\right) + f\left(-\frac{1}{\sqrt{3}}\right)$$

This equation is for two sampling points.

(c) Three sampling points formula:

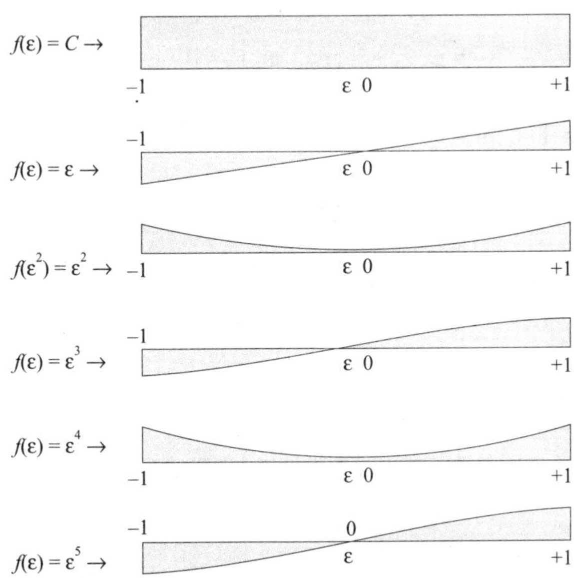

Fig. 7.16 Three sampling points

By the Gauss quadrature $(2 \times 3 - 1) = 5^{th}$ degree equation can be integrated. This equation consists of six terms as shown in Fig. 7.16.

(*a*) $f(\varepsilon) = C;\quad \int_{-1}^{1} f(\varepsilon)d\varepsilon = \int_{-1}^{1} C\, d\varepsilon = 2c$

$$= w_1 f(\varepsilon_1) + w_2 f(\varepsilon_2) + w_3 f(\varepsilon_3)$$

$$= w_1 C + w_2 C + w_3 C = 2c \qquad ...[1]$$

(*b*) $f(\varepsilon) = \varepsilon;\quad \int_{-1}^{1} f(\varepsilon)d\varepsilon = \int_{-1}^{1} \varepsilon\, d\varepsilon = 0$

$$= w_1 f(\varepsilon_1) + w_2 f(\varepsilon_2) + w_3 f(\varepsilon_3)$$

$$= w_1 \varepsilon_1 + w_2 \varepsilon_2 + w_3 \varepsilon_3 = 0 \qquad ...[2]$$

(*c*) $f(\varepsilon) = \varepsilon^2;\quad \int_{-1}^{1} f(\varepsilon)d\varepsilon = \int_{-1}^{1} \varepsilon^2 d\varepsilon = \frac{2}{3}$

$$= w_1 f(\varepsilon_1) + w_2 f(\varepsilon_2) + w_3 f(\varepsilon_3)$$

$$= w_1 \varepsilon_1^2 + w_2 \varepsilon_2^2 + w_3 \varepsilon_3^2 = \frac{2}{3} \qquad ...[3]$$

(*d*) $f(\varepsilon) = \varepsilon^3;\quad \int_{-1}^{1} f(\varepsilon)d\varepsilon = \int_{-1}^{1} \varepsilon^3 d\varepsilon = 0$

$$= w_1 f(\varepsilon_1) + w_2 f(\varepsilon_2) + w_3 f(\varepsilon_3)$$

$$= w_1 \varepsilon_1^3 + w_2 \varepsilon_2^3 + w_3 \varepsilon_3^3 = 0 \qquad ...[4]$$

(*e*) $f(\varepsilon) = \varepsilon^4;\quad \int_{-1}^{1} f(\varepsilon)d\varepsilon = \int_{-1}^{1} \varepsilon^4 d\varepsilon = \frac{2}{5}$

$$= w_1 f(\varepsilon_1) + w_2 f(\varepsilon_2) + w_3 f(\varepsilon_3)$$

$$= w_1 \varepsilon_1^4 + w_2 \varepsilon_2^4 + w_3 \varepsilon_3^4 = \frac{2}{5} \qquad ...[5]$$

(*f*) $f(\varepsilon) = \varepsilon^5;\quad \int_{-1}^{1} f(\varepsilon)d\varepsilon = \int_{-1}^{1} \varepsilon^5 d\varepsilon = 0$

$$= w_1 f_1(\varepsilon_1) + w_2 f(\varepsilon_2) + w_3 f(\varepsilon_3)$$

$$= w_1 \varepsilon_1^5 + w_2 \varepsilon_2^5 + w_3 \varepsilon_3^5 = 0 \qquad ...[6]$$

On solving equation [1] through equation [6], we get

$$w_1 = \frac{5}{9}$$

$$w_2 = \frac{8}{9}$$

$$w_3 = \frac{5}{9}$$

$$\varepsilon_1 = -\sqrt{\frac{3}{5}}$$

$$\varepsilon_2 = 0$$

$$\therefore \qquad \varepsilon_3 = +\sqrt{\frac{3}{5}}$$

$$\therefore \qquad \int_{-1}^{1} f(c)\, d\varepsilon = \frac{5}{9} f\left[-\sqrt{\frac{3}{5}}\right] + \frac{8}{9} f[0] + \frac{5}{9} f\left[+\sqrt{\frac{3}{5}}\right]$$

$$\int_{-1}^{1} f(\xi)\, d\xi = \frac{5}{9} f\left[-\sqrt{\frac{3}{5}}\right] + \frac{8}{9} f[0] + \frac{5}{9} f\left[+\sqrt{\frac{3}{5}}\right]$$

This equation is for three sampling points.

Table 7.1 Sampling points and their weights for Gaussian quadrature from (n = 1 to 4)

No. of Points	*Position of Points, ξ_1 (Roots)*	*Weights W_1*
1	0	2
2	0.577340269	1
	– 0.577340269	1
3	0	0.8888888
	0.774596669	0.5555555
	– 0.774596669	0.5555555
4	0.861136311	0.347854845
	– 0.861136311	0.347854845
	0.339981043	0.652145154
	– 0.339981043	0.652145154

(one-dimensional situations)

Use of Rodrigue's Formula: The sampling points and weighting coefficients which are on natural scale are constants and can be predetermined for the particular scheme such as two point or three point and so on. These base points can conveniently be found by solving for the roots of the equation as obtained from Rodrigue's formula. The equation for solving roots for n point integration is given as

$$P_n(x) = 0 = \frac{1}{2^n\, n_1} \frac{d^n}{dx^n} (x^2-1)^n \qquad ...[1]$$

or

$$P_n(x) = 0 = \frac{d}{dx^n} (x^2-1)^n \qquad ...[2]$$

For example, for two-point integration the Rodrigue's equation is

$$\frac{d^2}{dx^2} (x^2-1)^2 = 0 \qquad ...[3]$$

or,

$$3x^2-1 = 0$$

or,

$$x = \pm\sqrt{\frac{1}{3}}$$

For 3-point integration, Rodrigue's is obtained as

$$x(5x^2-3) = 0$$

$\Rightarrow$

$$x_1 = -\sqrt{3/5} = -x_3$$

and

$$x_2 = 0$$

Note: For all practical purposes two- or three-point integration is adequate. Higher order integrations may only be for academic interests.

Integration in FEA

In finite element analyses most of the problems are solved with elements upto quadratic variations and few with cubical ones. For all practical purposes these can adequately be solved with 2 point scheme, which is exact upto third order. Moreover, as the finite element method is based on principle of approximations with the help of parameters known at some selected points, it is convenient to take a numerical integration. There are available a number of methods for numerical integration but the one which suits most has to be selected.

Generally, in finite element analyses the following kind of integration is encountered.

$$I = \iiint f(x, y, z,)\, dx\, dy\, dz, \quad \text{(for three-dimensional cases)} \qquad ...[1]$$

$$I = \iint f(x, y)\, dx\, dy, \quad \text{(for two-dimensional cases)} \qquad ...[2]$$

$$I = \int f(x)\, dx, \quad \text{(for one-dimensional cases)} \qquad ...[3]$$

Integration using Gauss–Legendre scheme. In finite element analyses where the shape functions are taken in natural co-ordinates, which are free of scale, there is no need of transformation of limits. The limits in both the cases are same, *i.e.*, from –1 to +1. Just for the equivalence of $dx.\, dy.\, dz$, another factor as determinant of Jacobi matrix ($|J|$) comes as multiplier to $d\xi.d\eta.d\zeta$.

that is, $dx\,dy\,dz = |J|\,d\xi\,d\eta\,d\zeta$

Shape functions and their derivatives are evaluated at Gaussian point and the products $w_i f(\xi, \eta, \zeta)$ are summed up to get the integration.

Table 7.2 Abscissae and weight coefficients of the Gaussian Quadrature Formula (n = 1 to 10)

$\pm\xi$		$\int_{-1}^{1} f(x)dx = \sum_{j=1}^{n} w_i(\xi_j)$		W_i	
		$n = 1$			
0.00000				2.00000	
		$n = 2$			
0.57735	02691	89626	1.00000	00000	00000
		$n = 3$			
0.77459	66692	41483	0.55655	55555	55556
0.00000	00000	00000	0.88888	88888	88889
		$n = 4$			
0.86113	63115	94053	0.34785	48451	37454
0.33998	10435	84856	0.65214	51214	62546
		$n = 5$			
0.90617	98459	38664	0.23692	68850	56189
0.53846	93101	05683	0.47862	86704	99366
0.00000	00000	00000	0.56888	88888	88889
		$n = 6$			
0.93246	95142	03152	0.17132	44923	79170
0.66120	93864	66265	0.36076	15730	48139
0.23861	91860	83197	0.46791	39345	72691
		$n = 7$			
0.94910	79123	42759	0.12948	49661	68870
0.74143	11855	99394	0.27970	53914	89277
0.40584	51513	77397	0.38183	00505	05119
0.00000	00000	00000	0.41795	91836	73469
		$n = 8$			
0.96028	98564	97536	0.10122	85362	90376
0.79666	64774	13627	0.22238	10344	53374
0.52553	24099	16329	0.31370	66458	77887
0.18343	46424	95650	0.36268	37833	78362
		$n = 9$			
0.96816	02395	07826	0.08127	43883	61574
0.83603	11073	26636	0.18064	81606	94857
0.61337	14327	00590	0.26061	06964	02935
0.32425	34234	03809	0.31234	70770	40003
0.00000	00000	00000	0.33023	93550	01260
		$n = 10$			
0.97390	65285	17172	0.06667	13443	08688
0.86506	33666	88985	0.14935	13491	50581
0.67940	95682	99024	0.21908	63625	15982
0.43339	53941	29247	0.26926	67193	09996
0.14887	43389	81631	0.29552	48247	14753

Table 7.3A Numerical integration formula for triangles

Order	*Figure*	*Point*	*Triangular Coordinates*	*Weights*
Linear	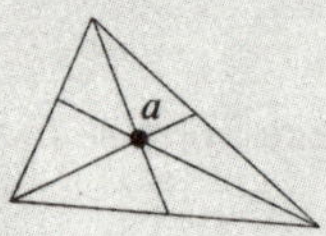	*a*	1/3, 1/3, 1/3	1
Quadratic	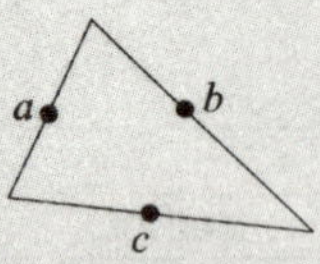	*a*	1/2, 1/2, 0	1/3
		b	0, 1/2, 1/2	1/3
		c	1/2, 0, 1/2	1/3
Cubic	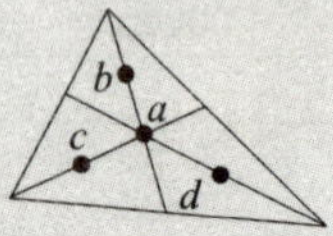	*a*	1/3, 1/3, 1/3	– 27/48
		b	0.6, 0.2, 0.2	25/48
		c	0.2, 0.6, 0.2	
		d	0.2, 0.2, 0.6	
Quintic	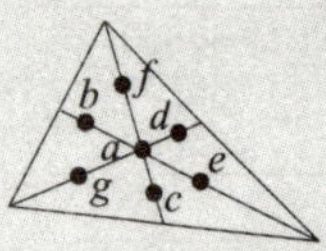	*a*	1/3, 1/3, 1/3	0.22500, 00000
		b	$\alpha_1, \beta_1, \beta_1$	0.13239, 41527
		c	$\beta_1, \alpha_1, \beta_1$	
		d	$\beta_1, \beta_1, \alpha_1$	
		e	$\alpha_2, \beta_2, \beta_2$	0.12593, 91805
		f	$\beta_2, \alpha_2, \beta_2$	
		g	$\beta_2, \beta_2, \alpha_2$	

Table 7.3B

Integration Order	*Degree of Precision*	ξ_i	ξ_2	*Weights*
3	2	$\xi_{11} = 1/6$	$\xi_{21} = \xi_{11}$	$W_1 = 1/3$
		$\xi_{12} = 2/3$	$\xi_{22} = \xi_{11}$	$W_2 = W_1$
		$\xi_{13} = \xi_{11}$	$\xi_{23} = \xi_{12}$	$W_3 = W_1$
7	5	$\xi_{11} = 0.101286507324$	$\xi_{21} = \xi_{11}$	$W_1 = 0.1259391$
		$\xi_{12} = 0.79742698535$	$\xi_{22} = \xi_{11}$	$W_2 = W_1$
		$\xi_{13} = \xi_{11}$	$\xi_{23} = \xi_{12}$	$W_3 = W_1$
		$\xi_{14} = 0.470142064105$	$\xi_{24} = \xi_{16}$	$W_4 = 0.1323941$
		$\xi_{15} = \xi_{14}$	$\xi_{25} = \xi_{14}$	$W_5 = W_4$
		$\xi_{16} = 0.059715871798$	$\xi_{26} = \xi_{14}$	$W_6 = W_4$
		$\xi_{17} = 1/3$	$\xi_{27} = \xi_{17}$	$W_7 = 0.225$

Example 1: Evaluate

$$I = \int_{-1}^{1} \left(3e^x + x^2 + \frac{1}{(x+2)} \right) dx$$

using one-point and two-point formulae.

Solution:

(i) *One-point formula:*

$$w_1 = 2; \xi_1 = 0; \; f(\xi) = 3e^{\xi} + \xi^2 + \frac{1}{(\xi + 2)}$$

$$\therefore \qquad I = \int_{-1}^{1} f(\xi)\, d\xi = w_1 f(\xi_1) = w_1 f(0) = 2 f(0) = 2\,[3e^0 + 0 + 0.5] = 7$$

$$\therefore \qquad I = 7$$

(ii) *Two-point formula:*

$$\left(w_1 = 1, w_2 = 1, \xi_1 = -\frac{1}{\sqrt{3}}; \; \xi_2 = \frac{1}{\sqrt{3}} \right)$$

$$\therefore \qquad I = \int_{-1}^{1} f(\xi) d\xi = w_1 f(\xi_1) + w_2 f(\xi_2)$$

$$= 1 \times f\left(\frac{1}{\sqrt{3}} \right) + 1 \times f\left(\frac{1}{\sqrt{3}} \right) = f\left(-\frac{1}{\sqrt{3}} \right) + f\left(\frac{1}{\sqrt{3}} \right)$$

or,

$$I = f(-0.57735) + f(+0.57735)$$

$$= \left[3e^{-0.57735} + (-0.57735)^2 + \frac{1}{(-0.57735 + 2)} \right]$$

$$+ \left[3e^{+0.57735} + (0.57735)^2 + \frac{1}{(0.57735 + 2)} \right]$$

$$= [1.6797 + 0.3333 + 0.7029] + [5.358 + 0.333 + 0.388]$$

$$\therefore \qquad I = 8.7952$$

and $\qquad I\,(\text{exact}) = 8.8165$

Example 2: Evaluate

$$\int_{-1}^{+1} (x^3 + 2x^2 + 1) dx$$

using the two-point Gauss formula.

Solution: Given:

$$I = \int_{a=-1}^{b=1} (x^3 + 2x^2 + 1)\, dx \qquad \text{...[1]}$$

Substituting

$$x = \frac{(b-a)}{2} \xi + \frac{(b+a)}{2}$$

$$= \frac{(1-1(-1))}{2}\xi + \frac{(1-1)}{2} = \xi$$

$\therefore$ $x = \xi;\ dx = d\xi$

$\therefore$ $I = \int_{-1}^{1} (\xi^3 + 2\xi^2 + 1) d\xi$

Hence, $f(\xi) = \xi^3 + 2\xi^2 + 1;\ \left(w_1 = w_2 = 1;\ \xi_1 = -\frac{1}{\sqrt{3}};\ \xi_2 = \frac{1}{\sqrt{3}} \right)$

$\therefore$ $f(\xi_1) = \left(-\frac{1}{\sqrt{3}}\right)^3 + 2\left(-\frac{1}{\sqrt{3}}\right)^2 + 1 = \left(-\frac{1}{3\sqrt{3}} + \frac{2}{3} + 1\right)$

$$f(\xi_2) = \left(\frac{1}{\sqrt{3}}\right)^3 + 2\left(\frac{1}{\sqrt{3}}\right)^2 + 1 = \left(\frac{1}{3\sqrt{3}} + \frac{2}{3} + 1\right)$$

$\therefore$ $I = w_1 f(\xi_1) + w_2 f(\xi_2) \quad = 1 \times \left(-\frac{1}{3\sqrt{3}} + \frac{2}{3} + 1\right) + 1 \times \left\{\frac{1}{3\sqrt{3}} + \frac{2}{3} + 1\right\}$

$\therefore$ $I = \frac{10}{3}$; and I (exact) $= \frac{10}{3}$

Example 3: Evaluate: $I = \int_0^{10} (3x^2 + 5x + x^3)\, dx$

by using Gauss quadrature formula

Solution: Given: $I = \int_{a=0}^{b=10} (3x^2 + 5x + x^3) dx$ ...[1]

The interval (0 – 10) may be changed to –1 to +1

by substituting $x = \frac{(b-a)}{2}\xi + \frac{(b+a)}{2}$ ...[2]

$$= \frac{(10-0)}{2}\xi + \frac{(10+0)}{2} = 5\xi + 5 \quad ...[3]$$

or $x = 5(1+\xi);\quad dx = 5d\xi$

$\therefore$ $I = \int_{-1}^{1} f(\xi)\, d\xi = \int_{-1}^{1} [3\{5(1+\xi)\}^2 + 5\{5(1+\xi)\} + \{5(1+\xi)\}^3]\, 5d\xi$

$$= \int_{-1}^{1} [75(1+\xi)^2 + 25(1+\xi) + 125(1+\xi)^3]\,(5)d\xi$$

or, $I = w_1 f(\xi_1) + w_2 f(\xi_2)$

where $f(\xi) = 125\,[3(1+\xi)^2 + (1+\xi) + 5(1+\xi)]$

$$f(\xi_1) = 125\,(0.5759 + 0.42265 + .37750)$$

$$= 125\,(1.33605)$$

$$f(\xi_2) = 125\,(7.46410 + 1.5775 + 19.62250)$$
$$= 125\,(28.66395)$$

which gives $I = 3750.00$

which is exact as can be compared with the value obtained through direct integration. This means the integration with two-point integration scheme is exact upto degree 3 (*i.e.*, $2n - 1$).

Example 4: Evaluate

$$I = \int_0^{10} (5x^4 + 20x^3 + 30x^2)dx \qquad ...[1]$$

using the 2-point Gauss formula.

Solution: $I = 160000.00$ (Exact)

Given: $\int_{a=0}^{b=10} (5x^4 + 20x^3 + 30x^2).dx$

Now, by the Gauss quadrature, the interval (0 – 10) may be changed to –1 to +1 by substituting

$$x = \frac{(b-a)}{2}\xi + \frac{(b+a)}{2} \qquad ...[2]$$

$$x = \frac{(10-0)}{2}\xi + \frac{(10+0)}{2} = 5\xi + 5 = 5\,(1+\xi)$$

$$\therefore \quad I = \int_{-1}^{1} f(\xi)d\xi = w_1 f(\xi_1) + w_2 f(\xi_2) = w_1F_1 + w_2F_2$$
$$= 5\,[\{5\,(1+\xi)^4\} + 20\,\{5\,(1+\xi)^3\} + 30\,\{5\,(1+\xi)^2\}.5d\xi]$$
$$= 625\,[25\cdot(1+\xi)^2 + 20\cdot(1+\xi) + 6]\cdot(1+\xi)^2\cdot d\xi$$

$$f(\xi_1) = 625\cdot[4.46583 + 8.453 + 6] \times 0.17863$$
$$f(\xi_2) = 625\cdot[62.20083 + 31.547 + 6) \times 2.48803$$

Weight factors (coefficients) $w_1 = w_2 = 1_3$

$\therefore$ $I = 157221.9$

The integration of 4th order function gives an error of 2778.1.

Example 5: Evaluate

$$\int_0^1 \frac{dx}{(1+x^2)}$$

$$\int_0^1 \frac{dx}{(1+x^2)} = \int_{a=1}^{b=1} \frac{dx}{(1+x^2)}$$

using Gauss three sampling point formula:

Solution:

$$I = \int_{-1}^{1} f(\xi)\,d\xi = \frac{5}{9} f(-0.7745) + \frac{8}{9} f(0) + \frac{5}{9} f(+0.7745). \qquad ...[1]$$

Given: $f(x) = \dfrac{1}{(1+x^2)}$, we know the transformation

$$x = \frac{(b-a)}{2}\xi + \frac{(b+a)}{2}$$

$$= \frac{(1-0)}{2}\xi + \frac{(1+0)}{2} = \frac{\xi}{2} + \frac{1}{2}$$

(i) when $\xi = -0.7745$

$$x = \frac{(1-0)}{2}(-0.7745) + \frac{(1+0)}{2} = 0.1125$$

$$g(x) = \frac{1}{(1+x^2)} = \frac{1}{1+(0.1125)^2} = 0.987447$$

(ii) when $\xi = 0$

$$x = \frac{(1-0)}{2}(0) + \frac{(1+0)}{2} = 0.5$$

$$\therefore \qquad g(x) = \frac{1}{(1+x^2)} = \frac{1}{1+(0.5)^2} = 0.8$$

(iii) when $\xi = +0.7745$

$$x = \frac{(1-0)}{2}(0.7745) + \frac{(1-0)}{2} = 0.88725$$

$$f(x) = \frac{1}{(1+x^2)} = \frac{1}{1+(0.88725)^2} = 0.5595305$$

$$\text{Area} = \frac{(b-a)}{2}\left[\frac{5}{9}f(-0.7745) + \frac{8}{9}f(0) + \frac{5}{9}f(+0.7745)\right]$$

$$= \frac{(1-0)}{2}\left[\frac{5}{9}\times 0.987447 + \frac{8}{9}\times 0.8 + \frac{5}{9}\times 0.5595305\right]$$

$$\text{Area} = 0.7852715$$

$$I(\text{exact}) = \frac{\pi}{4}$$

Example 6: Evaluate using Gauss two sampling point formula:

$$\int_0^1 \frac{1}{(1+x^2)}\,dx$$

Solution: We know that Gauss 2-point formula is given by

$$I = \int_{-1}^{1} f(\xi)d\xi = w_1 f(\xi_1) + w_2 f(\xi_2)$$

$$\left(w_1=1;\ w_2=1;\ \xi_1 = -\frac{1}{\sqrt{3}};\ \xi_2 = +\frac{1}{\sqrt{3}}\right)$$

Substituting,
$$x = \frac{(b-a)}{2}\xi + \frac{(b+a)}{2} \qquad \text{...[1]}$$

$$dx = \frac{(b-a)}{2}\, d\xi$$

Here,
$$I = \int_{a=0}^{b=1} \frac{1}{(1+x^2)}\, dx$$

Using [1],
$$x = \frac{(1-0)}{2}\,\xi + \frac{(1+0)}{2}$$

$$= \frac{(\xi+1)}{2}$$

and
$$dx = \frac{d\xi}{2}$$

∴
$$I = \int_{-1}^{1} \left(\frac{1}{2}\right) \frac{1}{(1+0.25(1+\xi)^2)}\, d\xi$$

or,
$$I = \frac{1}{2}\left[\frac{1}{1+0.25(-0.57735+1)^2}\right] + \frac{1}{2}\left[\frac{1}{1+0.25(-0.57735+1)^2}\right]$$

$$= \frac{1}{2}\,(\,0.95725 + 0.61651)$$

∴
$$I = 0.786875$$

$$I\ (\text{exact}) = \frac{\pi}{4}$$

Example 7: Evaluate $\int_1^3 \frac{dx}{x}$ using Gaussian three point formula.

Solution:
$$I = \int_{a=1}^{b=3} \frac{dx}{x} \qquad \text{...[1]}$$

The interval 1 – 3 may be changed to –1 to 1 by substituting

$$x = \frac{(b-a)}{2}\xi + \frac{(b+a)}{2} = \frac{(3-1)}{2}\xi + \frac{(3+1)}{2} \qquad \text{...[2]}$$

$$x = \xi + 2 \text{ and } dx = d\xi\,\frac{(b-a)}{2} = d\xi \qquad \text{...[3]}$$

$$I = \int_{-1}^{1} \frac{d\xi}{\xi + 2} = \int_{-1}^{1} f(\xi)\, d\xi \qquad \text{...[4]}$$

Three points may be chosen as : ξ_i –0.77459669, 0, +0. 77459669

$$w_i \qquad \frac{5}{9}, \qquad \frac{8}{9}, \qquad \frac{5}{9}$$

$$\therefore \qquad I = \int_{-1}^{1} f(\xi)\, d\xi = w_1 f(\xi_1) + w_2 f(\xi_2) + w_3 f(\xi_3)$$

$$= \left(\frac{5}{9}\right)\left[\frac{1}{2 - .77459669}\right] + \frac{8}{9}\left[\frac{1}{2}\right] + \frac{5}{9}\left[\frac{1}{2 + .77459669}\right]$$

$$\therefore \qquad I = 1.098039$$

whereas $\qquad I \text{ (exact)} = \log 3 = 1.098612289$

which gives an error of – 0.052%.

Example 8: Evaluate the integral

$$\int_0^1 \frac{dx}{1+x} \quad \text{using Gauss–Legendre three-point formula.}$$

Solution: Converting the interval (0, 1) to the interval (–1, 1).

Let $\qquad I = \int_{a=0}^{b=1} \frac{dx}{1+x}$

Writing $\qquad x = \frac{(b-a)}{2}\xi + \frac{(b+a)}{2}$

$$= \left(\frac{1-0}{2}\right)\xi + \frac{(1+0)}{2} = \frac{\xi}{2} + \frac{1}{2} = \frac{(\xi+1)}{2}$$

$$\therefore \qquad dx = \frac{d\xi}{2}$$

$$\therefore \qquad I = \int_0^1 \frac{dx}{(1+x)}$$

$$= \int_{-1}^{1} \frac{1}{1 + \frac{(\xi+1)}{2}} \cdot \frac{d\xi}{2}$$

$$= \int_{-1}^{1} \frac{d\xi}{(\xi+3)}$$

Using Gauss–Legendre three-point rule (corresponding to $n = 2$), we get

$$I = \frac{1}{9}\left[8\left(\frac{1}{0+3}\right) + 5\left(\frac{1}{3+\sqrt{3/5}}\right) + 5\left(\frac{1}{3-\sqrt{3/5}}\right)\right]$$

$$= \frac{131}{189} = 0.693122$$

$$I \text{ (exact)} = \log_2 = 0.693147$$

Example 9: Integrate by Gaussian quadrature ($n = 3$)

$$\int_1^2 \frac{dx}{1+x^3}$$

Solution: We have

$$I = \int_{a=1}^{b=2} \frac{dx}{1+x^3}$$

Converting the interval [1, 2] to the interval [–1, 1]

writing $$x = \frac{(b-a)}{2}\xi + \frac{(b+a)}{2} = \frac{(2-1)}{2}\xi + \frac{(2+1)}{2} = \frac{\xi}{2} + \frac{3}{2} = \frac{1}{2}(\xi + 3)$$

$\therefore$ $$I = \int_1^2 \frac{dx}{1+x^3}$$

$$= \frac{1}{2}\int_{-1}^{1} \frac{d\xi}{1+[(\xi+3)/2]^3}$$

Using the Gauss-Legendre four-point formula.

$$\int_{-1}^{1} f(\xi)\, d\xi = 0.652145\,[f(0.339981) + f(-0.339981)] + 0.347855\,[f(0.861136) + f(-0.861136)]$$

we obtain:

$$I = \frac{1}{2}[\,0.652145\,(\,0.176760 + 0.298268) + 0.347855\,(0.122020 + 0.449824)]$$

$$= 0.254353$$

Example 10: Evaluate $I = \int_1^2 \frac{2x dx}{1+x^4}$, using the Gauss-Legendre 1-point, 2-point and 3-point quadrature rules. Compare with the exact solution

$$I = \tan^{-1}(4) - (\pi/4)$$

Solution: To use the Gauss-Legendre rule, we convert the interval (1, 2) to the interval (–1, 1) in

$$I = \int_{a=1}^{b=2} \frac{2x dx}{1+x^4} \qquad ...[1]$$

writing, $$x = \frac{(b-a)}{2}\xi + \frac{(b+a)}{2}$$

$$= \frac{(2-1)}{2}\xi + \frac{(2+1)}{2} = \frac{\xi}{2} + \frac{3}{2} = \frac{(\xi+3)}{2}$$

$\therefore$ $$dx = \frac{d\xi}{2}$$

$\therefore$ $$I = \int_{-1}^{1} \frac{8(\xi+3)\,d\xi}{[16+(\xi+3)^4]} \int_{-1}^{1} f(\xi)\,d\xi \qquad ...[2]$$

Using the 1-point rule, we get

$$I = 2f(0) = 2\left[\frac{24}{16+81}\right] = 0.4948$$

Using the 2-point rule, we get

$$I = f\left(-\frac{1}{\sqrt{3}}\right) + f\left(\frac{1}{\sqrt{3}}\right)$$

$$= 0.3842 + 0.1592 + 0.5434$$

Using the 3-point rule, we get

$$I = \frac{1}{9}\left[5f\left(-\sqrt{\frac{3}{5}}\right) + 8f(0) + 5f\left(\sqrt{\frac{3}{5}}\right)\right]$$

$$= \frac{1}{9}[5\,(0.4393) + 8\,(0.2474) + 5\,(0.1379)]$$

$$I = 0.5406$$

$$I\,(\text{exact}) = 0.5404$$

Example 11: Evaluate $I = \int_0^{\pi} \sin x\, dx$ using the Gauss-Legendre quadrature formula.

Solution: Given: $$\text{I} = \int_{a=0}^{b=\pi} \sin x\, dx \qquad ...[1]$$

The interval 0 to π may be changed to -1 to 1

by substituting: $$x = \frac{(b-a)}{2}\xi + \frac{(b+a)}{2} = \frac{(\pi-0)}{2}\xi + \frac{(\pi+0)}{2} = \frac{\pi\xi}{2} + \frac{\pi}{2} = (\xi+1)\frac{\pi}{2}$$

$\therefore$ $$x = \frac{\pi}{2}(\xi+1); \quad dx = \frac{\pi}{2}\,d\xi$$

Thus $$I = \frac{\pi}{2}\int_{-1}^{1} f(\xi)\,d\xi = \frac{\pi}{2}\int_{-1}^{1} \sin\left[\frac{\pi}{2}(\xi+1)\right]d\xi$$

where $$f(\xi) = \sin\left[\frac{\pi}{2}(\xi + 1)\right]$$

Using three-point Gauss-Legendre formula, we get

$$I = \frac{\pi}{2}\left[\frac{5}{9}\sin\frac{\pi}{2}\left(-\sqrt{\frac{3}{5}}+1\right)+\frac{8}{9}\sin\frac{\pi}{2}(0+1)+\frac{5}{9}\sin\frac{\pi}{2}\left(\sqrt{\frac{3}{5}}+1\right)\right]$$

$$= 2.0014 \approx 2$$

Example 12: Evaluate

$$\int_2^3 \frac{\cos 2x}{1+\sin x}\,dx$$

using Gauss-Legendre two and three-point integration rules.

Solution:

Given: $$I = \int_{a=2}^{b=3} \frac{\cos 2x}{1+\sin x}\,dx \qquad \text{...[1]}$$

The interval (2 to 3) may be charged to –1 to +1 by substituting

$$x = \frac{(b-a)}{2}\xi + \frac{(b+a)}{2} = \frac{(3-2)}{2}\xi + \frac{(3+2)}{2} = \frac{1}{2}\xi + \frac{5}{2} \text{ in } I$$

$\therefore$ $$x = \frac{(\xi+5)}{2}$$

$\therefore$ $$I = \frac{1}{2}\int_{-1}^{+1}\frac{\cos(\xi+5)}{1+(\xi+5)/2}$$

Using the Gauss-Legendre 2-point formula

$$\int_{-1}^{1} f(\xi)\,d\xi = f\left(\frac{1}{\sqrt{3}}\right) + f\left(-\frac{1}{\sqrt{3}}\right)$$

we obtain:

$$I = \frac{1}{2}[0.56558356 - 0.15856672]$$

$$= 0.20350842$$

Using the Gauss-Legendre three-point formula

$$\int_{-1}^{1} f(\xi)\,d\xi = \frac{1}{9}\left[5+\left(-\sqrt{\frac{3}{5}}\right)+8f(0)+5f\left(\sqrt{\frac{3}{5}}\right)\right]$$

we obtain:

$$I = \frac{1}{18}[-1.26018516 + 1.41966658 + 3.48936887]$$

$$I = 0.20271391$$

Example 13: Use two-point Gaussian quadrature to find the deflection of a cantilever beam at the free end when the beam is subjected to uniformly distributed load as shown in the Fig. 7.17.

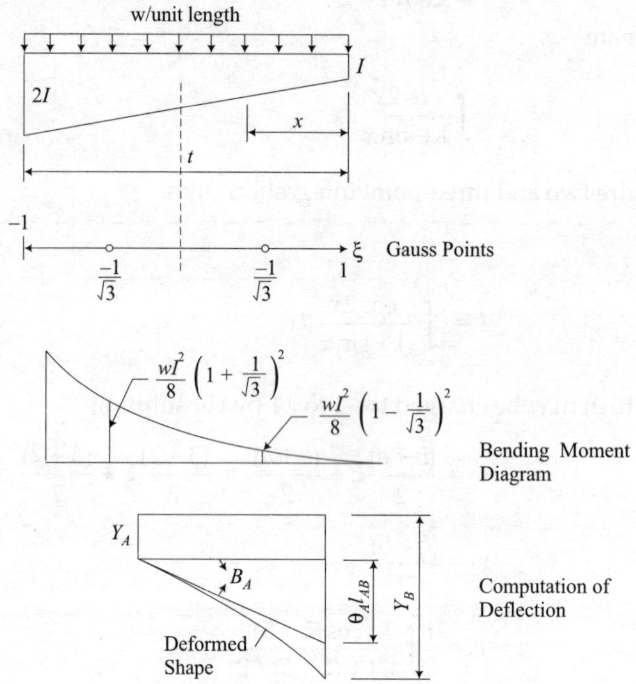

Fig. 7.17 Deflection of non-prismatic beam

Solution:

$$\theta_B = \theta_A + \int_A^B \frac{M}{EI} dx \qquad ...[1]$$

$$y_B = y_A + \theta_A \, l_{AB} + \int_A^B \frac{M_x \, d_x}{EI} \qquad ...[2]$$

(Since M and I are varying, and x is measured from B, at which deflection is required, we use Gaussian quadrature.)

For the cantilever beam $y_A = \theta_A = 0$

$$y_B = \int_0^1 \frac{Mx \, dx}{EI} = \frac{l}{2}\int_{-1}^{1} f(\xi) \, d\xi \qquad ...[3]$$

where $f(\xi) = M/EI$. Using two point Gaussian quadrature $\left(\xi = \frac{1}{\sqrt{3}}\right)$ on either side from centre and measuring X from B

$$f\left(\xi = -\frac{1}{\sqrt{3}}\right) = f\left(X = \frac{l}{2} - \frac{l}{2}\frac{1}{\sqrt{3}}\right)$$

$$= \frac{W\,(0.21113)^2\, l^2}{2E\,1.2113^2\, I}(0.2113\, l) \quad \text{...[4]}$$

$$f\left(\xi = \frac{1}{\sqrt{3}}\right) = f\left(X = \frac{l}{2} + \frac{l}{2}\frac{1}{\sqrt{3}}\right)$$

$$= \frac{W\,(0.7886)^2\, l^2}{2E\,1.7886\, I}(0.7886\, l) \quad \text{...[5]}$$

$$W_1 = W_2 = 1$$

$$Y_2 = \left[f\left(\xi = \frac{1}{\sqrt{3}}\right) + f\left(\xi = -\frac{1}{\sqrt{3}}\right)\right] l/2 \quad \text{...[6]}$$

$$= 0.07049\; wl^4/EI$$

Example 14: Use Gauss-Legendre numerical integration to integrate the body force term for the one-dimensional linear finite element assumed as a rod of elastic material fixed at both ends with constant cross-sectional area and length of 3 L with uniform body force loading f.

Solution: The matrix equation to be evaluated is written as: $A\int_0^L \begin{Bmatrix} (x-L)/L \\ x/L \end{Bmatrix} f\,dx$ ...[1]

$$I = \int_{-1}^{+1} f(\xi)\, d\xi = \sum_{i=1}^{n} w_i f(\xi_i) \quad \text{...[2]}$$

Using two sampling point Gauss formula ($n = 2$) in equation [2]

i.e., $$I = \int_{-1}^{1} f(\xi)\, d\xi = w_1 f(\xi_1) + w_2 f(\xi_2)$$

$$\left(w_1 = 1;\; w_2 = 1;\; \xi_1 = -\frac{1}{\sqrt{3}};\; \xi_2 = \frac{1}{\sqrt{3}}\right)$$

G. I. is $$A\int_{a=0}^{b=L} \begin{Bmatrix} \dfrac{(x-L)}{2} \\ \dfrac{x}{L} \end{Bmatrix} f dx$$

Substituting $$x = \frac{(b-a)}{2}\xi + \frac{(b+a)}{2}$$

$$x = \frac{(L-0)}{2}\xi + \frac{(L+0)}{2}$$

$$x = \frac{\xi L}{2} + \frac{L}{2} = \frac{(\xi L + L)}{2}$$

$\therefore$ $$dx = L\frac{d\xi}{2}$$

$\therefore$ G. I. becomes:

$$I = A\,f\,L\int_{-1}^{1}\begin{Bmatrix}(1-\xi)/4\\(1+\xi)/4\end{Bmatrix}d\xi \quad ...[3]$$

$$= \left(\frac{A\,f\,L}{4}\right)\{w_1 f(\xi_1) + w_2 f(\xi_2)\}$$

$$= \left(\frac{A\,f\,L}{4}\right)\left\{1\times f\left(-\frac{1}{\sqrt{3}}\right) + 1\times f\left(\frac{1}{\sqrt{3}}\right)\right\}$$

$$= \left(\frac{A\,f\,L}{4}\right)\left\{[1-(0.577735)]\,(1) + [1-(-\,0.577735)]\right\} \quad ...[1]$$

$$= \left(\frac{A\,f\,L}{2}\right). \qquad I\,(\text{exact}) = \left(\frac{A\,f\,L}{2}\right).$$

In the two-dimensional situation, in determining the stiffness matrix of an isoparametric quadrilateral, we are concerned with evaluating definite integrals of the form

$$I = \int_{-1}^{1}\int_{-1}^{1} F(\xi,\eta)\; d\xi\, d\eta \quad ...[1]$$

$$I = \int_{-1}^{1}\left(\int_{-1}^{1} F(\xi_1 \eta)\, d\xi\, d\eta = \int_{-1}^{1} (\Sigma\, W_i(\xi_i \eta)\, d\eta \quad ...[2]$$

$$= \sum_{i=1}^{\eta_i} W_i \sum_{j=1}^{\eta_i} W_j\, F(\xi_i, \eta_j)$$

$$= \sum_{i=1}^{\eta_i}\sum_{i=1}^{\eta_i} W_i W_j\; F(\xi_i, \eta_j)$$

In this expression it is assumed that there are η_j sampling points in the ξ direction and η_i in the direction η, there are consequently $\eta_i \times \eta_j$ sampling points in all.

In the three-dimensional situation the properties of an isoparametric brick element are to be evaluated. Then the integrals of the form

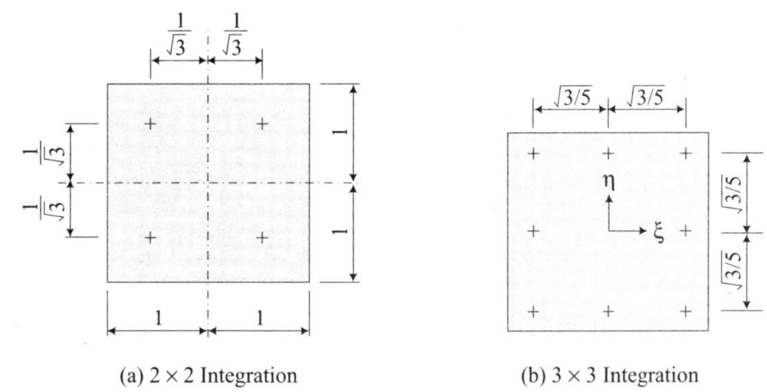

Fig. 7.18

$$I = \int_{-1}^{1}\int_{-1}^{1}\int_{-1}^{1} F(\xi, \eta, \rho)\, d\xi\, d\eta\, d\zeta \qquad \text{...[3]}$$

arise and can be calculated numerically as

$$I = \sum_{l=1}^{n_l}\sum_{j=1}^{n_j}\sum_{k=1}^{n_k} W_i\, W_j\, W_k\, F(\xi_k, \eta_j, \zeta_i) \qquad \text{...[4]}$$

For two-dimensional problem $n = 2$ means $2 \times 2 = 4$ Gaussian points and for three-dimensional problems it works out to be $2 \times 2 \times 2 = 8$. Thus,

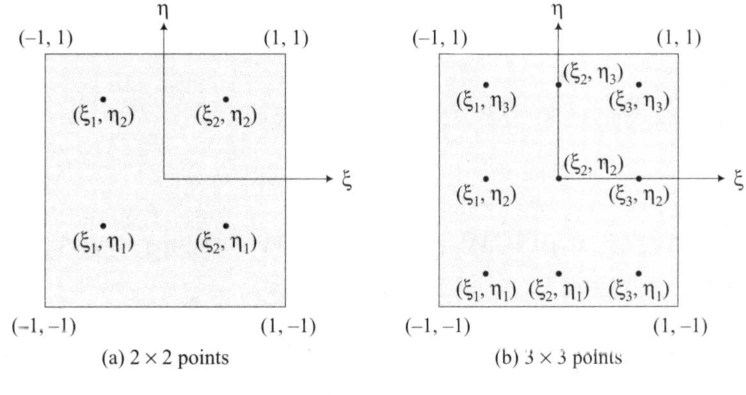

Fig. 7.19

Example 15: Evaluate $\int_{-1}^{1}\int_{-1}^{1} x^2y^2\,dxdy$ by using Gauss quadrature formula.

Solution: We convert the given double integral in terms of ξ and η by the substitutions

$$x = \frac{(b-a)}{2}\xi + \frac{(b+a)}{2} = \frac{[1-(-1)]}{2}\xi + \frac{(1-1)}{2}$$

$$= \frac{2}{2}\xi + 0 = \xi$$

$$(\therefore dx = d\xi) \quad \text{...[1]}$$

and

$$y = \frac{(d-c)}{2}\eta + \frac{(d+c)}{2} = \frac{[1-(-1)]}{2}\eta + \frac{(1-1)}{2} = \eta$$

$$(\therefore dy = d\eta) \quad \text{...[2]}$$

$$\therefore \quad I = \int_{a=-1}^{b=1}\int_{c=-1}^{d=1} x^2y^2\,dxdy = \int_{\eta=-1}^{\eta=1}\int_{\xi=-1}^{\xi=1} \xi^2\eta^2\,d\xi\,d\eta$$

$$(a = -1, b = 1, c = -1, d = 1)$$

$$= \int_{\eta=-1}^{1}\int_{\xi=-1}^{1} f(\xi, \eta)\,d\xi\,d\eta$$

where $f(\xi, \eta) = \xi^2\eta^2$ ...[3]

Both x and y in the G. I. have quadrate terms. If $n = 1$, then polynomial of maximum order $(2\times 1-1)$ or 1 can be evaluated exactly.

So, $n = 1$ is insufficient in the present case. If $n = 2$, then the polynomial of maximum order $(2\times 2-1)$ or 3 can be evaluated exactly. So for performing the above integration, a 2-point Gaussian quadrature for both x and y is employed.

$$I = \sum_{i=1}^{m}\sum_{j=1}^{n} W_iW_j F(\xi_i, \eta_j)$$

Here, $m = n = 2$

$$\therefore \quad I = \{W_1W_1 F(\xi_1, \eta_1) + W_1W_2 F(\xi_1, \eta_2) + W_2W_1 F(\xi_2, \eta_1) + W_2W_2 F(\xi_2, \eta_2)\} \quad \text{...[4]}$$

$$\left(\xi_1 = \frac{1}{\sqrt{3}};\ \eta_1 = -\frac{1}{\sqrt{3}};\ \xi_2 = -\frac{1}{\sqrt{3}};\ \eta_2 = -\frac{1}{\sqrt{3}}\right)$$

$$\left\{\begin{aligned} F(\xi_1,\eta_1) &= \xi_1^2\ \eta_1^2 = \left(\frac{1}{\sqrt{3}}\right)^2\left(-\frac{1}{\sqrt{3}}\right)^2 = \frac{1}{3}\times\frac{1}{3} = \frac{1}{9} \\ F(\xi_1,\eta_2) &= \xi_1^2\ \eta_2^2 = \left(\frac{1}{\sqrt{3}}\right)^2\left(-\frac{1}{\sqrt{3}}\right)^2 = \frac{1}{3}\times\frac{1}{3} = \frac{1}{9} \\ F(\xi_2,\eta_2) &= \xi_2^2\ \eta_2^2 = \left(-\frac{1}{\sqrt{3}}\right)^2\left(-\frac{1}{\sqrt{3}}\right)^2 = \frac{1}{3}\times\frac{1}{3} = \frac{1}{9} \\ F(\xi_2,\eta_1) &= \xi_2^2\ \eta_1^2 = \left(-\frac{1}{\sqrt{3}}\right)^2\left(-\frac{1}{\sqrt{3}}\right)^2 = \frac{1}{3}\times\frac{1}{3} = \frac{1}{9} \end{aligned}\right\}$$

$$\therefore \quad \left\{\begin{aligned} &(W_1 = W_2 = 1) \\ &I = 1\times 1\left(\frac{1}{9}\right) + 1\times 1\left(\frac{1}{9}\right) + 1\times 1\left(\frac{1}{9}\right) + 1\times 1\left(\frac{1}{9}\right) \\ &\quad = \frac{4}{9} \qquad \text{(using equation [4].} \end{aligned}\right\}$$

Example 16: Evaluate the integral

$$I = \int_{y=c=4}^{y=d=6}\int_{x=a=-2}^{x=b=2} (1-x)^2\,(4-y)^2\,dx\,dy \qquad \text{...[1]}$$

Converting the integral in terms of ξ and η as

$$X = \frac{(b-a)}{2},\xi + \frac{(b+a)}{2} = 2\,\xi + 0;\ dx = 2\,d\,\xi$$

$$Y = \frac{(d-c)}{2}\eta + \frac{(d+c)}{2} = \eta + 5\,;\ dy = d\,\eta \qquad [2]$$

$$I = 2\int_{\eta=-1}^{\eta=1}\int_{\xi=-1}^{1} (1-2\xi)^2\,(4-\eta-5)^2\,d\xi\,d\,\eta$$

$$= 2\int_{\eta=-1}^{1}\int_{\xi=-1}^{1} f(\xi,\eta)\,d\,\xi,\,d\,\eta \qquad \text{...[3]}$$

where $\qquad f(\xi,\eta) = (1-2\,\xi)^2\,(1+\eta)^2 \qquad$ ...[4]

Let us use two-point Gaussian quadrature function values at four points (1, 1), (1, 2), (2, 1) and (2, 2) as shown in Fig. 7.20 and compute.

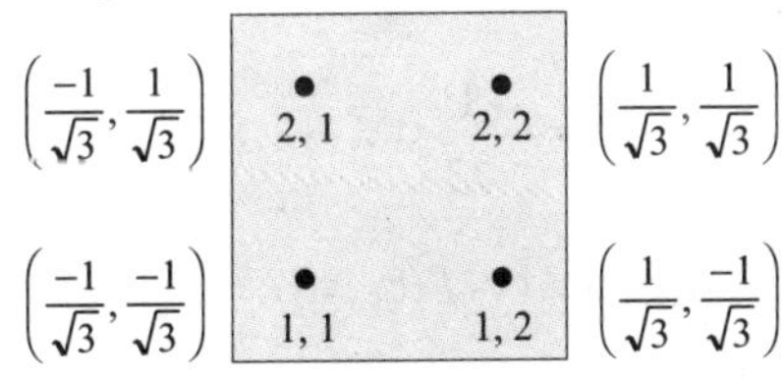

Fig. 7.20

$$\left\{\begin{aligned}
&\xi_1 = \frac{1}{\sqrt{3}};\ \eta_1 = -\frac{1}{\sqrt{3}};\ \xi_2 = -\frac{1}{\sqrt{3}};\ \eta_2 = -\frac{1}{\sqrt{3}}\\
&f(\xi_1, \eta_1) = (1-2\sqrt{3})^2\left(1-\frac{1}{\sqrt{3}}\right)^2 = 0.02394 \times 0.17862 = 0.004276162\\
&f(\xi_1, \eta_2) = \left(1+\frac{2}{\sqrt{3}}\right)^2\left(1-\frac{1}{\sqrt{3}}\right)^2 = 4.64288 \times 0.17862 = 0.829311225\\
&f(\xi_2, \eta_2) = \left(1+\frac{2}{\sqrt{3}}\right)^2\left(1+\frac{1}{\sqrt{3}}\right)^2 = 4.64288 \times 2.488 = 11.55148544\\
&f(\xi_2, \eta_1) = \left(1-\frac{2}{\sqrt{3}}\right)^2\left(1+\frac{1}{\sqrt{3}}\right)^2 = 0.02394 \times 2.488 = 0.05956272
\end{aligned}\right\}$$

$$I = \sum_{i=1}^{m}\sum_{j=1}^{n} W_i W_j f(\xi_i, \eta_j), \qquad (m = n = 2),$$

$$I = 2[W_1 W_1 f(\xi_1, \eta_1) + W_1 W_2 f(\xi_1, \eta_2) + W_2 W_1 f(\xi_2, \eta_1) + W_2 W_2 f(\xi_2, \eta_2)]$$

$\therefore$ $I = 2[1 \times 1 \times (0.004276162) + 1 \times 1\,(0.829311225) + 1 \times 1\,(11.55148544) + 1 \times 1\,(0.05956272)]$

$= 2\,[12.444635547]$

$= 24.\,88927108 \approx 24.88$ agrees with the calculated value 24.88.

Example 17: Using Gaussian quadrature, evaluate the following by two-point formula

$$\int_{-1}^{1}\int_{-1}^{1} (\xi^2 + 2\,\eta\,\xi + \eta^3)\, d\xi\, d\eta$$

[V.T.U. July/Aug. 2003, 6th sem. (B.E. Mech.)]

Solution : Given :

$$I = \int_{-1}^{1}\int_{-1}^{1} (\xi^2 + 2\eta\xi + \eta^3) d\xi\, d\eta \qquad ...[1]$$

Performing the above integration, a 2-point Gaussian quadrature for both ξ and η is employed.

$$\therefore \qquad I = \sum_{i=1}^{n}\sum_{j=1}^{n} W_i\, W_j\, F(\xi_i, \eta_j)$$

Here, $m = n = 2$

$$\left(\xi_1 = \frac{1}{\sqrt{3}};\ \eta_1 = -\frac{1}{\sqrt{3}};\ \xi_2 = -\frac{1}{\sqrt{3}};\ \eta_2 = -\frac{1}{\sqrt{3}}, W_1 = 1, W_2 = 1\right)$$

or,

$$I = W_1 W_1\, F(\xi_1, \eta_1) + W_1 W_2\, F(\xi_1, \eta_2) + W_2 W_1\, F(\xi_2, \eta_1) + W_2\, W_2\, F(\xi_2, \eta_2) \qquad ...[2]$$

Now, $$F(\xi_1, \eta_1) = (\xi_1^{\,2} + 2\eta_1\,\xi_1 + \eta_1^{\,3}) = \left[\left(\frac{1}{\sqrt{3}}\right)^2 + 2\left(-\frac{1}{\sqrt{3}}\right)\left(\frac{1}{\sqrt{3}}\right) + \left(-\frac{1}{\sqrt{3}}\right)^3\right]$$

$$= \left(\frac{1}{3} - \frac{2}{3} - \frac{1}{3\sqrt{3}}\right) = \left(\frac{\sqrt{3} - 2\sqrt{3} - 1}{3\sqrt{3}}\right)$$

$$= \frac{-\left(\sqrt{3} + 1\right)}{3\sqrt{3}} = \left(-\frac{1}{3} - \frac{1}{3\sqrt{3}}\right)$$

$$= -\left(\frac{1}{3} + \frac{1}{3\sqrt{3}}\right)$$

$$F(\xi_1, \eta_2) = (\xi_1^{\,2} + 2\xi_1\,\eta_2 + \eta_2^{\,3}) = \left[\left(\frac{1}{3} - \frac{2}{3} - \frac{1}{3\sqrt{3}}\right)\right] = -\left(\frac{1}{3} + \frac{1}{3\sqrt{3}}\right)$$

$$F(\xi_2, \eta_2) = (\xi_2^{\,2} + 2\xi_2\,\eta_2 + \eta_2^{\,3}) = \left[\left(\frac{1}{3} + \frac{2}{3} - \frac{1}{3\sqrt{3}}\right)\right] = \left(1 - \frac{1}{3\sqrt{3}}\right)$$

$$F(\xi_2, \eta_1) = (\xi_2^{\,2} + 2\xi_2\,\eta_1 + \eta_1^{\,3}) = \left[\left(\frac{1}{3} + \frac{2}{3} - \frac{1}{3\sqrt{3}}\right)\right] = \left(1 - \frac{1}{3\sqrt{3}}\right)$$

Hence, by using equation (2) we obtain

$$I = 1 \times 1 \times \left(-\frac{1}{3} - \frac{1}{3\sqrt{3}}\right) + 1 \times 1 \times \left(-\frac{1}{3} - \frac{1}{3\sqrt{3}}\right) + 1 \times 1 \times \left(1 - \frac{1}{3\sqrt{3}}\right)$$

$$+ 1 \times 1 \times \left(1 - \frac{1}{3\sqrt{3}}\right)$$

$$= \left(-\frac{1}{3} - \frac{1}{3\sqrt{3}}\right) - \frac{1}{3} - \frac{1}{3\sqrt{3}} + \left(1 - \frac{1}{3\sqrt{3}}\right) + \left(1 - \frac{1}{3\sqrt{3}}\right)$$

$$= \left(2 - \frac{2}{3}\right) - \left(\frac{1}{3\sqrt{3}} + \frac{1}{3\sqrt{3}} + \frac{1}{3\sqrt{3}} + \frac{1}{3\sqrt{3}}\right)$$

$$= \frac{4}{3} - \frac{4}{3\sqrt{3}}$$

$$= \frac{4}{3}\left(1 - \frac{1}{\sqrt{3}}\right)$$

$$= \frac{4}{3}\,(1 - 0.5773502691) = \frac{4}{3}\,(0.422649731)$$

$$\cong (1.333)\,(0.4226)$$
$$\cong 0.5633258$$

Example 18: Using three-point Gaussian quadrature find $\int xdA; \int xy\,dA$ for a triangular element whose vertices are (1, 1) (3, 2) and (2, 3) ($A = 3/2$)

where
$$x = x_1\xi_1 + x_2\xi_2 + x_3\xi_3$$
$$= (x_1 - x_3)\,\xi_1 + (x_2 - x_3)\xi_2 + x_3$$
$$= 2 - \xi_1 + \xi_2$$

Solution: Referring table for triangular elements

$$\int x\,dA = \frac{A}{3}\left(2-\frac{1}{6}+\frac{1}{6}\right)+\frac{A}{3}\left(2-\frac{2}{3}+\frac{1}{6}\right)+\frac{A}{3}\left(2-\frac{1}{6}+\frac{2}{3}\right) = 2A = 3 \qquad ...[1]$$

$$y = (y_1 - y_3)\,\xi_1 + (y_2 - y_1)\,\xi_2 + y_3$$
$$= -2\xi_1 - \xi_2 + 3$$

$$\therefore \qquad \int xydA = \int (2-\xi_1+\xi_2)\,(-2\xi_1-\xi_2+3)\,dA$$

$$= \frac{A}{3}\left\{\left(2-\frac{1}{6}+\frac{1}{6}\right)\left(-\frac{1}{3}-\frac{1}{6}+3\right)+\left(2-\frac{2}{3}+\frac{1}{6}\right)\left(-\frac{4}{3}-\frac{1}{6}+3\right)\right.$$
$$\left.+\left(2-\frac{1}{6}+\frac{2}{3}\right)\left(-\frac{1}{3}-\frac{2}{3}+3\right)\right\} \qquad ...[2]$$

$$= \frac{A}{3}\,\{5 + 2.25 + 5\} = 4.0833A \qquad ...[3]$$
$$= 6.12495$$

Example 19: Use Gauss-Legendre quadratures with $n = 2$ to numerically integrate

$$\int_0^a\int_0^b xy\,dx\,dy = \frac{a^2b^2}{4}$$

Solution: Converting the integral in terms of ξ and η as:

$$x = \frac{(b-a)}{2}\xi + \frac{(b+a)}{2}$$

$$= \frac{(a-0)}{2}\xi + \frac{(a+0)}{2} = \frac{a\xi}{2} + \frac{a}{2} = \frac{a(\xi+1)}{2}$$

$$\therefore \qquad dx = a\frac{d\xi}{2};$$

$$y = \frac{(d-c)}{2}\eta + \frac{(d+c)}{2} = \frac{(b-0)}{2}\eta + \frac{(b+0)}{2} = \frac{b\eta}{2} + \frac{b}{2} = \frac{b(\eta+1)}{2}$$

$$\therefore \qquad dy = \frac{bd\eta}{2}$$

$\therefore$ G.I. becomes

$$\int_{-1}^{+1} \frac{a}{2}(\xi+1)\,\frac{b}{2}(\eta+1)\,\frac{ab}{4}\,d\xi\,d\eta$$

$$I = \frac{a^2b^2}{16}\sum_{i=1}^{2}\sum_{j=1}^{2} W_i W_j (\xi_i + 1)(\eta_j + 1)$$

$$= \frac{a^2b^2}{16}\left[\sum_{i=1}^{2}\sum_{j=1}^{2} W_i W_j\, f(\xi_i, \eta_j)\right] \qquad ...[1]$$

where $f(\xi_i, \eta_j) = (\xi_i + 1)(\eta_j + 1)$

$$\left(\xi_1 = \frac{1}{\sqrt{3}};\ \eta_1 = -\frac{1}{\sqrt{3}},\ \xi_2 = -\frac{1}{\sqrt{3}},\ \eta_2 = -\frac{1}{\sqrt{3}}\right)$$

$$W_1 = 1;\ W_2 = 1$$

$$f(\xi_1, \eta_1) = (\xi_1 + 1)(\eta_1 + 1) = \left(\frac{1}{\sqrt{3}} + 1\right)\left(-\frac{1}{\sqrt{3}} + 1\right) = 1 - \left(\frac{1}{\sqrt{3}}\right)^2 = \left(1 - \frac{1}{\sqrt{3}}\right) = \frac{2}{3}$$

$$f(\xi_1, \eta_2) = (\xi_1 + 1)(\eta_2 + 1) = \left(\frac{1}{\sqrt{3}} + 1\right)\left(-\frac{1}{\sqrt{3}} + 1\right) = \left(\frac{2}{3}\right)$$

$$f(\xi_2, \eta_2) = (\xi_2 + 1)(\eta_2 + 1) = \left(-\frac{1}{\sqrt{3}} + 1\right)\left(-\frac{1}{\sqrt{3}} + 1\right) = \left(\frac{1}{3} - \frac{2}{\sqrt{3}} + 1\right)$$

$$f(\xi_2, \eta_1) = (\xi_2 + 1)(\eta_1 + 1) = \left(-\frac{1}{\sqrt{3}} + 1\right)\left(-\frac{1}{\sqrt{3}} + 1\right) = \left(\frac{1}{3} - \frac{2}{\sqrt{3}} + 1\right)$$

$$I = \frac{a^2b^2}{16}\,[W_1 W_1 f(\xi_1, \eta_1) + W_1W_2 f(\xi_1, \eta_2) + W_2W_1 f(\xi_2, \eta_1) + W_2W_2 f(\xi_2, \eta_2)]$$

$$= \frac{a^2b^2}{16}\left[1\times1\times\left(\frac{2}{3}\right) + 1\times1\times\left(\frac{2}{3}\right) + 1\times1\times\left(\frac{1}{3} - \frac{2}{\sqrt{3}} + 1\right) + 1\times1\times\left(\frac{1}{3} - \frac{2}{\sqrt{3}} + 1\right)\right]$$

$$= \frac{a^2b^2}{16}\left[\frac{2}{3} + \frac{2}{3} + \frac{1}{3} - \frac{2}{\sqrt{3}} + 1 + \frac{1}{3} - \frac{2}{\sqrt{3}} + 1\right]$$

$$= \frac{a^2b^2}{16}\left[2 - \frac{4}{\sqrt{3}} + 2\right]$$

$$= \frac{a^2b^2\,(4)}{16}\left[1 - \frac{1}{\sqrt{3}}\right] = \frac{a^2b^2}{4}\left(1 - \frac{1}{\sqrt{3}}\right)$$

$$I\text{ (exact)} = \frac{a^2b^2}{4}$$

7.4 HIGHER ORDER ELEMENTS

In order to get accurate results for stresses with a constant strain discretization one has to use a large number of elements.

That is, accuracy of calculation increases if higher order elements are used. Limitations on use of number of elements come from the total degrees of freedom the computer can handle . The limitation may be due to cost of computation time also. Hence, to use higher order elements we have to use less number of such elements.

Even with a fine grid one will face a problem of interpreting the result. Therefore, considerable effort has been devoted to develop "refined elements", *i.e.* elements having linear, quadratic or higher order strain expressions.

Definition: Any element with a second order or higher interpolation function is called the higher order element or refined elements. Within this element the variable will behave non-linearly. Quadratic, cubic are a few examples of higher order elements.

Shape Functions

The shape function is defined as the one whose value at a particular node i is unity and at other nodes zero. It is designated by the capital letter N_i. The subscript 'i' indicates the node number to which it belongs. Shape functions are usually obtained in terms of natural coordinate which is also known as intrinsic coordinate or local coordinate or rarely as normal coordinate.

It is termed as interpolation function because it is on the basis of this function that field variable (displacements etc.) can be determined by interpolation at any point inside the element or at any point on the boundary. It is called shape function because it is dependent on the shape of the element.

There is one shape function corresponding to each degree of freedom of the element. (*i.e.,* one shape function is associated with each node of the element). The shape functions may be polynomials of any order.

Polynomial Shape Functions

Case (1) (One-dimensional polynomial shape function): Polynomial shape function of nth order in one dimension is given by

$$u\,(x) = \alpha_1 + \alpha_2\,x + \alpha_3\,x^2 + \cdots + \alpha_{n+1}x^n \quad ...[1]$$

Its matrix form :

$$u = [G]\{\alpha\} \quad ...[2]$$

where $$[G] = [1, x, x^2, ...x^n]$$

and $$\{\alpha\}^T = [\alpha_1\ \alpha_2\ \alpha_3.....\ \alpha_{n+1}]$$

Hence, there are $m = n + 1$ terms in a one-dimensional nth order complete polynomial. ...[3]

Case (2) (Two-dimensional polynomial shape function): Two-dimensional polynomial model is

$$u\ (x, y) = \alpha_1 + \alpha_2 x + \alpha_3 y + \alpha_4 x^2 + \alpha_5 xy + \alpha_6 y^6 + \alpha_7 x^3 ... + \alpha_m y^n \quad ...[4]$$

$$v\ (x, y) = \alpha_{m+1} + \alpha_{m+2} x + \alpha_{m+3} y + \cdots + \alpha_{2m} y^n \quad ...[5]$$

or $$\{\delta\} = \begin{Bmatrix} u(x, y) \\ v(x, y) \end{Bmatrix} = [G]\{\alpha\}$$

$$= \begin{bmatrix} G_1 & 0 \\ 0 & G_1 \end{bmatrix} \{\alpha\} \quad ...[6]$$

where $$G_1 = [1\ x\ y\ x^2\ xy\ y^2\ x^3\ ...y^n\ ]$$

$$\{\alpha\}^T = [\alpha_1\ \alpha_2\ \alpha_3\ \alpha_4\ ....\alpha_{2m}]$$

It is seen that in two-dimensional problems total number of terms, m, in a complete n^{th} degree polynomial is

$$\boxed{m = \frac{(n+1)(n+2)}{2}} \quad ...[7]$$

For $n = 1$, we get first order complete polynomial

i.e., $$m = \frac{(1+1)(1+2)}{2} = 3$$

$\Rightarrow$ the first three terms are $\alpha_1 + \alpha_2 x + \alpha_3 y$

Similarly, for $n = 2$, $$m = \frac{(2+1)(2+2)}{2} = 6$$

and we know the first six terms are $\alpha_1 + \alpha_2 x + \alpha_3 y + \alpha_4 x^2 + \alpha_5 x\,y + \alpha_6 y^2$

In two-dimensional stress analysis basic field variable is displacement given by u and v, where

$$\boxed{\begin{aligned} u &= \Sigma\, N_i u_i \\ v &= \Sigma\, N_i v_i \end{aligned}} \quad ...[1]$$

where summation is the number of nodes of the element.

Illustration: For three-noded triangular element, displacement at $P(x, y)$ is

$$u = \Sigma\, N_i u_i = N_1 u_1 + N_2 u_2 + N_3 u_3$$

$$v = \Sigma\, N_i v_i = N_1 v_1 + N_2 v_2 + N_3 v_3$$

We can write the above in the matrix form as:

or
$$\begin{Bmatrix} u \\ v \end{Bmatrix} = \begin{bmatrix} N & 0 \\ 0 & N \end{bmatrix} \begin{Bmatrix} u_1 \\ u_2 \\ u_3 \\ v_1 \\ v_2 \\ v_3 \end{Bmatrix} = \begin{bmatrix} N_1 & N_2 & N_3 & 0 & 0 & 0 \\ 0 & 0 & 0 & N_1 & N_2 & N_3 \end{bmatrix} \begin{Bmatrix} u_1 \\ u_2 \\ u_3 \\ v_1 \\ v_2 \\ v_3 \end{Bmatrix}$$

or
$$\underset{2\times1}{\{\alpha\}} = \underset{2\times6}{[N]}\underset{6\times1}{\{\alpha\}_e} \qquad \text{...[2]}$$

where α is the displacement at any point in the element, $[N]$ shape function and $\{\alpha\}_e$ is a vector of nodal displacements.

Similarly, in case of 6-noded triangular element

$$\underset{2\times1}{\{\alpha\}} = \underset{2\times12}{[N]}\underset{12\times1}{\{\alpha\}_e} \qquad \text{...[3]}$$

In case of 4-noded rectangular element

$$\underset{2\times1}{\{\alpha\}} = \underset{2\times8}{[N]}\underset{8\times1}{\{\alpha\}_e} \qquad \text{...[4]}$$

Another convenient way to remember complete two-dimensional polynomial is to select/check the terms for parameter variation model, the terms are arranged in a triangular form as shown below. This triangle is known as Pascal's triangle.

															Constant	1
							1								Constant	1
						x		y							Linear	3
					x^2		xy		y^2						Quadratic	6
				x^3		x^2y		xy^2		y^3					Cubic	10
			x^4		x^3y		x^2y^2		xy^3		y^4				Quartic	15
		x^5		x^4y		x^3y^2		x^2y^3		xy^4		y^5			Quintic	21
	x^6		x^5y		x^4y^2		x^3y^3		x^2y^4		xy^5		y^6		Hexadic	28
	⋮		⋮		⋮		⋮		⋮		⋮		⋮			
x^n			…		…		x^ny^n		…		…			y^n		

PASCAL'S TRIANGLE

In order to choose terms for any element the counting of terms begins at top and goes downward. Depending upon type of variation required for the element the terms are taken up as equal to the number of nodes in the element.

(i) Quadratic along x, cubic along y. (ii) Linear along x, cubic along y.

Fig. 7.21 Some elements in two dimension

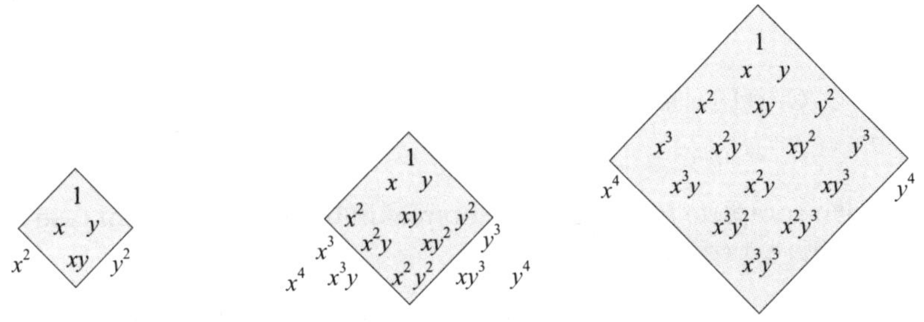

(a) Bilinear interpolation (b) Biquadratic interpolation (c) Bicubic expansion

Fig. 7.22

For three-dimensional element or parametric model the triangle for the terms can be constructed in similar manner. The displacement models for following 3 D elements can be given as:

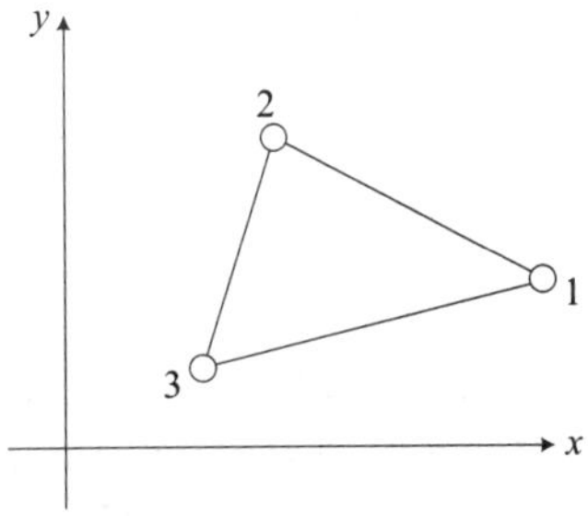

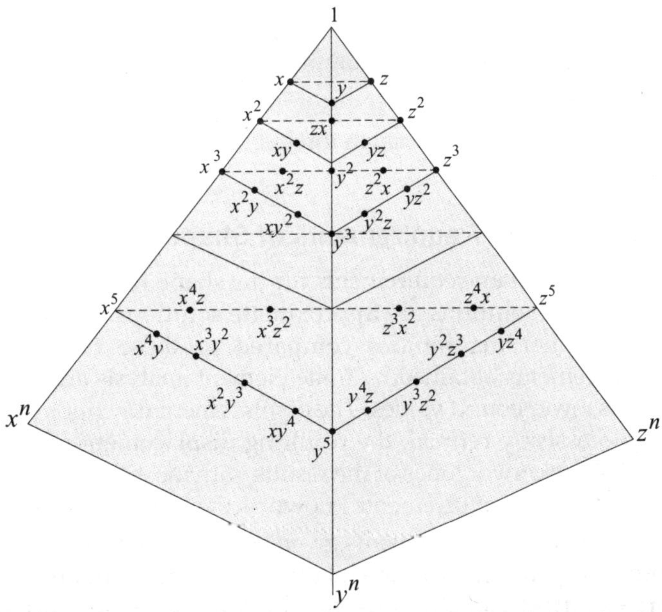

Fig. 7.23 Pascal's Tetrahedron

Case (3) (Three-dimensional polynomial shape functions): Three-dimensional shape function of nth order complete polynomial is given by

$$u\,(x, y, z) = \alpha_1 + \alpha_2 + \alpha_3 y + \alpha_4 z + \alpha_5 x^2 + \cdots + \alpha_m x^{n-1} z$$

$$v\,(x, y, z) = \alpha_{m+1} + \alpha_{m+2} x + \alpha_{m+3} y + \alpha_{m+4} z + \alpha_{m+5} x^2 + \cdots + \alpha_{2m} x^{n-1} z$$

$$w\,(x, y, z) = \alpha_{2m+1} + \alpha_{2m+2} x + \alpha_{2m+3} y + \alpha_{2m+4} z + \cdots \alpha_{3m} x^{n-1} z \qquad \text{...[1]}$$

or
$$\delta\,(x, y, z) = \begin{Bmatrix} u(x,y,z) \\ v(x,y,z) \\ w(x,y,z) \end{Bmatrix} = [G]\{\alpha\} = \begin{bmatrix} G_1 & 0 & 0 \\ 0 & G_1 & 0 \\ 0 & 0 & G_1 \end{bmatrix} \{\alpha\} \qquad \text{...[2]}$$

where
$$G_1 = [\,1\ x\ y\ z\ x^2\ xyy^2\ yzz^2\ zx....z^n z^{n-1} x..zx^{n-1}]$$

and
$$\{\alpha\}^T = [\alpha_1\ \alpha_2\ \alpha_3.....\alpha_{3m}]$$

It is seen that a complete nth order polynomial in three-dimensional case is having number of terms m given by the expression

$$m = \frac{(n+1)(n+2)(n+3)}{6} \qquad \text{...[3]}$$

when
$$n = 1,\ m = \frac{(1+1)(1+2)(1+3)}{6} = 4$$

i.e.,
$$\alpha_1 + \alpha_2 x + \alpha_3 y + \alpha_4 z$$

for $n = 2$,
$$m = \frac{(2+1)(2+2)(2+3)}{6} = 10$$

Hence, the second degree complete polynomial is

$$\alpha_1 + \alpha_2 x + \alpha_3 y + \alpha_4 z + \alpha_5 x^2 + \alpha_6 xy + \alpha_7 y^2 + \alpha_8 yz + \alpha_9 z^2 + \alpha_{10} zx$$

Complete polynomial in three dimensions may be expressed conveniently by a tetrahedron as shown in Fig. 7.23.

Convergence Requirements of Shape Functions

There are certain requirements for the shape functions of an element that will now be presented. Numerical solutions are approximate solutions. Stiffness coefficients for a displacement model have higher magnitudes compared to those for the exact solutions. In other words the displacements obtained by finite element analysis are lesser than the exact values. Thus the *FEM* yields lower bound values. The displacement functions must be chosen so that as the element mesh is successively refined, the resulting displacement solution converges to the correct answer. In ensure the convergence of the results with mesh refinement, the displacement approximation must satisfy certain requirements known as the 'convergence requirements' (or criteria).

Hence, the requirements given below must be met if we are to be assured of convergence of the process as the elements are increased in number and decreased in size. Thus :

(1) The displacements must be compatible between adjacent elements: Compatibility conditions refer to the continuity of displacements throughout the structure.

i.e., different parts of this structure must fit together without discontinuity at all stages of loading.

The 2nd part of this requirement, *compatibility* at interfaces, means that since the elements discretized have imaginary boundaries and they are the part of the same continuum behaving in union and are not behaving as separate bodies, the displacement or strains of the interfaces must be equal and in same direction if obtained from displacement model of element one and/or from displacement model of other adjacent elements. There should be no deformation causing openings or overlaps or separations at the element interfaces.

Broadly we can write that the displacement model must show continuity of deflected shape within and over interfaces of elements. This condition is called 'compatibility conditions'.

To describe the degree of compatibility achieved by interpolation function (or shape function) at the element interfaces, we have a specific notation in common use.

(i) We say that we have C^0 continuity when the field variable only (and none of its derivatives) maintains continuity at an interelement interface .

(ii) We say that we have C^1 continuity when the field variable and its first derivatives are continuous at element interfaces.

(iii) C^2 continuity indicates that the field variable, its first derivatives, and its second derivatives are all continuous at boundary interfaces, and so on.

Now, we can say that, for assurance of finite element convergence for a functional having P as the highest order of a derivative, completeness requires C^P continuity while compatibility requires C^{P-1} continuity".

Suppose that the displacement and its partial derivatives upto one order less than the highest order derivative appearing in strain energy function is continuous.

In plane stress and strain problem, it is sufficient if continuity of displacement is satisfied, since strain energy function includes only first order derivatives of the displacements.

$$(\text{strain energy} = \frac{1}{2}\ \text{stress} \times \text{strain})$$

This implies it is enough if C^0 continuity is ensured in plane stress and strain problems. In case

of flexure problems $\left(i.e., \frac{1}{2}\frac{M^2}{EI}\right)$

where $\left(M = -EI\frac{d^2w}{dx^2}\right)$.

(2) Rigid body modes must be represented: The chosen displacement function (model) should include the rigid body modes of the displacement of the element. This suggests that the rigid body displacements at the modes must not induce any straining of the element. In order to avoid complicated expressions, the terms pertaining to rigid body modes have not been sometimes considered in the displacement function. That is, when the nodal degrees of freedom are given values corresponding to rigid body motion, they must exhibit zero strain and zero nodal forces. In the displacement model (function) there should be a term which permits all points on the element to experience the same displacement.

Example (1): In the displacement function $u = \alpha_1 + \alpha_2 x + \alpha_3 y$; the term α_1 provides for the rigid body displacement.

This implies that to satisfy the requirement of rigid body displacement, there should be constant term in the shape function selected.

Example (2): In case of *CST*, the strain–displacement relation is given by

$$[\varepsilon] = [B]\begin{Bmatrix} u \\ v \end{Bmatrix} \qquad \text{...[1]}$$

where

$$[B] = \frac{1}{2\Delta}\begin{bmatrix} y_{23} & y_{31} & y_{12} & 0 & 0 & 0 \\ 0 & 0 & 0 & x_{32} & x_{13} & x_{12} \\ x_{32} & x_{13} & x_{21} & y_{23} & y_{31} & y_{12} \end{bmatrix} \qquad \text{...[2]}$$

For rigid body displacement

$$\left.\begin{aligned} u_1 &= u_2 = u_3 = 1 \\ \varepsilon_x &= y_{23} + y_{31} + y_{12} \\ &= (y_2 - y_3) + (y_3 - y_1) + (y_1 - y_2) = 0 \end{aligned}\right\} \qquad \text{...[3]}$$

Similarly we can prove that the other strains are also zero.

(3) The displacement field must represent constant strain states of the element: If at any time constant strains or constant temperature gradient occur within an element, the displacement model should be able to provide for it. The displacement model consists in it the linear terms $\alpha_2 x$, $\alpha_3 y$ etc. These terms indicate the constant strain states if the element size is progressively decreased to infinitesimal size.

In the displacement model,

$$u = \alpha_1 + \alpha_2 x + \alpha_3 y + \alpha_4 x^2 + \cdots + \alpha_m y^n$$

$$v = \alpha_{m+1} + \alpha_{m+2} x + \alpha_{m+3} x^2 + \cdots + \alpha_{2m} y^n$$

α_2 and α_{m+2} provide for uniform strain ε_x.

α_3 and α_{m+3} provide for uniform strain ε_y.

When a system is subdivided into large number of small elements, the nodes of the element may come quite close to one another and in such cases the variation of strains/gradient goes negligibly small giving rise to a constant strain or temperature gradient.

Within the element, therefore, the parametric model must be capable of calculating constant strains if it occurs. This corresponds to the terms $\alpha_2, \alpha_3, \alpha_4, \alpha_{m+2}, \alpha_{m+3}, \alpha_{m+4}, \alpha_{2m+2}, \alpha_{2m+3}, \alpha_{2m+4}$ in the equations above.

Consider, a small element Δx, where the strains at the two opposite faces are as shown in Fig. 7.24. Over a length Δx, the increment in strain can be given by $\dfrac{\partial ex}{\partial x} \cdot \Delta x$

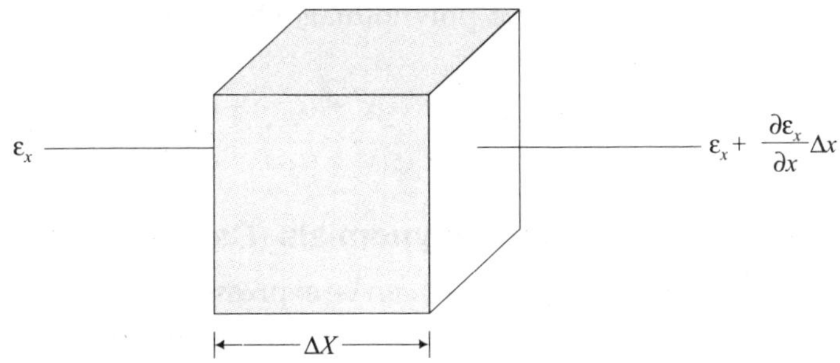

Fig. 7.24

$\therefore$ If $$\Delta x \to 0, \frac{\partial}{\partial x}\varepsilon_x . \Delta x \to 0$$

and the strains on the opposite faces become equal to ε_x, *i.e.*, a constant strain may exist over the element, constant strain conditions in other directions can also be explained in similar manner.

(4) Geometric isotropy: By interchanging x and y, the displacement pattern (model) does not change. Hence no problem will be encountered in the formation of stiffness matrix. However, it may be noted that all the terms (six terms upto quadratic variation) in the Pascal triangle are not considered. On the other hand, it is a balanced one since x^2 and its counterpart y^2 in the Pascal triangle are omitted. An additional consideration in the selection of polynomial shape function for the displacement model is that the pattern should be independent of the orientation of the local coordinate system. This property is known as geometric isotropy.

(5) Spatial isotropy: The displacement model (function) must satisfy the requirement of Spatial isotropy or Geometric invariance. This means that the displacement pattern, should be independent of the local coordinate system. This problem arises in two-dimensional elements. Constant strain condition in the element fulfills the requirement. In that case, displacement fields can be represented as a linear polynomial.

Example :
$$u = \alpha_0 + \alpha_1 x + \alpha_2 y$$
$$v = \alpha_3 + \alpha_4 x + \alpha_5 y$$

After a careful study of these two polynomials we ascertain that the displaced pattern is not biased towards a particular axis.

This property is known as Geometric isotropy. There are two simple guidelines to construct polynomial series with the desired property of isotropy :

(1). Polynomials of order n that are complete, have geometric isotropy

(2) Polynomials of order n that are not complete, yet contain appropriate terms to preserve 'symmetry', have geometric isotropy.

Example: Suppose we wish to construct a cubic polynomial expression for an element that has eight nodal values assigned to it. In this situation, we have to drop two terms from the complete cubic polynomial which contains 10 terms. To maintain geometric isotropy drop only terms that occur in symmetric pairs i.e., x^3, y^3 or x^2y, xy^2.

Thus, the acceptable eight term cubic polynomials shape functions exhibiting geometric isotropy are:

$$\alpha_1 + \alpha_2 x + \alpha_3 y + \alpha_4 x^2 + \alpha_5 xy + \alpha_6 y^2 + \alpha_7 x^2 y + \alpha_8 xy^2$$

and

$$\alpha_1 + \alpha_2 x + \alpha_3 y + \alpha_4 x^2 + \alpha_5 xy + \alpha_6 y^2 + \alpha_7 x^3 + \alpha_8 y^3$$

Derivation of Shape Functions Using Polynomials (Cartesian Coordinates)

The unknown coefficients, α's, of the polynomials can be expressed in terms of the displacements of the nodes of the element, called as nodal degrees of freedom. Initially shape functions were derived in terms of Cartesian coordinates, polynomial functions were used in this. After natural coordinates were identified and their advantage was noticed researchers started deriving shape functions in terms of natural coordinates.

Generalized Coordinate Approach

In this case, polynomials with number of constants exactly equal to nodal degrees of freedom of the element are selected. Care is taken to see that geometry isotropy is not lost. Using nodal values number of equations equal to number of constants in the polynomials are formed and then the constants found. Then the shape functions are identified.

A polynomial is the most common form of displacement model. This form is widely used because this form (1) is easy to handle mathematics of polynomial of any desired order (2) permits a recognisable approximation to the solution.

Example :

$$u = u\,(x) = \alpha_1 + \alpha_2 x + \alpha_3 x^2 + \cdots + \alpha_{n+1} x^n \qquad \text{...[1]}$$

is a polynomial of order n and $\alpha_1, \ldots \alpha_{n+1}$ are called generalised coordinates. The approximation of the polynomial in comparison with the true solution taking one, two and three terms is shown in Fig. 7.25.

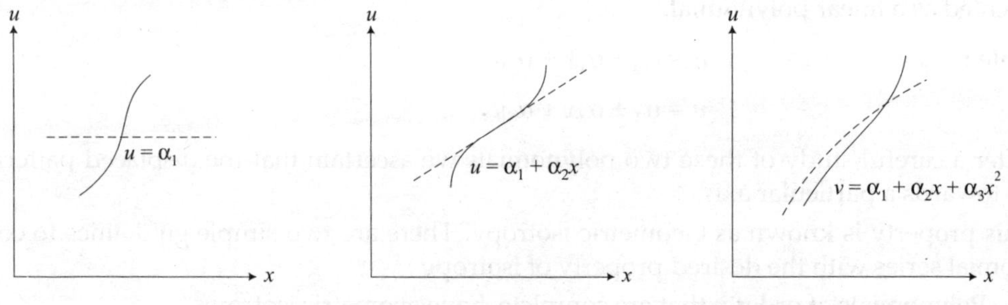

Fig. 7.25 Polynomial approximation to one dimension

Examples

1. Using polynomial method, find shape functions for two-noded line element.

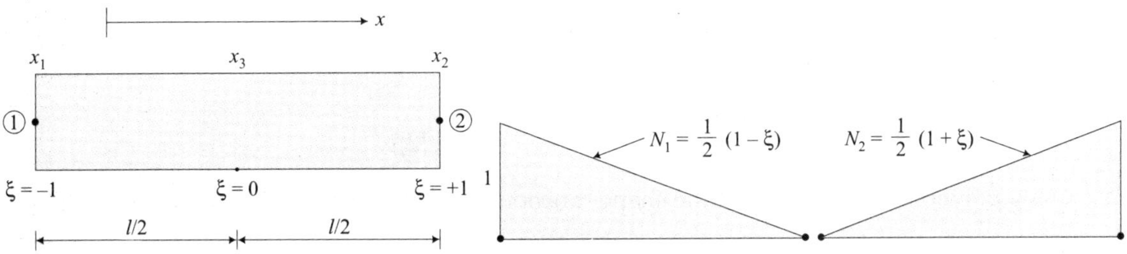

(a) Two-noded line element **(b)** Shape function for node 1 **(c)** Shape function for node 2

Fig. 7.26

Solution:

Fig. 7.26(a) shows the typical line element. In this case the origin is in the middle.

The simple natural coordinate is defined as:

$$\boxed{\xi = \frac{x - x_3}{(l/2)}}$$

when,	$x = x_1;\ \xi = \dfrac{x_1 - x_3}{(l/2)} = -\dfrac{l/2}{l/2} = -1$
when,	$x = x_3;\ \xi = \dfrac{x_3 - x_3}{(l/2)} = 0.$
when,	$x = x_2;\ \xi = \dfrac{x_2 - x_3}{(l/2)} = +\dfrac{l/2}{l/2} = +1$

Node 1:

when $\quad \xi = -1;\ N_1 = 1$

when $\quad \xi = +1;\ N_1 = 0$ ...[1]

$\therefore \quad N_1 = a_0 + a_1\xi$ ...[2]

$$\Rightarrow \quad \left.\begin{aligned} 1 &= a_0 - a_1 \\ 0 &= a_0 + a_1 \\ a_0 &= -a_1 \end{aligned}\right\} \quad \text{using [1]}$$

Solving,

$$a_0 = \frac{1}{2};\ a_1 = -\frac{1}{2}$$

$$\therefore \quad \boxed{N = \frac{(1-\xi)}{2}}, \quad \text{(using [2])}$$

There were two conditions. Therefore, a linear polynomial with two constants is assumed.

Node 2: when $\quad \xi = -1;\ N_2 = 0$

$\xi = +1;\ N_2 = 1$ ...[1]

Now, let $\quad N_2 = a_0 + a_1\,\xi$ ...[2]

Substituting the conditions from equation [1] into equation [2], we get

$$\left.\begin{aligned} 0 &= a_0 - a_1 \\ 1 &= a_0 + a_1 \end{aligned}\right\} \quad \text{...[3]}$$

Upon solving, $a_0 = \dfrac{1}{2}; a_1 = \dfrac{1}{2}$

$\therefore$ $\boxed{N_2 = \dfrac{(1+\xi)}{2}},$ (using [2])

2. Using polynomial method, find the shape function for the three-noded axial (linear) element.

Solution:

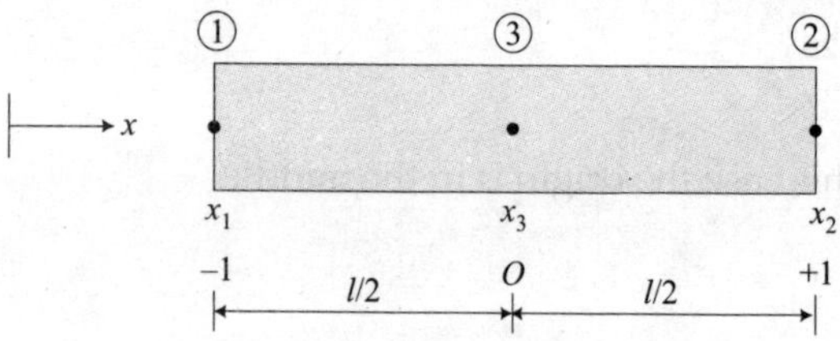

Fig. 7.27(a) Given element

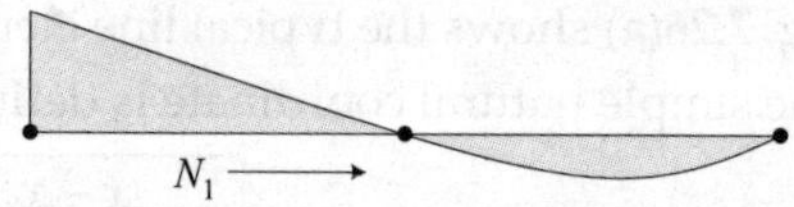

Fig. 7.27(b) Shape function N_1

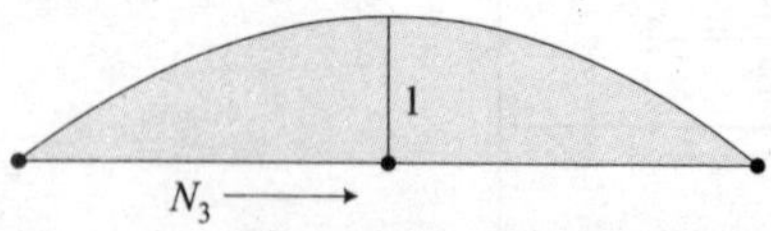

Fig. 7.27(c) Shape function N_3

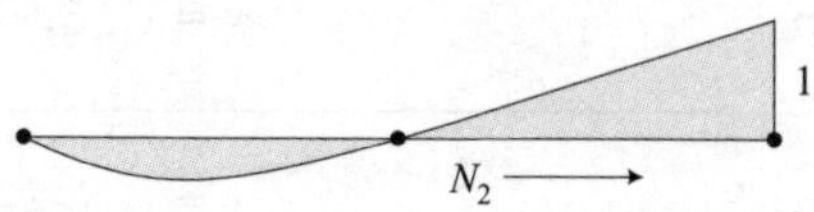

7.27(d) Shape function N_2

Figure 7.27(a) shows the typical element. Nodes (1) and (2) are called external nodes where connection is made to other adjoining elements. Node (3) is known as interval node which is introduced to improve the stiffness characteristic of the element. This element is known as higher order or refined element.

In this case, the origin is in the middle

$$\xi = \frac{(x - x_3)}{(l/2)} \text{ holds good.}$$

Node 1: We find that there are 3 conditions available.

They are:

$$\left\{\begin{aligned} \xi &= -1\,;\ N_1 = 1 \\ \xi &= 0\,;\ N_1 = 0 \\ \xi &= +1;\ N_1 = 0 \end{aligned}\right\}$$

$\therefore$ $\boxed{N = a_1 + a_2\,\xi + a_3\,\xi^2}$ ...[1]

$$\left\{\begin{array}{l} 1 = a_1 - a_2 + a_3 \\ 0 = a_1 + a_2 0 + a_3 0 \\ +1 = a_1 + a_2 + a_3 \end{array}\right\} \quad \begin{array}{r} ...[2] \\ ...[3] \\ ...[4] \end{array}$$

Solution by using determinants:

On solving for a_1 and a_2, we get

$$a_1 = 0; a_2 = -\frac{1}{2}; a_3 = +\frac{1}{2}$$

$$\frac{a_1}{\begin{vmatrix} -1 & 1 & -1 \\ 0 & 0 & 0 \\ 1 & 1 & -1 \end{vmatrix}} = \frac{-a_2}{\begin{vmatrix} 1 & 1 & -1 \\ 1 & 0 & 0 \\ 1 & 1 & -1 \end{vmatrix}} = \frac{a_3}{\begin{vmatrix} 1 & -1 & -1 \\ 1 & 0 & 0 \\ 1 & 1 & -1 \end{vmatrix}} = \frac{-1}{\begin{vmatrix} 1 & -1 & 1 \\ 1 & 0 & 0 \\ 1 & 1 & 1 \end{vmatrix}}$$

or, $$\frac{a_1}{\Delta_1} = \frac{-a_2}{\Delta_2} = \frac{a_3}{\Delta_3} = \frac{-1}{\Delta} \quad ...[1]$$

$$\Delta_1 \equiv \begin{vmatrix} -1 & 1 & -1 \\ 0 & 0 & 0 \\ 1 & 1 & -1 \end{vmatrix} \rightarrow R_3 \rightarrow R_3 + R_1$$

$$= \begin{vmatrix} 1 & 1 & -1 \\ 0 & 0 & 0 \\ 1 & 2 & -2 \end{vmatrix} = -1[0 - 0] - 1\,[0 - 0] \quad 1\,[0] = 0; \text{etc.}$$

$$N_1 = a_1 + a_2\xi + a_3\xi^2$$

$$\Rightarrow \qquad \boxed{N_1 = \frac{\xi}{2}(\xi - 1)}$$

This is shown in Fig. 7.27(b).

Node 2: $$N_2 = a_1 + a_2\,\xi + a_3\xi^2$$

Available conditions are :

$$\left\{\begin{array}{l} \text{when} \quad \xi = -1; \; N_2 = 0 \\ \text{when} \quad \xi = 0; \; N_2 = 0 \\ \text{when} \quad \xi = 1; \; N_2 = 1 \end{array}\right.$$

Substituting in N_2 and solving, we get

$$a_1 = 0; a_2 = \frac{1}{2}; a_3 = \frac{1}{2}$$

$$\Rightarrow \qquad \boxed{\therefore N_2 = \frac{\xi}{2}(\xi + 1)}$$

This is shown in Fig. 7.27(d).

Node 3:
$$N_3 = a_1 + a_2\xi + a_3\xi^2$$

Available conditions are:

$$\begin{cases} \text{when} \quad \xi = -1; N_3 = 0 \\ \text{when} \quad \xi = 0; N_3 = 1 \\ \text{when} \quad \xi = 1; N_3 = 0 \end{cases}$$

Substituting in N_3 and solving, we get

$$a_1 = 1; a_2 = 0 \text{ and } a_3 = -1$$

$$\therefore \qquad \boxed{N_3 = 1 - \xi^2}$$

The same is plotted in Fig. 7.27(c).

3. Using polynomial functions (generalised coordinates), find shape functions for two- noded bar/truss elements.

Solution:

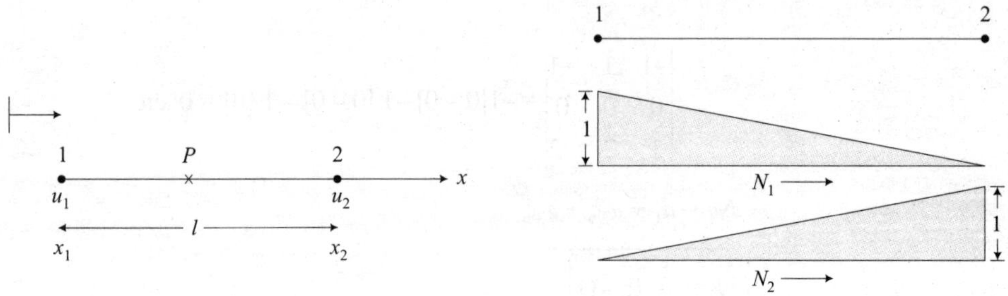

Fig. 7.28 Bar/truss element with two nodes

The bar/truss element will be able to carry either a tensile or compressive force when used in a system or assemblage. In this case total nodal unknowns are displacements u_1 and u_2 along x-axis. We choose a polynomial with only two constants to represent displacement at any point in the elements.

Let
$$u = \alpha_1 + \alpha_2 x \qquad \text{...[1]}$$

where α_1, α_2 are generalised coordinates. The matrix form of [1] is

$$u = [1 \quad x]\begin{Bmatrix}\alpha_1 \\ \alpha_2\end{Bmatrix}$$

$u = u_1$ at node [1] and $u = u_2$ at node [2].

We have
$$[\delta] = \begin{Bmatrix} u_1 \\ u_2 \end{Bmatrix} = \begin{bmatrix} 1 & x_2 \\ 1 & x_2 \end{bmatrix} \begin{Bmatrix} \alpha_1 \\ \alpha_2 \end{Bmatrix}$$

$$\begin{Bmatrix} \alpha_1 \\ \alpha_2 \end{Bmatrix} = \begin{bmatrix} 1 & x_1 \\ 1 & x_2 \end{bmatrix}^{-1} \begin{Bmatrix} u_1 \\ u_2 \end{Bmatrix}$$

$$= \frac{1}{(x_2 - x_1)} = \begin{bmatrix} x_2 & -1 \\ -x_1 & 1 \end{bmatrix}^T \begin{Bmatrix} u_1 \\ u_2 \end{Bmatrix}$$

$$= \frac{1}{l} \begin{bmatrix} x_2 & -x_1 \\ -1 & 1 \end{bmatrix} \begin{Bmatrix} u_1 \\ u_2 \end{Bmatrix}$$

$\therefore$
$$u = [1 \quad x] \begin{Bmatrix} \alpha_1 \\ \alpha_2 \end{Bmatrix}$$

$$= [1 \quad x] \frac{1}{l} \begin{bmatrix} x_2 & -x_1 \\ -1 & 1 \end{bmatrix} \begin{Bmatrix} u_1 \\ u_2 \end{Bmatrix}$$

$$= \frac{1}{l} [x_2 - x \quad -x_1 + x] \begin{Bmatrix} u_1 \\ u_2 \end{Bmatrix}$$

$$= \begin{bmatrix} \dfrac{x_2 - x}{l} & \dfrac{x - x_1}{l} \end{bmatrix} \begin{Bmatrix} u_1 \\ u_2 \end{Bmatrix}$$

$$= [N_1 \quad N_2] \begin{Bmatrix} u_1 \\ u_2 \end{Bmatrix}$$

$$= N_1 u_1 + N_2 u_2$$

where
$$N_1 = \frac{x_2 - x}{l}; \; N_2 = \frac{x - x_1}{l}$$

Thus, the shape function $[N]$ is

$$[N] = [N_1 \; N_2] = \begin{bmatrix} \dfrac{x_2 - x}{l} & \dfrac{x - x_1}{l} \end{bmatrix}$$

The shape functions N_1 and N_2 are as shown in Fig. 7.28.

4. Using polynomial functions (generalised coordinates) find the displacement transformation and shape functions for a three-noded truss element.

Solution: By data, the element has got three nodes of which two exterior nodes are having connection made to other elements and one internal node. There are three degrees of freedom. The given displacement function (field) has three generalised coordinates and so let us select

$$u = \alpha_0 + \alpha_1 x + \alpha_2 x^2 \qquad \text{...[1]}$$

as the displacement function.

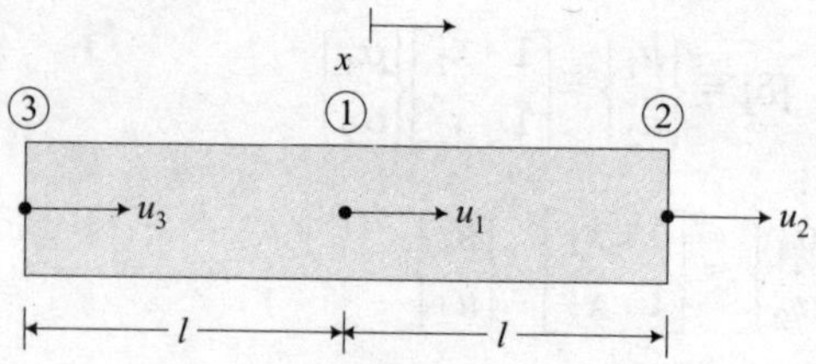

Fig. 7.29

These generalised coordinates can be evaluated using the three nodal degrees of freedom u_1, u_2 and u_3.

Since the origin is in the middle [Fig. 7.29] of the element

when $x = 0; \quad \therefore \; u_1 = \alpha_0$ ...[2]

when $x = +l; \quad u = u_2$

$\Rightarrow \quad u_2 = \alpha_0 + \alpha_1 l + \alpha_2 l^2$ ...[3]

when $x = -l; u = u_3$

$\Rightarrow \quad u_3 = \alpha_0 - \alpha_1 l + \alpha_2 l^2$ ...[4]

Writing equations [2], [3] and [4] in matrix form, we get

$$\begin{Bmatrix} u_1 \\ u_2 \\ u_3 \end{Bmatrix} = \begin{bmatrix} 1 & 0 & 0 \\ 1 & l & l^2 \\ 1 & -l & l^2 \end{bmatrix} \begin{Bmatrix} \alpha_0 \\ \alpha_1 \\ \alpha_2 \end{Bmatrix} \quad ...[5]$$

$$\downarrow \qquad \downarrow \qquad \downarrow$$

$$u^* \qquad C \qquad \alpha$$

where [C] = connectivity matrix.

Inverting the same, we get the displacement transformation matrix $[C]^{-1}$

$$[C]^{-1} = \frac{1}{2l^3}\begin{bmatrix} 2l^3 & 0 & 0 \\ 1 & l^2 & -l^2 \\ -2l & l & l \end{bmatrix}$$

Putting equation [1] in matrix form

$$[u] = [1 \;\; x \;\; x^2] \begin{Bmatrix} \alpha_0 \\ \alpha_1 \\ \alpha_2 \end{Bmatrix}$$

$$\downarrow \qquad \downarrow$$

$$P \qquad A$$

The shape function matrix $[N]$ is given by the expression

$$[N] = [P]\,[C]^{-1}$$

$$\therefore \qquad [N] = [1 \;\; x \;\; x^2]\begin{bmatrix} 2l^3 & 0 & 0 \\ 0 & l^2 & -l^2 \\ -2l & l & l \end{bmatrix}\frac{1}{2l^3}$$

Multiplying the above, we set the shape functions as indicated below :

$$N_1 = \left(1 - \frac{x^2}{l^2}\right)$$

$$N_2 = \left(\frac{x}{2l} + \frac{x^2}{2l^2}\right)$$

$$N_3 = \left(\frac{-x}{2l} + \frac{x^2}{2l^2}\right)$$

and the displacement function is $[u] = [N]\,\{u^*\} = N_1u_1 + N_2u_2 + N_3u_3$.

5. Using polynomial functions (generalised coordinates) determine shape function for a two-noded beam element.

Solution: The beam element has got four degrees of freedom. Eventhough there are two nodes, there will be four shape functions.

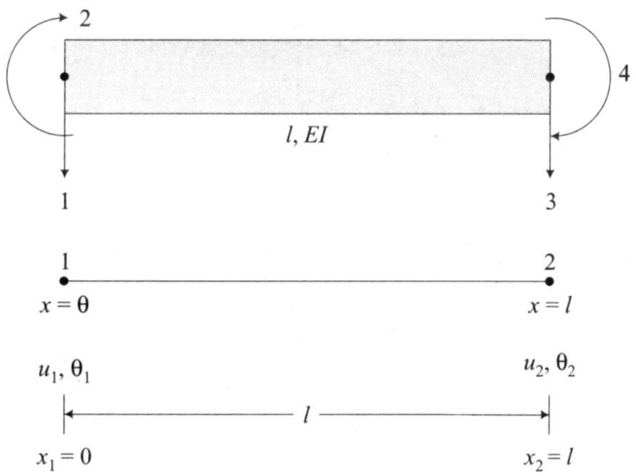

Fig. 7.30

Here C^1- continuity is to be satisfied, since the strain energy contains second differentiation term $\dfrac{d^2w}{dx^2}$ and so at each node, unknowns are the displacements and slopes, *i.e.*,

$$[\delta] = \begin{Bmatrix} w_1 \\ \theta_1 \\ w_2 \\ \theta_2 \end{Bmatrix}$$

where $$\theta_1 = \frac{\partial w_1}{\partial x}; \quad \theta_2 = \frac{\partial w_2}{\partial x}$$

Since there are four degrees of freedom we select polynomials with four constants, the displacement field will be:

$$w = \alpha_1 + \alpha_2 x + \alpha_3 x^2 + \alpha_4 x^3 \qquad ...[1]$$

[1] satisfies compatibility and completeness requirement. Now $\theta = \dfrac{\partial w}{\partial x} = \alpha_2 + 2\alpha_3 x + 3\alpha_4 x^2$. For the sake of convenience, we select local coordinate system

$$x_1 = 0; x_2 = l$$

$\therefore$
$$w_1 = \alpha_1; \ \theta_1 = \alpha_2$$
$$w_2 = \alpha_1 + \alpha_2 l + \alpha_3 l^2 + \alpha_4 l^3$$
$$\theta_2 = \alpha_2 + 2\alpha_3 l + 3\alpha_4 l^2$$

i.e.,
$$\{\delta\} = \begin{Bmatrix} w_1 \\ \theta_1 \\ w_2 \\ \theta_2 \end{Bmatrix} = \begin{bmatrix} 1 & 0 & 0 & 0 \\ 0 & 1 & 0 & 0 \\ 1 & l & l^2 & l^3 \\ 0 & 1 & 2l & 3l^2 \end{bmatrix} \begin{Bmatrix} \alpha_1 \\ \alpha_2 \\ \alpha_3 \\ \alpha_4 \end{Bmatrix}$$

$\therefore$
$$\begin{Bmatrix} \alpha_1 \\ \alpha_2 \\ \alpha_3 \\ \alpha_4 \end{Bmatrix} = \begin{bmatrix} 1 & 0 & 0 & 0 \\ 0 & 1 & 0 & 0 \\ 1 & l & l^2 & l^3 \\ 0 & 1 & 2l & 3l^2 \end{bmatrix} \begin{Bmatrix} w_1 \\ \theta_1 \\ w_2 \\ \theta_2 \end{Bmatrix}$$

$$= \frac{1}{3l^4 - 2l^4} \begin{bmatrix} l^4 & 0 & -3l^2 & 2l \\ 0 & l^4 & -2l^2 & l^2 \\ 0 & 0 & 3l^2 & -2l \\ 0 & 0 & -l^3 & l^2 \end{bmatrix} \begin{Bmatrix} w_1 \\ \theta_1 \\ w_2 \\ \theta_2 \end{Bmatrix}$$

$$= \begin{bmatrix} 1 & 0 & 0 & 0 \\ 0 & 1 & 0 & 0 \\ \frac{-3}{l^2} & \frac{-2}{l} & \frac{3}{l^2} & \frac{-1}{l} \\ \frac{2}{l^3} & \frac{1}{l^2} & \frac{-2}{l^3} & \frac{1}{l^2} \end{bmatrix} \begin{Bmatrix} w_1 \\ \theta_1 \\ w_2 \\ \theta_2 \end{Bmatrix}$$

$\therefore \quad w = \alpha_1 + \alpha_2 x + \alpha_3 x^2 + \alpha_4 x^3$

$$= [1 \;\; x \;\; x^2 \;\; x^3] \begin{Bmatrix} \alpha_1 \\ \alpha_2 \\ \alpha_3 \\ \alpha_4 \end{Bmatrix}$$

$$= [1 \;\; x \;\; x^2 \;\; x^3] \begin{bmatrix} 1 & 0 & 0 & 0 \\ 0 & 1 & 0 & 0 \\ \frac{-3}{l^2} & \frac{-2}{l} & \frac{3}{l^2} & \frac{-1}{l} \\ \frac{2}{l^3} & \frac{1}{l^2} & \frac{-2}{l^3} & \frac{1}{l^2} \end{bmatrix} \begin{Bmatrix} w_1 \\ \theta_1 \\ w_2 \\ \theta_2 \end{Bmatrix}$$

$$= \left[\left(1 - \frac{3x^2}{l^2} + \frac{2x^3}{l^3}\right)\left(x - \frac{2x^2}{l} + \frac{x^3}{l^2}\right)\left(\frac{3x^2}{l^2} - \frac{2x^3}{l^3}\right)\left(-\frac{x^2}{l} + \frac{x^3}{l^2}\right)\right] \begin{Bmatrix} w_1 \\ \theta_1 \\ w_2 \\ \theta_2 \end{Bmatrix}$$

$$= [N_1 \;\; N_2 \;\; N_3 \;\; N_4] \{\delta\}_e = [N] \{\delta\}_e$$

where $\quad [N] = [N_1 \;\; N_2 \;\; N_3 \;\; N_4]$

and $\left[\left(N_1 = 1 - \frac{3x^2}{l^2} + \frac{2x^3}{l^3}\right); \left(N_2 = x - \frac{2x^2}{l} + \frac{x^3}{l^2}\right); \left(N_3 = \frac{3x^2}{l^2} - \frac{2x^3}{l^3}\right); \left(N_4 = -\frac{x^2}{l} + \frac{x^3}{l^2}\right)\right]$

Variation of these functions is shown:

At node ①; ($N_1 = 1$)

$$N_2 = N_3 = N_4 = 0; \text{ and } \frac{\partial N_2}{\partial x} = 1, \frac{\partial N_1}{\partial x} = \frac{\partial N_3}{\partial x} = \frac{\partial N_4}{\partial x} = 0$$

Similarly at node ②

$$N_1 = N_2 = N_4 = 0; N_3 = 1 \text{ and } \frac{\partial N_1}{\partial x} = \frac{\partial N_2}{\partial x} = \frac{\partial N_3}{\partial x} = 0 \text{ and } \frac{\partial N_4}{\partial x} = 1.$$

6. Use polynomial functions (generalised coordinates) to find the shape functions for the constant strain triangle (*CST*).

Solution: The simplest element available in the analysis of plane elasticity problems is the 3-noded triangular element. The element is also referred to as constant strain triangle, because it will be seen in the derivation of element stiffness matrix that the strain remains constant at any point within the element.

A *CST* element is as shown in Fig. 7.31.

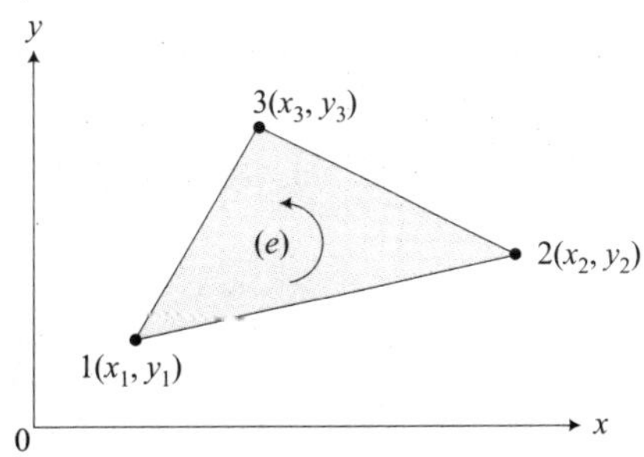

Fig. 7.31

Let the nodal variables be u_1, u_2, u_3, v_1, v_2 and v_3

or $\{\phi\}^T = [u_1 \ u_2 \ u_3 \ v_1 \ v_2 \ v_3]$

From the consideration of compatibility and completeness the following displacement model is selected.

$$\left\{\begin{matrix} u = \alpha_1 + \alpha_2 x + \alpha_3 y \\ v = \alpha_4 + \alpha_5 x + \alpha_6 y \end{matrix}\right\} \quad \text{...[1]}$$

$$\therefore \quad u_1 = \alpha_1 + \alpha_2 x_1 + \alpha_3 y_1$$

$$u_2 = \alpha_1 + \alpha_2 x_2 + \alpha_3 y_2$$

$$u_3 = \alpha_1 + \alpha_2 x_3 + \alpha_3 y_3$$

i. e.

$$\begin{Bmatrix} u_1 \\ u_2 \\ u_3 \end{Bmatrix} = \begin{bmatrix} 1 & x_1 & y_1 \\ 1 & x_2 & y_2 \\ 1 & x_3 & y_3 \end{bmatrix} \begin{Bmatrix} \alpha_1 \\ \alpha_2 \\ \alpha_3 \end{Bmatrix}$$

$$\therefore \quad \begin{Bmatrix} \alpha_1 \\ \alpha_2 \\ \alpha_3 \end{Bmatrix} = \begin{bmatrix} 1 & x_1 & y_1 \\ 1 & x_2 & y_2 \\ 1 & x_3 & y_3 \end{bmatrix} = \begin{Bmatrix} u_1 \\ u_2 \\ u_3 \end{Bmatrix}$$

Now,

$$\begin{vmatrix} 1 & x_1 & y_1 \\ 1 & x_2 & y_2 \\ 1 & x_3 & y_3 \end{vmatrix} = \begin{vmatrix} 1 & 1 & 1 \\ x_1 & x_2 & x_3 \\ y_1 & y_2 & y_3 \end{vmatrix} = 2A$$

$$\Rightarrow \quad A = \frac{1}{2} \begin{vmatrix} 1 & x_1 & y_1 \\ 1 & x_2 & y_2 \\ 1 & x_3 & y_3 \end{vmatrix}$$

where A is the area of triangle with vertices at (x_1, y_1); (x_2, y_2) and (x_3, y_3), *i.e.*, area of the element.

$$\therefore \quad \begin{Bmatrix} \alpha_1 \\ \alpha_2 \\ \alpha_3 \end{Bmatrix} = \frac{1}{2A} \begin{bmatrix} x_2 y_3 - x_3 y_2 & y_2 - y_1 & x_3 - x_2 \\ x_3 y_1 - x_1 y_3 & y_3 - y_1 & x_1 - x_3 \\ x_1 y_2 - x_2 y_1 & y_1 - y_2 & x_2 - x_1 \end{bmatrix}^T \begin{Bmatrix} u_1 \\ u_2 \\ u_3 \end{Bmatrix}$$

$$= \frac{1}{2A} \begin{bmatrix} a_1 & b_1 & c_1 \\ a_2 & b_2 & c_2 \\ a_3 & b_3 & c_3 \end{bmatrix} \begin{Bmatrix} u_1 \\ u_2 \\ u_3 \end{Bmatrix}$$

$$= \frac{1}{2A} \begin{bmatrix} a_1 & a_2 & a_3 \\ b_1 & b_2 & b_3 \\ c_1 & c_2 & c_3 \end{bmatrix} \begin{Bmatrix} u_1 \\ u_2 \\ u_3 \end{Bmatrix} \quad \text{...[2]}$$

where

$$\left.\begin{array}{lll} a_1 = x_2y_3 - x_3y_2; & a_2 = x_3y_1 - x_1y_3; & a_3 = x_1y_2 - x_2y_1 \\ b_1 = y_2 - y_3 & b_2 = y_3 - y_1 & b_3 = y_1 - y_2 \\ c_1 = x_3 - x_2 & c_2 = x_1 - x_3 & c_3 = x_2 - x_1 \end{array}\right\} \quad ...[3]$$

$\therefore$

$$u = \alpha_1 + \alpha_2 x + \alpha_3 y$$

$$= [1 \ x \ y] \begin{Bmatrix} \alpha_1 \\ \alpha_2 \\ \alpha_3 \end{Bmatrix} = [1 \ x \ y] \frac{1}{2A} \begin{bmatrix} a_1 & a_2 & a_3 \\ b_1 & b_2 & b_3 \\ c_1 & c_2 & c_3 \end{bmatrix} \begin{Bmatrix} u_1 \\ u_2 \\ u_3 \end{Bmatrix}$$

$$= \left[\frac{a_1 + b_1x + c_1y}{2A} \ \frac{a_2 + b_2x + c_2y}{2A} \ \frac{a_3 + b_3x + c_3y}{2A} \right] \begin{Bmatrix} u_1 \\ u_2 \\ u_3 \end{Bmatrix} \quad ...[4]$$

On simplifying equation [4], we obtain

$$\left\{ \begin{array}{l} u = u^{(e)}(x, y) = [N_1(x, y) \quad N_2(x, y) \quad N_3(x, y)] \begin{Bmatrix} u_1 \\ u_2 \\ u_3 \end{Bmatrix} \\ = N_1(x, y)\, u_1 + N_2(x, y)\, u_2 + N_3(x, y) u_3 \\ = N^{(e)} \phi^{(e)} \end{array} \right\} \quad ...[5]$$

where

$$\left\{ \begin{array}{l} N^{(e)} = [N_1 \, N_2 \, N_3]; \ \phi^{(e)} = [u_1 \ u_2 \ u_3]^T \\ N_1(x, y) = \dfrac{1}{2A}(a_1 + b_1x + c_1y); \\ N_2(x, y) = \dfrac{1}{2A}(a_2 + b_2x + c_2y); \\ N_3(x, y) = \dfrac{1}{2A}(a_3 + b_3x + c_3y) \end{array} \right\}$$

Similarly,

$$v = [N_1(x, y)\, N_2(x, y)\, N_3(x, y)] \begin{Bmatrix} v_1 \\ v_2 \\ v_3 \end{Bmatrix}$$

The functions $N_i = N_i(x, y)$, $i = 1, 2, 3$ are called shape functions and defined in the element (e). It is easily verified from equation [3] that

$$N_i(x_i, y_j) = \delta_{ij} \quad [6]$$

where δ_{ij} is a Kronecker delta,

$$\delta_{ij} = 1, \ i = j; \ \delta_{ij} = 0, \ i \neq j$$

$$\begin{Bmatrix} u \\ v \end{Bmatrix} = \begin{Bmatrix} u(x, y) \\ v(x, y) \end{Bmatrix} = \begin{bmatrix} N_1 & N_2 & N_3 & 0 & 0 & 0 \\ 0 & 0 & 0 & N_1 & N_2 & N_3 \end{bmatrix} \begin{Bmatrix} u_1 \\ u_2 \\ u_3 \\ v_1 \\ v_2 \\ v_3 \end{Bmatrix}$$

$$= \begin{bmatrix} N & 0 \\ 0 & N \end{bmatrix} \{\phi^{(e)}\} \qquad \text{...[7]}$$

Natural coordinates for triangular elements: Representation of the natural coordinates (or area coordinates or local coordinates) at any point $P(x, y)$ belongs to the triangular element.

Let (e) be a triangular element and a point $P(x, y) \in (e)$ can be associated with (L_1, L_2, L_3) defined by:

$$L_1 = \frac{\text{area } P23}{\text{area } 123}, \quad L_2 = \frac{\text{area } P31}{\text{area } 123}, \quad L_3 = \frac{\text{area } P12}{\text{area } 123} \qquad \text{...[1]}$$

as shown in the Fig. 7.32.

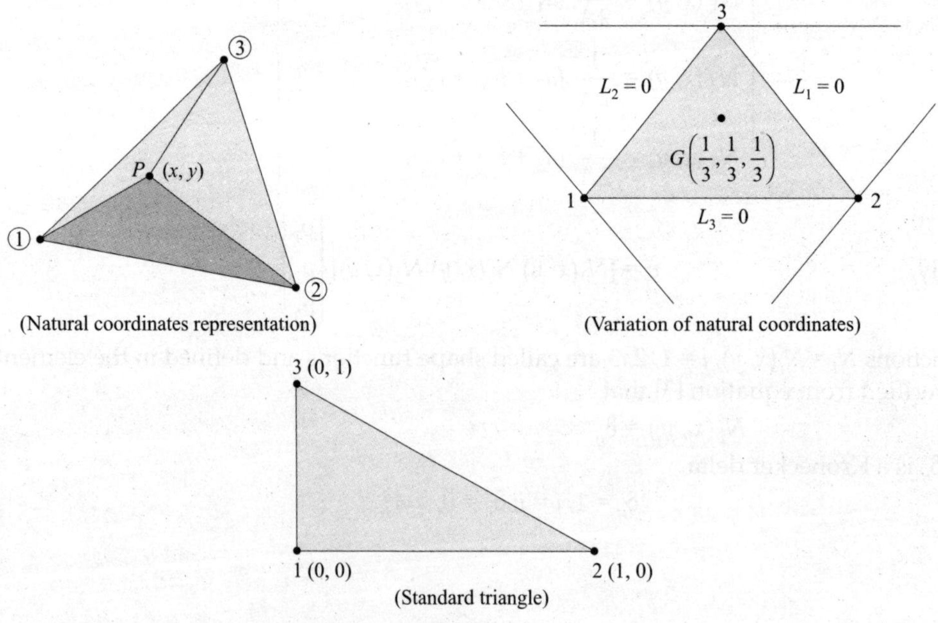

Fig. 7.32

Examples:

1. Using natural coordinate system, determine the shape functions for a two-noded bar element.

Solution: Since this element has only two nodal values, we select only linear function in natural coordinates.

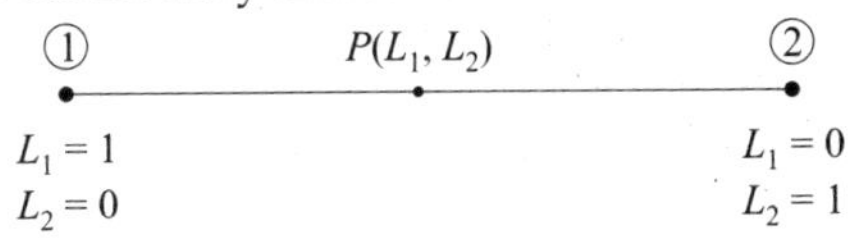

Let $$u = \alpha_1 L_1 + \alpha_2 L_2$$

$$= [L_1 \; L_2] \begin{bmatrix} \alpha_1 \\ \alpha_2 \end{bmatrix}$$

[Now, $L_1 = 1$; $L_2 = 0$ at node (1) and $L_1 = 0$; $L_2 = 1$ at node (2)]

$\therefore$ $$\begin{bmatrix} \alpha_1 \\ \alpha_2 \end{bmatrix} = \begin{bmatrix} 1 & 0 \\ 0 & 1 \end{bmatrix}^{-1} \begin{Bmatrix} u_1 \\ u_2 \end{Bmatrix}$$

$$= \frac{1}{1} \begin{bmatrix} 1 & 0 \\ 0 & 1 \end{bmatrix} \begin{Bmatrix} u_1 \\ u_2 \end{Bmatrix}$$

$\therefore$ $$u = [\, L_1 \; L_2 \,] \begin{Bmatrix} \alpha_1 \\ \alpha_2 \end{Bmatrix}$$

$$= [L_1 \; L_2] \begin{bmatrix} 1 & 0 \\ 0 & 1 \end{bmatrix} \begin{Bmatrix} u_1 \\ u_2 \end{Bmatrix} = [L_1 \; L_2] \begin{Bmatrix} u_1 \\ u_2 \end{Bmatrix}$$

Since $$u = [N_1 \; N_2] \begin{Bmatrix} u_1 \\ u_2 \end{Bmatrix} \quad \text{(by definition of shape function)}$$

Hence, $$N_1 = L_1 \text{ and } N_2 = L_2$$

2. Using natural coordinate system, derive the expression for shape function for a two-noded bar element taking natural coordinate ξ as varying from –1 to +1.

Solution: The bar element in the natural coordinate ξ varies from –1 to +1 as shown in Fig. 7.33.

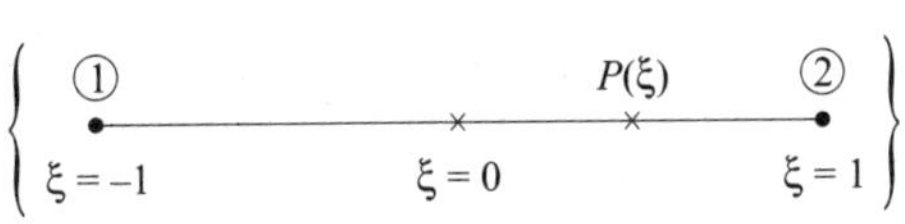

Fig. 7.33

Let us assume a linear polynomial

Now, $$u = \alpha_1 + \alpha_2 \xi$$

$$= [1 \;\; \xi] \begin{Bmatrix} \alpha_1 \\ \alpha_2 \end{Bmatrix} \qquad \text{...[1]}$$

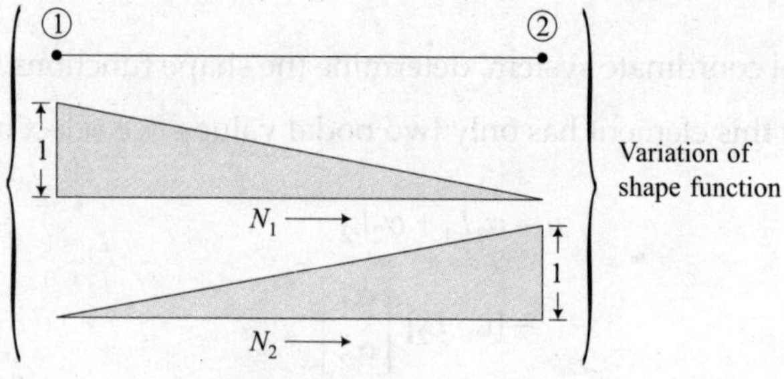

Fig. 7.34

$$\begin{Bmatrix} u_1 \\ u_2 \end{Bmatrix} = \begin{bmatrix} 1 & -1 \\ 1 & 1 \end{bmatrix} \begin{Bmatrix} \alpha_1 \\ \alpha_2 \end{Bmatrix}; \text{ since } \xi = -1 \text{ at node (1) and } \xi = 1 \text{ at node (2)}$$

$$\therefore \quad \begin{Bmatrix} \alpha_1 \\ \alpha_2 \end{Bmatrix} = \begin{bmatrix} 1 & -1 \\ 1 & 1 \end{bmatrix}^{-1} \begin{Bmatrix} u_1 \\ u_2 \end{Bmatrix}$$

$$= \frac{1}{1+1} = \begin{bmatrix} 1 & -1 \\ +1 & 1 \end{bmatrix}^{-1} \begin{Bmatrix} u_1 \\ u_2 \end{Bmatrix}$$

$$= \frac{1}{2} \begin{bmatrix} 1 & 1 \\ -1 & 1 \end{bmatrix} \begin{Bmatrix} u_1 \\ u_2 \end{Bmatrix}$$

$$\therefore \quad u = [1 \quad \xi] \begin{Bmatrix} \alpha_1 \\ \alpha_2 \end{Bmatrix} = [1 \quad \xi] \frac{1}{2} \begin{bmatrix} 1 & 1 \\ -1 & 1 \end{bmatrix} \begin{Bmatrix} u_1 \\ u_2 \end{Bmatrix}$$

$$= \frac{1}{2} [1 - \xi \quad 1 + \xi] \begin{Bmatrix} u_1 \\ u_2 \end{Bmatrix}$$

$$= \begin{bmatrix} \frac{1-\xi}{2} & \frac{1+\xi}{2} \end{bmatrix} \begin{Bmatrix} u_1 \\ u_2 \end{Bmatrix} = [N_1 \; N_2] \begin{Bmatrix} u_1 \\ u_2 \end{Bmatrix}$$

where $$N_1 = \frac{1-\xi}{2}; \quad N_2 = \frac{1+\xi}{2} \qquad \text{...[2]}$$

3. Using natural coordinates, find the shape functions for a constant strain triangular (*CST*) element.

Solution: A constant strain triangular element (*CST*) of arbitrary shape with nodes at the corners denoted as 1, 2 and is shown in Fig. 7.35. The natural coordinates of 1, 2 and 3 are (1, 0, 0), (0, 1, 0) and (0, 0, 1) respectively.

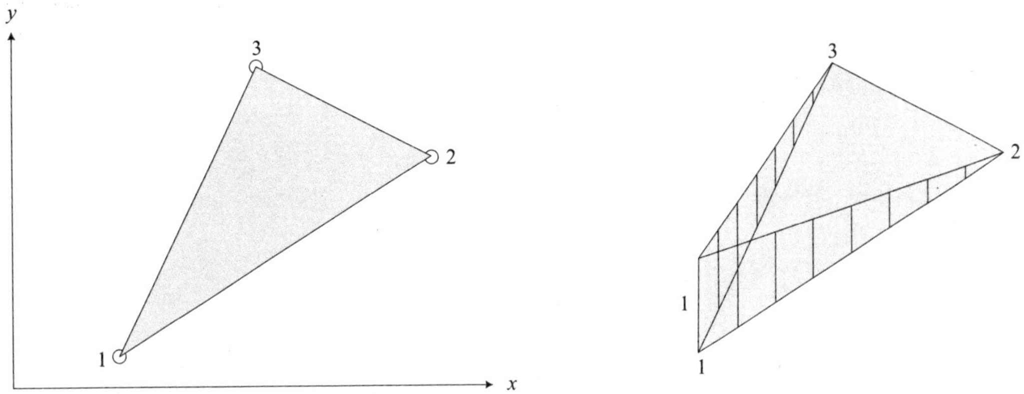

(a) Typical CST element

(b) Variation of N_1

Fig. 7.35

The linear displacement variation of a triangular element shown in Fig. 7.35 can be expressed in natural coordinates as,

$$u = \alpha_1 L_1 + \alpha_2 L_2 + \alpha_3 L_3 \qquad \text{...[1]}$$

i.e.,

$$u = \{\phi_2\}^T \{\alpha\} \qquad \text{...[2]}$$

where $\{\phi_2\}^T = [L_1 \;\; L_2 \;\; L_3]$ and $\{\alpha\}^T = [\alpha_1 \;\; \alpha_2 \;\; \alpha_3]$

Substituting the natural coordinates for nodes 1, 2 and 3, the u-components of nodal displacements in [2]
we obtain:

$$u = [L_1 \;\; L_2 \;\; L_3] \begin{Bmatrix} \alpha_1 \\ \alpha_2 \\ \alpha_3 \end{Bmatrix}$$

$$\therefore \quad \begin{Bmatrix} u_1 \\ u_2 \\ u_3 \end{Bmatrix} = \begin{bmatrix} 1 & 0 & 0 \\ 0 & 1 & 0 \\ 0 & 0 & 1 \end{bmatrix} \begin{Bmatrix} \alpha_1 \\ \alpha_2 \\ \alpha_3 \end{Bmatrix}$$

($\because$ $L_1 = (1, 0, 0)$; $L_2 = (0, 1, 0)$ and $L_3 = (0, 0, 1)$)

$$\{\alpha\} = \begin{bmatrix} 1 & 0 & 0 \\ 0 & 1 & 0 \\ 0 & 0 & 1 \end{bmatrix}^{-1} \begin{Bmatrix} u_1 \\ u_2 \\ u_3 \end{Bmatrix}$$

$$= \begin{bmatrix} 1 & 0 & 0 \\ 0 & 1 & 0 \\ 0 & 0 & 1 \end{bmatrix} \begin{Bmatrix} u_1 \\ u_2 \\ u_3 \end{Bmatrix}$$

$$\therefore \qquad u = [L_1 \ L_2 \ L_3] \begin{Bmatrix} \alpha_1 \\ \alpha_2 \\ \alpha_3 \end{Bmatrix}$$

$$= [L_1 \ L_2 \ L_3] \begin{bmatrix} 1 & 0 & 0 \\ 0 & 1 & 0 \\ 0 & 0 & 1 \end{bmatrix} \begin{Bmatrix} u_1 \\ u_2 \\ u_3 \end{Bmatrix}$$

$$= [L_1 \ L_2 \ L_3] \begin{Bmatrix} \mu_1 \\ \mu_2 \\ \mu_3 \end{Bmatrix}$$

$$= [N_1 \ N_2 \ N_3] \begin{Bmatrix} \mu_1 \\ \mu_2 \\ \mu_3 \end{Bmatrix}$$

Similarly, we can show that

$$v = L_1 v_1 + L_2 v_2 + L_3 v_3$$
$$= N_1 v_1 + N_2 v_2 + N_3 v_3$$
$$= [N_1 \ N_2 \ N_3] \begin{Bmatrix} v_1 \\ v_2 \\ v_3 \end{Bmatrix}$$

$$\therefore \qquad \begin{Bmatrix} u \\ v \end{Bmatrix} = \begin{bmatrix} L_1 & L_2 & L_3 & 0 & 0 & 0 \\ 0 & 0 & 0 & N_1 & N_2 & N_3 \end{bmatrix} \{\delta\}_e = \begin{bmatrix} N & 0 \\ 0 & N \end{bmatrix} \{\delta\}_e$$

where $\qquad N_1 = L_1; N_2 = L_2$

and $\qquad N_3 = L_3;$

Variation of N_1 is shown in Fig. 7.35(b).

4. Derive shape functions in terms of global coordinates for a three-node triangular element with nodes 1, 2 and 3 located at (x_1, y_1), (x_2, y_2) and (x_3, y_3) respectively, in the global system.

Solution: Let us assume an interpolation function in the form

$$u = c_1 + c_2 x + c_3 y \qquad \text{...[1]}$$

Equation [1] in matrix form is,

$$u = [1 \ x \ y] \begin{Bmatrix} c_1 \\ c_2 \\ c_3 \end{Bmatrix}$$

or, $$u = [\alpha] \{C\} \qquad \text{...[2]}$$

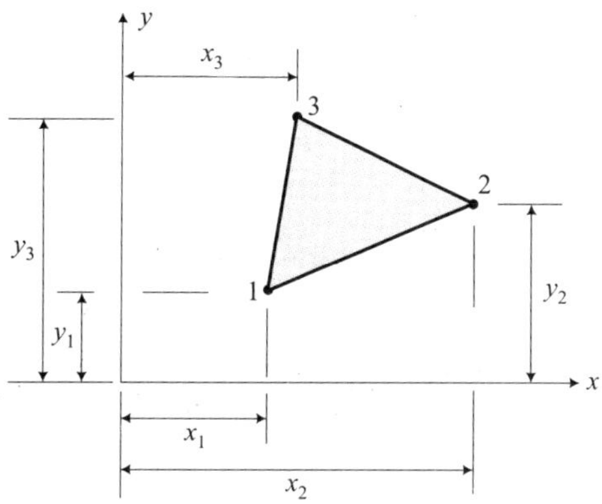

Fig. 7.36

The boundary conditions are in terms of nodal point values of u :

$$u\,(x_1, y_1) = u_1; \quad u(x_2, y_2) = u_2; \quad u\,(x_3, y_3) = u_3 \qquad ...[3]$$

Substituting equation [3] into [1] to obtain three equations that can be solved for c_1, c_2 and c_3:

$$\begin{Bmatrix} u_1 = c_1 + c_2x_1 + c_3y_1 \\ u_2 = c_1 + c_2x_2 + c_3y_2 \\ u_3 = c_1 + c_2x_3 + c_3y_3 \end{Bmatrix} \quad \text{or} \quad \begin{Bmatrix} u_1 \\ u_2 \\ u_3 \end{Bmatrix} = \begin{bmatrix} 1 & x_1 & y_1 \\ 1 & x_2 & y_2 \\ 1 & x_3 & y_3 \end{bmatrix} \begin{Bmatrix} c_1 \\ c_2 \\ c_3 \end{Bmatrix} \qquad ...[4]$$

Now, [4] in matrix format $\{u\} = [x]\,\{c\}$ ...[5]

Solving for $\{c\}$ as: $\{c\} = [x]^{-1}\,\{u\}$ ...[6]

Substituting from equation [6] into equation [2]

$$\begin{aligned} u &= [\alpha]\,\{c\} \\ &= [\alpha]\,[x]^{-1}\,\{u\} \\ &= [N]\,\{u\} \end{aligned} \qquad ...[7]$$

The shape functions are the product of the first two matrices on the r.h.s of equation [7]

$$[N] = [\alpha]\,[x]^{-1}$$

or $$[N_1\ N_2\ N_3] = [1\ x\ y]\,[x]^{-1} \qquad ...[8]$$

solving equation [8] gives

$$N_1 = [(x_2y_3 - x_3, y_2) + x(y_2 - y_3) + y\,(x_3 - x_2)]\,/2A$$

$$N_2 = [(x_3y_1 - x_1, y_3) + x(y_3 - y_1) + y\,(x_1 - x_3)]\,/2A$$

$$N_3 = [(x_1y_2 - x_2, y_1) + x(y_1 - y_2) + y\,(x_2 - x_1)]\,/2A$$

where $$A = \frac{1}{2} \det \begin{bmatrix} 1 & x_1 & y_1 \\ 1 & x_2 & y_2 \\ 1 & x_3 & y_3 \end{bmatrix}$$

A is the area of the triangular element.

5. Using natural coordinate system, determine the shape function for linear strain triangular element (*LST* element).

Solution: Considering Pascal's triangle, the displacements in x and y directions vary in the quadratic form (one order higher than the variation of strain). As there are six nodal values we have to pick a polynomial with six constants. Let us choose the shape function with quadratic terms in natural coordinate system as:

$$u = \alpha_1 L_1^2 + \alpha_2 L_2^2 + \alpha_3 L_3^2 + \alpha_4 L_1 L_2 + \alpha_5 L_2 L_3 + \alpha_6 L_3 L_1 \quad \text{...[1]}$$

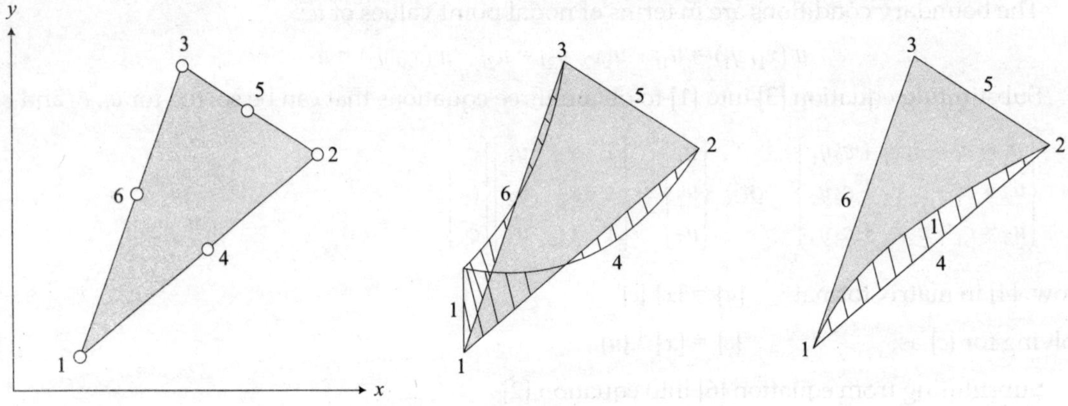

(a) Typical LST element (b) Variation of N_1 (c) Variation of N_4

Fig. 7.37

$$\begin{Bmatrix} u_1 \\ u_2 \\ u_3 \\ u_4 \\ u_5 \\ u_6 \end{Bmatrix} = \begin{bmatrix} 1 & 0 & 0 & 0 & 0 & 0 \\ 0 & 1 & 0 & 0 & 0 & 0 \\ 0 & 0 & 1 & 0 & 0 & 0 \\ \frac{1}{4} & \frac{1}{4} & 0 & \frac{1}{4} & 0 & 0 \\ 0 & \frac{1}{4} & \frac{1}{4} & 0 & \frac{1}{4} & 0 \\ \frac{1}{4} & 0 & \frac{1}{4} & 0 & 0 & \frac{1}{4} \end{bmatrix} \begin{Bmatrix} \alpha_1 \\ \alpha_2 \\ \alpha_3 \\ \alpha_4 \\ \alpha_5 \\ \alpha_6 \end{Bmatrix}$$

The natural coordinates at nodes 1 to 6 are:

(1, 0, 0)	(½, ½, 0)
(0, 1, 0)	(0, ½, ½)
(0, 0, 1)	(½, 0, ½)

We have, $\{u\}_e = [A]\{\alpha\}$

where $\{u\}_e$ = nodal displacement vector in x-direction,

$\{\alpha\}$ = generalised coordinates vector (constant in polynomials)

and
$$[A^{-1}] = \begin{bmatrix} 1 & 0 & 0 & 0 & 0 & 0 \\ 0 & 1 & 0 & 0 & 0 & 0 \\ 0 & 0 & 1 & 0 & 0 & 0 \\ -1 & -1 & 0 & 4 & 0 & 0 \\ 0 & -1 & -1 & 0 & 4 & 0 \\ -1 & 0 & -1 & 0 & 0 & 4 \end{bmatrix}$$

$\therefore$ $\{\alpha\} = [A]^{-1}\{u\}_e$

and we have by [1]

$$u = \alpha_1 L_1^2 + \alpha_2 L_2^2 + \alpha_3 L_3^2 + \alpha_4 L_1 L_2 + \alpha_5 L_2 L_3 + \alpha_6 L_3 L_1$$

$$= [L_1^2 \; L_2^2 \; L_3^2 \; L_1 \; L_2 \; L_2 \; L_3 \; L_3 \; L_1]\begin{Bmatrix} \alpha_1 \\ \alpha_2 \\ \alpha_3 \\ \alpha_4 \\ \alpha_5 \\ \alpha_6 \end{Bmatrix}$$

$$= [L_1^2 \; L_2^2 \; L_3^2 \; L_1 \; L_2 \; L_2 \; L_3 \; L_3 \; L_1][A]^{-1}\{u\}_e$$

$$= [L_1^2 \; L_2^2 \; L_3^2 \; L_1 \; L_2 \; L_2 \; L_3 \; L_3 \; L_1]\begin{bmatrix} 1 & 0 & 0 & 0 & 0 & 0 \\ 0 & 1 & 0 & 0 & 0 & 0 \\ 0 & 0 & 1 & 0 & 0 & 0 \\ -1 & -1 & 0 & 4 & 0 & 0 \\ 0 & -1 & -1 & 0 & 4 & 0 \\ -1 & 0 & -1 & 0 & 0 & 4 \end{bmatrix}$$

$$= [(L_1^2 - L_1 L_2 - L_3 L_1), (L_2^2 - L_1 L_2 - L_2 L_3), (L_3^2 - L_2 L_3 - L_3 L_1), 4 L_1 L_2, 4 L_2 L_3, 4 L_3 L_1]\{u\}_e$$

$$= [N_1 \; N_2 \; N_3 \; N_4 \; N_5 \; N_6 \{u\}_e$$

where
$$N_1 = L_1^2 - L_1 L_2 - L_3 L_1$$
$$N_2 = L_2^2 - L_1 L_2 - L_2 L_3$$

$$N_3 = L_3^2 - L_2L_3 - L_3 L_1$$

$$N_4 = 4\, L_1L_2$$

$$N_5 = 4\, L_2L_3$$

and $$N_6 = 4\, L_3L_1$$

Now, $$N_1 = L_1^2 - L_1L_2 - L_3L_1$$

$$= L_1 (L_1 - L_2 - L_3)$$

$$= L_1 [L_1 - (1 - L_1)] = L_1 (2L_1 - 1)$$

$(\because L_1 + L_2 + L_3 = 1)$

Similarly $$N_2 = L_2 (2L_2 - 1)$$

and $$N_3 = L_3 (2L_3 - 1)$$

Now, it can be easily proved that

$$v = [N_1\, N_2\, N_3\, N_4\, N_5\, N_6]\, \{v\}_e$$

$$\therefore \quad \begin{bmatrix} u \\ v \end{bmatrix} = \begin{bmatrix} N & 0 \\ 0 & N \end{bmatrix} \{v\}_e \quad ...[2]$$

where $$\{\delta\}_e^T = [u_1 u_2\, u_3\, u_4\, u_5\, u_6\, v_1\, v_2\, v_3\, v_4\, v_5\, v_6]$$

and $$[N] = [N_1\, N_2\, N_3\, N_4\, N_5\, N_6]\, \{\delta\}_e$$

$$N_1 = L_1 (2L_1 - 1);$$

$$N_2 = L_2 (2L_2 - 1);$$

$$N_3 = L_3 (2L_3 - 1);$$

$$N_4 = 4L_1L_2$$

$$N_5 = 4L_2L_3$$

$$N_6 = 4L_3L_1$$

The variations of N_1 and N_4 are as shown in Fig. 7.37.

6. Using natural coordinates, find the shape functions for 4-noded rectangular element.

Solution: We descretize the domain R into rectangular elements by drawing lines parallel to the x and y axes. The natural coordinates for a rectangular element shown in Fig. 7.38 are defined by

$$\xi = \frac{x - x_c}{a} \text{ and } \eta = \frac{y - y_c}{a} \quad ...[1]$$

where x_c and y_c are the coordinates of two origin and $2a \times 2b$ is the size of the element.

Let us assume the value of the function over the element as :

$$\boxed{u = \alpha_1 + \alpha_2\xi + \alpha_3\eta + \alpha_4\xi\eta} \quad ...[1]$$

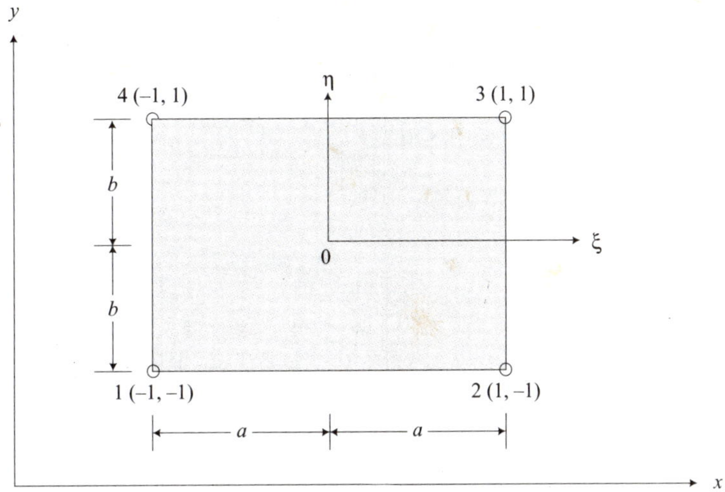

Fig. 7.38 Typical 4-noded rectangular element

The nodal displacements $\{u\}_e$ can be obtained as:

$$\{u\}_e = \begin{Bmatrix} u_1 \\ u_2 \\ u_3 \\ u_4 \end{Bmatrix} = \begin{bmatrix} 1 & -1 & -1 & 1 \\ 1 & 1 & -1 & -1 \\ 1 & 1 & 1 & 1 \\ 1 & -1 & 1 & -1 \end{bmatrix} \begin{Bmatrix} \alpha_1 \\ \alpha_2 \\ \alpha_3 \\ \alpha_4 \end{Bmatrix} \qquad \text{...[2]}$$

$$\therefore \qquad \begin{Bmatrix} \alpha_1 \\ \alpha_2 \\ \alpha_3 \\ \alpha_4 \end{Bmatrix} = \begin{bmatrix} 1 & -1 & -1 & 1 \\ 1 & 1 & -1 & -1 \\ 1 & 1 & 1 & 1 \\ 1 & -1 & 1 & -1 \end{bmatrix}^{-1} \begin{Bmatrix} u_1 \\ u_2 \\ u_3 \\ u_4 \end{Bmatrix}$$

or in matrix format:

$$\{\alpha\} = [A]^{-1} \{u\}_e$$

where $[A^{-1}]$ can be shown to be equal to

$$[A]^{-1} = \begin{bmatrix} \frac{1}{4} & \frac{1}{4} & \frac{1}{4} & \frac{1}{4} \\ -\frac{1}{4} & \frac{1}{4} & \frac{1}{4} & -\frac{1}{4} \\ -\frac{1}{4} & -\frac{1}{4} & \frac{1}{4} & \frac{1}{4} \\ \frac{1}{4} & -\frac{1}{4} & \frac{1}{4} & -\frac{1}{4} \end{bmatrix}$$

$$\therefore \qquad u = \alpha_1 + \alpha_2\xi + \alpha_3\eta + \alpha_4\xi\eta$$

$$= [1 \;\; \xi \;\; \eta \;\; \xi\eta] \begin{Bmatrix} \alpha_1 \\ \alpha_2 \\ \alpha_3 \\ \alpha_4 \end{Bmatrix}$$

$$= [1 \;\; \xi \;\; \eta \;\; \xi\eta] \begin{bmatrix} \frac{1}{4} & \frac{1}{4} & \frac{1}{4} & \frac{1}{4} \\ -\frac{1}{4} & \frac{1}{4} & \frac{1}{4} & -\frac{1}{4} \\ -\frac{1}{4} & -\frac{1}{4} & \frac{1}{4} & \frac{1}{4} \\ \frac{1}{4} & -\frac{1}{4} & \frac{1}{4} & -\frac{1}{4} \end{bmatrix} \{u\}_e \qquad \text{...[1]}$$

$$\{u\} = \frac{1}{4} \begin{Bmatrix} (1-\xi)\,(1-\eta) \\ (1+\xi)\,(1-\eta) \\ (1+\xi)\,(1+\eta) \\ (1-\xi)\,(1+\eta) \end{Bmatrix} \{u\}_e \qquad \text{...[2]}$$

$$= \begin{bmatrix} \frac{(1-\xi)\,(1-\eta)}{4} \\ \frac{(1+\xi)\,(1-\eta)}{4} \\ \frac{(1+\xi)\,(1+\eta)}{4} \\ \frac{(1-\xi)\,(1+\eta)}{4} \end{bmatrix} \{u\}_e \qquad \text{...[3]}$$

$$= \begin{Bmatrix} N_1 \\ N_2 \\ N_3 \\ N_4 \end{Bmatrix} \{u\}_e = [N]\,\{u\}_e \qquad \text{...[4]}$$

where
$$\begin{Bmatrix} N_1 = \dfrac{(1-\xi)(1-\eta)}{4}; \\ N_2 = \dfrac{(1+\xi)(1-\eta)}{4}; \\ N_3 = \dfrac{(1+\xi)(1+\eta)}{4}; \\ N_4 = \dfrac{(1-\xi)(1+\eta)}{4} \end{Bmatrix}$$

we can write:

$$N_i = \frac{1}{4}(1+\xi\,\xi_i)(1+\eta\,\eta_i),\ (i = 1, 2, 3 \text{ and } 4) \qquad \text{...[5]}$$

Similarly,
$$V = [N_1\ N_2\ N_3\ N_4]\begin{Bmatrix} v_1 \\ v_2 \\ v_3 \\ v_4 \end{Bmatrix}$$

$\therefore$
$$\begin{Bmatrix} u \\ v \end{Bmatrix} = \begin{bmatrix} N_1 & N_2 & N_3 & N_4 & 0 & 0 & 0 & 0 \\ 0 & 0 & 0 & 0 & N_1 & N_2 & N_3 & N_4 \end{bmatrix}\{\delta\}_e$$

$$N_i = \begin{cases} 1, \text{ at node } i \\ 0, \text{ at all other nodes} \end{cases}$$

7. Using natural coordinates, find the shape function for quadratic rectangular element.

Solution: Let us assume a complete third order polynomial in the form:

$$u = \alpha_1 + \alpha_2\xi + \alpha_3\eta + \alpha_4\xi^2 + \alpha_5\xi\eta + \alpha_6\eta^2 + \alpha_7\xi^2\eta + \alpha_8\xi\eta^2 \qquad \text{...[1]}$$

(dropping ξ^3 and η^3 terms, as [1] has to maintain geometric isotropy).

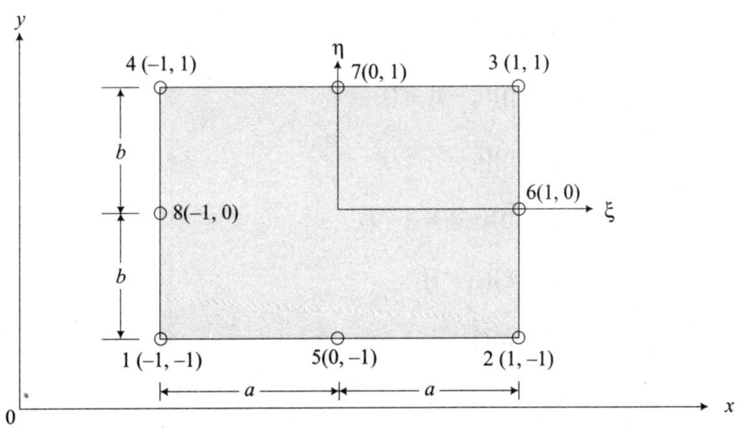

Fig. 7.39 Typical quadratic triangular element

Since the geometric invariance is achieved if the polynomial includes all the terms *i.e.*, the polynomial is a complete one. However, invariance may be achieved if the polynomial is "balanced" in case all the terms cannot be included.

For the given element, there are 8 nodal values for u and 8 for v.

The nodal displacements $\{u\}_e$ can be obtained by substituting the coordinates for the nodes as

$$\{u\}_e = \begin{Bmatrix} u_1 \\ u_2 \\ u_3 \\ u_4 \\ u_5 \\ u_6 \\ u_7 \\ u_8 \end{Bmatrix} = \begin{bmatrix} 1 & -1 & -1 & 1 & 1 & 1 & -1 & -1 \\ 1 & 1 & -1 & 1 & -1 & 1 & -1 & 1 \\ 1 & 1 & 1 & 1 & 1 & 1 & 1 & 1 \\ 1 & -1 & 1 & 1 & -1 & 1 & 1 & -1 \\ 1 & 0 & -1 & 0 & 0 & 1 & 0 & 0 \\ 1 & 1 & 0 & 1 & 0 & 0 & 0 & 0 \\ 1 & 0 & 1 & 0 & 0 & 1 & 0 & 0 \\ 1 & -1 & 0 & 1 & 0 & 0 & 0 & 0 \end{bmatrix} \begin{Bmatrix} \alpha_1 \\ \alpha_2 \\ \alpha_3 \\ \alpha_4 \\ \alpha_5 \\ \alpha_6 \\ \alpha_7 \\ \alpha_8 \end{Bmatrix} \quad \text{...[2]}$$

$$\{u\}_e = [A]\{\alpha\} \quad \text{...[3]}$$

or

$$\{\alpha\} = [A]^{-1}\{u\}_e \quad \text{....[4]}$$

where $[A]^{-1}$ can be shown to be equal to

$$[A]^{-1} = \frac{1}{4}\begin{bmatrix} -1 & -1 & -1 & -1 & 2 & 2 & 2 & 2 \\ 0 & 0 & 0 & 0 & 0 & 2 & 0 & -2 \\ 0 & 0 & 0 & 0 & -2 & 0 & 2 & 0 \\ 1 & 1 & 1 & 1 & -2 & 0 & -2 & 0 \\ 1 & -1 & 1 & -1 & 0 & 0 & 0 & 0 \\ 1 & 1 & 1 & 1 & 0 & -2 & 0 & -2 \\ -1 & -1 & 1 & 1 & 2 & 0 & -2 & 0 \\ -1 & 1 & 1 & -1 & 0 & -2 & 0 & 2 \end{bmatrix} \quad \text{....[5]}$$

but,

$$u = [1 \;\; \xi \;\; \eta \;\; \xi^2 \;\; \xi\eta \;\; \eta^2 \;\; \xi^2\eta \;\; \xi\eta^2]\{\alpha\}$$

$$= [1 \;\; \xi \;\; \eta \;\; \xi^2 \;\; \xi\eta \;\; \eta^2 \;\; \xi^2\eta \;\; \xi\eta^2][A]^{-1}\{u\}_e \qquad \text{(Using [4])}$$

$$\begin{Bmatrix} \frac{1}{4}(1-\xi)(1-\eta)(-\xi-\eta)(-\xi-\eta-1) \\ \frac{1}{4}(1+\xi)(1-\eta)(\xi-\eta-1) \\ \frac{1}{4}(1+\xi)(1+\eta)(\xi+\eta-1) \\ \frac{1}{4}(1-\xi)(1+\eta)(-\xi+\eta-1) \\ \frac{1}{2}(1+\xi)(1-\xi)(\eta-1) \\ \frac{1}{2}(1+\xi)(1+\eta)(1-\eta) \\ \frac{1}{2}(1+\xi)(1-\xi)(\eta+1) \\ \frac{1}{2}(1-\xi)(1+\eta)(1-\eta) \end{Bmatrix} \quad \text{...[6]}$$

$$= \begin{Bmatrix} N_1 \\ N_2 \\ N_3 \\ N_4 \\ N_5 \\ N_6 \\ N_7 \\ N_8 \end{Bmatrix} \{u\}_e \qquad \text{...[7]}$$

where $N_1, ..., N_8$ are as defined in equation [6].

For corner nodes : $(i = 1, 2, 3, 4)$

$$N_i = \frac{1}{4}(1 + \xi\xi_i)(1 + \eta\eta_i)(\xi\xi_i + \eta\eta_i - 1) \qquad \text{...[8]}$$

For midside nodes

If $\xi_i = 0$, then $N_i = \frac{1}{2}(1 - \xi^2)(1 + \eta\eta_i)$; (*i.e.*, for 5, 7)

If $\eta_i = 0$, then $N_i = \frac{1}{2}(1 - \eta^2)(1 + \xi\xi_i)$; (*i.e.*, for 6, 8)

Thus,

$$\begin{Bmatrix} u \\ v \end{Bmatrix} = \begin{bmatrix} \underset{1\times 8}{N} & \underset{1\times 8}{0} \\ \underset{1\times 8}{0} & \underset{1\times 8}{N} \end{bmatrix} \begin{Bmatrix} u_e \\ v_e \end{Bmatrix}$$

(N_i = 1 at node i and is zero at all other nodes. Variation is quadratic.)

Shape Functions for Serendipity Family Elements

Consider the elements shown in Fig. 7.40 with nodal points located only on the boundary. These elements are rectangles with sides parallel to x and y axes. This family of elements are called as "serendipity elements" after the famous princes of Serendip noted for their chance discoveries. Serendipity means the faculty of making happily chance finds.

Zienkiewicz called these elements as "Serendip family".

From the Fig. 7.40, it may be observed that progressively increasing number of nodes is placed on the element boundary. The variation of function on the edges is to ensure continuity such that it is linear quadratic.

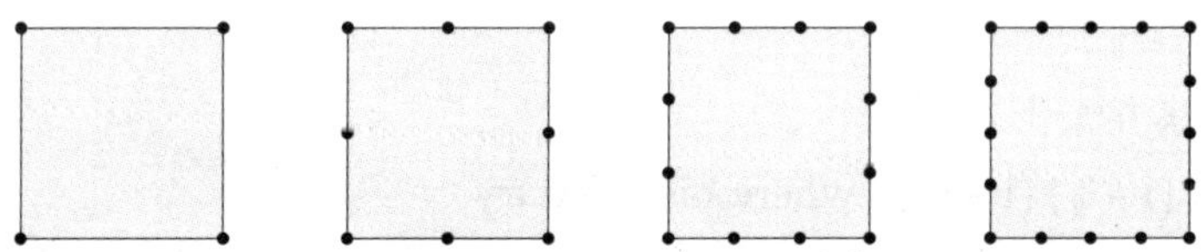

(a) Linear (b) Quadratic (c) Cubic (d) Quartic

Fig. 7.40 Rectangular element. 'Serendip' family

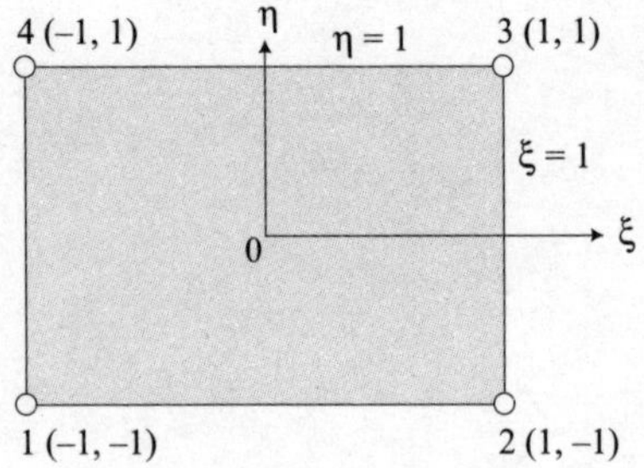

Fig. 7.41 Natural coordinate system (Rectangular element)

Serendipity elements

1. Shape functions for 4-noded rectangular element

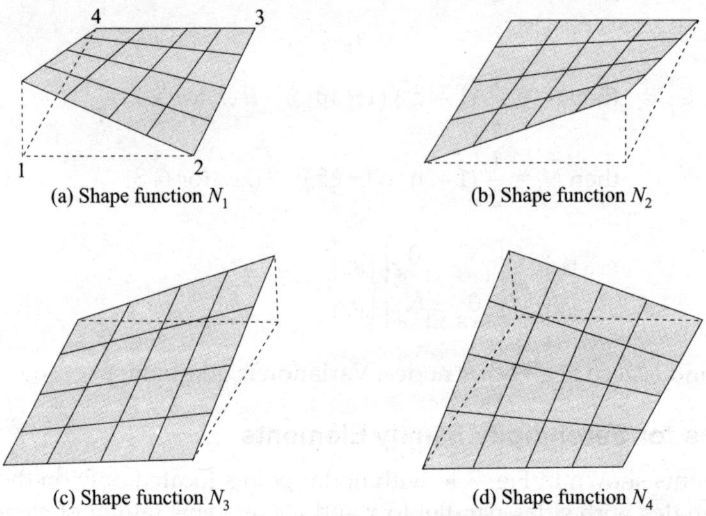

Fig. 7.42 Shape function

The shape functions are shown in Fig. 7.42 (*a* – *d*).

N_1 has to satisfy the conditions

(*a*) along $\xi = 1$, $N_1 = 0$

(*b*) along $\eta = 1$, $N_1 = 0$ and

(*c*) at $\xi = -1, \eta = -1$, $N_1 = 1$.

Let $N_1 = k\,(1 - \xi)\,(1 - \eta)$, where k is arbitrary.

Conditions (*a*) and (*b*) are satisfied.

Condition (*c*) yields

$$1 = k\,(1 + 1\,(1 + 1)$$

$\Rightarrow$ $$k = \frac{1}{4}$$

$\therefore$ $$N_1 = \frac{(1-\xi)(1-\eta)}{4}$$

Similarly, $$N_2 = \frac{(1+\xi)(1-\eta)}{4}$$

$$N_3 = \frac{(1+\xi)(1+\eta)}{4}$$

and $$N_4 = \frac{(1-\xi)(1-\eta)}{4}$$

or, $$N_i = \frac{1}{4}\,(1 + \xi_i\,\xi)\,(1 + \eta_i\,\eta)$$

or, $$\{N\} = \frac{1}{4}\begin{Bmatrix}(1-\xi)\,(1-\eta)\\(1+\xi)\,(1-\eta)\\(1+\xi)\,(1+\eta)\\(1-\xi)\,(1+\eta)\end{Bmatrix}$$

are the required shape functions.

2. **Derivation of element stiffness for linear serendipity rectangle**

The displacement in x and y directions can be written in terms of nodal displacements as

$$\begin{Bmatrix}\bar{\text{u}}\\\bar{\text{v}}\end{Bmatrix} = \begin{Bmatrix}\{N\}^T & \{0\}\\\{0\} & \{N\}^T\end{Bmatrix}\begin{Bmatrix}\underline{u}\\\underline{v}\end{Bmatrix} \quad ...[1]$$

Now, strain–displacement relation is written as:

$$\{\bar{\in}\} = \begin{bmatrix}\frac{\partial}{\partial x}\{N\}^T & \{0\}\\\{0\} & \frac{\partial}{\partial y}\{N\}^T\\\frac{\partial}{\partial y}\{N\}^T & \frac{\partial}{\partial x}\{N\}^T\end{bmatrix}\begin{Bmatrix}\underline{u}\\\underline{v}\end{Bmatrix} \quad ...[2]$$

where $$\frac{\partial}{\partial x} = \frac{\partial}{\partial \xi}\frac{\partial \xi}{\partial x} = \frac{1}{a}\frac{\partial}{\partial \xi} \quad ...[3]$$

$$\frac{\partial}{\partial y} = \frac{\partial}{\partial \eta}\frac{\partial \eta}{\partial y} = \frac{1}{b}\frac{\partial}{\partial \eta}$$

or $$\{\bar{\epsilon}\} = [B]\begin{Bmatrix} \underline{U} \\ \underline{V} \end{Bmatrix} \qquad ...[4]$$

and $$[K] = \int_V [B]^T [C]\,[B]\,dv = abt \int_{-1}^{1}\int_{-1}^{1} [B]^T [C]\,[B]\,d\xi\,d\eta \qquad ...[5]$$

In the above case $$\varepsilon_x = \frac{1}{4a} < -(1-\eta),\ (1-\eta),\ (1+\eta),\ -(1+\eta) > \begin{Bmatrix} u_1 \\ u_2 \\ u_3 \\ u_4 \end{Bmatrix} \qquad ...[6]$$

$$[B] = \begin{bmatrix} \frac{1}{a}\frac{\partial}{\partial \xi}\{N\}^T & \{0\} \\ \{0\} & \frac{1}{b}\frac{\partial}{\partial \eta}\{N\}^T \\ \frac{1}{b}\frac{\partial}{\partial \eta}\{N^*\}^T & \frac{1}{a}\frac{\partial}{\partial \xi}\{N^*\}^T \end{bmatrix} \qquad ...[7]$$

where N^* are constant equal to its value at centroid.

The derivation of stiffness and consistent load vector is exactly same as triangular element except that $ab\ d\xi\ d\eta = dx\ dy$. Hence

$$t\int B^T CB\,dA = abt \int_{\xi=-1}^{1}\int_{\eta=-1}^{1} B^T CB\,d\xi\,d\eta. \qquad ...[8]$$

Higher Order Serendipity Elements

Quadratic edge displacement

1. Derivation of sape functions for quadratic serendipity family element.

In Pascal's triangle we can write a polynomial over the element (which consists of 8 nodes) as:

$$\tilde{U} = a_1 + a_2x + a_3y + a_4\,x^2 + a_3xy + a_6y^2 + a_7x^2y + a_8xy^2$$

$$\tilde{V} = a_a + a_{10}x + a_{11}y + a_{12}x^2 + a_{13}xy + a_{14}y^2 + a_{15}x^2y + a_{16}xy^2$$

or for side nodes where $\xi = 0$

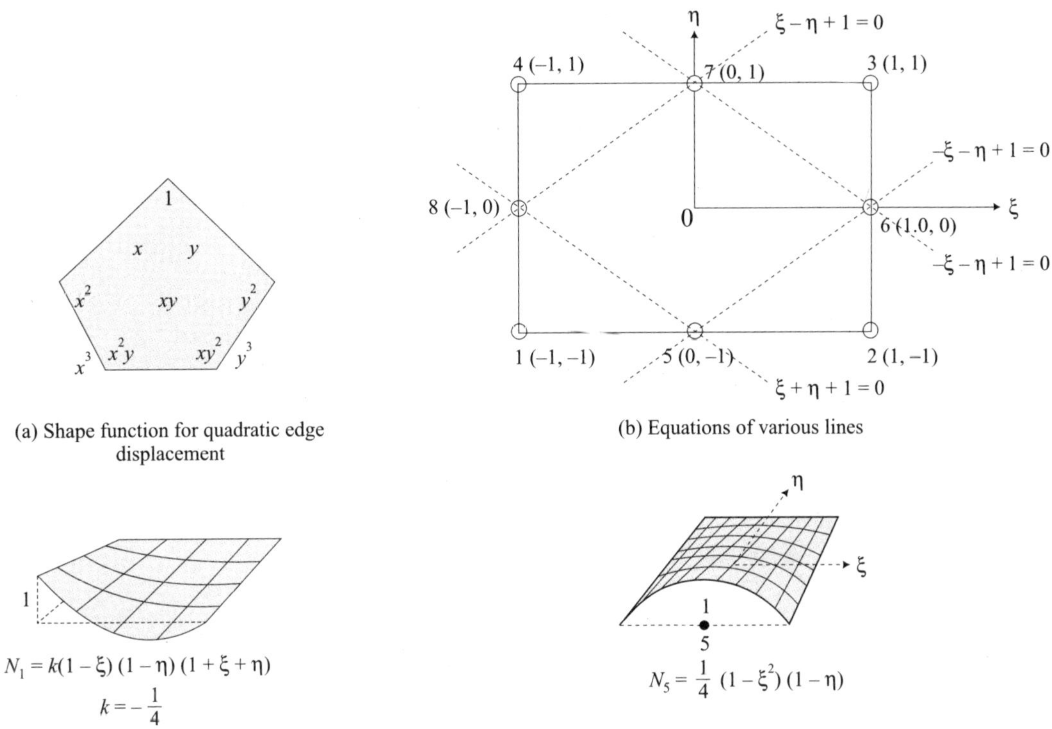

(a) Shape function for quadratic edge displacement

(b) Equations of various lines

(c) Shape function N_1

(d) Shape function N_5

Fig. 7.43

The conditions to be satisfied by N_1 are,

(a) along line $\xi = 1$, $N_1 = 0$

(b) along line $\eta = 1$, $N_1 = 0$

(c) along line 5 – 8, $N_1 = 0$

$$\xi + \eta + 1 = 0.$$

(d) At $\xi = -1$ and $\eta = -1$, $N_1 = 1$

Hence let

$$N_1 = C(1 - \xi)(1 - \eta)(1 + \xi + \eta)$$

It satisfies the requirements *a, b, c*. In other words it ensues $N_1 = 0$ at nodes 2, 3, ...8. From condition (*d*),

$$1 = C(1 + 1)(1 + 1)(1 - 1 - 1) \quad \therefore C = -\frac{1}{4}$$

$\therefore$ For the corner node

$$N_1 = -\frac{(1-\xi)(1-\eta)(1+\xi+\eta)}{4}$$

Similarly, we can show that

$$N_2 = -\frac{(1+\xi)(1-\eta)(1-\xi+\eta)}{4}$$

$$N_3 = -\frac{(1+\xi)(1+\eta)(1-\xi-\eta)}{4}$$

and $$N_4 = -\frac{(1-\xi)(1+\eta)(1+\xi-\eta)}{4}$$

for mid-side node 5 the conditions to be satisfied are

(*a*) Along $\xi = 1$ $\quad N_5 = 0$

(*b*) Along $\eta = 1$ $\quad N_5 = 0$

(*c*) Along $\xi = -1$ $\quad N_5 = 0$

(*d*) At node 5 where $\xi = 0, \eta = -1,\ N_5 = 1$

∴ Let $\quad N_5 = C\,(1-\xi)(1-\eta)\,(1+\xi)$

This form satisfies $N_1 = 0$ at all nodes other than node 5. From the condition '*d*' we get $1 = C\,(1-0)(1+1)\,(1+0)$

$$\therefore \quad C = \frac{1}{2}$$

$$\therefore \quad N_5 = \frac{1}{2}\,(1-\xi)\,(1-\eta)\,(1+\xi) = \frac{(1-\xi^2)(1-\eta)}{2}$$

Similarly, it may be shown that

$$N_6 = \frac{(1+\xi)(1-\eta^2)}{2}$$

$$N_7 = \frac{(1-\xi^2)(1+\eta)}{2}$$

$$N_8 = \frac{(1-\xi)(1-\eta^2)}{2}$$

Hence, let

$N_1 = C\,(1-\xi)\,(1-\eta)$, where C is arbitrary constant.

Conditions (*a*) and (*b*) are satisfied. Condition (*c*) gives,

$$1 = C\,(1+1)\,(1+1) \text{ or } C = \frac{1}{4}$$

$$\therefore \quad N_1 = \frac{(1-\xi)(1-\eta)}{4}$$

On the same lines we can get,

$$N_2 = \frac{(1+\xi)(1-\eta)}{4}$$

$$N_3 = \frac{(1+\xi)(1+\eta)}{4}$$

and $$N_4 = \frac{(1-\xi)(1-\eta)}{4}$$

Now we can write the shape function vector as

$$\{N\} = \frac{1}{4}\begin{Bmatrix} (1-\xi)(1-\eta)(-\eta-\xi-1) \\ (1+\xi)(1-\eta)(-\eta+\xi-1) \\ (1+\xi)(1+\eta)(\eta+\xi-1) \\ (1-\xi)(1+\eta)(\eta-\xi-1) \\ (1-\xi^2)(1-\eta) \\ (1-\eta^2)(1+\xi) \\ (1-\xi^2)(1+\eta) \\ (1-\eta^2)(1-\xi) \end{Bmatrix}$$

Cubic Edge Displacements

2. Derivation of the shape functions for cubic serendipity family element.

Solution: The displacement U in x direction is written (from Pascal's triangle shown in Fig. 7.44(a)) as:

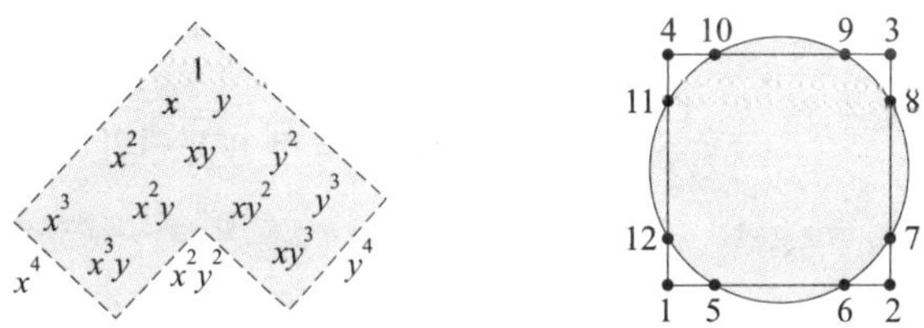

(a) Pascal triangle (b)

Fig. 7.44

$$\tilde{U} = a_1 + a_2x + a_3y + a_4x^2 + a_5xy + a_6y^2 + a_7x^3 + a_8x^2y + a_9xy^2 + a_{10}y^3 + a_{11}x^3y + a_{12}xy^3$$

$$\begin{Bmatrix} N_1 = 0 \text{ for all nodes except 1 and is 1 for node 1.} \\ N_1 = 0 \text{ is satisfied for nodes 2, 7, 8, 3 if } 1-\xi = 0 \\ N_1 = 0 \text{ is satisfied for nodes 3, 9, 10, 4 if } 1-\eta = 0 \end{Bmatrix}$$

The points 5, 6, 7, 8, 9, 10, 11, 12 lie on the circle shown in Fig. 7.44(b). The radius of this circle = OA.

$$= \sqrt{\left\{\frac{1}{3}\right\}^2 + 1} \text{ since } A\left(\frac{1}{3}, 1\right) \text{ and } O = (0, 0) = \sqrt{\frac{10}{3}}$$

The equation of the circle is

$$\xi^2 + \eta^2 = \frac{10}{3}$$

$$\xi^2 + \eta^2 - \frac{10}{3} = 0$$

Satisfies $N_1 = 0$ for nodes 5 to 12.

$N_1 = C(1 - \xi)(1 - \eta)\left(\xi^2 + \eta^2 - \frac{10}{9}\right)$ satisfies $N_1 = 0$ for all nodes except for node 1.

For node 1 $\quad N_1 = 1.$

$$\therefore \quad 1 = (1 - \xi_1)(1 - \eta_1)\left(\xi_1^2 + \eta_1^2 - \frac{10}{9}\right)$$

But we know $\quad \xi_1 = \eta_1 = -1$

$$\therefore \quad 1 = 2 \times 2 \times \left(1 + 1 - \frac{10}{9}\right) C = \frac{32}{9} C$$

or $$C = \frac{9}{32}$$

$$\therefore \quad N_1 = \frac{9}{32}(1 - \xi)(1 - \eta)\left(\xi^2 + \eta^2 - \frac{10}{9}\right)$$

$$= \frac{1}{32}(1 - \xi)(1 - \eta)\{9(\xi^2 + \eta^2) - 10\}$$

Along the similar lines, we get,

$$\left\{\begin{aligned} N_2 &= \frac{1}{32}(1 + \xi)(1 - \eta)\{9(\xi^2 + \eta^2) - 10\} \\ N_3 &= \frac{1}{32}(1 + \xi)(1 + \eta)\{9(\xi^2 + \eta^2) - 10\} \\ N_4 &= \frac{1}{32}(1 - \xi)(1 + \eta)\{9(\xi^2 + \eta^2) - 10\} \end{aligned}\right\}$$

From midside node 5,

$1 - \xi = 0$ ensures $N_5 = 0$ at nodes 2, 7, 8, 3

$1 - \eta = 0$ ensures $N_5 = 0$ at nodes 3, 9, 10, 4

$1 + \xi = 0$ ensures $N_5 = 0$ at nodes 4, 11, 12, 1

$1 - 3\xi = 0$ ensures $N_5 = 0$ at node 6.

$\therefore$ Let $\quad N_5 = C(1 - \xi)(1 - \eta)(1 + \xi)(1 - 3\xi)$

At node 5, $\quad \xi = -\frac{1}{3}, \eta = -1$ and $N_5 = 1$

$$\therefore \qquad 1 = C \times \frac{4}{3} \times 2 \times \frac{2}{3} \times 2$$

$$C = \frac{9}{32}$$

$$\therefore \qquad N_5 = \frac{9}{32} (1 - \xi)(1 - \eta)(1 + \xi)(1 - 3\xi)$$

$$= \frac{9}{32} (1 - \xi^2)(1 - \eta)(1 - 3\xi)$$

Similarly, the shape functions for all the twelve nodes may be written as:

$$[N] = \frac{1}{32} \left\{ \begin{matrix} (1-\xi) & (1-\eta) & [9(\xi^2+\eta^2)-10] \\ (1+\xi) & (1-\eta) & [9(\xi^2+\eta^2)-10] \\ (1+\xi) & (1+\eta) & [9(\xi^2+\eta^2)-10] \\ (1-\xi) & (1+\eta) & [9(\xi^2+\eta^2)-10] \\ 9(1-\xi^2) & (1-\eta) & (1-3\xi) \\ 9(1-\xi^2) & (1-\eta) & (1+3\xi) \\ 9(1-\eta^2) & (1+\xi) & (1-3\eta) \\ 9(1-\eta^2) & (1+\xi) & (1+3\eta) \\ 9(1-\xi^2) & (1+\eta) & (1+3\xi) \\ 9(1-\xi^2) & (1+\eta) & (1-3\xi) \\ 9(1-\eta^2) & (1-\xi) & (1+3\eta) \\ 9(1-\eta^2) & (1-\xi) & (1-3\eta) \end{matrix} \right\}$$

It is not immediately obvious that these elements contain rigid body modes. Constant strain states must be checked.

Hermite Polynomials as Shape Functions

If we wish C^1or higher continuity between elements for a one-dimensional element, we may use the Hermite polynomial element. Hermite polynomial in one dimension is denoted as $H^n(x)$. It is a polynomial of order $2n + 1$.

Thus, $H^1(x)$ is a first order polynomial which is cubic in x.

$H^2(x)$ is a second order polynomial and is of 5th order.

Shape functions for a cubic polynomial that must satisfy conditions on as well as notation at each end of the beam. Hermite's polynomial is a natural choice for deriving the shape function.

We can also use the following notations:

$H^n_{mi}(x)$; n = order of the Hermitian function

where $\quad m$ = order of derivatives; i = node number.

Illustration

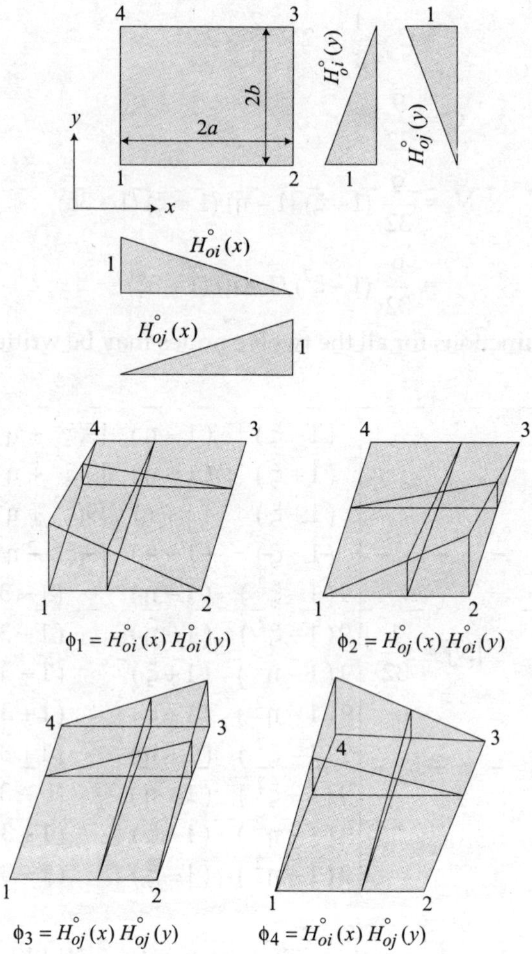

Fig. 7.45 First order Hermitian element

(1): Fig. 7.45

$$\overline{U} = [N_1 \; N_2 \; N_3 \; N_4] \begin{Bmatrix} U_1 \\ U_2 \\ U_3 \\ U_4 \end{Bmatrix}$$

where
$$H^o_{oi}(x) = (1 - x/2a)\,;\; H^o_{oj}(x) = x/2a$$
$$H^o_{oi}(y) = (1 - y/2b)\,;\; H^o_{oj}(y) = y/2b$$

The displacements in x and y directions may be written in terms of nodal displacements Fig. 7.46(a).

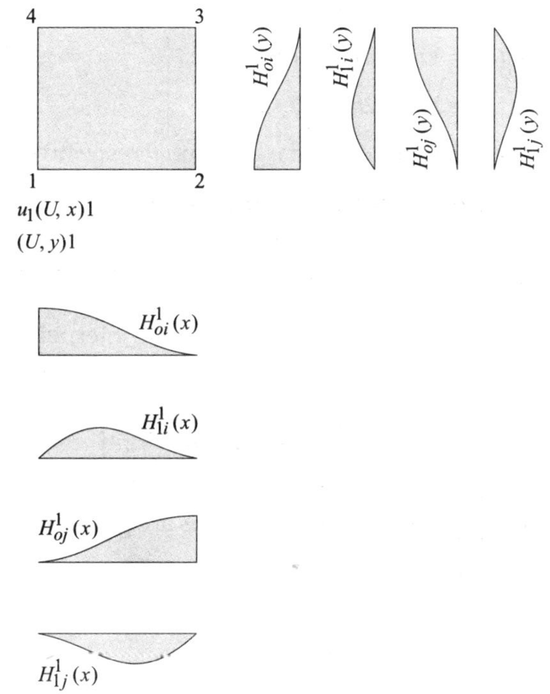

Fig. 7.46(a) Second order Hermitian element and shape functions

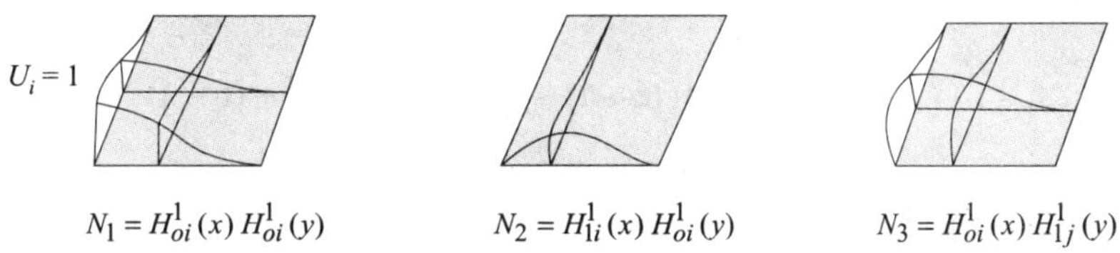

Fig. 7.46(b) Shape functions N_1, N_2, N_3 for second order Hermitian Element

$$\tilde{U} = [N_1\, N_2 \ldots N_{12}] \begin{Bmatrix} U_1 \\ (U, x)_1 \\ (U, y)_1 \\ \vdots \\ \vdots \\ (U, y)_4 \end{Bmatrix}$$

where $N_1 \ldots N_{12}$ are the shape functions which are shown in Fig. 7.46(b).

It is to be noted for a rectangle

$$\left\{\begin{array}{l} H^1_{oi}(x) = 1 + 2\,(x/2a)^3 - 3\,(x/2a)^2 \\ H^1_{ij}(x) = x\,(1 - x/2a)^2 \\ H^1_{oj}(x) = 3\,(x/2a)^2 - 2\,(x/2a)^3 \\ \mathrm{H}^1_{ij}(x) = 2a\,(x/2a)^3 - (x/2a)^2 \end{array}\right\}$$

To get the shape function $H(y)$ replace x and a by y and b respectively.

Cubic hermite shape functions for rectangular element

The function and its first order partial derivatives $u, \dfrac{\partial u}{\partial x}, \dfrac{\partial u}{\partial y}$ are specified at the four corner nodes of the rectangular element (e) as shown in Fig. 7.47. Satisfying the interpolation conditions, we get the cubic Hermite polynomial.

$$u^{(e)}(x, y) = \sum_{i=1}^{4}\left[N_i(x, y)\,u_i + H_i(x, y)\left(\frac{\partial u}{\partial x}\right)_i + K_i(x, y)\left(\frac{\partial u}{\partial y}\right)_i\right] \qquad ...[1]$$

where the shape functions N_i, H_i and K_i in ξ, η coordinates are given by

$$N_1(\xi, \eta) = \frac{1}{8}(\eta - 1)(\xi - 1)\left[\frac{1}{2}(\eta - 1)(\xi - 1) - \frac{1}{2}(\eta + 1)(\xi + 1) - (\eta + 1)(\eta - 1) - (\xi + 1)(\xi - 1)\right]$$

$$H_1(\xi, \eta) = \frac{a}{8}(\eta - 1)(\xi - 1)^2(\xi + 1)$$

$$K_1(\xi, \eta) = -\frac{b}{8}(\eta - 1)^2(\xi - 1)(\eta + 1)$$

$$N_2(\xi, \eta) = \frac{1}{8}(\eta - 1)(\xi + 1)\left[\frac{1}{2}(\eta - 1)(\xi + 1) - \frac{1}{2}(\eta + 1)(\xi - 1) + (\eta + 1)(\eta - 1) + (\xi + 1)(\xi - 1)\right]$$

$$H_2(\xi, \eta) = \frac{a}{8}(\eta - 1)(\xi + 1)^2(\xi - 1)$$

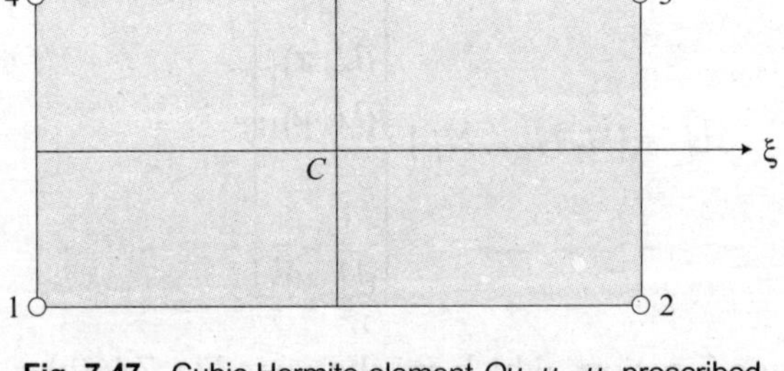

Fig. 7.47 Cubic Hermite element Ou, u_x, u_y prescribed

$$K_2(\xi, \eta) = \frac{b}{8}(\eta - 1)^2(\xi + 1)(\eta + 1)$$

$$N_3(\xi, \eta) = \frac{1}{8}(\eta + 1)(\xi + 1)\left[\frac{1}{2}(\eta+1)(\xi+1) - \frac{1}{2}(\xi-1)(\eta-1) - (\eta+1)(\eta-1) - (\xi+1)(\xi-1)\right]$$

$$H_3(\xi, \eta) = \frac{a}{8}(\eta + 1)(\xi + 1)^2(\xi - 1)$$

$$K_3(\xi, \eta) = \frac{b}{8}(\eta + 1)^2(\xi + 1)(\eta - 1)$$

$$N_4(\xi, \eta) = \frac{1}{8}(\eta + 1)(\xi - 1)\left[\frac{1}{2}(\eta + 1)(\xi - 1) - \frac{1}{2}(\eta - 1)(\xi + 1) + (\eta + 1)(\eta - 1) + (\xi + 1)(\xi - 1)\right]$$

$$H_4(\xi, \eta) = \frac{a}{8}(\eta + 1)(\xi - 1)^2(\xi + 1)$$

$$K_4(\xi, \eta) = -\frac{b}{8}(\eta + 1)^2(\xi - 1)(\eta - 1)$$

where $a = \frac{1}{2}$ side 12 and $b = \frac{1}{2}$ side 23.

Construction of Shape Functions Using Lagrangian Polynomials

Lagrangian interpolation provides an easy means of *generating shape functions.*

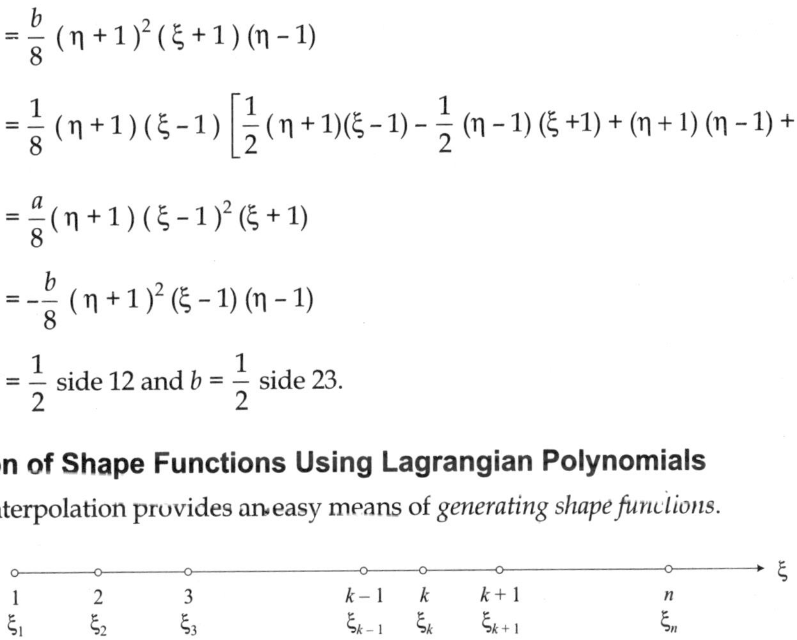

Fig. 7.48 Intervals for Lagrangian interpolation

Let us consider the linear element as shown in Fig. 7.48, a line.

It is divided into equally spaced points 1, 2, 3, ..., k, ... n.

The coordinates (Natural) of these points are ξ_1, ξ_2, ξ_3, ... ξ_k, ... ξ_n.

The Lagrange polynomials satisfy the shape function properties given by equation

i.e., $N_i = 1$ at node ξ_i

$N_i = 0$ at node ξ_j $(i \neq j)$

For different nodes, they are

$$N_1 = \frac{(\xi - \xi_2)(\xi - \xi_3)\ldots(\xi - \xi_n)}{(\xi_1 - \xi_2)(\xi_1 - \xi_3)\ldots(\xi_1 - \xi_n)}$$

$$N_k = \frac{(\xi - \xi_1)(\xi - \xi_2)\ldots(\xi - \xi_{k-1})(\xi - \xi_{k+1})\ldots(\xi - \xi_n)}{(\xi_k - \xi_1)(\xi_k - \xi_2)\ldots(\xi_k - \xi_{k-1})(\xi_k - \xi_{k+1})\ldots(\xi_k - \xi_n)} \quad \ldots[1]$$

Note: In two dimensions, the shape function is given by $N_{ij} = N_i N_j$ where N_i and N_j are one-dimensional values along their respective ones as given by eqn. [1].

Quadratic Shape Function

A quadratic shape function is achieved by the quadratic variation within the element by introducing a third node within the element called internal node.

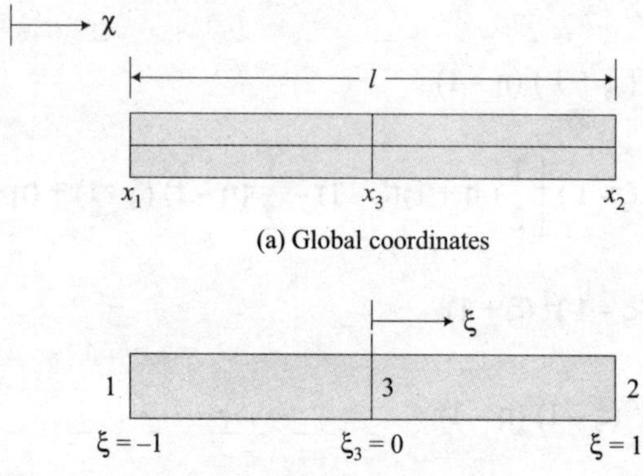

(a) Global coordinates

(b) Natural coordinates

Fig. 7.49

The shape function for:

Node (1)

$$N_1 = \frac{(\xi - \xi_2)(\xi - \xi_3)}{(\xi_1 - \xi_2)(\xi_1 - \xi_3)}$$

$$\left.\begin{aligned} \xi_2 &= +1 \\ \xi_3 &= 0 \\ \xi_1 &= -1 \end{aligned}\right| \quad \therefore \quad N_1 = \frac{(\xi - 1)(\xi - 0)}{(-1-1)(-1-0)}$$

$$= \frac{\xi(\xi - 1)}{2}$$

$$\therefore \quad \boxed{N_1 = \frac{\xi(\xi - 1)}{2}}$$

Node (2)

$$N_2 = \frac{(\xi - \xi_1)(\xi - \xi_3)}{(\xi_2 - \xi_1)(\xi_2 - \xi_3)}$$

$$= \frac{(\xi + 1)(\xi - 0)}{(1+1)(1-0)} = \frac{\xi(1 + \xi)}{2}$$

$$\therefore \quad \boxed{N_2 = \frac{\xi(1 + \xi)}{2}}$$

Node (3)

$$N_3 = \frac{(\xi - \xi_1)(\xi - \xi_2)}{(\xi_3 - \xi_1)(\xi_3 - \xi_2)}$$

$$= \frac{(\xi + 1)(\xi - 1)}{(0 + 1)(0 - 1)} = (1 + \xi)(1 - \xi) = (1 - \xi^2)$$

$$\therefore \qquad \boxed{N_3 = (1 - \xi^2)}$$

N_1, N_2, N_3 are the required shape functions. They are called *Lagrange shape functions*.

Shape Functions for the 4-Noded Linear (Axial) Element

What are Cubic Shape Functions?

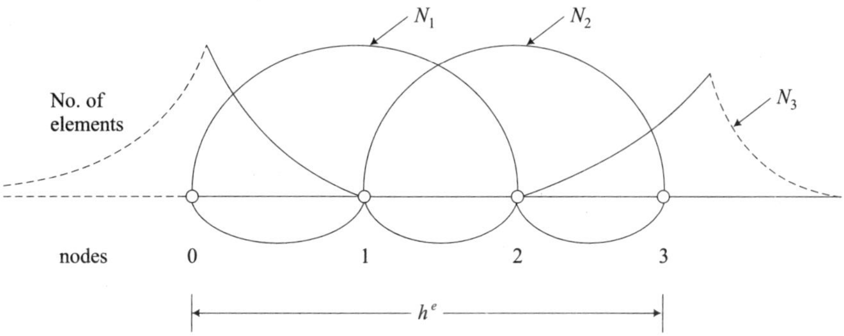

Fig. 7.50 Cubic forms

The shape functions are:

N_0 at node 0

N_1 at node 1

N_2 at node 2

N_3 at node 3

Node $[0] \rightarrow N_0$, where

$$N_0 = \frac{(\xi - \xi_1)(\xi - \xi_2)(\xi - \xi_3)}{(\xi_0 - \xi_1)(\xi_0 - \xi_2)(\xi_0 - \xi_3)}$$

$$= \frac{\left(\xi + \frac{1}{3}\right)\left(\xi - \frac{1}{3}\right)(\xi - 1)}{\left(-1 + \frac{1}{3}\right)\left(-1 - \frac{1}{3}\right)(-1 - 1)}$$

$\xi_0 = -1$

$\xi_1 = -\frac{1}{3}$

$\xi_2 = \frac{1}{3}$

$\xi_3 = 1$

$$= \frac{\left(\xi + \frac{1}{3}\right)\left(\xi - \frac{1}{3}\right)(\xi - 1)}{\left(-\frac{2}{3}\right)\left(-\frac{4}{3}\right)(-2)}$$

$$= -\frac{9}{16}\left(\xi + \frac{1}{3}\right)\left(\xi - \frac{1}{3}\right)(\xi - 1)$$

$$\boxed{N_0 = -\frac{9}{16}\left(\xi + \frac{1}{3}\right)\left(\xi - \frac{1}{3}\right)(\xi - 1)}$$

Node [1] $\rightarrow N_1$

$$N_1 = \frac{(\xi - \xi_0)(\xi - \xi_2)(\xi - \xi_3)}{(\xi_1 - \xi_0)(\xi_1 - \xi_2)(\xi_1 - \xi_3)}$$

$\xi_0 = -1$

$\xi_1 = -\frac{1}{3}$

$\xi_2 = \frac{1}{3}$

$\xi_3 = 1$

$$= \frac{(\xi + 1)\left(\xi - \frac{1}{3}\right)(\xi - 1)}{\left(-\frac{1}{3} + 1\right)\left(-\frac{1}{3} - \frac{1}{3}\right)(-\frac{1}{3} - 1)}$$

$$= \frac{(\xi + 1)\left(\xi - \frac{1}{3}\right)(\xi - 1)}{\left(\frac{2}{3}\right)\left(-\frac{2}{3}\right)\left(-\frac{4}{3}\right)}$$

$$= \frac{27}{16}(\xi + 1)\left(\xi - \frac{1}{3}\right)(\xi - 1)$$

$$\boxed{N_1 = +\frac{27}{16}(\xi + 1)\left(\xi - \frac{1}{3}\right)(\xi - 1)}$$

Node [2] $\rightarrow N_2$.

$$N_2 = \frac{(\xi - \xi_0)(\xi - \xi_3)(\xi - \xi_1)}{(\xi_2 - \xi_0)(\xi_2 - \xi_1)(\xi_2 - \xi_3)}$$

$$= \frac{(\xi + 1)(\xi - 1)\left(\xi + \frac{1}{3}\right)}{\left(\frac{1}{3} + 1\right)\left(\frac{1}{3} + \frac{1}{3}\right)\left(\frac{1}{3} - 1\right)}$$

$$= \frac{(\xi + 1)\left(\xi + \frac{1}{3}\right)(\xi - 1)}{\left(\frac{4}{3}\right)\left(\frac{2}{3}\right)\left(-\frac{2}{3}\right)}$$

$$\therefore \quad \boxed{N_2 = -\frac{27}{16}(\xi+1)\left(\xi+\frac{1}{3}\right)(\xi-1)}$$

Node [3] $\rightarrow N_3$.

$$N_3 = \frac{(\xi-\xi_0)(\xi-\xi_1)(\xi-\xi_2)}{(\xi_3-\xi_0)(\xi_3-\xi_1)(\xi_3-\xi_2)}$$

$\xi_0 = -1$

$\xi_1 = -\frac{1}{3}$

$\xi_2 = \frac{1}{3}$

$\xi_3 = 1$

$$= \frac{(\xi+1)\left(\xi+\frac{1}{3}\right)\left(\xi-\frac{1}{3}\right)}{(1+1)\left(1+\frac{1}{3}\right)\left(1-\frac{1}{3}\right)}$$

$$= \frac{9}{16}(\xi+1)\left(\xi+\frac{1}{3}\right)\left(\xi-\frac{1}{3}\right)$$

$$\therefore \quad \boxed{N_3 = \frac{9}{16}(\xi+1)\left(\xi+\frac{1}{3}\right)\left(\xi-\frac{1}{3}\right)}$$

Two-dimensional Elements

One-dimensional concepts will be extended to two-dimensions (quadratic model).

Shape functions for the 6-noded [Quadratic Model] triangular element using natural or local coordinates.

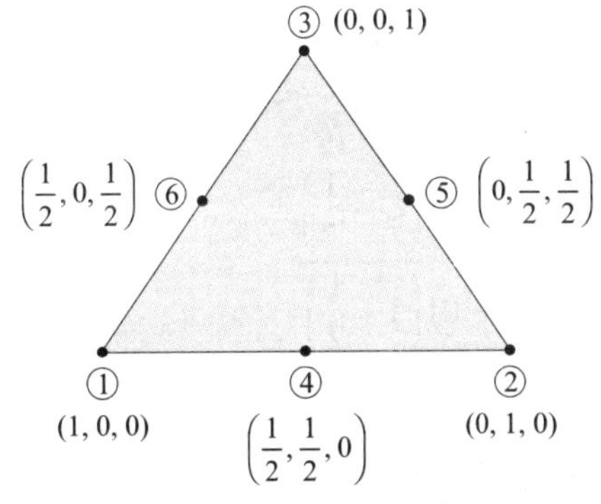

Fig. 7.51 Triangular element quadratic model

The L_1 coordinates for nodes 1, 4 and 2 are $\left(1, \frac{1}{2}, 0\right)$ respectively,

A Lagrange shape function can be constructed as follows.

Node (1): $\boxed{\xi_1 = 1, \xi_4 = \frac{1}{2}, \xi_2 = 0,}$

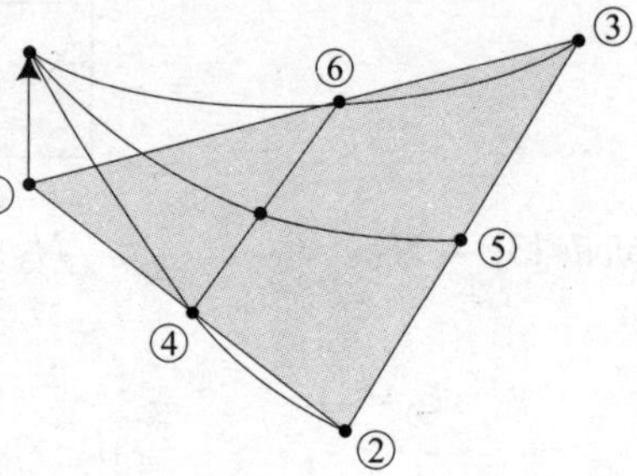

$$\therefore \qquad N_1 = \frac{(\xi - \xi_2)(\xi - \xi_4)}{(\xi_1 - \xi_2)(\xi_1 - \xi_4)}$$

$$= \frac{(\xi - 0)\left(\xi - \frac{1}{2}\right)}{(1-0)\left(1 - \frac{1}{2}\right)}$$

$$= \xi\,(2\,\xi - 1)$$

$$\therefore \qquad N_1 = \xi\,(2\xi - 1)$$

$$\xi = L_1 = \text{Area coordinate of the system}$$

$$\boxed{N_1 = L_1\,(2L_1 - 1)}$$

It can be seen that the value of this function at all other nodes *except one is zero*. Shape function pattern for node (1) = N_1.

Node (2): The L_2 coordinates for the nodes 2, 5 and 3 are $\left(1, \frac{1}{2}, 0\right)$ respectively.

$$\xi_2 = 1, \xi_5 = \frac{1}{2}, \xi_3 = 0.$$

A Lagrangian shape function can be constructed as:

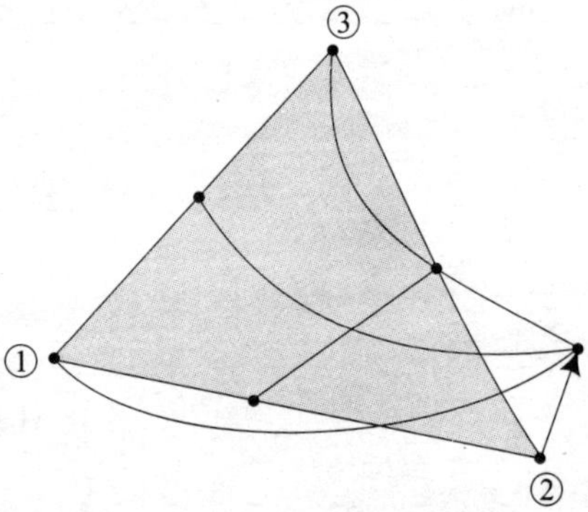

$$N_2 = \frac{(\xi - \xi_3)(\xi - \xi_5)}{(\xi_2 - \xi_3)(\xi_2 - \xi_5)}$$

$$= \frac{(\xi - 0)\left(\xi - \frac{1}{2}\right)}{(1-0)\left(1 - \frac{1}{2}\right)}$$

$$= \frac{\xi\left(\xi - \frac{1}{2}\right)}{\frac{1}{2}} = \xi(2\xi - 1)$$

Let $\qquad \xi = L_2 = \text{area coordinate of the system.}$

$$\boxed{\therefore N_2 = L_2\,(2L_2 - 1)}$$

Node (3): The L_3 coordinate for nodes 1, 6 and 3 are 0, $\frac{1}{2}$, 1 respectively. Shape functions for node 3 = N_3.

A Lagrangian shape function can be constructed as follows:

$$\xi_1 = 0, \xi_6 = \frac{1}{2}, \xi_3 = 1$$

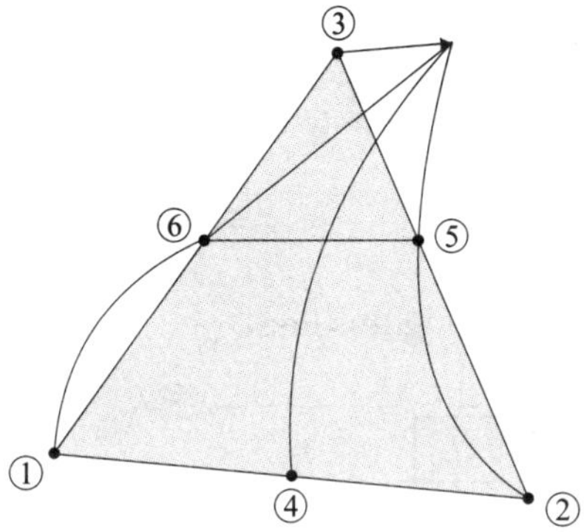

Fig. 7.52 A shape function for node 3 = N_3

$$N_3 = \frac{(\xi - \xi_1)(\xi - \xi_6)}{(\xi_3 - \xi_1)(\xi_3 - \xi_6)}$$

$$= \frac{(\xi - 0)\left(\xi - \frac{1}{2}\right)}{(1-0)\left(1 - \frac{1}{2}\right)} = \frac{\xi\left(\xi - \frac{1}{2}\right)}{\frac{1}{2}}$$

$$= \xi\,(2\xi - 1)$$

Let $\xi = L_3$ = Area coordinates of the system

$$\boxed{N_3 = L_3\,(2L_3 - 1)}$$

Node (4): L_1 coordinate for nodes 1, 4, 2 are $\left(1, \frac{1}{2}, 0\right)$ respectively.

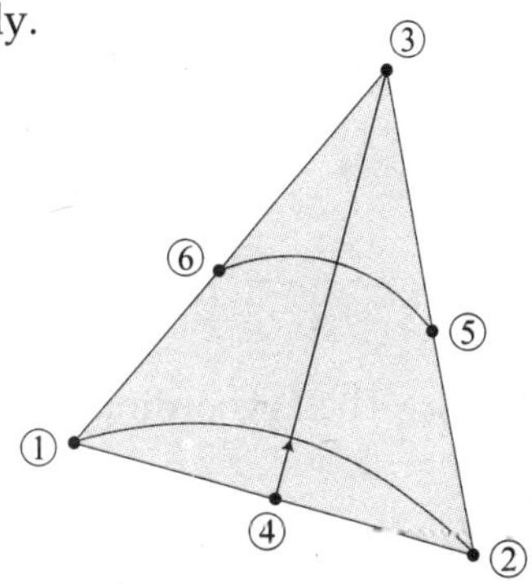

Fig. 7.53 Shape function for node 4 = N_4

so, $$N_4 = \frac{(\xi - \xi_1)(\xi - \xi_2)}{(\xi_4 - \xi_1)(\xi_4 - \xi_2)}$$

$\xi_1 = 1$	$\xi_4 = \frac{1}{2}$	$\xi_2 = 0$

The pattern is same as shown.

$\therefore$ $$N_4 = \frac{(\xi - 1)(\xi - 0)}{\left(\frac{1}{2} - 1\right)\left(\frac{1}{2} - 0\right)}$$

$$= -4\xi(\xi - 1)$$

$$= 4\xi(1 - \xi)$$

Let $\qquad \xi = L_1 =$ Area coordinate of the system

$\therefore \qquad N_4 = 4L_1(1 - L_1)$

Along the line containing the nodes 1, 4, 2

$$L_1 + L_2 + L_3 = 1$$

$$L_1 + L_2 + 0 = 1$$

Since $\qquad L_3 = 0$ along this line

$\therefore \qquad L_2 = L - L_1$

$\therefore \qquad \boxed{N_4 = 4L_1L_2}$

Node (5): The L_2 coordinates for nodes 2, 5, 3 are $\left(1, \frac{1}{2}, 0\right)$ respectively.

$$\therefore \qquad N_5 = \frac{(\xi - \xi_2)(\xi - \xi_3)}{(\xi_5 - \xi_2)(\xi_5 - \xi_3)}$$

$$\boxed{\xi_2 = 1, \xi_5 = \frac{1}{2}, \xi_3 = 0}$$

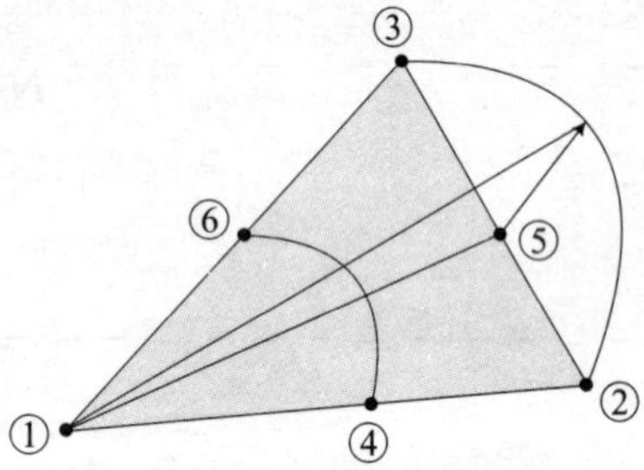

Fig. 7.54

$$N_5 = \frac{(\xi - 1)(\xi - 0)}{\left(\frac{1}{2} - 1\right)\left(\frac{1}{2} - 0\right)} = -4\xi(\xi - 1)$$

$$= 4\xi(1 - \xi)$$

$$= 4L_2(1 - L_2) = 4L_2L_3$$

Along the line containing the nodes 2, 5, 3

$$L_1 + L_2 + L_3 = 1$$

$$0 + L_2 + L_3 = 1$$

$$L_3 = 1 - L_2$$

$\therefore \qquad \boxed{N_5 = 4L_2L_3}$

Let $\qquad \xi = L_2 =$ Area coordinate of the system

Node (6): The L_3 coordinates for the nodes 3, 6, 1 are $\left(1, \frac{1}{2}, 0\right)$ respectively.

$$\boxed{\xi_3 = 1} \; \boxed{\xi_6 = \frac{1}{2}} \; \boxed{\xi_1 = 0}$$

$$N_6 = \frac{(\xi - \xi_1)(\xi - \xi_3)}{(\xi_6 - \xi_1)(\xi_6 - \xi_3)}$$

$$= \frac{(\xi - 0)(\xi - 1)}{\left(\frac{1}{2} - 0\right)\left(\frac{1}{2} - 1\right)}$$

$$= -4\xi(\xi - 1)$$

$$= 4\xi(\xi - 1)$$

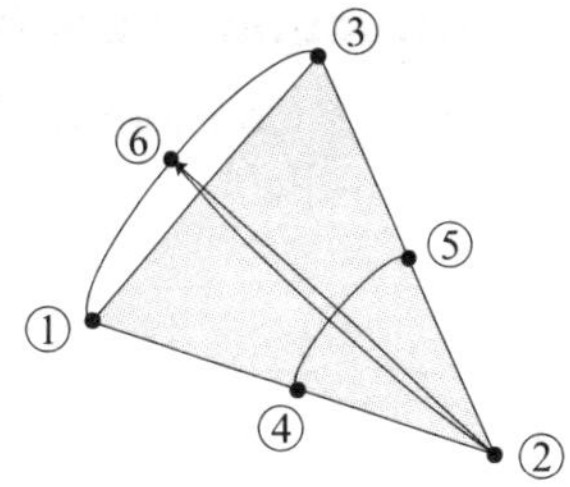

Fig. 7.55

Let $\xi = L_3 =$ Area coordinate of the system.

$$N_6 = 4\,L_3\,(1 - L_3)$$

Along the line containing the nodes 3, 6, 1

$$L_1 + L_2 + L_3 = 1$$

$$L_1 + 0 + L_3 = 0$$

$$L_3 = 1 - L_1$$

$\therefore$ $N_6 = 4\,L_3 L_1 = 4L_1 L_3$

$\therefore$ $\boxed{N_6 = 4L_1 L_2}$

we have $N_1 = L_1(2L_1 - 1)$

$$N_2 = L_2(2L_2 - 1)$$

$$N_3 = L_3(2L_3 - 1)$$

$$N_4 = 4L_1\,L_2$$

$$N_5 = 4L_2\,L_3$$

$$N_6 = 4L_1\,L_3$$

Shape functions for the cubic model: (Δ element)

Consider a cubic model (Triangular element cubic model).

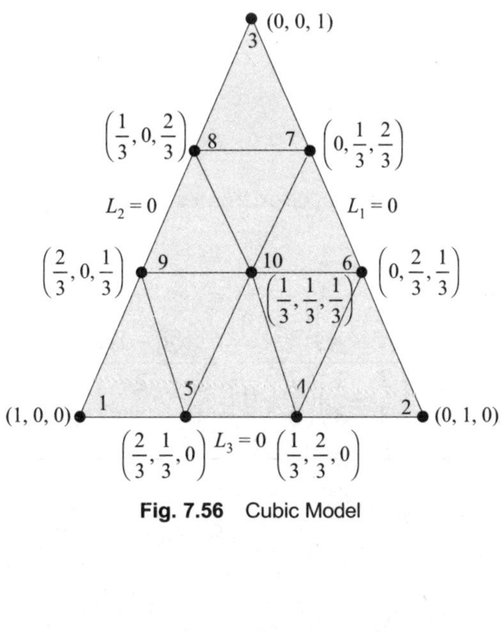

Fig. 7.56 Cubic Model

Construction of Shape Functions

Node (1): Consider the line having nodes 1, 4, 5 and 2, then

$$N_1 = \frac{(\xi - \xi_4)(\xi - \xi_5)(\xi - \xi_2)}{(\xi_1 - \xi_4)(\xi_1 - \xi_5)(\xi_1 - \xi_2)}$$

where $$\boxed{\xi_1 = 1 \;\; \xi_2 = 0 \;\; \xi_4 = \frac{2}{3} \;\; \xi_5 = \frac{1}{3}}$$

$\therefore$ $$N_1 = \frac{\left(\xi - \frac{2}{3}\right)\left(\xi - \frac{1}{3}\right)(\xi - 0)}{\left(1 - \frac{2}{3}\right)\left(1 - \frac{1}{3}\right)(1 - 0)}$$

$$= \frac{(3\xi - 2)(3\xi - 1)\xi}{2}$$

Let $$\xi = L_1$$

$\therefore$ $$\boxed{N_1 = \frac{(3L_1 - 2)(3L_1 - 1)L_1}{2}}$$

Similarly $$N_2 = \frac{(3L_2 - 2)(3L_2 - 1)L_2}{2}$$

$$N_3 = \frac{(3L_3 - 2)(3L_3 - 1)L_3}{2}$$

Node [4]: Consider the case having nodes 4, 10 and 7,

then, $$C_1 = \frac{(\xi - \xi_{10})(\xi - \xi_7)}{(\xi_4 - \xi_{10})(\xi_4 - \xi_7)}$$

$$\boxed{\begin{aligned} \xi_4 &= \frac{2}{3} \\ \xi_{10} &= \frac{1}{3} \\ \xi_7 &= 0 \end{aligned}} \quad L_1 \text{ coordinates}$$

$\therefore$ $$C_1 = \frac{\left(\xi - \frac{1}{3}\right)(\xi - 0)}{\left(\frac{2}{3} - \frac{1}{3}\right)\left(\frac{2}{3} - 0\right)} = \frac{(3\xi - 1)\xi}{2}$$

Let $$\xi = L_1$$

$$\therefore \qquad C_1 = \frac{(3L_1 - 1)L_1}{2} \qquad \text{[i]}$$

Again, consider the line having nodes 4, 9,

then
$$C_2 = \frac{(\xi - \xi_9)}{(\xi_4 - \xi_9)}$$

$$\xi_9 = 0$$
$$\xi_4 = \frac{1}{3} \qquad L_2 \text{ ordinals}$$

$$\therefore \qquad C_2 = \frac{(\xi - 0)}{\left(\frac{1}{3} - 0\right)} = 3\xi$$

Let
$$\xi = L_2$$
$$\therefore \qquad C_2 = 3L_2$$

Now,
$$N_4 = C_1 \times C_2 = \frac{3L_2 3L_1(3L_1 - 1)}{2}$$

$$\therefore \qquad \boxed{N_4 = \frac{9L_1L_2(3L_1 - 1)}{2}}$$

Along the similar steps shape functions for all other side nodes can be constructed.

Node [10] (internal node): At the node 10, three lines are intersecting. For any one line shape function can be found and then multiplied as shown below:

Let L_1 coordinates along the line having nodes 4, 10, 7.

$$C_1 = \frac{(\xi - \xi_4)(\xi - \xi_7)}{(\xi_{10} - \xi_4)(\xi_{10} - \xi_7)}$$

where
$$\boxed{\begin{aligned} \xi_4 &= \frac{2}{3} \\ \xi_{10} &= \frac{1}{3} \\ \xi_7 &= 0 \end{aligned}}$$

$$\therefore \qquad C_1 = \frac{\left(\xi - \frac{2}{3}\right)(\xi - 0)}{\left(\frac{1}{3} - \frac{2}{3}\right)\left(\frac{1}{3} - 0\right)} = -3\xi(3\xi - 2)$$

Let $\xi = L_1$

$\therefore$ $C_1 = -3L_1(3\,L_1 - 2)$

Similarly $C_2 = -3L_2(3L_2 - 2)$

$C_3 = -\,3L_3(3L_3 - 2)$

$\therefore$ $N_{10} = -3 \times 3 \times 3 \times (3L_1 - 2)\,(3L_2 - 2)\,(3L_3 - 2)L_1\,L_2\,L_3$

$= -27\,L_1\,L_2\,L_3\,(3L_1 - 2)(3L_2 - 2)\,(3L_3 - 2)$

At node 10

$$(3L_1 - 2) = -1$$

$$(3L_2 - 2) = -1$$

$$(3L_3 - 2) = -1$$

$\therefore$ $\boxed{N_{10} = 27\,L_1 L_2\,L_3}$

It may be verified that this is the shape function for node 10.

Shape Function for a 4-noded Rectangular Element.

Shape function for a rectangular element in $x - y$ plane assuming a function

$$u = C_1 + C_2x + C_3y + C_4xy$$

over the element as shown in Fig. 7.57 write the result

$$u = N_1u_1 + N_2u_2 + N_3u_3 + N_4u_4$$

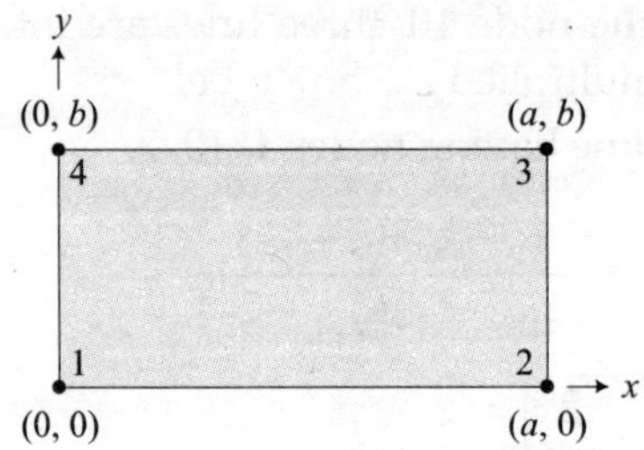

Fig. 7.57

The boundary conditions are

$$\left.\begin{aligned} u\,(0, 0) &= u_1 \\ u\,(a, 0) &= u_2 \\ u\,(a, b) &= u_3 \\ u\,(0, b) &= u_4 \end{aligned}\right\} \qquad \text{...[1]}$$

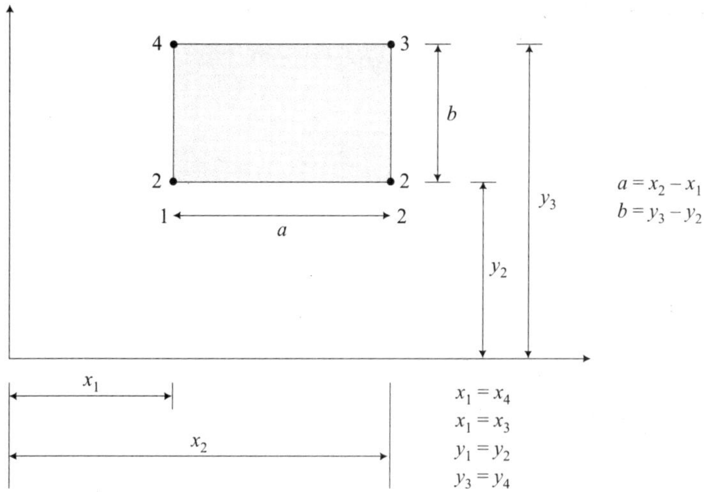

Fig. 7.58

Substituting equation [1] into equation:

$$u = C_1 + C_2 x + C_3 y + C_4 xy \quad ...[2a]$$

yields

$$\left.\begin{aligned} u_1 &= C_1 \\ u_2 &= C_1 + C_2 a \\ u_3 &= C_1 + C_2 u + C_3 b + C_4 ab \\ u_4 &= C_1 + C_3 b \end{aligned}\right\} \quad [2b]$$

Solving for C_1, C_2, C_3, C_4, we get

$$\boxed{C_1 = u_1,\ C_2 = \frac{u_2 - u_1}{a},\ C_3 = \frac{u_4 - u_1}{b},\ C_4 = \frac{u_1 - u_2 + u_3 - u_4}{ab}} \quad ...[3]$$

Substituting equation [3] into equation [2*a*] and rearrange, to get the form of equation

$$u = N_1 u_1 + N_2 u_2 + N_3 u_3 + N_4 u_4 \quad ...[4]$$

$$u = u_1 + \left(\frac{u_2 - u_1}{a}\right) x + \left(\frac{u_4 - u_1}{b}\right) y + \left(\frac{u_1 - u_2 + u_3 - u_4}{ab}\right) xy$$

$$= \frac{ab \cdot u_1 + x \cdot b \cdot u_2 - xbu_1 + y \cdot a \cdot u_4 - y \cdot a \cdot u_1 + xy \cdot u_1 + xyu_3 - xyu_3 - xyu_4}{ab}$$

$$= \frac{(ab - xb - ya + xy)u_1}{ab} + \frac{(xb - xy)u_2}{ab} + \frac{(xy)u_3}{ab} + \frac{(ay - xy)u_4}{ab}$$

$$\Rightarrow \quad u = \frac{(a - x)(b - y)}{ab} u_1 + \frac{x(b - y)}{ab} u_2 + \frac{(xy)}{ab} u_3 + \frac{y(a - x)}{ab} u_4 \quad ...[5]$$

Compare the coefficients of u_1, u_2, u_3 and u_4 in eqns. [4] & [5], we get

$$N_1 = \frac{(a-x)(b-y)}{ab}$$

$$N_2 = \frac{x(b-y)}{ab}$$

$$N_3 = \frac{(xy)}{ab}$$

$$N_4 = \frac{y(a-x)}{ab}$$

are the shape functions ...[6]

Example 1. Derive the shape functions for the six-node element as shown in Fig. 7.59.

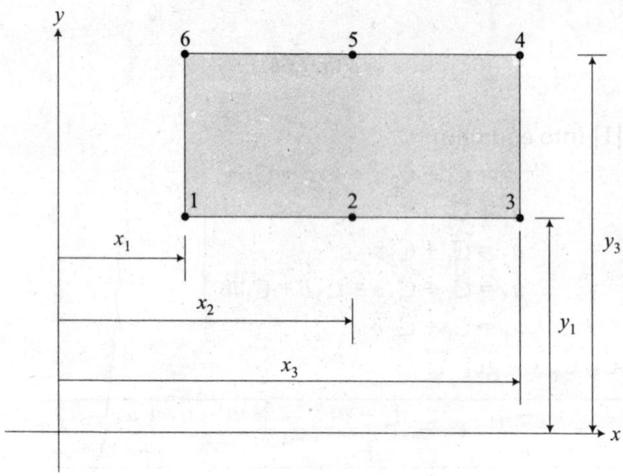

Fig. 7.59

The Lagrange shape functions are constructed as follows:

We can combine two- and 3-node interpolation formulas using Lagrange interpolation polynomial

$$N_1 = \left(\frac{(x-x_2)(x-x_3)}{(x_1-x_2)(x_1-x_3)}\right)\left(\frac{y-y_2}{y_1-y_2}\right)$$

using $N_1 = L_1 x\, L_1 y$.

Similarly,

$$N_2 = \left(\frac{(x-x_1)(x-x_3)}{(x_2-x_1)(x_2-x_1)}\right)\left(\frac{y-y_2}{y_1-y_2}\right)$$

using $N_2 = L_2 x\, L_2 y$.

$$N_3 = \left(\frac{(x - x_1)(x - x_2)}{(x_3 - x_1)(x_3 - x_2)}\right)\left(\frac{y - y_2}{y_1 - y_2}\right)$$

where ($N_3 = L_3x\, L_3y$).

Similarly,
$$N_4 = \left(\frac{(x - x_1)(x - x_2)}{(x_3 - x_1)(x_3 - x_2)}\right)\left(\frac{y - y_1}{y_2 - y_1}\right)$$

$$N_5 = \left(\frac{(x - x_1)(x - x_3)}{(x_2 - x_1)(x_2 - x_3)}\right)\left(\frac{y - y_1}{y_2 - y_1}\right)$$

and
$$N_6 = \left(\frac{(x - x_2)(x - x_3)}{(x_1 - x_2)(x_1 - x_3)}\right)\left(\frac{y - y_1}{y_2 - y_1}\right)$$

Example 2. Derive shape functions for a nine-node element using ξ, η coordinates and the node numbering sequence shown in Fig. 7.60.

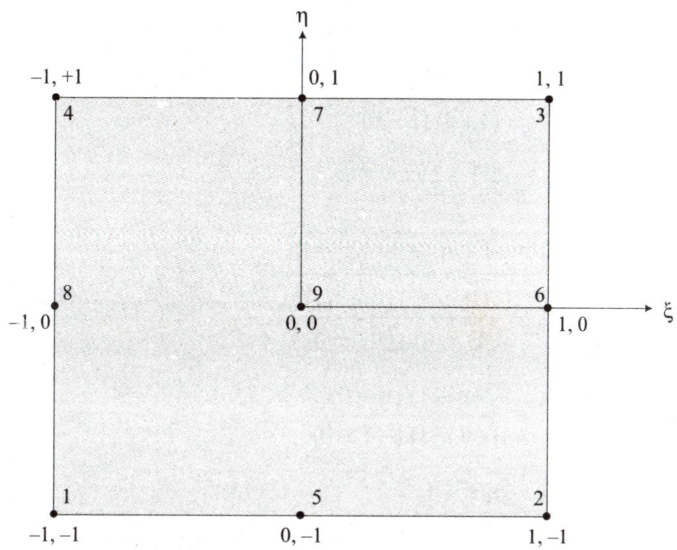

Fig. 7.60

Shape functions for 9-node element are

Node (1) $\boxed{\xi_1 = -1, \xi_2 = 1, \xi_5 = 0}$

$$N_{1\xi} = \frac{(\xi - \xi_2)(\xi - \xi_5)}{(\xi_1 - \xi_2)(\xi_1 - \xi_5)}$$

$$= \frac{(\xi - 1)(\xi - 0)}{(-1 - 1)(-1 - 0)}$$

$$= \frac{\xi(\xi-1)}{2}$$

$$\boxed{\eta_1 = -1, \quad \eta_4 = 1, \quad \eta_8 = 0}$$

$$N_{1\eta} = \frac{(\eta-\eta_4)(\eta-\eta_8)}{(\eta_1-\eta_4)(\eta_1-\eta_8)} = \frac{(\eta-1)(\eta-0)}{(-1-1)(-1-0)} = \frac{\eta(\eta-1)}{2}$$

$$N_1 = N_{1\xi} N_{1\eta}$$

$$= \frac{\xi(\xi-1)}{2} \cdot \frac{\eta(\eta-1)}{2}$$

$$\boxed{N_1 = \frac{\xi\eta(\xi-1)(\eta-1)}{4}}$$

Node (2)

$$\boxed{\xi_2 = 1, \ \xi_1 = -1, \ \xi_5 = 0}$$

$\therefore$
$$N_{2\xi} = \frac{(\xi-\xi_1)(\xi-\xi_5)}{(\xi_2-\xi_1)(\xi_2-\xi_5)}$$

$$= \frac{(\xi+1)(\xi-0)}{(1+1)(1-0)}$$

$$= \frac{\xi(1+\xi)}{2}$$

$$\boxed{\eta_2 = -1 \ | \ \eta_3 = 1 \ | \ \eta_6 = 0}$$

and
$$N_{2\eta} = \frac{(\eta-\eta_3)(\eta-\eta_6)}{(\eta_2-\eta_3)(\eta_2-\eta_6)}$$

$$= \frac{(\eta-1)(\eta-0)}{(-1-1)(-1-0)}$$

$$= \frac{\eta(\eta-1)}{2}$$

$\therefore$
$$N_2 = N_{2\xi}\, \eta_{2\eta}$$

$$= \frac{\xi(\xi+1)}{2} \cdot \frac{(\eta-1)\eta}{2}$$

$$= \frac{\xi\eta(\xi+1)(\eta-1)}{4}$$

Similarly,
$$N_3 = \frac{\xi\eta(\xi+1)(\eta+1)}{4}$$

$$N_4 = \frac{\xi\eta(\xi-1)(\eta+1)}{4}$$

$$N_5 = \frac{-\eta(\xi+1)(\xi-1)(\eta-1)}{2}$$

$$N_6 = \frac{-\xi(\xi+1)(\eta+1)(\eta-1)}{2}$$

$$N_7 = \frac{-\eta(\xi+1)(\xi-1)(\eta+1)}{2}$$

$$N_8 = \frac{-\xi(\xi-1)(\eta+1)(\eta-1)}{2}$$

$$N_9 = (\xi+1)(\xi-1)(\eta+1)(\eta-1)$$

$$N_{9\xi} = \frac{(\xi-\xi_6)(\xi-\xi_8)}{(\xi_9-\xi_6)(\xi_9-\xi_8)}$$

$$= \frac{(\xi-1)(\xi+1)}{(0-1)(0+1)} = \frac{(\xi-1)(\xi+1)}{-1}$$

$$N_{9\eta} = \frac{(\eta-1)(\eta+1)}{(0-1)(0+1)} = \frac{(\eta-1)(\eta+1)}{-1}$$

$$N_9 = N_{9\xi}\,N_{9\eta} = (\xi+1)(\xi-1)(\eta+1)(\eta-1)$$

Example 3. Derive shape functions for an 8-node isoparametric element in ξ, η coordinates using the numbering sequence as shown in Fig. 7.61.

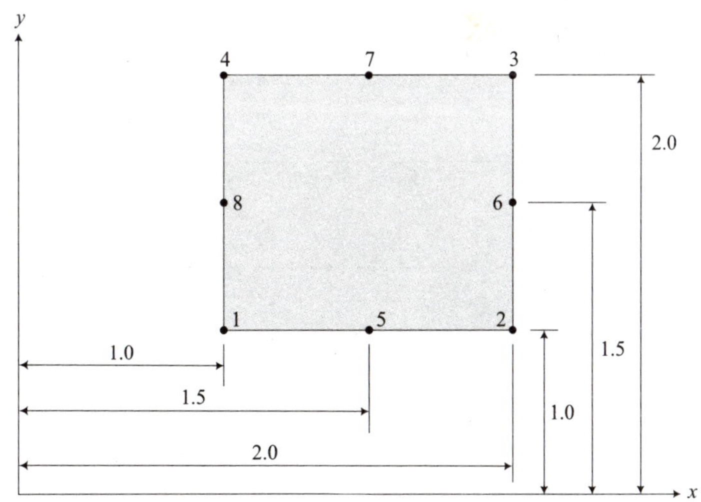

Fig. 7.61

The shape functions are:

$$N_1 = \frac{(1-\xi)(1-\eta)(-\xi-\eta-1)}{4}$$

$$N_2 = \frac{(1+\xi)(1-\eta)(\xi-\eta-1)}{4}$$

$$N_3 = \frac{(1+\xi)(1+\eta)(\xi+\eta-1)}{4}$$

$$N_4 = \frac{(1-\xi)(1+\eta)(-\xi+\eta-1)}{4}$$

$$N_5 = \frac{(1-\xi^2)(1-\eta)}{2}$$

$$N_6 = \frac{(1+\xi)(1-\eta^2)}{2}$$

$$N_7 = \frac{(1-\xi^2)(1+\eta)}{2}$$

$$N_8 = \frac{(1-\xi)(1-\eta^2)}{2}$$

Example 4. Derive the shape functions for a 12-node rectangular element as shown in Fig. 7.62.

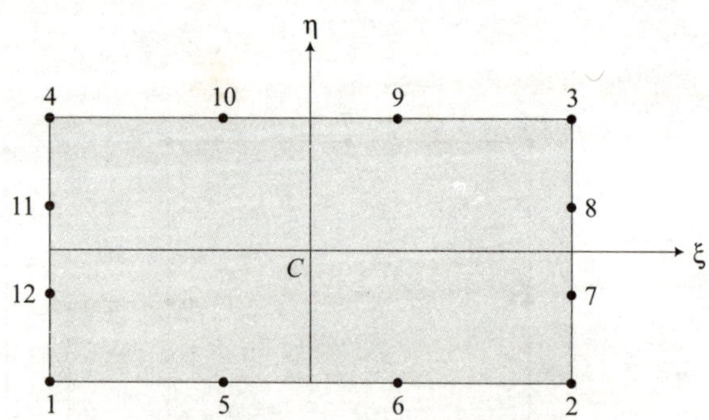

Fig. 7.62 Rectangular element with twelve nodes

The piecewise cubic polynomial is given by

$$u^{(e)}(x, y) = \sum_{i=1}^{12} N_i(x, y)\, u_i \qquad ...[1]$$

where the shape functions N_i in ξ, η coordinates become at corner nodes $i = 1, 2, 3, 4$

$$N_i(\xi, \eta) = \frac{1}{32}(1+\xi\xi_i)(1+\eta\eta_i)\,[9(\xi^2+\eta^2)-10]$$

on midside nodes $\quad i = 7, 8, 11, 12$

$$N_i(\xi, \eta) = \frac{9}{32}(1 + \xi\xi_i)(1 - \eta^2)[1 + 9\eta\eta_i)$$

and on midside nodes $\quad i = 5, 6, 9, 10$

$$N_i(\xi, \eta) = \frac{9}{32}(1 + \eta\eta_i)(1 - \xi^2)(1 + 9\xi\xi_i) \qquad ...[2]$$

EXERCISES

1. Using isoparametric concept, formulate the element stiffness matrix for a unidimensional two-noded element with constant cross-section A and Young's modulus E.

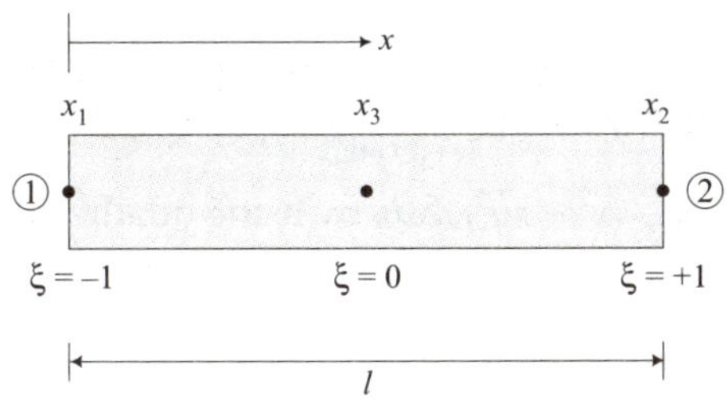

Fig. P7.1

$$\left(\textbf{Ans}: K = \frac{AE}{l}\begin{bmatrix} 1 & -1 \\ -1 & 1 \end{bmatrix}\right)$$

2. (i) Explain the isoparametric concept in finite element analysis.
   (ii) State and explain the three basic laws on which isoparametric concept is developed.
3. Explain the terms isoparametric, subparametric and superparametric elements.
4. Write short notes on
   (a) Uniqueness of mapping of isoparametric elements.
   (b) Jacobian matrix.
5. For the quadratic triangle, derive the stiffness matrix of the element.
6. Detemine the shape functions for isoparametric cubic element.
7. Write the shape functions of the quadratic triangle and verify the shape function properties.
8. Write the interpolation function for a line segment having 3 nodes in terms of natural coordinates.
9. For a cubic triangle, write the shape functions in terms of area coordinates and verify the properties of the shape function .
10. Verify the convergence criteria for the 3-noded triangle where shape functions have been expressed in terms of area coordinates.

11. For isoparametric quadratic element develop expressions for the nodal forces due to the presence of body forces.
12. Construct a displacement field parallel to x-axis (*i.e.*, u) for the element:

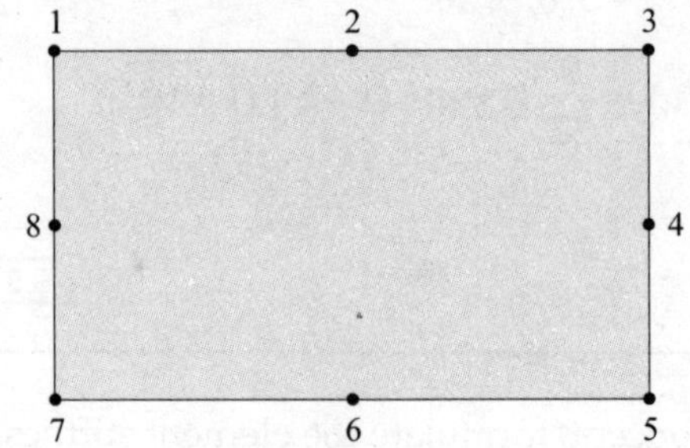

Fig. P7.12

13. For the derivation of the stiffness matrix of the rectangular element by using isoparametric element, evaluate explicitly the stiffness matrix of the element.
14. Evaluate the integral $\iint_A (x^2 + xy^2)\, dxdy$ by using quadrature, where A denotes the region shown in Fig. P8 .14.

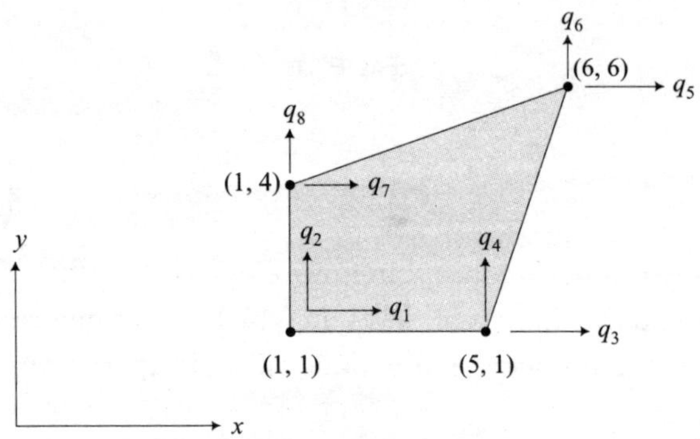

Fig. P7.14

8

Dynamic Analysis

Dynamic analysis of structures by the finite element method are not straightforward as static ones are. However, the method can be extended without much difficulty to the dynamic analysis of structures. The procedure is similar. The mass matrix for dynamic analysis is to be determined.

The original development of the finite element procedure occurred in the aircraft industry in 1956 in an effort to solve the equations for the dynamic response of complex aircraft structures. The finite element technique, as applied to the continuous structure, reduces to a system having multiple degrees of freedom. Dynamic analysis is of two types: free vibration analysis and the response of the structure to forced vibrations; damping may or may not be considered depending on its importance. (Damping in fact varies widely depending upon composition, thermal treatment, stress history , frequency and material processing etc.). For proper accounting of damping loss of energy concept is suitable using a loss factor n^s. (Ref. Lazan B. J. Damping of Materials and Members in Structural Mechanics, 1968, Pergamon Press, New York).

The external force that acts as a function of time in forced vibrations. Depending upon the nature of time variation of this force, the problem can be categorised in four types: harmonic, periodic, transient and random. In structural dynamics, coordinates serve to establish stiffness matrix and assist in the formulation of governing differential equation of motion.

Free Vibration Analysis

Let us assume a solution of the type $x = ae_0^{p.t}$ ...[1]

This gives $[M]\,[X]\,[P^2] + [C]\,[X]\,[P] + [K]\,[X] = 0$ ...[2]

Multiply both sides by $[x]^T$

$[X]^T\,[M]\,[X]\,[P]^2 + [X]\,[C]\,[X]\,[P] + [X]^T\,[K]\,[X] = [0]$ ...[3]

(Put $[X]^T\,[M]\,[X] = [LL]$) ...[4]

and $[X]^I\,[C]\,[X] = [C'_m]$) ...[5]

Hence, $[P^2] + [LL]^{-1}[C'_m]\,[P] + [LL]^{-1}\,[X]^T\,[K]\,[X] = \{0\}$ ...[6]

This orthogonalisation diagonalises $[M]$ and $[K]$ matrices. $[LL]^{-1} \{X\}^T [K] \{X\}$ gives the eigen values straight in the form of a diagonal matrix with diagonal elements as its eigen values for different nodes.

Definition: (Free vibration) If a solid body, such as an engineering structure, is deformed elastically and suddenly released, it tends to vibrate about its equilibrium position. This periodic motion due to the restoring strain energy is called free vibration. The number of cycles per unit time is called the frequency and the maximum displacement from the equilibium position is the amplitude.

Damping Matrix [*C*]

Evaluation of correct damping has always been a problem. There's not an adequate technique to frame it. For working solution damping is considered, per Rayleigh, as a linear function of mass and stiffness matrices.

i.e. $$[C] = A\,[M] + B\,[K] \qquad ...[1]$$

or $$[X]^T [C] [X] = A\,[X^T] [M] [X] + B\,[X]^T [K] [X] \qquad ...[2]$$

or $$[LL]^{-1} [X]^T [C] [X] = [A] + B\,[w^2]$$

or $$[C_m] = [A] + B.\,[w^2] \qquad ...[3]$$

where $$[C_m] = [LL]^{-1} [X]^T [C] [X] \qquad [4]$$

and $$w^2 = [LL]^{-1} [X]^T [K] [X] \qquad [5]$$

or $$[C_m] = [A + B.w^2] \qquad [6]$$

That is , this way [C] also is diagonalised by assumptions. It is taken only as a certain percentage of the eigen values. There are cases like aerodynamic vibrations where a negative damping due to wind acts and the matrix [C] = [C] + [C] may be a full matrix.

Free Undamped Vibration Analysis

The general expression for the damping matrix of the element can be shown to be

$$[C]_e = \int [N]^T \mu [N] dx \qquad ...[1]$$

where μ is some numerical value of the viscous damping . Generation of the damping matrix creates difficulties, mainly because of the damping mechanism and the damping levels in the structures. As such it will not be discussed here.

Now, if all the elements are assembled, the final equations will be

$$[M] \{\ddot{X}\} + [C] \{\ddot{X}\} + [K] \{X\} = \{P\} \qquad ...[2]$$

If free undamped vibration is considered, then equation [2] becomes

$$[M] \{\ddot{X}\} + [K] \{X\} = \{0\} \qquad ...[3]$$

Let us assume a solution of the type $\{X\} = a\, e^{ipt}\, \{\psi\}$ ...[4]

so, $$\{\ddot{X}\} = -ap^2 e^{ipt}\, \{\psi\}$$ ...[5]

Combining eqns. [3] to [5] results

$$-P^2\, [M]\, \{\psi\} + [K]\, \{\psi\} = 0$$

or $$[K]^{-1}\, [M]\, \{\psi\} = w^2\, [I]\, \{\psi\}.$$ ...[6]

where $w^2 = 1/P^2$ and $[I]$ is an identity matrix

Equation [6] can be written as

$$\left[\overline{D}\right] \{\psi\} = w^2\, [I]\, \{\psi\}$$ ...[7]

where $\left[\overline{D}\right] = [K]^{-1}[M]$ ...[8]

Equation [7] is a typical eigen value problem. There are various techniques available for their solution.

A non-trivial solution of equation [7] will be

$$\det \left\| \left[\overline{D}\right] - w^2\, [I] \right\| = 0$$ ...[9]

Formulation: Element mass matrices for 1-D element

Dynamic problems may be solved using Hamilton's principle in mechanics which states that a conservative mechanical system moves in such a way as to minimise the integral involving the potential and kinetic energies. That is, among the set of all admissible configurations of the system the actual motion makes the quantity

$$I = \int_{t_1}^{t_2} [T - U - V]\, dt$$ ...[1]

Stationary assuming that the configuration of the system is prescribed at the limits t_1 and t_2 is normally stated as

$$\delta \int_{t_1}^{t_2} L\, dt = 0$$ ...[2]

where $L = T - U - V$ is called the Lagrangian. T, U, V are the kinetic, the strain energies & potential of the applied loads, respectively.

Mass Matrices

In dynamic analysis of structures, in addition to the stiffness of the structure, a property essential in the dynamic behaviour is the inertia or mass, which is represented by a mass matrix.

The mass matrices can be classified into two types: (*a*) consistent mass and (*b*) lumped mass. Consistent mass is derived from the variational principle and in lumped mass, the mass is assumed to be concentrated at the node.

(a) Consistent Mass

By the application of variational principle (Hamilton's principle), we can define Lagrangian.

$$L = T - U - V \qquad ...[1]$$

Assuming
$$\begin{Bmatrix} \{u\} = [N]\{q\} \\ \{\varepsilon\} = [B]\{q\} \end{Bmatrix} \qquad ...[2]$$

we obtain:

$$[M] = \int_{t_1}^{t_2} \{\delta q\}^T \int_V B^T CB\, dV \{q\} - \{\delta \dot{q}\}^T \int_V \rho\, [N]^T [N]\, dV\, \{\dot{q}\}$$

$$- \{\delta q\}^T \int_V [N]^T \{F\} dV - \{\delta q\}^T \int_{S_1} [N]^T \{T\} ds_1$$

(On integration of the second term by parts w.r.t time), we get

$$[M] = \int_{t_1}^{t_2} \{\delta \dot{q}\}^T \int_V \rho [N]^T [N] dV\, \{q\}\, dt - [\{\delta q\}^T \int_V \rho [N]^T [N]\, dV\, \{\dot{q}\}]_{t_1}^{t_2}$$

$$- \int_{t_1}^{t_2} \{\delta^2 q\}^T \int \rho\, [N]^T [N] dV\, \{\dot{q}\}\, dt \qquad ...[3]$$

By Hamiltons's principle, the tentative displacement configuration must satisfy given conditions at times t_1 and t_2. So the first term of r.h.s. of [3] vanishes. Substituting remaining terms into equation

we find:

$$\delta q\,(t_1) = 0$$

$$\delta q\,(t_2) = 0 \qquad ...[4]$$

$$\Rightarrow [M] = \int_{t_1}^{t_2} \{\delta q\}^T \{\int_V \rho\, [N]^T [N]\, dV\, \{\ddot{q}\} + \int_V [B]^T [C]\, [B]\, dV\, \{q\} - \int_V [N]^T \{F\} dV - \int_{S_1} [N]^T \{T\}\, ds_1\}\, dt = 0 \qquad ...[5]$$

Since q is arbitrary, the expression in the parenthesis must vanish. Hence we obtain equation of motion for the element

$$[M]\,\{\ddot{q}\} + K\,\{q\} = \{Q\} \qquad ...[6]$$

K and Q are the element stiffness matrix and load vector defined in the usual manner. M is the consistent mass defined by

$$[M] = \int_v \rho\, [N]^T [N] dV \qquad ...[7]$$

The general expression for element mass matrix is defined to be

$$m^e = \rho \int_e [N]^T [N]\, dV \qquad ...[8]$$

where ρ is the density of the structure.

(b) Lumped Mass Matrices

The mass matrix for a structural element is where the mass is distributed accurately through the element.

The mass matrix is given by the equation

$$[M] = \int_V \rho [N]^T [N] dV$$

is consistent mass matrix. The opposite being lumped mass matrix, where it is assumed that the total element mass in each direction is distributed equally between the nodes.

Example (1): For the truss element, the lumped mass approach gives a mass matrix:

$$m^e = \frac{\rho A_e l_e}{2}\begin{bmatrix}1 & 0 & 0 & 0\\ 0 & 1 & 0 & 0\\ 0 & 0 & 1 & 0\\ 0 & 0 & 0 & 1\end{bmatrix}$$

Example (2): For the beam element, the lumped mass matrix is :

$$m^e = \frac{\rho A_e l_e}{2}\begin{bmatrix}1 & 0 & 0 & 0\\ 0 & 0 & 0 & 0\\ 0 & 0 & 1 & 0\\ 0 & 0 & 0 & 0\end{bmatrix}$$

Note: The mass of the length, area or volume tributory to a particular node is considered to be concentrated at the node. The lumped mass technique is easier to handle since only diagonal elements are involved. Consistent mass matrices yield more accurate results for flexural elements such as beams.

Forced Vibration

By using Hamilton's principle, dynamic problems may be solved. The functional for this principle is the Lagrangian defined as

$$L = T - U - V \qquad \text{...[1]}$$

where T is the kinetic energy, U is the strain energy and V is the potential of the applied loads. For linear elastic body, the Lagrangian 'L' is expressed in matrix form as:

$$L = \frac{1}{2}\left[\left[\int \rho \{\dot{u}\}^T \{\dot{u}\} - \{\varepsilon\}^T [c]\{\varepsilon\} + 2\{u\}^T \{F\}\right] dV + \int_S \{u\}^T [T] ds\right] \qquad \text{...[2]}$$

Hamilton's principle may be stated as follows

$$\int_{t_1}^{t_2} L dt = 0 \qquad \text{...[3]}$$

Element Mass Matrices for 1-*D* Element

(i) One-dimensional bar element: Element Mass Matrix, m^e

A bar element is a member subjected to axial bloads only. The governing equations describe the behaviour of such a body. A number of bar type finite elements of increasing order such as linear, quadratic and cubic are available.

$$\{u\}^T = [u_1 \; u_2]$$
$$N = [N_1 \; N_2]$$
$$N_1 = \frac{1-\xi}{2}$$
$$N_2 = \frac{1+\xi}{2}$$
$$dx = \frac{l_e}{2} d\xi$$

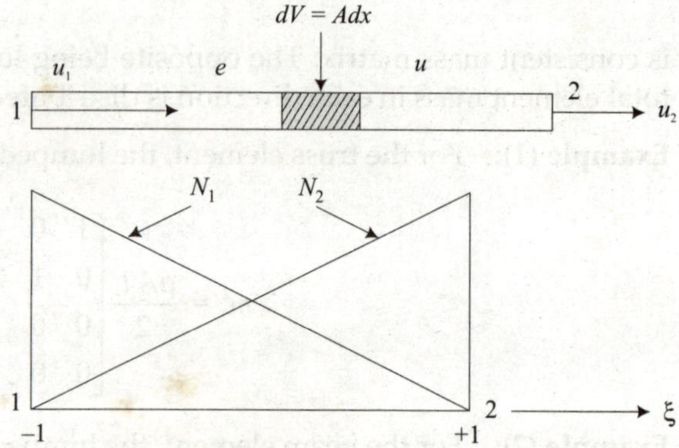

Bar Element

Now, element mass matrix is given by

$$m^e = \rho \int_e N^T N dv$$

$$= \rho \int_e N^T N A dx; \quad (\because dV = Adx)$$

$$= \frac{\rho A_e l_e}{2} \int_{-1}^{+1} N^T N d\xi \quad (\because dx = \frac{l_e}{2} d\xi)$$

$$= \frac{\rho A_e l_e}{2} \int_{-1}^{+1} \begin{bmatrix} N_1^2 & N_1 N_2 \\ N_2 N_1 & N_2^2 \end{bmatrix} d\xi$$

$$= \frac{\rho A_e l_e}{2} \begin{bmatrix} \int_{-1}^{+1} N_1^2 d\xi & \int_{-1}^{+1} N_1 N_2 d\xi \\ \int_{-1}^{+1} N_2 N_1 d\xi & \int_{-1}^{+1} N_2^2 d\xi \end{bmatrix}$$

$$= \frac{\rho A_e l_e}{2} \begin{bmatrix} \frac{1}{4}\int_{-1}^{+1} (1-\xi)^2 d\xi & \frac{1}{4}\int_{-1}^{+1} (1-\xi^2) d\xi \\ \frac{1}{4}\int_{-1}^{+1} (1-\xi^2) d\xi & \frac{1}{4}\int_{-1}^{+1} (1+\xi)^2 d\xi \end{bmatrix}$$

$$= \frac{\rho A_e l_e}{2}\begin{bmatrix} \frac{2}{3} & \frac{1}{3} \\ \frac{1}{3} & \frac{2}{3} \end{bmatrix}$$

$$= \frac{\rho A_e l_e}{6}\begin{bmatrix} 2 & 1 \\ 1 & 2 \end{bmatrix}$$

Remember: For the bar element, the element mass matrix [EMM] is:

$$m^e = \frac{\rho A_e l_e}{6}\begin{bmatrix} 2 & 1 \\ 1 & 2 \end{bmatrix}$$

(ii) Truss Element: Element mass matrix, m^e

The truss elements are parts of a truss structure linked together by pinned joints that transmit only axial force to the elements. Unlike an axial element, a truss element is two-dimensional *i.e.*, it can displace both in x and y directions. Thus, we have two degrees of freedom at each node and totally four degrees of freedom for a truss element. This type of element will be able to carry either a tensile or a compressive force, when used in a system or assemblage.

For the truss element shown in Fig. (8.1),

$$\alpha^T = [\alpha_1 \ \alpha_2]$$

$$u^T = [u_1 \ u_2 \ u_3 \ u_4]$$

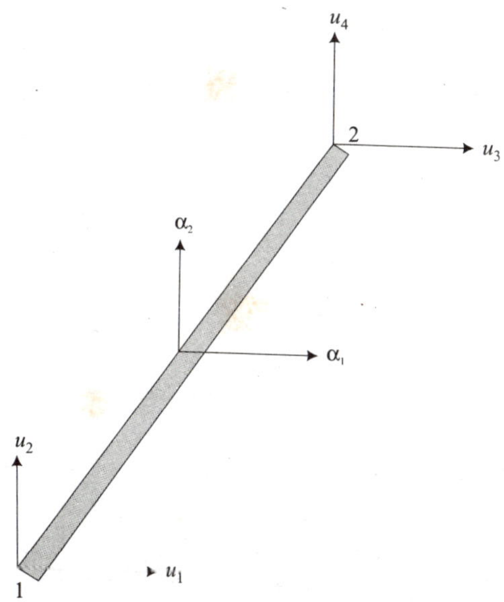

Fig. 8.1 Truss element

$$N = \begin{bmatrix} N_1 & 0 & N_2 & 0 \\ 0 & N_1 & 0 & N_2 \end{bmatrix}$$

where $$N_1 = \frac{1-\xi}{2}\,;\, N_2 = \frac{1+\xi}{2}$$

and ξ is defined from –1 to +1.

Now, element mass matrix is given by

$$m^e = \rho \int_e N^T N dV$$

$$= \rho \int_e N^T N\, A dx \; (\therefore dV = A dx)$$

$$m^e = \frac{\rho A_e l_e}{2} \int_{-1}^{+1} N^T N d\xi\,;\, \{\therefore dx = \frac{l_e}{2} d\xi\}$$

$$= \frac{\rho A_e l_e}{2} \int_{-1}^{+1} \begin{bmatrix} N_1 & 0 \\ 0 & N_1 \\ N_2 & 0 \\ 0 & N_2 \end{bmatrix} \begin{bmatrix} N_1 & 0 & N_2 & 0 \\ 0 & N_1 & 0 & N_2 \end{bmatrix} d\xi$$

$$= \frac{\rho A_e l_e}{2} \int_{-1}^{+1} \begin{bmatrix} N_1^2 + 0 & 0+0 & N_1 N_2 + 0 & 0+0 \\ 0+0 & 0+N_1^2 & 0+0 & 0+N_1 N_2 \\ N_2 N_1 + 0 & 0+0 & N_2^2 + 0 & 0+0 \\ 0+0 & 0+N_2 N_1 & 0+0 & 0+N_2^2 \end{bmatrix} d\xi$$

$$= \frac{\rho A_e l_e}{2} \begin{bmatrix} \frac{1}{4}\int_{-1}^{+1}(1-\xi)^2 d\xi & 0 & \frac{1}{4}\int_{-1}^{+1}(1-\xi^2) d\xi & 0 \\ 0 & \frac{1}{4}\int_{-1}^{+1}(1-\xi)^2 d\xi & 0 & \frac{1}{4}\int_{-1}^{+1}(1-\xi^2) d\xi \\ \frac{1}{4}\int_{-1}^{+1}(1-\xi^2) d\xi & 0 & \frac{1}{4}\int_{-1}^{+1}(1+\xi)^2 d\xi & 0 \\ 0 & \frac{1}{4}\int_{-1}^{+1}(1-\xi^2) d\xi & 0 & \frac{1}{4}\int_{-1}^{+1}(1+\xi)^2 d\xi \end{bmatrix}$$

$$= \frac{\rho A_e l_e}{2} \begin{bmatrix} \frac{2}{3} & 0 & \frac{1}{3} & 0 \\ 0 & \frac{2}{3} & 0 & \frac{1}{3} \\ \frac{1}{3} & 0 & \frac{2}{3} & 0 \\ 0 & \frac{1}{3} & 0 & \frac{2}{3} \end{bmatrix}$$

$$= \frac{\rho A_e l_e}{2 \times 3} \begin{bmatrix} 2 & 0 & 1 & 0 \\ 0 & 2 & 0 & 1 \\ 1 & 0 & 2 & 0 \\ 0 & 1 & 0 & 2 \end{bmatrix}$$

Remember: For the truss element, the element mass matrix (EMM) is :

$$m^e = \frac{\rho A_e l_e}{6} \begin{bmatrix} 2 & 0 & 1 & 0 \\ 0 & 2 & 0 & 1 \\ 1 & 0 & 2 & 0 \\ 0 & 1 & 0 & 2 \end{bmatrix}$$

Element Mass Matrix for Beam Element

For the beam element shown in Fig. 8.2, we make use of the Hermite shape functions $y = Hq$ to find the element mass matrix.

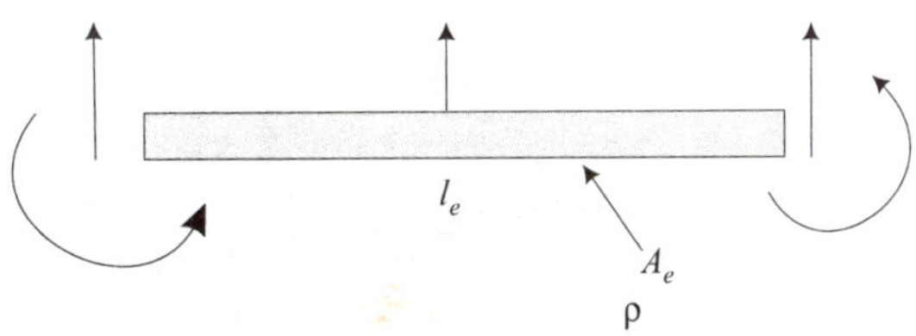

Fig. 8.2 Beam element

$$m^e = \int_{-1}^{+1} H^T H \rho \, A_e \frac{l_e}{2} d\xi \qquad \text{...[1]}$$

$$m^e = \frac{\rho A_e l_e}{2} \int_{-1}^{+1} H^T H \, d\xi$$

or
$$m^e = \frac{\rho A_e l_e}{2}\int_{-1}^{+1}\begin{bmatrix} H_1 \\ \frac{l_e}{2}H_2 \\ H_3 \\ \frac{l_e}{2}H_4 \end{bmatrix}\left[H_1,\ \frac{l_e}{2}H_2,\ H_3,\ \frac{l_e}{2}H_4\right]d\xi$$

We know that

$$H_1 = \frac{1}{4}\,(2 - 3\xi + \xi^3);$$

$$H_2 = \frac{1}{4}\,(1 - \xi - \xi^2 + \xi^3);$$

$$H_3 = \frac{1}{4}\,(2 + 3\xi - \xi^3);$$

$$H_4 = \frac{1}{4}\,(-1 - \xi + \xi^2 + \xi^3);$$

$$\int_{-1}^{+1}\xi^2 d\xi = \frac{2}{3};\ \int_{-1}^{+1}\xi d\xi = 0;\ \int_{-1}^{+1} d\xi = 2$$

$$H_1 = \frac{1}{4}\,(2 - 3\xi + \xi^3);\ H_1^2 = \frac{1}{16}\,(2 - 3\xi + \xi^3)^2$$

$$\int_{-1}^{+1} H_1^2\, d\xi = \left(\frac{26}{35}\right) = \frac{6}{6}\left(\frac{26}{35}\right) = \frac{156}{210}\ \text{(after integration)}$$

$$m^e = \frac{\rho A_e l_e}{2}\int_{-1}^{+1}\begin{bmatrix} H_1^2, & \frac{l_e}{2}H_1H_2, & H_1H_3, & \frac{l_e}{2}H_1H_4 \\ H_1H_2\frac{l_e}{2}, & \left(\frac{l_e}{2}\right)^2 H_2^2, & \frac{l_eH_2H_3}{2}, & \left(\frac{l_e}{2}\right)^2 H_2H_4 \\ H_3H_1, & \frac{l_e}{2}H_3H_2, & H_3^2, & \frac{l_e}{2}H_3H_4 \\ \frac{l_e}{2}H_4H_1, & \left(\frac{l_e}{2}\right)^2 H_4H_2, & \frac{l_e}{2}H_3H_4, & \left(\frac{l_e}{2}\right)^2 H_4^2 \end{bmatrix}d\xi$$

On integrating, we get

$$m^e = \frac{\rho A_e l_e}{420}\begin{bmatrix} 156 & 22l_e & 54 & -13l_e \\ & 4l_e^2 & 13l_e & -3l_e^2 \\ & & 156 & -22l_e \\ \textit{Symmetric} & & & 4l_e^2 \end{bmatrix}$$

Element Mass Matrix for Frame Element

The mass matrix for frame element is given by

$$m^{e'} = \begin{bmatrix} 2a & 0 & 0 & a & 0 & 0 \\ & 156b & 22l_e b & 0 & 54b & -13l_e b \\ & & 4l_e^2 b & 0 & 13l_e b & -3l_e^2 b \\ & & & 2a & 0 & 0 \\ & & & & 156b & -22l_e b \\ Symmetric & & & & & 4l_e^2 b \end{bmatrix} \quad ...[1]$$

where $a = \dfrac{\rho A_e l_e}{6}$ and $b = \dfrac{\rho A_e l_e}{420}$

Mass matrix m^e in the global system :

$$m^e = L^T m^{e'} L \quad ...[2]$$

Element Mass Matrix for Axis Symmetric Triangular Element

A triangular axi-symmetric element has three nodes and at each node there are two degrees of freedom (q_1, q_2) etc. representing the radial and axial displacements u and w respectively.
These displacements can be expressed as

$$\mathbf{u}^T = [u\ w]^T = Nq$$

The vectors q and N are similar to those for the triangular element given by the equation $q^T = [q_1\ q_2 .... q_6]$ and N is the shape function matrix given by

$$N = \begin{bmatrix} N_1 & 0 & N_2 & 0 & N_3 & 0 \\ 0 & N_1 & 0 & N_2 & 0 & N_3 \end{bmatrix}$$

we have,

$$m^e = \int_e \rho N^T N dV = \int_e \rho N^T N \, 2\pi \, r \, dA \quad ...[1]$$

since

$$r = N_1 r_1 + N_2 r_2 + N_3 r_3, \text{ we have}$$

$$m^e = 2\pi\rho \int_e (N_1 r_1 + N_2 r_2 + N_3 r_3) N^T N dA$$

Note that

$$\int_e N_1^3 \, dA = \frac{2A_e}{20}, \int_e N_1^2 N_2 \, dA = \frac{2A_e}{60}, \int_e N_1 N_2 N_3 \, dA = \frac{2A_e}{120}$$

we get

$$m^e = \frac{\pi\rho A_e}{10}\begin{bmatrix} \frac{4}{3}r_1 + 2\bar{r} & 0 & 2\bar{r} - \frac{r_3}{3} & 0 & 2r - \frac{r_2}{3} & 0 \\ & \frac{4}{3}r_1 + 2\bar{r} & 0 & 2\bar{r} - \frac{r_3}{3} & 0 & 2\bar{r} - \frac{r_2}{3} \\ & & \frac{4}{3}r_2 + 2\bar{r} & 0 & 2\bar{r} - \frac{r_1}{3} & 0 \\ & & & \frac{4}{3}r_2 + 2\bar{r} & 0 & 2\bar{r} - \frac{r_1}{3} \\ & & & & \frac{4}{3}r_3 + 2\bar{r} & 0 \\ & & & & & \frac{4}{3}r_3 + 2\bar{r} \end{bmatrix} \quad ...[2]$$

where $\bar{r} = \dfrac{r_1 + r_2 + r_3}{3}$ ...[3]

Element Mass Matrix for CST Element

If at any time constant strains or constant temperature gradient occur within an element, the displacement model should be able to provide for it. The triangular element is most versatile in that it can fit into any irregular boundary – the curved boundaries are to be approximated by a series of straight lines. This element is referred to as constant strain triangle (CST), because it will be seen in the derivation of element stiffness matrix that the strains remain constant at any point within the element.

For the plane stress, plane strain conditions for the CST element is as shown in Fig. 8.3.

$$\phi^T = [u\ v]$$

$$q^T = [q_1\ q_2 \ldots q_6] \quad ...[1]$$

$$N = \begin{bmatrix} N_1 & 0 & N_2 & 0 & N_3 & 0 \\ 0 & N_1 & 0 & N_2 & 0 & N_3 \end{bmatrix}$$

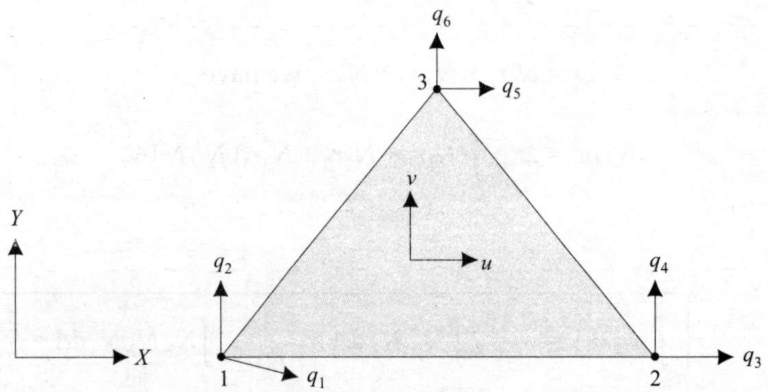

Fig. 8.3 CST element

The element mass matrix m^e is given by

$$m^e = \rho t_e \int_e N^T N dA = \rho t_e \int_e \begin{bmatrix} N_1 & 0 \\ 0 & N_1 \\ N_2 & 0 \\ 0 & N_2 \\ N_3 & 0 \\ 0 & N_3 \end{bmatrix} \begin{bmatrix} N_1 & 0 & N_2 & 0 & N_3 & 0 \\ 0 & N_1 & 0 & N_2 & 0 & N_3 \end{bmatrix} dA$$

$$\int_e N_1^2 dA = \frac{1}{6} A_e, \int_e N_1 N_2 dA = \frac{1}{12} A_e,$$

we have,

$$m^e = \frac{\rho t_e A_e}{12} \begin{bmatrix} 2 & 0 & 1 & 0 & 1 & 0 \\ 0 & 2 & 0 & 1 & 0 & 1 \\ 1 & 0 & 2 & 0 & 1 & 0 \\ 0 & 1 & 0 & 2 & 0 & 1 \\ 1 & 0 & 1 & 0 & 2 & 0 \\ 0 & 1 & 0 & 1 & 0 & 2 \end{bmatrix} \quad ...[2]$$

Element Mass Matrix (m^e) for the Quadrilateral Element for Plane Stress and Plane Strain

The general quadrilateral element is four sided with two degrees of freedom at each node. The nodal displacement vector for this case takes the form,

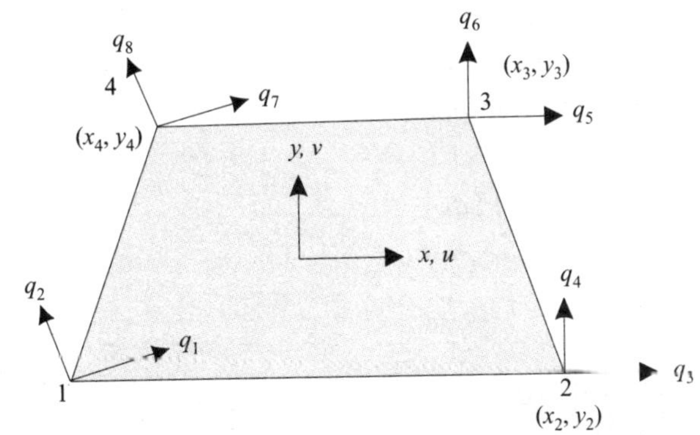

Fig. 8.4

$$q = [q_1\, q_2\, q_3\, q_4\, q_5\, q_6\, q_7\, q_8]^T \qquad \text{...[1]}$$

$$\phi^T = [u\; v]$$

$$N = \begin{bmatrix} N_1 & 0 & N_2 & 0 & N_3 & 0 & N_4 & 0 \\ 0 & N_1 & 0 & N_2 & 0 & N_3 & 0 & N_4 \end{bmatrix} \qquad \text{...[2]}$$

The mass matrix is given by

$$m^e = \rho t_e \int_{-1}^{1}\int_{-1}^{1} N^T N \det J \, d\xi \, d\eta \qquad \text{...[3]}$$

This integral needs to be evaluated by numerical integration.

Element Mass Matrix for the Tetrahedral Element

The element mass matrix of the element is given by

$$m^e = \frac{\rho V_e}{20} \begin{bmatrix} 2 & 0 & 0 & 1 & 0 & 0 & 1 & 0 & 0 & 1 & 0 & 0 \\ & 2 & 0 & 0 & 1 & 0 & 0 & 1 & 0 & 0 & 1 & 0 \\ & & 2 & 0 & 0 & 1 & 0 & 0 & 1 & 0 & 0 & 1 \\ & & & 2 & 0 & 0 & 1 & 0 & 0 & 1 & 0 & 0 \\ & & & & 2 & 0 & 0 & 1 & 0 & 0 & 1 & 0 \\ & & & & & 2 & 0 & 0 & 1 & 0 & 0 & 1 \\ & & & & & & 2 & 0 & 0 & 1 & 0 & 0 \\ & & & & & & & 2 & 0 & 0 & 1 & 0 \\ & & \text{Symmetric} & & & & & & 2 & 0 & 0 & 1 \\ & & & & & & & & & 2 & 0 & 0 \\ & & & & & & & & & & 2 & 0 \\ & & & & & & & & & & & 2 \end{bmatrix}$$

For the tetrahedral element, $\mathbf{U}^T = [u r w]$ ...[2]

The necessary condition for a functional:

$$I = \int_{t_1}^{t_2} L\,dt \qquad \text{...[3]}$$

to be maximum or minimum

is that
$$\frac{\partial L}{\partial q_i} - \frac{d}{dt}\left(\frac{\partial L}{\partial \dot{q}_i}\right) = 0 \qquad \text{...[4]}$$

$$i = 1 \text{ to } n.$$

where t = 1 to t = 2 represent time interval for the state of motion of a body and $L = L(q_1, q_2, ..., q_n; \dot{q}_1, \dot{q}_2, ... \dot{q}_n); \dot{q}_i = \frac{d}{dt} q_i$

Illustration: Consider a spring mass system (Fig. 8.5a). The dynamic equilibrium equation may be written as

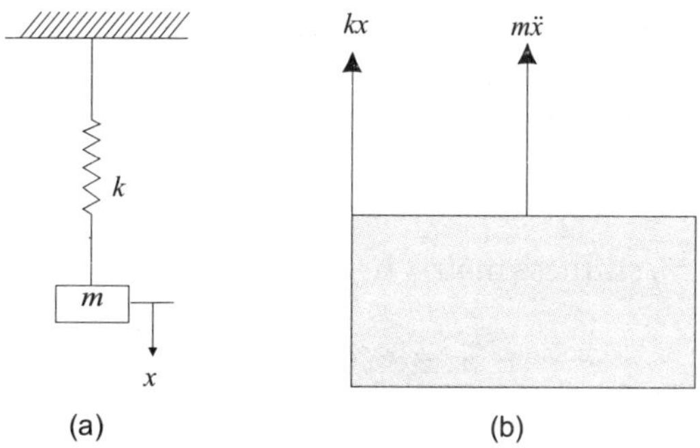

Fig. 8.5

$$m\ddot{x} + kx = 0\,;\ (\ddot{x} = \frac{d^2x}{dt^2}) \qquad ...[1]$$

$$\Rightarrow \qquad \ddot{x} + (^k/_m)x = 0 \qquad ...[2]$$

$$\Rightarrow \qquad \ddot{x} + w^2 x = 0$$

$\Rightarrow$ $x = A \sin wt + B \cos wt$ is the complete solution. ...[3]

[3] is a simple harmonic motion. Knowing the initial conditions, we can find x in terms of the equilibrium equation when there is damping present as is given by Fig. 8.5(b).

$$\boxed{m\,\ddot{x} + C\,\dot{x} + kx = 0;} \qquad ...[4]$$

where C = damping coefficient

or, $\boxed{L = T - \Pi = T - U - V = m\ddot{x} + C\dot{x} + kx = 0}$

Hence, using,

$$\frac{\partial L}{\partial x} - \frac{d}{dt}\left(\frac{\partial L}{\partial \dot{x}}\right) = 0$$

$$\Rightarrow \qquad k - \frac{d}{dt}(C) = 0$$

$$\Rightarrow \qquad k = 0\,;$$

$\therefore$ [4] becomes

$$m\ddot{x} + C\dot{x} = 0$$

or which will be written as:

$$[m \;\; C]\begin{Bmatrix}\ddot{x}\\ \dot{x}\end{Bmatrix} = 0$$

which is of the form

$$M\,\ddot{x} + C\,\dot{x} = 0$$

where M is the mass matrix, k is the stiffness matrix, and $\ddot{x}$ and $\dot{x}$ are vectors representing accelerations and velocities.

Example: Consider a spring mass system (Fig. 8.6). The Lagrangian is given by

$$L = \frac{1}{2}m_1\dot{x}_1^2 + \frac{1}{2}m_2\,\dot{x}_2^2 - \frac{1}{2}c_1x_1^2 - \frac{1}{2}c_2\,(x_2 - x_1)^2 = T - U - V.$$

Compute the mass matrix M, stiffness matrix k.

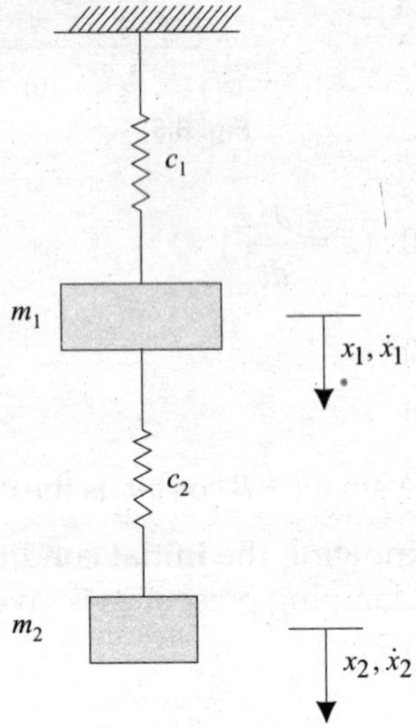

Fig. 8.6

Solution: Given $\qquad L = T - \Pi = T - U - V$

$$= \frac{1}{2}\,m_1\dot{x}_1^2 + \frac{1}{2}\,m_2\dot{x}_2^2 - \frac{1}{2}c_1x_1^2 - \frac{1}{2}c_2\,(x_2 - x_1)^2 = 0$$

using the equations of motion :

$$\frac{\partial L}{\partial x_1} - \frac{d}{dt}\left(\frac{\partial L}{\partial \dot{x}_1}\right) = 0; \text{ and } \frac{\partial L}{\partial x_2} - \frac{d}{dt}\left(\frac{\partial L}{\partial \dot{x}_2}\right) = 0$$

we obtain: $\qquad m_1\ddot{x}_1 + C_1\,x_1 - C_2\,(x_2 - x_1) = 0 \qquad$ ...[1]

and $\qquad m_2\ddot{x}_2 + C_2\,(x_2 - x_1) = 0 \qquad$ ...[2]

Equations [1] and [2] can be written in the matrix form

$$\begin{bmatrix} m_1 & 0 \\ 0 & m_2 \end{bmatrix} \begin{Bmatrix} \ddot{x}_1 \\ \ddot{x}_2 \end{Bmatrix} + \begin{bmatrix} (c_1 + c_2) & -c_2 \\ -c_2 & c_2 \end{bmatrix} \begin{Bmatrix} x_1 \\ x_2 \end{Bmatrix} = 0$$

$$\downarrow \quad \downarrow \quad \downarrow \quad \downarrow$$

$$M \quad \ddot{x} \quad + \quad k \quad x \quad = 0$$

or $$M\ddot{x} + kx = 0 \qquad \text{...[3]}$$

where M is mass matrix, K is the stiffness matrix, $\ddot{x} = \dfrac{d^2x}{dt^2}$ = accelerations, and x = displacements.

Computation of Eigen Value and Eigen Vector

The dynamic analysis of the generalised problem in free vibration consists of evaluating an eigen value λ (= w^2), which is a measure of frequency vibrations, together with the corresponding eigen vector U which satisfies:

$$\boxed{KU = \lambda MU} \qquad \text{...[1]}$$

where K and M are symmetric matrices, also K is positive definite.

Properties

P_1: Corresponding to every eigen vector $U_1, U_2, \ldots U_n$, there exists a set of scalars satisfying

$$KU_i = \lambda_i MU_i \qquad \text{...[2]}$$

(where $0 \le \lambda_1 \le \lambda_2 \le \ldots \le \lambda_n$).

P_2: Eigen vectors are orthogonal w.r.t. both the stiffness and mass matrices.

i.e. $$\boxed{\begin{matrix} U_i^T MU_j = 0\,;\, if\, i \ne j \\ U_i^T KU_j = 0\,;\, if\, i \ne j \end{matrix}} \quad \text{(orthogonal)} \qquad \text{...[3]}$$

The lengths of U_i are given by

$$\boxed{U_i^T MU_i = 1} \quad \text{(normalised)} \qquad \text{...[4]}$$

From equation [4], we find

$$\boxed{U_i^T KU_i = \lambda_i} \qquad \text{...[5]}$$

Procedure

The procedure underlying the computation of eigen values and eigen vectors can be classified as

(1) Characteristic polynomial technique

(2) Guyan reduction.

(1) By definition, we have

$$KU = \lambda MU$$

$$\Rightarrow \quad (K - \lambda M)\,U = 0 \qquad \text{...[1]}$$

$$\Rightarrow \qquad \det (K - \lambda M) = 0$$

$$\Rightarrow \qquad |K - \lambda M| = 0 \text{ (For non-trivial eigen vector)}$$

$|k - \lambda M|$ is called the characteristic polynomial in λ.

Example (1): Determine the eigen values for the truss system shown in Fig. 8.7.

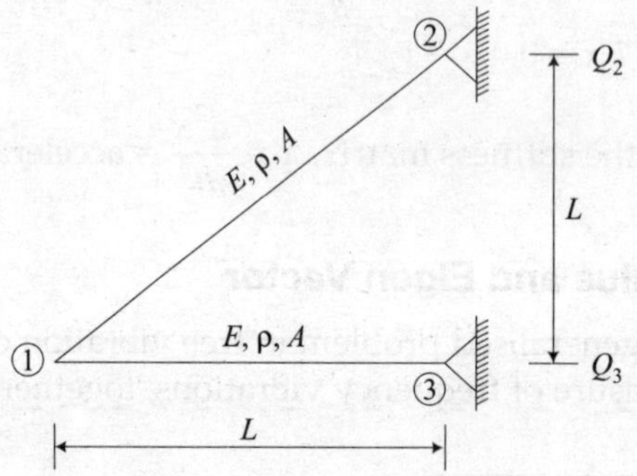

Fig. 8.7 Two bar truss for vibration analysis

Solution: Gathering the stiffness and mass matrix values corresponding to the degrees of freedom U

$$\left\{\frac{EA}{L}\begin{bmatrix} 1+\dfrac{1}{2\sqrt{2}} & \dfrac{1}{2\sqrt{2}} \\ \dfrac{1}{2\sqrt{2}} & \dfrac{1}{2\sqrt{2}} \end{bmatrix} = \lambda\frac{\rho AL}{6}\begin{bmatrix} 2+\sqrt{2} & \sqrt{2} \\ \sqrt{2} & \sqrt{2} \end{bmatrix}\right\}\begin{Bmatrix} U_2 \\ U_3 \end{Bmatrix} = \{0\}$$

Assume $\qquad \lambda = w^2 \rho AL^2/6EA$

$$8\lambda^2 - 6\lambda + 1 = 0$$

On solving, we get

$$\lambda_1 = \frac{1}{4};\ \lambda_2 = \frac{1}{2} \text{ are the eigen values}$$

$$\Rightarrow \qquad \left.\begin{aligned} w_1 &= \frac{\sqrt{1.5}}{L}\sqrt{\frac{E}{\rho}} \\ w_2 &= \frac{\sqrt{3}}{L}\sqrt{\frac{E}{\rho}} \end{aligned}\right\} \text{ are the two natural frequencies.}$$

Example (2): The axial displacement is written in terms of nodal displacements as shown in Fig. 8.8(a).

$$U = [1 - \xi, \xi]\begin{Bmatrix} u_i \\ u_j \end{Bmatrix}$$

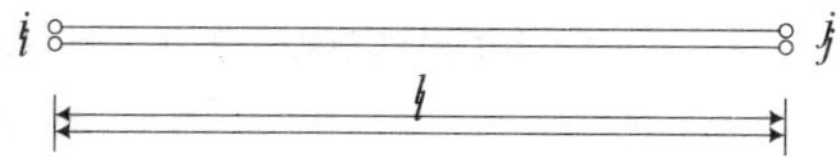

(a) Pin jointed truss element

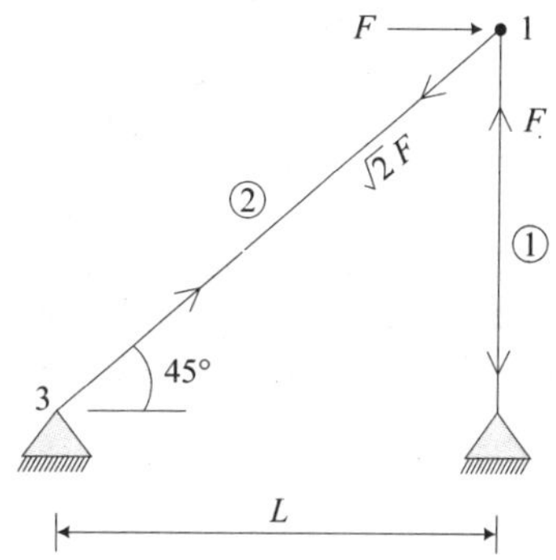

(b) Pin jointed truss.

Fig. 8.8

The consistent mass matrix is given by

$$\rho\int N^T N dV = \rho\int\begin{bmatrix} N_1 \\ N_2 \end{bmatrix}[N_1\ N_2]dV$$

$$= \rho\int\begin{bmatrix} (1-\xi) \\ (\xi) \end{bmatrix}[(1-\xi)\ (\xi)]dV$$

$$[m] = \rho LA\begin{bmatrix} \frac{1}{3} & \frac{1}{6} \\ \frac{1}{6} & \frac{1}{3} \end{bmatrix}$$

The lumped mass is given by

$$m = \rho LA\begin{bmatrix} \frac{1}{2} & 0 \\ 0 & \frac{1}{2} \end{bmatrix}$$

The two-dimensional pin-jointed structure is shown in Fig. 8.8(b).

Element 1

$$[K_1] = \frac{EA}{L}\begin{bmatrix} 0 & 0 \\ 0 & 1 \end{bmatrix}$$

$$[M_1] = \rho LA\begin{bmatrix} \frac{1}{3} & \frac{1}{6} \\ \frac{1}{6} & \frac{1}{3} \end{bmatrix}$$

Element 2

$$[K_2] = \frac{EA}{L}\begin{bmatrix} 0.3536 & 0.3536 \\ 0.3536 & 0.3536 \end{bmatrix}$$

$$[M_2] = \frac{AL}{6}\begin{bmatrix} 1.414 & 0 \\ 0 & 4.242 \end{bmatrix}$$

Let us assume $U_i = A_i \sin wt$, the det. is given by

$$\det |K - w^2 M| = 0$$

$$\Rightarrow \qquad \frac{EA}{L}\left\|\begin{bmatrix} 0.3536 & 0.3536 \\ 0.3536 & 1.3536 \end{bmatrix}\right\| - \frac{\rho ALw^2}{6}\left\|\begin{bmatrix} 3.414 & 1 \\ 1 & 6.242 \end{bmatrix}\right\| = 0$$

Assuming $\qquad \lambda = \rho\, AL^2w^2/6EA$

The characteristic equation is written as

$$20.31\,\lambda^2 - 6.1128\,\lambda + 0.3536 = 0$$

On solving, we get

$$\lambda_1 = 0.079\,;\ \lambda_2 = 0.221$$

Hence, the fundamental frequency is given by

$$w = \frac{0.648}{L}\sqrt{E/\rho}$$

Example (3): Consider a simply supported beam shown in Fig. 8.9 represented by a single element. The boundary conditions are $V_1 = V_2 = 0$ and so the possible degrees of freedom are θ_1 and θ_2 only. Obtain K and M matrices of order 2×2.

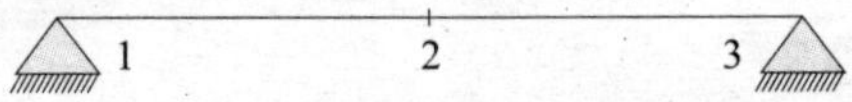

Fig. 8.9 Free vibration of a simply supported beam

Solution: We have $\| [K] - w^2[M] \| = 0$

(using consistent mass)

Define: $\lambda = \rho A L^2 w^2/420\,EI$

We get

$$\frac{EI}{L}\begin{Vmatrix} L(4-4\lambda) & L(2+3\lambda) \\ L(2+3\lambda) & L(4-4\lambda) \end{Vmatrix} = 0$$

On solving

$$\Rightarrow \qquad \lambda_1 = \frac{2}{7}\,;\ \lambda_2 = 6;$$

These natural frequencies are given by

$$w_1 = 10.9544\,\sqrt{EI/\rho AL^4}$$

$$w_2 = 50.2\,\sqrt{EI/\rho AL^4}$$

and the exact solutions are found to be

$$\left.\begin{aligned} w_1 &= 9.8696\,\sqrt{EI/\rho AL^4} \\ w_2 &= 39.478\,\sqrt{EI/\rho AL^4} \end{aligned}\right\} \qquad \text{...[1]}$$

For a two-element modeling of the whole beam and consistency symmetry, we get

$$\begin{Vmatrix} L^2(4-4\lambda) & -L(6+13\lambda)/2 \\ -L(6+13\lambda)/2 & (12-156\lambda) \end{Vmatrix} = 0$$

The characteristic equation is

$$455\,\lambda^2 - 828\,\lambda + 12 = 0$$

Solving

$$\left.\begin{aligned} w_1 &= 9.9086\sqrt{EI/\rho AL^4} \\ w_2 &= 43.818\sqrt{EI/\rho AL^4} \end{aligned}\right\} \qquad \text{...[2]}$$

The values obtained in [2] are an improvement over those obtained in [1].

Suppose a uniform shaft having one end fixed, the other end free is divided into two equal elements and the node numbers are as shown in Fig. 8.10.

Then, the complete mass matrix [M] being

$$[M] = \frac{\bar{I}L}{12}\begin{bmatrix} 2 & 1 \\ 1 & 4 \end{bmatrix}$$

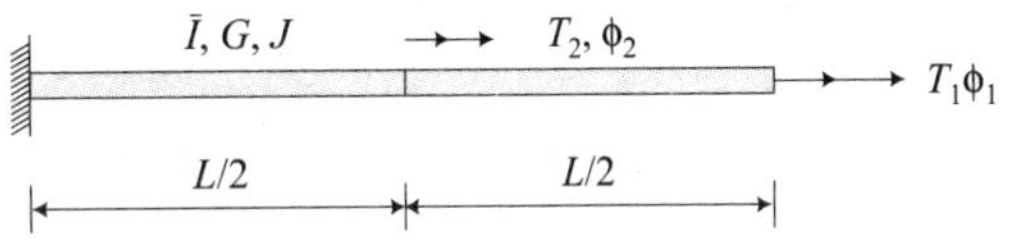

Fig. 8.10

and

$$[K] = \frac{2GJ}{L}\begin{bmatrix} 1 & -1 \\ -1 & 2 \end{bmatrix}$$

$\therefore$

$$\left[\bar{D}\right] = \frac{L}{2GJ}\begin{bmatrix} 2 & 1 \\ 1 & 1 \end{bmatrix}\frac{\bar{I}L}{12}\begin{bmatrix} 2 & 1 \\ 1 & 4 \end{bmatrix} = \frac{\bar{I}L^2}{24GJ}\begin{bmatrix} 5 & 6 \\ 3 & 5 \end{bmatrix}$$

Equation $\quad \det\left\|\left[\bar{D}\right] - w^2[I]\right\| = 0$

then becomes after putting $\quad d = \dfrac{\bar{I}L_1^2}{24GJ}$

Let

$$\begin{Vmatrix} 5d - w^2 & 6d \\ 3d & 5d - w^2 \end{Vmatrix} = 0$$

or, $\quad (5d - w^2)^2 = 18d^2$

or, $\quad 5d - w^2 = \pm 4.2426d$

or, $\quad w^2 = .7574d,\ 9.2426d.$

$\therefore$ The natural frequencies are

$$p^2 = 2.597\frac{GJ}{\bar{I}L^2};$$

$$p^2 = 31.687\frac{GJ}{\bar{I}L^2}$$

The computation of eigen values and eigen vectors for a set of equation is far more expensive than the computation of displacements for a set of stiffness equations of the same size. Hence, the reduction of the size of the eigen value equations of motion in the free vibration is far more essential than the reduction of the stiffness equations in static analysis.

Assume the equations of motion may be arranged and partitioned as

$$\left(\begin{bmatrix} k_{rr} & k_{ro} \\ k_{or} & k_{oo} \end{bmatrix} - w^2 \begin{bmatrix} M_{rr} & M_{ro} \\ M_{or} & M_{oo} \end{bmatrix}\right)\begin{Bmatrix} Q_r \\ Q_o \end{Bmatrix} = \begin{Bmatrix} F_r \\ 0 \end{Bmatrix} \quad \text{...[1]}$$

where Q_r = retained set and Q_o = omitted set.
Disregarding the inertia force, we have

$$\begin{bmatrix} k_{rr} & k_{ro} \\ k_{or} & k_{oo} \end{bmatrix}\begin{Bmatrix} Q_r \\ Q_o \end{Bmatrix} = \begin{Bmatrix} F_r \\ 0 \end{Bmatrix} \quad \text{...[2]}$$

We can solve for Q_o in terms of Q_r as:

$$\boxed{Q_o = -k_{oo}^{-1} k_{or} Q_r} \quad \text{...[3]}$$

and substituting in equation [2], we get

$$(k_{rr} - k_{ro} k_{oo}^{-1} k_{or})\, Q_r = F_r \quad \text{...[4]}$$

The above equation can also be arrived at by using transformation

$$\begin{Bmatrix} Q_r \\ Q_o \end{Bmatrix} = \begin{bmatrix} [I] \\ -[k_{oo}]^{-1} \quad [k_{or}] \end{bmatrix}\{Q_r\} = [T]\,\{Q_r\}$$

$$= [T]^T [k]\,[T]\{Q_r\} = \{F_r\} \quad \text{...[5]}$$

or $$[k_r]\,\{Q_r\} = [F_r] \quad \text{...[6]}$$

where $$[k_r] = [T]^T\,[k][T] \quad \text{...[7]}$$

Using the same procedure the mass matrix corresponding to retained set (Q_r) is

$$[M_r] = [T]^T\,[M]\,[T] \quad \text{...[8]}$$

and eigen value problem now becomes

$$([k_r] - w^2\,[M_r]\,\{Q_r\} = \{F_r\} = \{0\}$$

This reduction method is due to GUYAN.

Example: Determine the lowest eigen value for the following equations using retained set Q_r and omitted set Q_o of degrees of freedom.

$$\left\{\left[\begin{array}{cc|c} 3 & 1 & 0 \\ 2 & 4 & -1 \\ \hline 0 & -1 & 1 \end{array}\right] - \lambda \begin{bmatrix} 1 & 0 & 0 \\ 0 & 1 & 0 \\ 0 & 0 & 1 \end{bmatrix}\right\}\begin{Bmatrix} X_1 \\ X_2 \\ X_3 \end{Bmatrix} = \{0\}$$

Solution: Let us denote $\begin{Bmatrix} X_1 \\ X_2 \end{Bmatrix}$ as $\{X_r\}$; $X_3 = X_o$

$$[T] = \begin{bmatrix} 2 & 0 \\ 0 & 1 \\ -k_o^{-1} & k_{or} \end{bmatrix} \text{ where } \begin{matrix} [k_{rr}] = \begin{bmatrix} 3 & 1 \\ 1 & 4 \end{bmatrix} \\ [k_{ro}] = \begin{bmatrix} 0 \\ -1 \end{bmatrix} \end{matrix}$$

$$[k_{or}] = [0\ -1];\ [k_{oo}] = 1$$

$$[T] = \begin{bmatrix} 1 & 0 \\ 0 & 1 \\ 0 & -1 \end{bmatrix};\ T^T kT = \begin{bmatrix} 3 & 1 \\ 1 & 5 \end{bmatrix} = [k_r];$$

$$T^T MT = \begin{bmatrix} 1 & 0 \\ 0 & 2 \end{bmatrix} = [M_r]$$

$$\det \left| \begin{bmatrix} 3 & 1 \\ 1 & 5 \end{bmatrix} - \lambda \begin{bmatrix} 1 & 0 \\ 0 & 2 \end{bmatrix} \right| = 0$$

$$2\lambda^2 - 11\lambda + 14 = 0$$

$$\Rightarrow \qquad \lambda = 2 \text{ ; and } \lambda = 3.5$$

whereas the solutions using the whole matrix are 1.58 and 2.1.

9

Axi-Symmetric Elements

9.1 INTRODUCTION

Many problems in stress analysis which are of practical interest involve solids of resolution (axi-symmetric solids) subject to axially symmetric loading. All structures are three-dimensional. However, we render them two-dimensional for the purpose of analysis using some conditions. The mathematical problem of axi-symmetric stress analysis is very similar to that of plane stress/strain since once again the situation is two-dimensional formulations. A primary difference is that the volume integration must be replaced with $dV = 2\pi r dr dz$.

For example, a thin structure is treated as the plane stress problem whereas a long cylindrical body is assumed to be in plane strain conditions. But there are certain structures, which by their geometry become one dimensionless and make the analysis simpler. These structures have symmetry about an axis and are called axi-symmetric structures.

Hence, a three-dimensional axi-symmetric structure reduces to a simple two-dimensional problem. Axial symmetry is a condition where the properties of the radial sections of a body of revolution are identical. Normally, the axis of symmetry will be identified with z-axis of the Cartesian coordinate system. And any point in the plane perpendicular to the z-axis will be represented by the polar coordinates (r, θ). Because of symmetry all deformations and stresses are independent of coordinate Q. Thus the problem is reduced as two-dimensional in the coordinates r and z.

9.2 AXI-SYMMETRIC FORMULATION

In order to consider the axi-symmetric case of the total potential energy, the total potential energy expression can be modified to consider the axis-symmetric case and is

$$\Pi = \frac{1}{2}\int_0^{2\pi}\int_{Area} \sigma^T \varepsilon dA d\theta - \int_0^{2\pi}\int_{Area} u^T f r dA d\theta - \int_0^{2\pi}\int_L u^T T r dl d\theta - \sum_i u_i^T P_i \qquad ...[1]$$

where the various types of loads act as shown in Fig. 9.1.

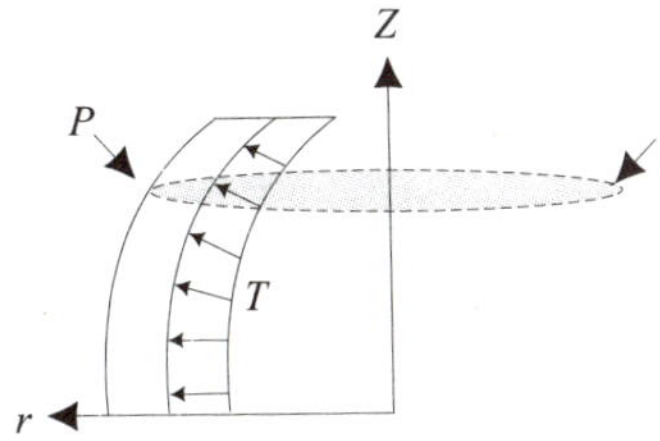

Fig. 9.1

In the above equation [1], P_i is the concentrated force acting over a circumference of a circle as shown by the dotted circle and not simply at a point.

Examples of axi-symmetric structural elements

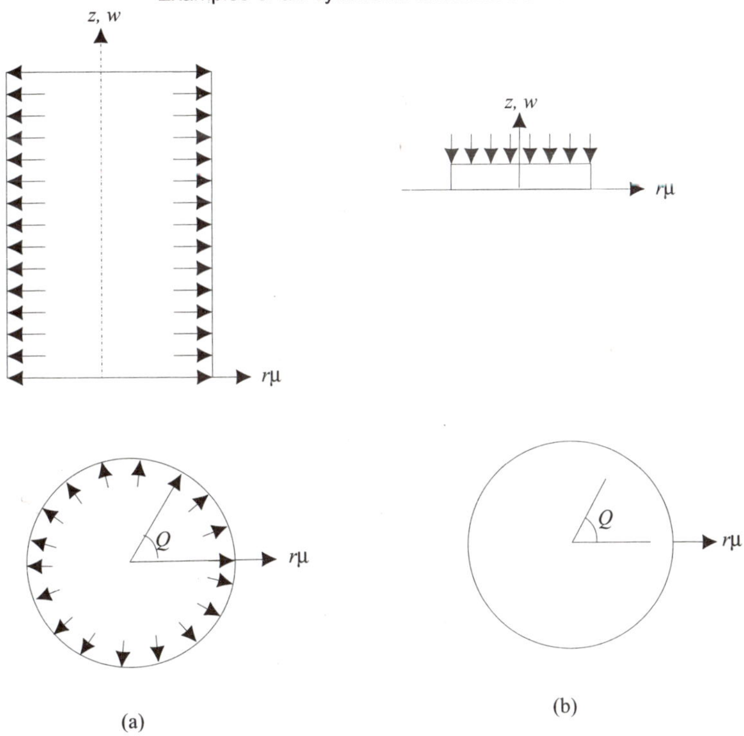

Fig. 9.2

If such elements are subjected to axis-symmetric loadings like uniform internal or external pressures, then there exits symmetry about any axis. Because of symmetry, the stress components are independent of the angular (θ) coordinate.

In these cases $r = r_{r\theta} = r_{\theta z} = \tau_{\theta z} = 0$

i.e., all derivatives w.r.t θ vanish.

An axi-symmetric element is shown in Fig. 9.3.

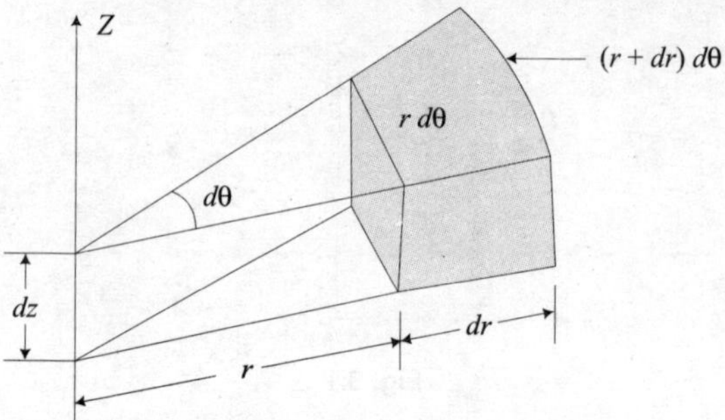

Fig. 9.3

The deformation of the element can be considered in the r-θ as well as r-z planes shown in Fig. 9.4. The radial strain, ε , is given by

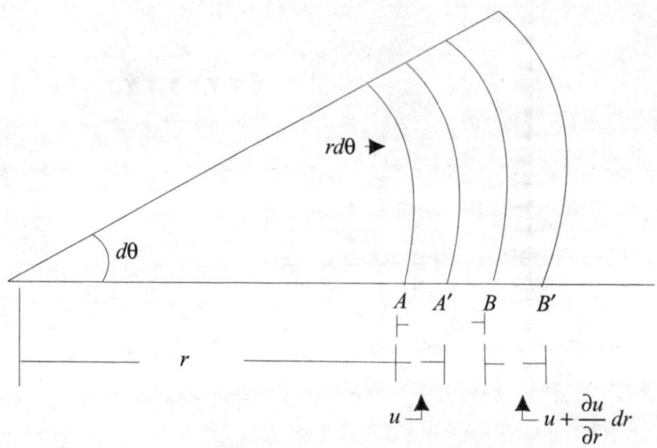

Fig. 9.4 r-θ plane

$$\varepsilon_r = \frac{u + \dfrac{\partial u}{\partial r} dr - u}{dr} = \frac{\partial u}{\partial r} \qquad \text{...[1]}$$

The tangential strain, ε_θ is given by

$$\varepsilon_\theta = \frac{(r + u)d\theta - rd\theta}{r\, d\theta} = \frac{u}{r} \qquad \text{...[2]}$$

In the r-z plane, the deformation will be as shown in Fig. 9.5.

$$\text{The axial strain } \varepsilon_z = \frac{w + \dfrac{\partial w}{\partial z}dz - w}{dz} = \frac{\partial w}{\partial z} \quad \text{...[3]}$$

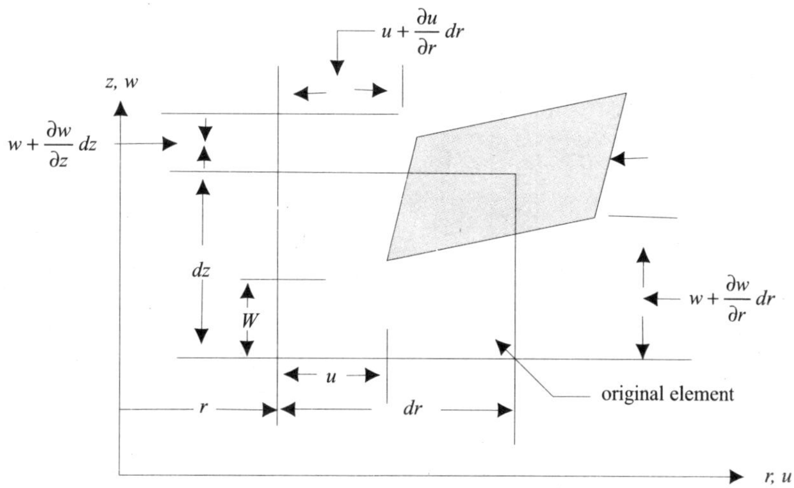

Fig. 9.5

$$\text{The shear strain, } \gamma_{rz} = \frac{\partial u}{\partial z} + \frac{\partial w}{\partial r} \quad \text{...[4]}$$

Thus, for the axi-symmetric case the strain vector is given by

$$\varepsilon = \begin{Bmatrix} \varepsilon_r \\ \varepsilon_\theta \\ \varepsilon_z \\ \gamma_{rz} \end{Bmatrix} = \begin{Bmatrix} \dfrac{\partial u}{\partial r} \\ \dfrac{u}{r} \\ \dfrac{\partial w}{\partial z} \\ \dfrac{\partial u}{\partial z} + \dfrac{\partial w}{\partial r} \end{Bmatrix} = \begin{bmatrix} \dfrac{\partial u}{\partial r} & \dfrac{u}{r} & \dfrac{\partial w}{\partial z} & \dfrac{\partial u}{\partial z} + \dfrac{\partial w}{\partial r} \end{bmatrix}^T \quad \text{...[5]}$$

Stress-strain relation for axi-symmetric elements

From Hooke's law we find the strains are related to stress as

$$\varepsilon_r = \frac{\sigma_r}{E} - \mu\frac{\sigma_\theta}{E} - \mu\frac{\sigma_z}{E} \quad \text{...[6a]}$$

$$\varepsilon_z = \frac{\sigma_z}{E} - \mu\frac{\sigma_\theta}{E} - \mu\frac{\sigma_r}{E} \quad \text{...[6b]}$$

$$\varepsilon_\theta = \frac{\sigma_\theta}{E} - \mu\frac{\sigma_r}{E} - \mu\frac{\sigma_z}{E} \quad \text{...[6c]}$$

$$\gamma_{rz} = \frac{2(1+\mu)}{E}\tau_{rz} \quad \text{...[6d]}$$

The above can be put in the form

$$\sigma = \mathbf{D}\,\varepsilon \quad \text{...[7]}$$

where,

$$\sigma = \begin{Bmatrix} \sigma_r \\ \sigma_z \\ \sigma_\theta \\ \tau_{rz} \end{Bmatrix} = [\sigma_r\ \sigma_z\ \sigma_\theta\ \tau_{rz}]^T \quad \text{...[8]}$$

and

$$\mathbf{D} = \frac{E}{(1+\mu)(1-2\mu)}\begin{bmatrix} 1-\mu & \mu & \mu & 0 \\ \mu & 1-\mu & \mu & 0 \\ \mu & \mu & 1-\mu & 0 \\ 0 & 0 & 0 & \frac{1-2\mu}{2} \end{bmatrix} \quad \text{...[9]}$$

9.3 FORMULATION OF AXI-SYMMETRIC TRIANGULAR ELEMENT

The region R is divided into elements as shown in Fig. 9.6. Here each element is a triangle with nodes i, j, k numbered anticlockwise. Though each element is completely represented by the area in the rz plane, in reality, it is a ring shaped solid of revolution obtained by revolving the triangle about the z-axis.

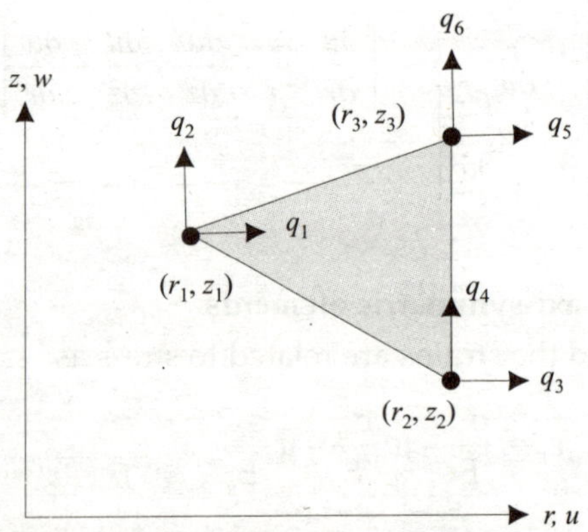

Fig. 9.6

An arbitrary element (*e*) is shown in Fig. 9.7, which is an axisymmetric triangular element and has three nodes and at each node there are two degrees of freedom (q_1, q_2) etc., representing the radial and axial displacements u and w respectively. We note here that the odd subscripts show the radial displacement and even refer to the axial displacements. These displacements can be expressed as:

$$\mathbf{u} = [u \;\; w]^T = \mathbf{N}q$$

where N is a shape function given by

$$N = \begin{bmatrix} N_1 & 0 & N_2 & 0 & N_3 & 0 \\ 0 & N_1 & 0 & N_2 & 0 & N_3 \end{bmatrix} \qquad ...[2]$$

and
$$q = \begin{Bmatrix} q_1 \\ q_2 \\ q_3 \\ q_4 \\ q_5 \\ q_6 \end{Bmatrix} \qquad ...[3]$$

Note that $N_1 = 1$ at node 1 and $N_1 = 0$ at nodes 2 and 3 etc. We can take for the shape functions the following expressions $N_1 = \xi$; $N_2 = \eta$; $N_3 = 1 - \xi - \eta$.

Hence, we can write the expressions for the displacements u and w as

$$\begin{cases} u = \xi q_1 + \eta q_3 + (1 - \xi - \eta) q_5 & ...[4u] \\ w = \xi q_2 + \eta q_4 + (1 - \xi - \eta) q_6 & ...[4b] \end{cases}$$

Jacobian is nothing but the matrix containing the derivatives of the global coordinates with respect to the natural coordinates of a particular element. Thus

$$\begin{Bmatrix} \dfrac{\partial u}{\partial \xi} \\ \dfrac{\partial u}{\partial \eta} \end{Bmatrix} = \begin{Bmatrix} \dfrac{\partial r}{\partial \xi} & \dfrac{\partial z}{\partial \xi} \\ \dfrac{\partial r}{\partial \eta} & \dfrac{\partial z}{\partial \eta} \end{Bmatrix} \begin{Bmatrix} \dfrac{\partial u}{\partial r} \\ \dfrac{\partial u}{\partial z} \end{Bmatrix} = J \begin{Bmatrix} \dfrac{\partial u}{\partial r} \\ \dfrac{\partial u}{\partial z} \end{Bmatrix} \qquad ...[5]$$

Similarly,

$$\begin{Bmatrix} \dfrac{\partial w}{\partial \xi} \\ \dfrac{\partial w}{\partial \eta} \end{Bmatrix} = \begin{Bmatrix} \dfrac{\partial r}{\partial \xi} & \dfrac{\partial z}{\partial \xi} \\ \dfrac{\partial r}{\partial \eta} & \dfrac{\partial z}{\partial \eta} \end{Bmatrix} \begin{Bmatrix} \dfrac{\partial w}{\partial r} \\ \dfrac{\partial w}{\partial z} \end{Bmatrix} = J \begin{Bmatrix} \dfrac{\partial w}{\partial r} \\ \dfrac{\partial w}{\partial z} \end{Bmatrix} \qquad ...[6]$$

Thus,

$$\mathbf{J} = \begin{Bmatrix} \dfrac{\partial r}{\partial \xi} & \dfrac{\partial z}{\partial \xi} \\ \dfrac{\partial r}{\partial \eta} & \dfrac{\partial z}{\partial \eta} \end{Bmatrix} = \begin{bmatrix} (r_1 - r_3) & (z_1 - z_3) \\ (r_2 - r_3) & (z_1 - z_3) \end{bmatrix} \qquad ...[7]$$

The inverse relation is given as

$$\begin{Bmatrix} \dfrac{\partial u}{\partial r} \\ \dfrac{\partial u}{\partial z} \end{Bmatrix} = \mathbf{J}^{-1} \begin{Bmatrix} \dfrac{\partial u}{\partial \xi} \\ \dfrac{\partial u}{\partial \eta} \end{Bmatrix} \text{ and } \begin{Bmatrix} \dfrac{\partial w}{\partial r} \\ \dfrac{\partial w}{\partial z} \end{Bmatrix} = \mathbf{J}^{-1} \begin{Bmatrix} \dfrac{\partial w}{\partial \xi} \\ \dfrac{\partial w}{\partial \eta} \end{Bmatrix} \qquad ...[8]$$

where the inverse J^{-1} is given by

$$\mathbf{J}^{-1} = \frac{1}{\det J} \begin{bmatrix} (z_2 - z_3) & -(z_1 - z_3) \\ -(r_2 - r_3) & (r_1 - r_3) \end{bmatrix} \qquad ...[9]$$

Strain–displacement matrix (B) for the axi-symmetric element.

From equations
$$\varepsilon_r = \frac{u + \dfrac{\partial u}{\partial r} dr - u}{dr} = \frac{\partial u}{\partial r},$$

$$\varepsilon_\theta = \frac{(r+u)d\theta - rd\theta}{rd\theta} = \frac{u}{r},$$

$$\varepsilon_z = \frac{w + \dfrac{\partial w}{\partial z} dz - w}{dz} = \frac{\partial u}{\partial z}$$

$$\gamma_{rz} = \frac{\partial u}{\partial z} + \frac{\partial u}{\partial r}$$ and equation [8] we find that:

$$\varepsilon_r = \frac{\partial u}{\partial r} = \frac{1}{\det J} [(z_2 - z_3)(q_1 - q_5) - (z_1 - z_3)(q_3 - q_5)]$$

$$\varepsilon_r = \frac{\partial w}{\partial z} = \frac{1}{\det J} [-(r_2 - r_3)(q_2 - q_6) + (r_1 - r_3)(q_4 - q_6)]$$

$$\varepsilon_\theta = \frac{u}{r} = \frac{\xi q_1 + \eta q_3 + (1 - \xi - \eta) q_5}{r}$$

$$\gamma_{rz} = \frac{\partial u}{\partial z} + \frac{\partial w}{\partial r}$$

$$= \frac{-(r_2 - r_3)(q_1 - q_5) + (r_1 - r_3)(q_3 - q_5) + (z_2 - z_3)(q_2 - q_6) - (z_1 - z_3)(q_4 - q_6)}{\det J}$$

The above equations can be written in the form

$$\begin{Bmatrix} \varepsilon_r \\ \varepsilon_z \\ \varepsilon_{\theta r} \\ \gamma_{rz} \end{Bmatrix} = \begin{bmatrix} \frac{z_2 - z_3}{\det J} & 0 & \frac{z_3 - z_1}{\det J} & 0 & \frac{z_1 - z_2}{\det J} & 0 \\ 0 & \frac{r_3 - r_2}{\det J} & 0 & \frac{r_1 - r_3}{\det J} & 0 & \frac{r_2 - r_1}{\det J} \\ \frac{\xi}{r} & 0 & \frac{\eta}{r} & 0 & \frac{1-\xi-\eta}{r} & 0 \\ \frac{r_3 - r_2}{\det J} & \frac{z_2 - z_3}{\det J} & \frac{r_1 - r_3}{\det J} & \frac{z_3 - z_1}{\det J} & \frac{r_2 - r_1}{\det J} & \frac{z_1 - z_2}{\det J} \end{bmatrix}$$

or simply as

$$\varepsilon = \mathbf{B}q$$

where the strain-displacement matrix **B** has the form

$$B = \begin{bmatrix} \frac{z_2 - z_3}{\det J} & 0 & \frac{z_3 - z_1}{\det J} & 0 & \frac{z_1 - z_2}{\det J} & 0 \\ 0 & \frac{r_3 - r_2}{\det J} & 0 & \frac{r_1 - r_3}{\det J} & 0 & \frac{r_2 - r_1}{\det J} \\ \frac{\xi}{r} & 0 & \frac{\eta}{r} & 0 & \frac{1-\xi-\eta}{r} & 0 \\ \frac{r_3 - r_2}{\det J} & \frac{z_2 - z_3}{\det J} & \frac{r_1 - r_3}{\det J} & \frac{z_3 - z_1}{\det J} & \frac{r_2 - r_1}{\det J} & \frac{z_1 - z_2}{\det J} \end{bmatrix}$$

Potential Energy Approach

We have from potential energy expression:

$$\Pi = \frac{1}{2}\int_0^{2\pi}\int_{Area} \sigma^T \varepsilon dA d\theta - \int_0^{2\pi}\int_{Area} u^T f r dA d\theta - \int_0^{2\pi}\int_L u^T T r dl d\theta - \sum_i u_i^T P_i$$

By Hooke's law, we have

$$\Pi = \sum_{NE}\left[\frac{1}{2}\left(2\pi\int_e \varepsilon^T D\, \varepsilon\, r\, d\, A\right) - 2\pi\int_e u^T f\, r\, d\, A - 2\pi\int_e u^T\, T\, r\, dl\right] - \sum_i u_i^T P_i$$

The first term on the RHS gives the element strain energy, which can be written as

$$U_e = \frac{1}{2}\left(2\pi\int_e \varepsilon^T D \cdot \varepsilon\;\; r\, d\, A\right) = \frac{1}{2} q^T \left(2\pi\int_e B^T D\, B\, r\, d\, A\right) q$$

and the element stiffness matrix is given by

$$k^e = 2\pi\int_e \mathbf{B}^T \mathbf{D}\mathbf{B}\, r\, dA$$

Evaluation of the Matrix B by the Midpoint Rule

The matrix **B** is not only coordinates dependent but also involves the radial coordinate r explicitly in its coefficients. The evaluation being tricky can be evaluated in two ways.

(i) The simplest way of evaluating **B** is to calculate it at the centroid of the triangle where we can replace r by $\bar{r}$, where

$$\bar{r} = \frac{r_1 + r_2 + r_3}{3}$$

The value of **B** is then used as representative value for the triangle. In this method we can replace **B** by $\overline{\mathbf{B}}$ to show that **B** is evaluated at the centroid. In this the element stiffness matrix, given in eq. gets modified as

$$k^e = 2\pi \int_e B^T D B \, r \, dA = 2\pi \, \bar{r} \, \overline{B}^T D \, \overline{B} \int_e dA = 2\pi \, \bar{r} \, A_e \, \overline{B}^T D \, \overline{B}$$

where the area A_e of the element is as usual equal to half the determinant of **J**, *i.e.*,

$A_e = \frac{1}{2} \, |\det \mathbf{J}|$. This method is called the centroid of *midpoint rule.*

(ii) In anothers method is to replace r by $r = \xi r_1 + \eta \, r_2 + (1 - \xi - \eta) r_3$ and carry out the integration elaborately.

(1) Body force term

Work done by the body force is given by:

$$W_b = 2\pi \int_e u^T f \, rdA = 2\pi \int_e (u f_r + w f_z) \, rdA$$

$$= 2\pi \int_e [\, (\xi \, q_1 + \eta \, q_3 + (1 - \xi - \eta) q_5) \, f_r + (\xi \, q_2 + \eta \, q_4 + (1 - \xi - \eta) q_6) \, f_z] \, rdA$$

As in stiffness here also the quantities are evaluated at the centroid of the triangle and hence can be written as

$$W_b = 2\pi \int_e u^T fr \, dA = q^T f^e$$

The element body force vector f^e is given by

$$f^e = \frac{2\pi \bar{r} A_e}{3} \begin{Bmatrix} \bar{f}_r \\ \bar{f}_z \\ \bar{f}_r \\ \bar{f}_z \\ \bar{f}_r \\ \bar{f}_z \end{Bmatrix} = \frac{2\pi \bar{r} A_e}{3} \left[\bar{f}_r \;\; \bar{f}_z \;\; \bar{f}_r \;\; \bar{f}_z \;\; \bar{f}_r \;\; \bar{f}_z \right]^T$$

(2) Surface force term

Since the work done by the surface traction is

$$W_{ST} = 2\pi \int_e \boldsymbol{u}^T \boldsymbol{T} r\, dl = 2\pi \int_e (\boldsymbol{u}\, T_r + wT_z)\, r\, dA$$

where T_r and T_z are components of traction forces that act uniformly over the edge of the element (Fig. 9.8).

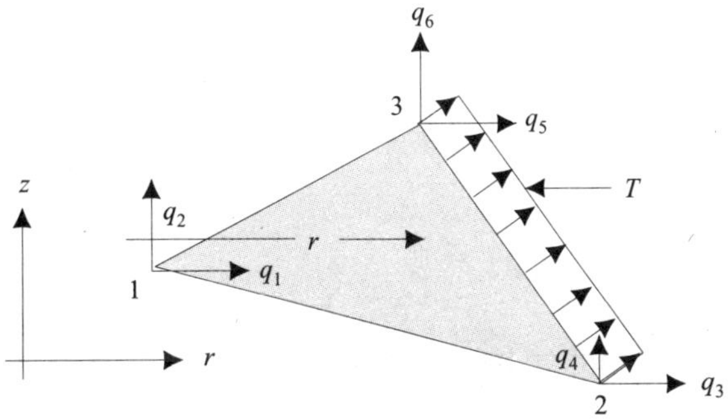

Fig. 9.8

For the loading shown, we can write

$$W_{ST} = 2\pi \int_e \boldsymbol{u}^T Tr\, dl = 2\pi \int_e (u\, T_r + w\, T_z)\, r\, dA = \boldsymbol{q}^T \boldsymbol{T}^e$$

where
$$\boldsymbol{q} = [q_3\, q_4\, q_5\, q_6]^T$$
$$\boldsymbol{T}^e = 2\pi\, l_{23}\, [\alpha T_r \;\; \alpha T_z \;\; \beta T_r \;\; \beta T_z]^T$$
$$\alpha = \frac{2r_2 + r_3}{6} \text{ and } \beta = \frac{r_2 + 2r_3}{6}$$
$$l_{23} = \sqrt{(r_2 - r_3)^2 + (z_2 - z_3)^2}$$

9.4 GALERKIN APPROACH

In Galerkin method, the consistent variation ϕ is expressed as,

$$\phi = N\,\psi \quad \text{...[1]}$$

where
$$\psi = [\psi_1, \psi_2, \ldots \psi_6]^T \quad \text{...[2]}$$

The corresponding $\varepsilon\,(\phi)$ is given by

$$\varepsilon\,(\phi) = B\psi \quad \text{...[3]}$$

The global vector of variations ψ is represented by

$$\psi = [\psi_1, \psi_2, \ldots \psi_N]^T \quad \text{...[4]}$$

Let us introduce the interpolated displacements into the Galerkin variational form:

$$2\pi \int \sigma^T \varepsilon(\phi) r dA - \left[2\pi \int_A \phi^T f r \, dA + 2\pi \int_L \phi^T T r \, dl + \Sigma \phi_i^T P_i \right] = 0 \qquad ...[5]$$

where $\phi = [\phi_r, \phi_z]^T$ ...[6]

The first term representing the internal virtual work gives

$$\text{Internal virtual work} = 2\pi \int_A \sigma^T \varepsilon(\phi) \, rdA$$

$$= \sum_e 2\pi \int_e q^T B^T D B \psi \, rdA$$

$$= \sum_e q^T k^e \psi \qquad ...[7]$$

where the element stiffness k^e is given by $k^e = 2\pi \, \bar{r} \, A_e \bar{B}^T D \bar{B}$ ...[8]

We note that k^e is symmetric.

$$\text{The internal virtual work} = \sum_e q^T k^e \psi = \Sigma \psi^T k^e q$$

$$= \psi^T k Q$$

where k is the global stiffness matrix.

The external virtual work = $\psi^T F$

using [6] which involves body force, surface fractions, and can be treated in the same way as in the potential energy approach, by replacing q and ψ.

Shape functions and the element stiffness matrix for an axisymmetric annular ring element as shown in Fig. 9.9.

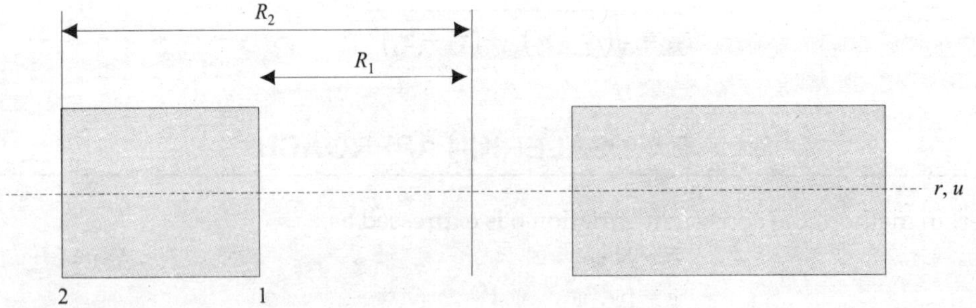

Fig. 9.9 Annular ring element

We know that in an axi-symmetric problem the radial displacement u depends on r.

$$u = \alpha_1 + \alpha_2 r$$

$$= [1 \;\; r] \begin{Bmatrix} \alpha_1 \\ \alpha_2 \end{Bmatrix} \qquad ...[1]$$

$$\begin{Bmatrix} u_1 \\ u_2 \end{Bmatrix} = \begin{bmatrix} 1 & r_1 \\ 1 & r_2 \end{bmatrix} \{\alpha\}$$

or

$$\{\alpha\} = \begin{bmatrix} 1 & r_1 \\ 1 & r_2 \end{bmatrix}^{-1} \begin{Bmatrix} u_1 \\ u_2 \end{Bmatrix}$$

$$= \frac{1}{(r_2 - r_1)} \begin{bmatrix} r_2 & -r_1 \\ 1 & -1 \end{bmatrix} \begin{Bmatrix} u_1 \\ u_2 \end{Bmatrix} \quad \text{...[2]}$$

Hence,

$$u = [1 \;\; r] \frac{1}{(r_2 - r_1)} \begin{bmatrix} r_2 & -r_1 \\ 1 & -1 \end{bmatrix} \begin{Bmatrix} u_1 \\ u_2 \end{Bmatrix} \quad \text{...[3]}$$

Now, the strain–displacement 'ε' is given by

$$\{\varepsilon\} = \begin{bmatrix} \dfrac{\partial u}{\partial r} \\ \dfrac{u}{r} \end{bmatrix} = \begin{Bmatrix} \varepsilon_1 \\ \varepsilon_2 \end{Bmatrix}$$

$$= \begin{bmatrix} 0 & 1 \\ \frac{1}{r} & 1 \end{bmatrix} \frac{1}{(r_2 - r_1)} \begin{bmatrix} r_2 & -r_1 \\ 1 & -1 \end{bmatrix} \begin{Bmatrix} u_1 \\ u_2 \end{Bmatrix} \quad \text{...[4]}$$

$$= [B] \begin{Bmatrix} u_1 \\ u_2 \end{Bmatrix}$$

$$\{\sigma\} = \begin{Bmatrix} \sigma_r \\ \sigma_\theta \end{Bmatrix} = \begin{bmatrix} 1 & r \\ r & 1 \end{bmatrix} \begin{Bmatrix} \varepsilon_r \\ \varepsilon_\theta \end{Bmatrix} = [C]\,[B] \begin{Bmatrix} u_1 \\ u_2 \end{Bmatrix} \quad \text{...[5]}$$

$$[K] = \int B^T CBdr \quad \text{...[6]}$$

$$= 2\pi t \int_{r_1}^{r_2} B^T CBrdr \quad \text{...[7]}$$

where t = thickness of the ring element.

Analyse a solid element in the form of a right circular cone under the action of self weight along its axis using single axis-symmetric ring element of triangular cross-section.

(i) Determine the shape function for the element.

(ii) Derive the components of vectors of nodal forces. Assume the height, radius, weight, density to be equal to 1.

Solution:

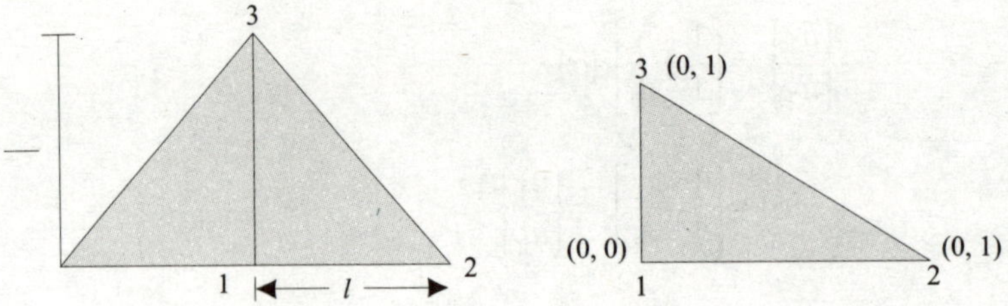

Fig. 9.10

$$\begin{Bmatrix} U \\ V \end{Bmatrix} = \begin{bmatrix} 1 & r & z & 0 & 0 & 0 \\ 0 & 0 & 0 & 1 & r & z \end{bmatrix} \begin{Bmatrix} \alpha_1 \\ \vdots \\ \alpha_6 \end{Bmatrix} \qquad ..[1]$$

$$\begin{Bmatrix} U_1 \\ \vdots \\ U_3 \\ V_1 \\ \vdots \\ V_3 \end{Bmatrix} = \begin{bmatrix} 1 & 0 & 0 & 0 & 0 & 0 \\ 1 & 1 & 0 & 0 & 0 & 0 \\ 1 & 0 & 1 & 0 & 0 & 0 \\ 0 & 0 & 0 & 1 & 0 & 0 \\ 0 & 0 & 0 & 1 & 1 & 0 \\ 0 & 0 & 0 & 1 & 0 & 1 \end{bmatrix} \{\alpha\} \qquad ...[2]$$

$$\{\alpha\} = \begin{bmatrix} 1 & 0 & 0 & 0 & 0 & 0 \\ -1 & 1 & 0 & 0 & 0 & 0 \\ -1 & 0 & 1 & 0 & 0 & 0 \\ 0 & 0 & 0 & 1 & 0 & 0 \\ 0 & 0 & 0 & -1 & 1 & 0 \\ 0 & 0 & 0 & -1 & 0 & 1 \end{bmatrix} \begin{Bmatrix} \underline{U} \\ \underline{V} \end{Bmatrix} \qquad ...[3]$$

$$\therefore \quad \begin{Bmatrix} U \\ V \end{Bmatrix} = \begin{bmatrix} (1-r-z) & r & z & 0 & 0 & 0 \\ 0 & 0 & 0 & (1-r-z) & r & z \end{bmatrix} \begin{Bmatrix} \underline{U} \\ \underline{V} \end{Bmatrix} \qquad ...[4]$$

The strain–displacement matrix is written as

$U_1 = U_3 = 0$ [B] matrix is reduced to 4×4 written as

$$\{\varepsilon\} = \begin{bmatrix} +1 & 0 & 0 & 0 \\ 0 & -1 & 0 & 1 \\ 0 & -1 & 1 & 0 \\ 1 & 0 & 0 & 0 \end{bmatrix} \begin{Bmatrix} \underline{U} \\ \underline{V} \end{Bmatrix} \qquad ...[5]$$

or

$$[B] = \begin{Bmatrix} \underline{U} \\ \underline{V} \end{Bmatrix}$$

$$[K] = 2\pi \int_{z=0}^{1} \int_{r=0}^{1} [B]^T [C][B] r dr\, dz \qquad ...[6]$$

$$\begin{Bmatrix} Q_x \\ Q_y \end{Bmatrix} = 2\pi \int_{z=0}^{1} \int_{r=0}^{1} N^T \begin{Bmatrix} 0 \\ -ug \end{Bmatrix} r dr dz \qquad ...[7]$$

The integration of the elements of [K] and the consistent load vector $\begin{Bmatrix} Q_x \\ Q_y \end{Bmatrix}$ are to be carried out using Table 9.1.

Example: A thick walled long cylinder of internal radius 70 mm and external radius 100 mm is subjected to an internal pressure of 4 Mpa. It is placed inside another rigid cylinder with internal radius equal to the external radius of the pressurized cylinder (Fig. 9.11). Given that E = 200 Gpa and Poisson's ratio μ = 0.3, determine the radial displacement at the inner edge.

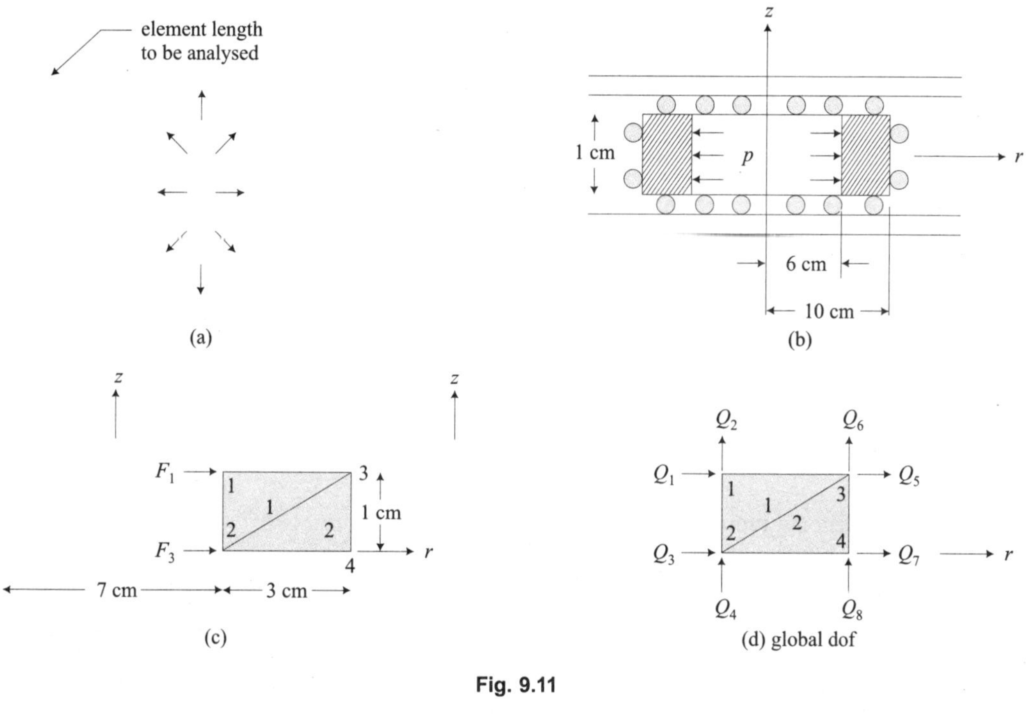

Fig. 9.11

Solution: We analyse the problem as follows:

Table 9.1

element	1	2	3 ← *local nodes*
1	1	2	3
2	1	4	3

The **D** matrix for both the elements are same and

$$\mathbf{D} = 10^5 \begin{bmatrix} 2.69 & 1.15 & 1.15 & 0 \\ 1.15 & 2.69 & 1.15 & 0 \\ 1.15 & 1.15 & 2.69 & 0 \\ 0 & 0 & 0 & 0.77 \end{bmatrix} Mpa$$

$$\det \mathbf{J} = \begin{vmatrix} r_1 - r_3 & z_1 - z_3 \\ r_2 - r_3 & z_2 - z_3 \end{vmatrix} = \begin{vmatrix} 7-10 & 1-1 \\ 7-10 & 0-1 \end{vmatrix} = 3 \text{ cm}^2, \quad \text{(After expension)}$$

Area of the element 1 = 1.5 cm^2

Let us calculate the **B** matrix for each element following the centroid method.

Now, for **element** 1, $\bar{r} = \frac{1}{3}(7 + 7 + 10) = 8$ cm, and

$$N_1 = N_2 = N_3 = \frac{1}{3}$$

Hence,

$$\bar{B}^1 = \begin{bmatrix} -0.33 & 0 & 0 & 0 & 0.33 & 0 \\ 0 & 1 & 0 & -1 & 0 & 0 \\ 0.0417 & 0 & 0.0417 & 0 & 0.0417 & 0 \\ 1 & -0.33 & -0.33 & 0 & 0 & 0.33 \end{bmatrix}$$

Hence,

$$\mathrm{D}\bar{B}^1 = 10^5 \begin{bmatrix} -0.849 & 1.15 & 0.048 & -1.15 & 0.945 & 0 \\ -0.335 & 2.69 & 0.048 & -2.69 & 0.431 & 0 \\ -0.271 & 1.15 & 0.112 & -1.15 & 0.495 & 0 \\ 0.77 & -0.257 & -0.257 & 0 & 0 & 0.257 \end{bmatrix}$$

The stiffness matrix is given by

$$K^1 = 2\pi \bar{r} A_e \bar{B}^T \mathbf{D} \bar{B}$$

$$\begin{matrix} 1 & 2 & 3 & 4 & 5 & 6 \end{matrix} \leftarrow \text{global } dof$$

$$= 10^5 \begin{bmatrix} 78.6 & -44.66 & -28.22 & -25.27 & -22.18 & 19.39 \\ & 209.4 & 10.11 & -202.91 & 32.51 & -6.49 \\ & & 54.84 & -3.62 & 1.58 & -6.49 \\ & & & 202.91 & -32.51 & 0 \\ & \text{Symmetric} & & & 25.34 & 0 \\ & & & & & 6.49 \end{bmatrix}$$

Element 2

$$\det \mathbf{J} = \begin{vmatrix} 7-10 & 0-1 \\ 10-10 & 0-1 \end{vmatrix} = 3 \text{ cm}^2, \quad \text{(After expension)}$$

$$= 1.5 \text{ cm}^2 \quad \text{(represents the area of the element)}$$

$$\bar{r}_2 = \frac{1}{3}(7 + 10 + 10) = 9$$

$$\Rightarrow \qquad N_1 = N_2 = N_3 = \frac{1}{3}$$

Hence,

$$\bar{B}^2 = \begin{bmatrix} -0.33 & 0 & 0.33 & 0 & 0 & 0 \\ 0 & 0 & 0 & -1 & 0 & 1 \\ 0.037 & 0 & 0.037 & 0 & 0.037 & 0 \\ 0 & -0.33 & -1 & 0.33 & 1 & 0 \end{bmatrix}$$

$$\mathbf{D}\bar{B}^2 = 10^5 \begin{bmatrix} -0.86 & 0 & 0.943 & -1.15 & 0.043 & 1.15 \\ -.34 & 0 & 0.426 & -2.69 & 0.043 & 2.69 \\ -.283 & 0 & 0.483 & -1.15 & 0.1 & 1.15 \\ 0 & -0.26 & -0.77 & 0.26 & 0.77 & 0 \end{bmatrix}$$

So, the element stiffness matrix is given by

$$k^2 = \underline{2\pi\ \bar{r}_2 A_e}\ \bar{B}^T \mathbf{D}\ \bar{B}$$

$$\begin{matrix} 3 & 4 & 7 & 8 & 5 & 6 \end{matrix} \leftarrow \text{global } dof$$

$$= 10^5 \begin{bmatrix} 23.76 & 0 & 25.12 & 28.85 & 0.85 & -28.85 \\ & 7.64 & 22.06 & -7.38 & -22.06 & 0 \\ & & 93.52 & 58.21 & -63.81 & 36.15 \\ & & & 23.51 & 18.16 & -228.27 \\ & & & & 68.74 & 3.65 \\ & & & & & 228.27 \end{bmatrix}$$

The loads are acting only on the nodes 1 and 2, are equal and

$$F_1 = F_3 = \frac{1}{2}(2\pi r l_e p) = \pi (7.0)(1.0)(4) = 8800\ N$$

After assemblaging the stiffness, we obtain the global stiffness matrix. Now, if we use the elimination approach then except 1 and 3 all other *dof* will be zero. This yields

$$10^5 \begin{bmatrix} 78.6 & -28.22 \\ -28.22 & 78.6 \end{bmatrix} \begin{Bmatrix} Q_1 \\ Q_3 \end{Bmatrix} = \begin{Bmatrix} 8800 \\ 8800 \end{Bmatrix}$$

from which $Q_1 = Q_3 = 0.17467 \times 10^{-2}$ cm.

EXERCISES

1. Explain the term 'axi-symmetric problems' and give constitutive law for such problems.
2. Develop the strain-displacement matrix **B** for the axi-symmetric element.
3. Develop the element stiffness matrix for the axi-symmetric eiement.
4. Write a note on the actual evaluation of the matrix **B**.
5. Write down the expression for the total potential energy for the axi-symmetric element.
6. Evaluate the body force term for the axi-symmetric element.
7. State the strain-displacement vector in cylindrical coordinates.
8. State the stress-strain relations for axi-symmetric case.

Appendix I

SUMMARY

Displacement Method

The term displacement method is used within the field of study known as theory of structures. The displacement-based finite element method can be regarded as an extension of the displacement method of analysis which had been used for many years in matrix analysis of beam and truss structures. The displacement method is often referred to as the stiffness method and predates the use of computers in structural analysis. The three-moment equation and the slope-deflection method were forerunners of the displacement method.

Nodal Degrees of Freedom

Each finite element has a fixed number of nodes and every node is capable of deformation. The deformation includes displacements, rotations, and/or strains. These are collectively known as nodal degrees of freedom. The nodal degrees of freedom, also called as nodal displacement parameters, normally refer to the displacement such as u, v and w and their first partial derivative (rotations) at a node, but very often may include other terms such as stresses, strains and second or even higher partial derivatives. The nodal degree of freedom is a variable that describes the behaviour of a node in an element.

Example (1). In one-dimensional axial element, each node has only one degree of freedom i.e., each node can move in the ± x-direction only.

Example (2). For the truss (axial) element shown in Fig. A.1 the dof is 5. This is also called kinematic degree of indeterminacy.

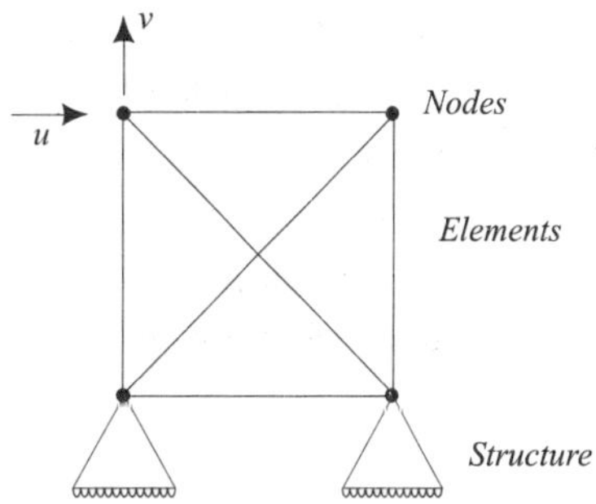

Fig. A.1 2-dimensional pin jointed truss

Different Coordinate Systems

In finite element formulation, the actual format of the displacement model would depend upon the coordinate system chosen. The different coordinate systems in use are:

1. Generalized coordinates
2. Natural coordinates.

These are of two types:

(i) Triangular coordinates

(ii) Quadrilateral coordinates.

1. Generalised Coordinates

In this system, displacement model (or parametric model) is taken in the form of a polynomial where unknown coefficients of the variables are named as generalized coordinates.

The general form of variation of any function can be described by a mathematical polynomial as

$$f(x) = A_1 + A_2x + A_3x^2 + a_4x^3 + a_5x^5 \qquad ...(1)$$

where A_1, A_2, A_3, ... are generalized coordinates.

We note that:

(i) In case of elements with two nodes (points) no variation other than a linear one is possible i.e., only a straight line can be determined uniquely with the help of two nodes,

(ii) In order to represent a square or a quadratic variation, 3 nodes are required, and

(iii) For cubic variation 4 nodes are required, and so on.

Therefore, according to the object of studies, number of nodes on each side of the element are chosen.

Example 1. Given: $A = \begin{bmatrix} 4 & 5 \\ 2 & 1 \end{bmatrix}$. Determine

(i) Inverse of matrix

(ii) Eigen values

[VTU, 6th Sem., J/A, 2003; (10 marks)]

Solution: Given: $A = \begin{bmatrix} 4 & 5 \\ 2 & 1 \end{bmatrix}$

Step (1): $I_n\,|A| = \begin{vmatrix} 4 & 5 \\ 2 & 1 \end{vmatrix}$,

the cofactor of 4 is 1

the cofactor of 5 is –2

the cofactor of 2 is –5

the cofactor of 1 is 4

∴ the matrix B formed of the cofactors of the elements of | A | is

$$B = \begin{bmatrix} 1 & -2 \\ -5 & 4 \end{bmatrix}$$

Now, Adj(A) = the transpose of the matrix B

$$= \begin{bmatrix} 1 & -5 \\ -2 & 4 \end{bmatrix};$$

Step (2): $|A| = 4 \times 1 - 5 \times 2 = 4 - 10 = -6$

Step (3): $$A^{-1} = \frac{1}{-6} \begin{bmatrix} 1 & -5 \\ -2 & 4 \end{bmatrix} = \begin{bmatrix} -\frac{1}{6} & \frac{5}{6} \\ \frac{2}{6} & -\frac{4}{6} \end{bmatrix}$$

Verification: $$AA^{-1} = A^{-1}A = \begin{bmatrix} 4 & 5 \\ 2 & 1 \end{bmatrix} \begin{bmatrix} -\frac{1}{6} & \frac{5}{6} \\ \frac{2}{6} & -\frac{4}{6} \end{bmatrix}$$

$$\boxed{AA^{-1} = A^{-1}A = I} = \begin{bmatrix} -\frac{4}{6} + \frac{10}{6} & \frac{20}{6} - \frac{20}{6} \\ -\frac{2}{6} + \frac{2}{6} & \frac{10}{6} - \frac{4}{6} \end{bmatrix} = \begin{bmatrix} 1 & 0 \\ 0 & 1 \end{bmatrix} = I$$

(ii) Given: $$A = \begin{bmatrix} 4 & 5 \\ 2 & 1 \end{bmatrix}$$

$$A - \lambda I = \begin{bmatrix} 4 & 5 \\ 2 & 1 \end{bmatrix} - \lambda \begin{bmatrix} 1 & 0 \\ 0 & 1 \end{bmatrix}$$

$$= \begin{bmatrix} 4 - \lambda & 5 \\ 2 & 1 - \lambda \end{bmatrix}$$

$$|A - \lambda I| = 0 \Rightarrow \begin{bmatrix} 4 - \lambda & 5 \\ 2 & 1 - \lambda \end{bmatrix} = 0$$

$$\Rightarrow (4 - \lambda)(1 - \lambda) - 10 = 0$$

$$\Rightarrow 4 - 5\lambda + \lambda^2 - 10 = 0$$

$$\Rightarrow \lambda^2 - 5\lambda - 6 = 0$$

$$\Rightarrow \lambda^2 - 6\lambda + \lambda - 6 = 0$$

$$\Rightarrow \lambda(\lambda - 6) + (\lambda - 6) - 0$$

$$\Rightarrow (\lambda + 1)(\lambda - 6) = 0$$

$$\Rightarrow \lambda = 6 \text{ and } \lambda = -1$$

Eigen values are $\lambda = 6$ and $\lambda = -1$

Example 2. Find the Eigen values of

$$A = \begin{bmatrix} 4 & -\sqrt{5} \\ -\sqrt{5} & 8 \end{bmatrix}$$

[VTU, 6th Sem., J/F, 2004; (5 marks)]

Solution:

Let
$$A = \begin{bmatrix} 4 & -\sqrt{5} \\ -\sqrt{5} & 8 \end{bmatrix}$$

$$A - \lambda I = \begin{bmatrix} 4 & -\sqrt{5} \\ -\sqrt{5} & 8 \end{bmatrix} - \lambda \begin{bmatrix} 1 & 0 \\ 0 & 1 \end{bmatrix}$$

$$= \begin{bmatrix} 4-\lambda & -\sqrt{5} \\ -\sqrt{5} & 8-\lambda \end{bmatrix}$$

$$\Rightarrow (4 - \lambda)(8 - \lambda) + 5 = 0$$

$$\Rightarrow 32 - 12\lambda + \lambda^2 - 5 = 0$$

$$\Rightarrow \lambda^2 - 12\lambda + 27 = 0$$

$$\Rightarrow \lambda^2 - 3\lambda - 9\lambda + 27 = 0$$

$$\Rightarrow \lambda(\lambda - 3) - 9(\lambda - 3) = 0$$

$$\Rightarrow (\lambda - 3)(\lambda - 9) = 0$$

$$\Rightarrow \lambda = 3, \lambda = 9$$

$\therefore$ Eigen values are $\lambda = 3$ and $\lambda = 9$

Example 3. Solve the following system of simultaneous equations by Gaussian elimination method.

$$2x_1 + x_2 + 3x_3 = 10$$

$$4x_1 + x_2 + x_3 = 5$$

$$3x_1 + 2x_2 + 3x_3 = 3$$

[VTU, 6th Sem., J/F, 2004; (10 marks)]

Solution: Consider the augmented matrix

or
$$[A/b] = \left[\begin{array}{ccc|c} 2 & 1 & 3 & 10 \\ 4 & 1 & 1 & 5 \\ 3 & 2 & 3 & 3 \end{array}\right]$$

$$\text{Applying} \quad \xrightarrow[R_3 \to 2R_3 - 3R_1]{R_2 \to R_2 - 2R_1} \begin{bmatrix} 2 & 1 & 3 & 10 \\ 0 & -1 & -5 & -15 \\ 0 & 1 & -3 & -24 \end{bmatrix}$$

$$\text{Applying} \quad \xrightarrow{R_3 \to R \;+ R_2} \begin{bmatrix} 2 & 1 & 3 & 10 \\ 0 & -1 & -5 & -15 \\ 0 & 0 & -8 & -39 \end{bmatrix}$$

Back substitution gives the solution

i.e.,
$$2x_1 + x_2 + 3x_3 = 10$$
$$-x_2 - 5x_3 = -15$$
$$-8x_3 = -39$$

$$\Rightarrow \quad x_3 = \frac{39}{8}; x_2 = +15 - 5\left(\frac{39}{8}\right)$$
$$= 15 - \frac{195}{8} = \frac{120 - 195}{8} = \left(-\frac{75}{8}\right)$$

$$\therefore \quad x_1 = \frac{1}{2}[10 - x_2 - 3x_3]$$
$$= \frac{1}{2}\left[10 + \frac{75}{8} - \frac{117}{8}\right]$$
$$= \frac{1}{16}[80 + 75 - 117] = \frac{1}{16}[38] = \left(\frac{19}{8}\right)$$

$$\Rightarrow \quad \left(x_3 = \frac{39}{8}; x_2 = -\frac{75}{8} \text{ and } x_1 = \frac{19}{8}\right)$$

Example 4. With an example explain Rayleigh-Ritz method.

[VTU, 6th Sem., J/A, 2003; (10 marks)]

Solution: Rayleigh-Ritz method:

A method for determining approximate solutions of functional equations through the expedient of replacing them by finite systems of equations. It is one of the most useful approximate methods stemming from variational considerations, wherein, for the present we employ the following approximate displacement field components for expressing the total energy:

$$\left\{\begin{aligned} u &= \Sigma a_i \phi_i(x, y, z) \text{ for } i = 1 \text{ to } m_1 \\ v &= \Sigma a_j \phi_j(x, y, z) \text{ for } j = m_1 + 1 \text{ to } m_2 \\ w &= \Sigma a_k \phi_k(x, y, z) \text{ for } k = m_2 + 1 \text{ to } m \end{aligned}\right\} \quad ...[1]$$

where a_i, a_j, and a_k are '3n' linearly independent parameters, which behave as generalized coordinates. The functions ϕ_i are usually taken as polynomials.

Substituting from the expression in [1] for displacement in strain displacements in stress-strain relations, we get the total potential energy

$$\Pi = \Pi\,(a_1, a_2, ..., a_{m1}, a_{m1+1} \ldots a_{m2+1} \ldots a_m)$$

But from the principle of minimum potential energy,

$$\frac{d\Pi}{da_i} = 0 \text{ for } i = 1 \text{ to } m \qquad \text{...[2]}$$

Equation [2] yields the solution of 'm' equation, and we get the values of 'a'. With these values of 'a_i's and 'ϕ_i's satisfying boundary conditions, the displacements are obtained. Then the strains and stresses can be assembled.

Example 5. Let us consider a simply supported beam loaded at centre as shown in Fig. A.2. Let us assume the trial function in the form of trigonometric series as (instead of infinite degrees of freedom select k degrees of freedom)

$$y` = \sum_{n=1}^{k} a_n \sin\left(\frac{n\pi}{l}\right)x \qquad \text{...[1]}$$

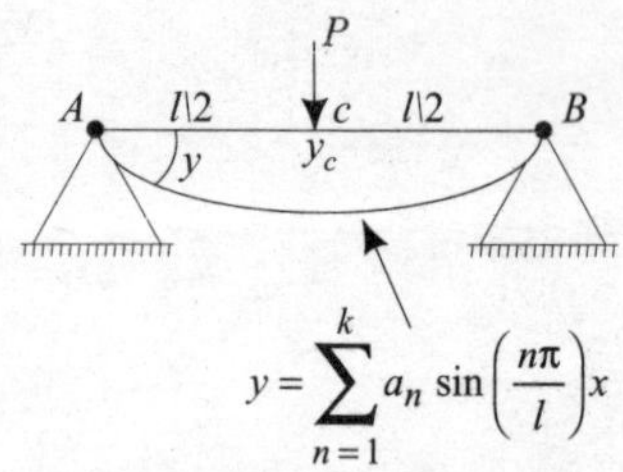

Fig. A.2

Compute the central deflection for the beam.

Solution: (i) *Central concentrated load*

The trial function assumed is $y = \sum_{n=1}^{k} a_n \sin\frac{n\pi}{l}x$. It may be seen that this trial function completely satisfies the geometric boundary conditions at the two ends.

$$y = \sum_{n=1}^{k} a_n \sin\left(\frac{n\pi}{l}\right)x \qquad \text{...[1]}$$

We know that

$$\left.\begin{aligned} &\int_0^l \sin\frac{n\pi}{l}x \sin\frac{m\pi}{l}x\,dx = 0;\ (m \neq n) \\ &\int_0^l \sin\frac{m\pi}{l}x \sin\frac{n\pi}{l}x\,dx = \frac{l}{2};\ (m = m) \end{aligned}\right\} \qquad \text{...[2]}$$

Given: $$y = \sum_{n=1}^{k} a_n \sin\left(\frac{n\pi}{l}\right)x$$

Now,
$$y' = \sum_{n=1}^{k} \frac{a_n \pi_n}{l} \cos\left(\frac{n\pi}{l}\right)x \qquad ...[3]$$

$$y'' = -\sum_{n=1}^{k} \frac{a_n^2 \pi^2 n^2}{l^2} \sin\left(\frac{n\pi}{l}\right)x \qquad ...[4]$$

$$\pi = \frac{1}{2}\int_0^l EI\,(y'')^2\,dx - Py_c$$

$$= \frac{1}{2}\int_0^l EI\left\{\sum_{n=1}^{k}\left[\frac{n^4\pi^4}{l^4}.a_n^2 \sin^2\left(\frac{n\pi}{l}\right)x\right]\right\}dx - P\sum_{n=1}^{k}\left[a_n \sin\frac{n\pi}{2}\right] \qquad ...[5]$$

$$\pi = V + U \quad (\therefore V = -Pa_n)$$

$$\pi = \frac{1}{4}\frac{EI\,\pi^4}{l^3}\sum_{n=1,3,5}^{k} n^4 a_n^2 - P\sum_{n=1,3,5} a_n \sin\frac{n\pi}{2} \qquad ...[6]$$

(U remains same as long as the trial function is unchanged)

Effecting variation invoking the stationary property

$$\delta\pi = \frac{d\pi}{da_n}\delta a_n = 0;\ \frac{\partial\pi}{\partial u_1} = 0;\ \frac{\partial\pi}{\partial a_3} = 0;\ \frac{\partial\pi}{\partial a_5} = 0 \qquad ...[7]$$

Equilibrium equations

$$\frac{\partial\pi}{\partial a_n} = \frac{EI\pi^4}{4l^3}n^4(2a_n) - P\sin\frac{n\pi}{2} = 0 \qquad ...[8]$$

$$a_n = \frac{2Pl^3}{\pi^4 EI}\frac{1}{n^4}\sin\frac{n\pi}{2}$$

Hence the approximate deflected shape is given by

$$y = \frac{2Pl^3}{\pi^4 EI}\sum_{n=1,3,5}^{k}\frac{1}{n^4}\sin\frac{n\pi}{2}\sin\frac{n\pi}{l}x \qquad ...[9]$$

The central deflection

$$y_{\max} = a_n = \frac{2Pl^3}{\pi^4 EI}\sum_{n=1,3,5}^{k}\frac{1}{n^4} \qquad ...[10]$$

$$\pi = -\frac{P^2 l^3}{\pi^4 EI}\sum_{n=1,3,5}^{k}\frac{1}{n^4} \qquad ...[11]$$

If we choose one term

$$y = \frac{2Pl^3}{\pi^4 EI} \sin \frac{\pi}{l} x;$$

$$\pi = \frac{-P^2 l^3}{\pi^4 EI} \quad \text{...[12]}$$

If we choose two terms

$$y = \frac{2Pl^3}{\pi^4 EI} \left[\sin \frac{\pi}{l} x - \frac{1}{81} \sin \frac{3\pi}{l} x \right] \quad \text{...[13]}$$

$$y_c = \frac{2Pl^3}{\pi^4 EI} \left[1 + \frac{1}{81} \right] \text{ and so on.} \quad \text{...[14]}$$

(ii) *Uniformly distributed load*

The strain energy will remain the same as before. The only difference is in the computation of potential energy V.

$$V = -\int_0^l (w\,dx) y$$

$$= -\int_0^l w a_n \sin \frac{\pi x}{l} dx$$

$$= -\frac{2\,w a_n l}{\pi}$$

$$\pi = \left(\frac{EI\pi^4 a_n^2}{4l^3} - \frac{2 w a_n l}{\pi} \right)$$

$$\delta\pi = \frac{d\pi}{da_n} \delta\pi = 0; \qquad (n = 1, 2, 3, ...)$$

$$\left[\frac{2EI\pi^4 a_n}{4l^3} - \frac{2wl}{\pi} \right] \delta a_n = 0$$

$$\therefore \qquad a_n = \frac{4}{\pi^5} \frac{wl^4}{EI} \approx \frac{5}{384} \frac{wl^4}{EI} \text{ (Exact)}$$

$\therefore$ The central deflection $y_{max} = a_n$.

Example 6. Given the potential energy for the bar Fig. A.3, with body force neglected is

$$\Pi = \frac{1}{2} \int_0^L EA \left(\frac{du}{dx} \right)^2 dx - 2u_1; \qquad [u_1 = u(x)]_{x=1}$$

Compute the stress in the bar.

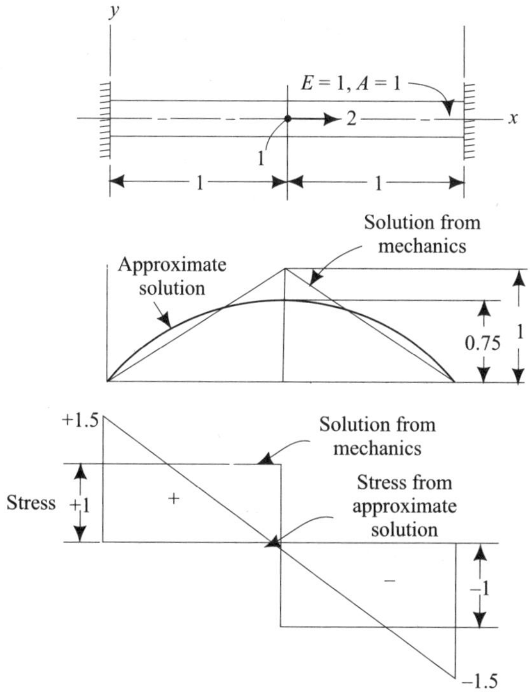

Fig. A.3

Solution: Suppose we are given a polynomial function

$$u = a_1 + a_2x + a_3x^2$$

with end condition $u(0) = 0$; and $u(2) = 0$.

Now,
$$\begin{Bmatrix} u(0) = 0 \Rightarrow 0 = a_1 \\ u(2) = 0 \Rightarrow 0 = a_1 + 2a_2 + 4a_3 \end{Bmatrix}$$

$$\Rightarrow a_2 = -2a_3$$

$$\Rightarrow u = a_3(-2x + x^2); \qquad u_1 = -a_3$$

Then,
$$\frac{du}{dx} = 2a_3\,(-1 + x)$$

$$\Pi = \frac{1}{2}\int_0^2 4\,a_3^2\,(-1+x)^2\,dx - 2(-a_3)$$

$$= 2\,a_3^2 \int_0^2 (1 - 2x + x^2)dx + 2a_3$$

$$= 2\,a_3^2\left(\frac{2}{3}\right) + 2a_3$$

Since $$\frac{\partial \pi}{\partial a_3} = 4a_3\left(\frac{2}{3}\right) + 2 = 0, \text{ resulting in,}$$

$$a_3 = -0.75;$$

$$u_1 = -a_3 = 0.75$$

The stress in the bar is given by

$$\sigma = E\frac{du}{dx} = 1.5\,(1 - x)$$

Appendix II

BEAM ELEMENTS

In the Finite element notation, we will define the beam element with general degrees of freedom as shown in Fig. (i). Beam elements are related directly to matrix analysis of structures.

The primary difference between the beam finite element and the finite elements of the previous chapters is the order of the governing differential equation. Beam theory is based upon a fourth-order ordinary differential equation. The elements whether one-or two-dimensional, can only transmit forces directly from one element to another through the nodes with nodes acting like a pin-joint. But in structures, we get into a situation where both forces and moments have to be considered. We require an element, which can transmit not only forces but also moment. To effect such a requirement, we should have elements with both translational and rotational degrees of freedom at their nodes. Beam, plate and shell elements all belong to this category. All these elements are also further categorized as thin and thick depending upon whether transverse shear strain is taken into account or not.

Some important assumptions used in Euler-Bernoulli beam theory are:

(i) Plane cross-sections before bending will remain plane even after bending.

(ii) All points lying on a normal to the neutral section will deflect the same amount in the transverse direction.

SHAPE FUNCTIONS

Let us choose a two noded beam element (Fig. (i)) in which nodal unknowns are displacement w and the slope $\frac{\partial w}{\partial x}$. Since the element has four degrees of freedom, we have to select the polynomial with only 4 constants, that is the first order (cubic) Hermitian polynomials as shape function.

(a) Natural co-ordinates varying from zero to one

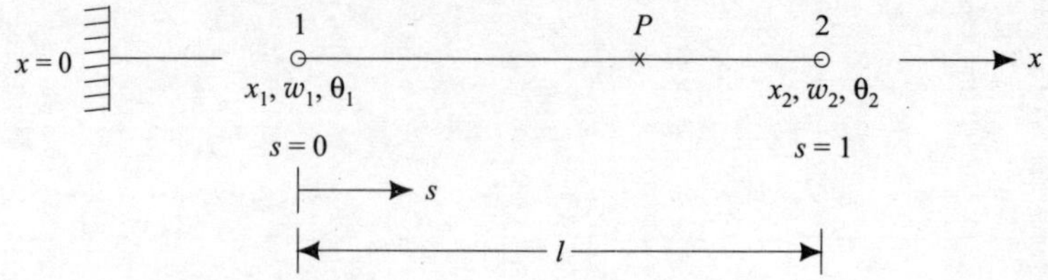

These are given as

$$H_{01}^{1}(s) = 1 - 3s^2 + 2s^3$$
$$H_{11}^{1}(s) = ls(s-1)^2$$
$$H_{02}^{1}(s) = s^2(3-2s)$$
$$H_{12}^{1}(s) = ls^2(s-1)$$

Where $s = \dfrac{x - x_1}{x_2 - x_1} = \dfrac{x - x_1}{l}$ and $l = x_2 - x_1$

The displacement nodel for the beam element is

$$w = [N]\{\delta\}_e$$

i.e.,
$$w = [N_1 \quad N_2 \quad N_3 \quad N_4]\begin{Bmatrix} w_1 \\ \theta_1 \\ w_2 \\ \theta_2 \end{Bmatrix}$$

where
$$\theta = \frac{\partial w}{\partial x}$$

Then
$$\frac{\partial w}{\partial x} = \left[\frac{\partial N_1}{\partial x} \quad \frac{\partial N_2}{\partial x} \quad \frac{\partial N_3}{\partial x} \quad \frac{\partial N_4}{\partial x}\right]\begin{Bmatrix} w_1 \\ \theta_1 \\ w_2 \\ \theta_2 \end{Bmatrix}$$

We select Hermitian order polynomial functions as shape function i.e.,

$$N_1 = H_{01}^{1}(s) = 1 - 3s^2 + 2s^3$$
$$N_2 = H_{11}^{1}(s) = ls(s-1)^2$$
$$N_3 = H_{02}^{1}(s) = s^2(s-2s)$$
$$N_4 = H_{12}^{1}(s) = ls^2(s-1)$$

Now it may be observed that, at node 1,

$$N_1 = 1,\ N_2 = N_3 = N_4 = 0$$

and

$$\frac{\partial N_1}{\partial s} = 0,\ \frac{\partial N_2}{\partial s} = 1,\ \frac{\partial N_3}{\partial s} = \frac{\partial N_4}{\partial s} = 0$$

Similarly at node 2,

$$N_1 = 0,\ N_2 = 0,\ N_3 = 1,\ N_4 = 0$$

$$\frac{\partial N_1}{\partial s} = \frac{\partial N_2}{\partial s} = \frac{\partial N_3}{\partial s} = 0 \text{ and } \frac{\partial N_4}{\partial s} = 1$$

is the required shape function.

(b) If we choose natural co-ordinate as shown in Fig (i),

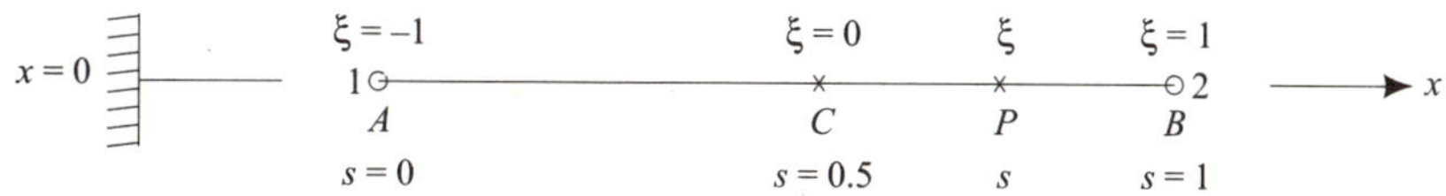

Fig. (i) Beam element with natural co-ordinates ξ varying from −1 to 1

then,

$$\xi = \frac{PC}{\frac{l}{2}} = \frac{2}{l}(AP - AC) = 2(s - 0.5) = 2s - 1$$

Substituting in equation we get, Hermite polynomial of first order as,

$$\left\{\begin{aligned} N_1 &= H_{01}^1 = \frac{2 - 3\xi + \xi^3}{4} \\ N_2 &= H_{11}^1 = (1 - \xi - \xi^2 + \xi^3)\frac{l}{8} \\ N_3 &= H_{02}^1 = \frac{1}{4}(2 + 3\xi - \xi^3) \\ N_4 &= H_{12}^1 = \frac{l}{8}(-1 - \xi + \xi^2 + \xi^3) \end{aligned}\right\} \quad \text{...[1]}$$

Hermitian shape functions in two dimensions may be constructed as multiplication of Hermitian functions in x and y directions. Thus.

$$N_1 = H_{01}^1(x)\ H_{01}^1(y)\ ;\quad N_2 = H_{11}^1(x)\ H_{11}^1(y)$$

$$N_3 = H_{02}^1(x)\ H_{02}^1(y)\ ;\quad N_4 = H_{12}^1(x)\ H_{12}^1(y)$$

(c) Consider the beam element with general degrees of freedom as shown in Fig (i)

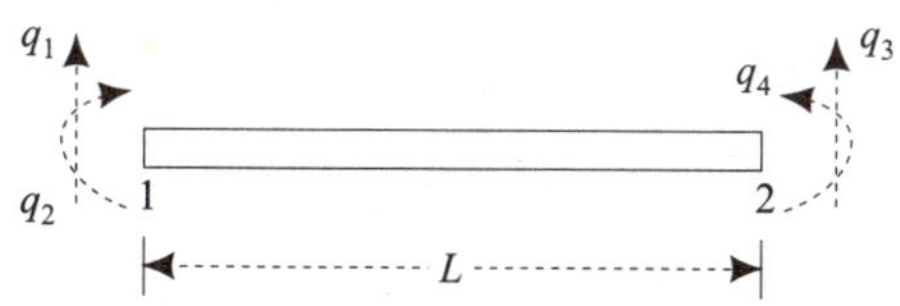

The local degrees of freedom are given by

$$q = \begin{Bmatrix} q_1 \\ q_2 \\ q_3 \\ q_4 \end{Bmatrix} = [q_1 \quad q_2 \quad q_3 \quad q_4]^T$$

Where the odd subscript refers to transverse displacements and the even to rotational displacements. The shape function for a beam element differs from that for an axial element since we have to take care of slopes at the ends. The shape functions are called the *Hermite shape function* which satisfy both nodal displacement and slope continuity requirements. The general expression can be assumed as

$$H_{mi}^n = A_i + B_i\xi + C_i\xi^2 + D_i\xi^3 \qquad [i = 1, 2, 3, 4]$$

In the above, the odd subscript refers to transverse displacement and even to rotation, i.e., $\left(\frac{dH}{d\xi}\right)_2$ refers to slope at node 1 and subscript 2 refers to global degree of freedom. Note that *i* refers to degrees of freedom and not the nodal number. Thus, we have the conditions as given in the Table 1 that must be satisfied by *H*.

Table 1

	H_1	$\left(\frac{dH}{d\xi}\right)_1$	H_2	$\left(\frac{dH}{d\xi}\right)_2$	H_3	$\left(\frac{dH}{d\xi}\right)_3$	H_4	$\left(\frac{dH}{d\xi}\right)_4$
$\xi = -1$	1	0	0	1	0	0	0	0
$\xi = +1$	0	0	0	0	1	0	0	1

Using the above table, it is found that

$$\left\{\begin{aligned} N_1 &= H_{01}^1 = \frac{1}{4}(2 - 3\xi + \xi^3) \\ N_2 &= H_{11}^1 = \frac{1}{4}(1 - \xi - \xi^2 + \xi^3) \\ N_3 &= H_{02}^1 = \frac{1}{4}(2 + 3\xi - \xi^3) \\ N_4 &= H_{12}^1 = \frac{1}{4}(-1 - \xi + \xi^2 + \xi^3) \end{aligned}\right\} \qquad \text{...[1]}$$

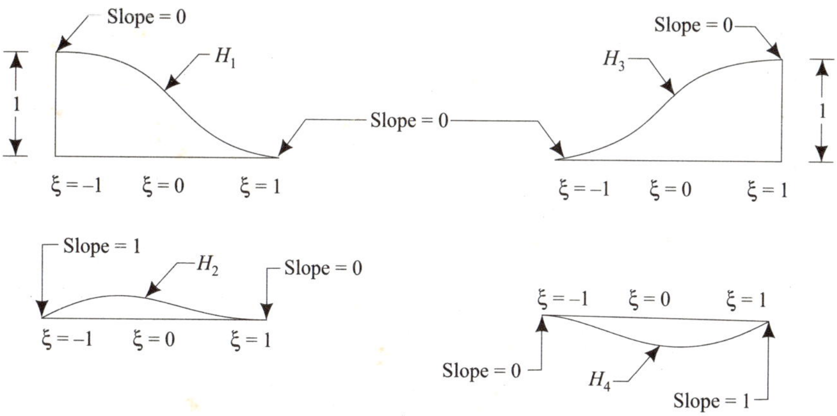

The shape functions are shown graphically in

Since y is the transverse displacement, we can represent the displacement at any point by the relation

$$y(\xi) = H_1 y_1 + H_2 \left(\frac{dy}{d\xi}\right)_{\text{node 1}} + H_3 y_2 + H_4 \left(\frac{dy}{d\xi}\right)_{\text{node 2}}$$

The x-ξ relation is given by

$$x = \frac{1-\xi}{2} x_1 + \frac{1+\xi}{2} x_2 = \frac{x_1 + x_2}{2} + \frac{x_2 - x_1}{2} \xi$$

Hence,

$$dx = \frac{x_2 - x_1}{2} d\xi = \frac{l_e}{2} d\xi$$

Where l_e is the element length. Then

$$\frac{dy}{d\xi} = \frac{dy}{dx}\frac{dx}{d\xi} = \frac{l_e}{2}\frac{dy}{dx}$$

Since

$$\left(\frac{dy}{dx}\right)_{\text{node 1}} = q_2 \text{ and } \left(\frac{dy}{dx}\right)_{\text{node 2}} = q_4$$

Eq. (240) gets modified as

$$y(\xi) = H_1 q_1 + \frac{l_e}{2} H_2 q_2 + H_3 q_3 + \frac{l_e}{2} H_4 q_4$$

The above can be put in matrix notation as

$$y = \mathbf{Hq}$$

Where

$$H = \left[H_1 \;\; \frac{l_e}{2} H_2 \;\; H_3 \;\; \frac{l_e}{2} H_4\right]$$